THEORY OF
INFORMATION
Fundamentality, Diversity and Unification

World Scientific Series in Information Studies — **Vol. 1**

THEORY OF
INFORMATION

Fundamentality, Diversity and Unification

Mark Burgin

University of California, Los Angeles, USA

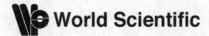

World Scientific

NEW JERSEY · LONDON · SINGAPORE · BEIJING · SHANGHAI · HONG KONG · TAIPEI · CHENNAI

Published by

World Scientific Publishing Co. Pte. Ltd.

5 Toh Tuck Link, Singapore 596224

USA office: 27 Warren Street, Suite 401-402, Hackensack, NJ 07601

UK office: 57 Shelton Street, Covent Garden, London WC2H 9HE

British Library Cataloguing-in-Publication Data
A catalogue record for this book is available from the British Library.

World Scientific Series in Information Studies – Vol. 1
THEORY OF INFORMATION
Fundamentality, Diversity and Unification

Copyright © 2010 by World Scientific Publishing Co. Pte. Ltd.

ISBN-13 978-981-283-548-2
ISBN-10 981-283-548-2

Printed in Singapore.

Preface

We set sail on this new sea because there is knowledge to be gained.

John F. Kennedy (1917–1963)

As a general rule the most successful man in life is the man who has the best information.

Benjamin Disraeli (1804–1881)

There is an ancient Greek legend about the mystic creature Sphinx who plagued the city of Thebes. The Sphinx would routinely ask the passerby's a riddle and whoever could not answer it would be snatched up and eaten by the beast. The hero Oedipus then went before the Sphinx and asked it to tell him the riddle, upon telling him Oedipus responded with the correct answer to the riddle. Hearing this, the Sphinx disappeared, bringing an end its threat to people.

Now people encounter a new mystic Sphinx that emerges from the technological development. This modern Sphinx asks the question what is information and how it functions. If people do not find a correct answer to this problem, technology threats to destroy humankind because people become more and more dependent on information technology. Really, the importance of information for society grows so fast that our time is called the "information age" (cf., for example, (Giuliano, 1983; Goguen, 1997; Mason, 1986; Crawford and Bray-Crawford, 1995; Stephanidis, 2000)). It is generally acknowledged that we have been living in the "information age", at least, since the middle of the 20th century. Information not only constitutes the very foundation of most industrial sectors, but more significantly has been now transformed into a primary tradable resource or commodity (Macgregor, 2005). In fact, there is no aspect of human experience that lies outside information influence.

For instance, Martin (1995) acknowledges that information functions as the lifeblood of society, writing:

Without an uninterrupted flow of the vital resource [information], society, as we know it, would quickly run into difficulties, with business and industry, education, leisure, travel, and communications, national and international affairs all vulnerable to disruption. In more advanced societies, this vulnerability is heightened by an increasing dependence on the enabling powers of information and communication technologies.

Information is all around us, as well as in us. Our senses collect and our brains filter, organize and process information every second of the day. Information makes our lives possible.

Since creation of the Internet, the volume of information available to people has grown exponentially. The World Wide Web, despite only being generally available since 1995, has thoroughly trampled all existing information media in its path to become one of the primary information delivery mechanisms (Duff, 2003).

An estimated 1-2 exabytes (1 exabyte is equal to 10^{18} bites) of new information is produced (and stored) each year. This includes all media of information storage: books, magazines, documents, the Internet, PCs, photographs, x-rays, TV, radio, music, CDs, DVDs, etc. It makes roughly 250 megabytes for every person on the planet. However, most of information is digital (93%, to be precise).

Printed materials are estimated as 240 terabytes (1 terabyte is equal to 10^{15} bites). This makes up less than a fraction of 1% of the total new information storage. However, there are about 7.5 billion office documents printed each year, as well as almost one million books, 40,000 journals, and 80,000 magazines.

At the same time, deficiencies with information quality impose significant costs on individual companies, organizations, and the whole economy, resulting in the estimated costs only to the US economy at $600 billion per year (cf., (Hill, 2004)).

During the last fifty years or so, the term *information* succeeded in pervading science, technology, humanities, economy, politics, as well as everyday life. Goverments, companies and some individuals spend vast fortunes to acquire necessary information. However, there are many problems with this term. As Goguen writes (1997), "we live in an "Age of Information," but it is an open scandal that there is no theory, nor even definition, of information that is both broad and precise enough to make such an assertion meaningful." Indeed, we are overwhelmed with myriad information from a wide spectrum of information sources, such as the World Wide Web, e-mails, images, speeches, documents, books, journals, etc. Moreover, as Loewenstein (1999) demonstrates, "information flow ... is the prime mover of life – molecular information flowing in circles brings forth the organization we call 'organism' and maintains it against the ever-present disorganizing pressures in the physics universe." Each living being on the Earth has the structure determined by information encoded in its DNA.

At the same time, our experience demonstrates that mundane understanding of the notion of information may be very misleading. For instance, many identify information and message. However, receiving a message that consists of a random sequence of letters, a person gets no information. Thus, there are messages without information. Another example that demonstrates that a message and information are distinct essences is a situation when one and same message gives different information to different people. So, as it often happens in science, to overcome limitations of the commonplace image of information, the main problem is to find or to build the right theory.

Looking into information science, we encounter a peculiar situation. On the one hand, it has a lot of theories, a diversity of results, and even a proclaimed success. Scientists have created a bulk of information theories: Shannon's statistical information theory, semantic information theory, dynamic information theory, qualitative information theory, Marschak's economical information theory, utility information theory, Fisher's statistical information theory, algorithmic information theory and so on. Researchers study information ecology (cf., for example, (Davenport, 1997) and information economics (cf., for example, (Marschak, 1959; 1964; Arrow, 1984; Godin, 2008)), created information algebra (cf., for example, (Burgin, 1997b; Kohlas, 2003)), information geometry (cf., for example, (Amari and Nagaoka, 1985)), information logic (van Rijsbergen, 1986; 1989; Demri and Orlowska,

1999), information calculus (van Rijsbergen and Laimas, 1996), physics of information (cf., for example, (Stonier, 1990; Siegfried, 2000; Pattee, 2006)), and philosophy of information (cf., for example, (Herold, 2004)), but still do not know what information is. Each year, dozens of books and thousands of papers are published on problems of information. On the other hand, as it is written in the introduction to the authoritative book "Information Policy and Scientific Research" (1986), *"Our main problem is that we do not really know what information is."*

Moreover, due to the huge diversity of phenomena that are considered under the name "information", some researchers have come to the conclusion that it is impossible to have an all-encompassing definition of information, as well as to build a unified theory of information (cf., for example, (Capuro, Fleissner, and Hofkirchner, 1999) or (Melik-Gaikazyan, 1997)).

The situation of insufficient knowledge on information is really critical, because society more and more relies on information processing. Computers are becoming smaller, cheaper, interconnected and ubiquitous, while the impact of the Internet grows continuously. Computers solve a multitude of problems for people. Embedded devices control cars and planes, ships and spacecrafts. Furth (1994) writes:

"In 1991 companies for the first time spent more on computing and communications gear — the capital goods of the new era — than on industrial, mining, farm, and construction machines. Info tech is now as vital, and often as intangible, as the air we breathe, which is filled with radio waves. In an automobile you notice the $675 worth of steel around you, but not the $782 worth of microelectronics."

The question "What is information?" is singled out as one of the most profound and pervasive problems for computer science. As Wing writes (2008), it reflects the kind of far-reaching issues that drive day-to-day research and researchers toward understanding and expanding the frontiers of computing.

As we see, the concept of information is one of the greatest ideas ever created by the mind of human being, and our task is to make this concept exact in the whole its entirety. For instance, Stonier (1996) studies information as a basic property of the universe.

Information processes dominate society, permeating into every sphere of life, and a tiny amount of information can cause destruction of the whole Earth because pushing a small button can start a nuclear war. Von Baeyer states (2004), *information is the essence of science*, while contemporary society is based on modern technology, which in turn, is based on science. Thus, it is vital to understand information as a phenomenon, to know regularities of information processes, and to use this knowledge to the benefit of humankind. Stonier (1991) and other researchers stress, information science is badly in need of an information theory.

Numerous attempts to achieve this knowledge and to build such a theory are reflected in the literature where information is considered. There are many excellent, good and not so good books on information. Every year dozen of books appear in which some issues of information treated. So, suggesting to a reader one more book on information, it is necessary to explain why this book is different from the diversity of those books on information that have already been published.

The main goal of this book is not to demonstrate a series of sophisticated theorems and not to describe one or two (even the most popular) ways of understanding and modeling the phenomenon of information. Other books will do this. This book is aimed at finding ways for a synthesized understanding the information phenomena and at building a general theory of information encapsulating other existing information theories in one unified theoretical system. With this goal in mind, after presenting the general theory of information, the main attention in the book is paid to the most popular information theories with the solid mathematical core. This allows the reader to understand how the general theory of information unifies existing knowledge in information studies.

From this perspective, the book has a three-fold purpose. The first aspiration is to present the main mathematically based directions in information theory. The second objective is analysis and synthesis of existing directions into a unified system based on a new approach that is called the *general theory of information*. The third aspiration of the book is to explain how this synthesis opens new kinds of possibilities for information technology, information sciences, computer science, knowledge engineering, psychology, linguistics, social sciences, and education.

At the beginning of the book, contemporary situation in information studies and information processing is analyzed. Problems and paradoxes related to theoretical issues of information and information processes are discussed. To develop a valid and efficient theory, we need a thorough analysis of information processes in nature, technology, and society. We provide such an analysis in the first chapter of this book.

Elaborating on this analysis, as well on synthesis of the core ideas in existing information theories, we make a natural step to the general theory of information, foundations of which are presented in the second chapter of this book. This theory provides a base for systematizing other information theories and for unification of our knowledge about information, information systems, and information processes. It makes possible to overcome arguments of those who argue the impossibility of a unified information theory.

According to Arthur Clarke (1962), the only way of discovering the limits of the possible is to venture a little way past them into the impossible. In the general theory of information, this is done by means of a parametric approach to information with a specific system parameter called an infological system. This provides for a better understanding, as well as for more efficient theory development and practical applications. Discoveries of new, unknown before types of information are portrayed in the book. In addition, applications of the general theory of information to different fields are demonstrated in the second chapter. Relations between knowledge and information are found and analyzed from the point of view of knowledge management. The function of information in education is made clear, changing our understanding of educational processes. The role of information in the functioning of an individual is illuminated, demonstrating existence of new kinds of information.

Next chapters contain the most important and developed approaches and special theories of information, which are presented from a general perspective given by the general theory of information. An emphasis is made on mathematically oriented theories. We consider such theories as statistical information theory (including Shannon's theory and its later development), Fisher's approach to information and its application to physics, algorithmic information theory, and different versions of semantic information theory and of qualitative information theory. A wide spectrum of information theories is presented, demonstrating a lot of research that goes beyond the most popular direction called Shannon's information, or communication, theory. The general theory of information does not only allow one to unify this variety of theoretical directions, but also it helps to enlighten applications and development of theoretical results.

The primary audience for the book includes undergraduate and graduate students, information technology professionals, and researchers in information sciences, computer science, knowledge engineering, cognitive sciences, philosophy, artificial intelligence, psychology, linguistics, bioinformatics, social sciences, and education. Experts in information technology will be exposed to tools that information theory suggests for their professional activity: in designing software, building computers, applying computers to solving various problems, and developing global and local networks.

In addition, a reader with an interest in science and technology can find here an explanation of information essence and functioning, as well as answers to the following questions:
- how information is related to knowledge and data,
- how information is modeled by mathematical structures,
- how these models are used to better understand computers and the Internet, cognition and education, communication and computation.

It is necessary to inform potential readers who would prefer to understand the essence without dealing with mathematical formulas that the book gives this opportunity. In spite that it has many formulas, it is possible to skip these formulas and still get a lot of information about the information phenomena. The exposition is constructed in such a way that it allows the reader to achieve several levels of understanding. It is possible to read without formulas and to achieve the first level. Understanding of basic formulas brings the reader to the second level. Mathematical results formulated in the form of propositions and theorems elevate the reader to the third level of understanding.

Acknowledgments

Many wonderful people have made contributions to my efforts with this work. I am especially grateful to the staff at World Scientific and especially, Ms. Tan Rok Ting, for their encouragement and help in bringing about this publication. I would like to thank the teachers and especially, my thesis advisor, A. G. Kurosh, who helped shape my scientific viewpoint and research style. In developing ideas in information theory, I have benefited from conversations with many friends and colleagues. Thus, I am grateful for the interest and helpful discussions with those who have communicated with me on these problems. I greatly appreciate advice and help of A. N. Kolmogorov from Moscow State University in the development of algorithmic information theory. I have also benefited from the discussions I had with Gordana Dodig-Crnkovic from Mälardalen University on general problems of information and with Frank Land from the London School of Economics and Political Science on knowledge management. Collaboration with Narayan Debnath from Winona State University in developing applications of algorithmic information theory to software engineering gave much to the development of ideas and concepts of the general theory of information. I would also like to thank the reviewers for their useful comments. Credit for my desire to write this book must go to my academic colleagues. Their questions and queries made significant contribution to my understanding of algorithms and computation. I would particularly like to thank many fine participants of the Jacob Marschak Interdisciplinary Colloquium on Mathematics in the Behavioral Sciences at UCLA and especially, Colloquium Director, Michael Intrilligator, for extensive and helpful discussions on problems of information that gave me much encouragement for further work in this direction. Comments and observations of participants of the Applied Mathematics Colloquium of the Department of Mathematics, Seminar of Theoretical Computer Science of the Department of Computer Science at UCLA, and the Internet discussion group on Foundations of Information Science (FIS) were useful in the development of my views on information. I would also like to thank the Departments of Mathematics and Computer Science in the School of Engineering at UCLA for providing space, equipment, and helpful discussions.

M. Burgin
Los Angeles

Contents

Chapter 1

Introduction

It is impossible to defeat an ignorant man by argument.
William McAdoo

... I had thought that the magic of the information age was that
it allowed us to know more,
but then I realized that the magic of the information age was that
it allowed us to know less.
David Brooks

We live in the world where information is everywhere. All knowledge is possible only because we receive, collect and produce information. People discovered existence of information and now talk of information is everywhere in our society. As Barwise and Seligman write (1997), in recent years, information became all the rage. The reason is that people are immersed in information, they cannot live without information and they are information systems themselves. The whole life is based on information processes as Loewenstein convincingly demonstrates in his book (1999). Information has become a key concept in sociology, political science, and the economics of the so-called information society. Thus, to better understand life, society, technology and many other things, we need to know what information is and how it behaves. Debons and Horne write (1997), if information science is to be a science of information, then some clear understanding of the object in question requires definition. That is why in the first section of this chapter, we show different approaches to defining and understanding information, demonstrating what conundrum exists in this area.

1

In the second section, we discuss the role of information in society. In the third section, we examine the role of information in nature. In the fourth section, technological aspects of information are considered. The last section of Introduction reveals the goals and structure of the whole book.

1.1. How Do We Know What Information Is?

> *Five paths to a single destination. What a waste.*
> *Better a labyrinth that leads everywhere and nowhere.*
> Umberto Eco, *Foucalt's Pendulum*

Etymologically the term information is a noun formed from the verb "*to inform*", which was borrowed in the 15th century from the Latin word "*informare*", which means "*to give form to*", "*to shape*", or "*to form*".

As Capurro writes (1978; 1991), "key theories of Greek ontology and epistemology based on the concepts of *typos, idéa* and *morphé* were at the origin of the Latin term *informatio*. These connotations were maintained throughout the Middle Ages but disappeared as scholastic ontology was superseded by modern science."

During the Renaissance the word "*to inform*" was synonymous to the word "*to instruct.*" Later its meaning extended essentially and it became one of the most important technical and scientific terms. One of the most common ways to define information is to describe it as one or more statements or facts that are received by people and that have some form of worth to the recipient (Losee, 1997).

A detailed exposition for the etymological and historical background of the meaning of the term "information" is given by Capurro (1978). Here we are mostly interested in the scientific meaning of this term.

The problem with the word *information* is that the concept seems so intuitive and pervasive that people do not think a lot about it. However, Martin (1995) writes:

What is information? ... although the question may appear rhetorical, there is a sense in which the answer is that nobody really knows.

Thus, academics have long arguments about what constitutes information and many tried to answer the question "What is Information?"

The outstanding American mathematician and cybernetician Norbert Wiener (1894–1964) was one of the first who considered information beyond its day-to-day usage. He was one of the founders of cybernetics as a scientific discipline. The aim was to bring together similar research efforts in communication engineering, psychology, sociology, biology and medicine. From the point of view of these disciplines, it is above all the quantity of information which, apart from message, amount of interference (noise) and coding technique, is to be accounted for. According to Wiener the transmission of information is only possible as a transmission of alternatives, for if only a message about one possible state is to be transmitted it is most easily done by not transmitting any information at all. Therefore he calls for the development of a statistical theory of the amount of information, a quantity that has natural affinities to entropy in statistical mechanics. While the amount of information of a system is a measure of the degree of order, the entropy of a system is one of the measures of the degree of disorder. However, this did not solve the question of a concept of information proper. Throughout his life Wiener attached special importance to finding an answer to this question. To this purpose he made use of the results of a long-term collaboration with medical scientists.

His research led Wiener to make the famous statement (cf., for example, (Wiener, 1961)):

"Information is information, not matter or energy."

Although it is not a definition of information, this statement contained the message that the actual objects used for communication, i.e., for conveying information, are less important than information itself.

Wiener also described information in a different way. He writes (1954) that information is a name for the content of what is exchanged with the outer world as we adjust to it, and make our adjustment felt upon it.

Chaitin (1999) developed this idea. He writes:

"The conventional view is that matter is primary, and that information, if it exists, emerges from the matter. But what if information is primary and matter is the secondary phenomenon! After all, the same information can have many different material representations in biology, in physics, and in psychology: DNA, RNA; DVD's, videotapes; long-term memory, short-term memory, nerve impulses, hormones. The material representation is irrelevant, what counts is the information itself. The same software can run on many machines.

INFORMATION is a really revolutionary new kind of concept, and recognition of this fact is one of the milestones of this age".

Even before Wiener, electrical engineers instead of providing a definition of information and looking for its intrinsic features, focused on measuring information, using the term to describe data transmission. The goal was to maximize information transmitted or received, or minimize noise, or both. In 1928, Hartley published a paper, called *Transmission of Information*. There he used the word information, and made explicitly clear the idea that information in this context was a measurable quantity. The suggested measure reflected only that the receiver was able to distinguish that one sequence of symbols had been sent rather than any other — quite regardless of any associated meaning or other psychological or semantic aspect the symbols might represent.

This venue of research successfully continued and since the time of Wiener's pioneering works, information science emerged giving birth to many information theories and producing a quantity of definitions of information. The birth of information theory is placed officially in 1948, when the outstanding American engineer and mathematician Claude Elwood Shannon (1916–2001) published his first epoch-making paper.

In this work, Shannon further developed works of Ralph Vinton Lyon Hartley (1888–1970) and Harry Nyquist (1889–1976), who introduced fundamental ideas related to the transmission of information in the context of the telegraph as a communications system.

Shannon himself applied the word *information* only in a descriptive sense to the output of an information source, and he stays resolutely within the framework of telecommunications, using the title *communication theory*. His followers renamed the theory *information theory*, and now, as Hajek writes, it is too late to revert to the name given by Shannon. Moreover, Shannon's theory is a kind of information theory as communication is information exchange. As this theory is built on statistical considerations, it is called *statistical information theory*.

Although Shannon called what he created a theory of communication and wrote not about information itself, but about quantity of information, his research made possible to elaborate some definitions of information.

In the 1950s and 1960s, Shannon's concept of information invaded various disciplines. Apart from the natural sciences and economics, it was mainly in the humanities, in particular, in cognitive psychology, that specialized definitions of information were elaborated (cf. (Gibson, 1966; Neisser, 1967; Seiffert, 1968; Attneave, 1974; Collins, 2007)). In addition, increasing efforts were made to develop a unified definition integrating all aspects of the problem that had already been investigated in individual disciplines. Important efforts for the development of a definition were made in semiotics, cybernetics and philosophy (cf. (MacKay, 1956; Shreider, 1967; Nauta 1970; Titze, 1971)). However, apart from a variety of verbal definitions and vague hints for the development of new information theories, these efforts have yielded little and have certainly not led to a generally accepted definition.

For the majority of people, the most popular idea is that information is a message or communication. But a message is not information because the same message can contain a lot of information for one person and no information for another person.

The most utilized scientific definition of information (cf., for example, (Hartley, 1928) or (Ursul, 1971)) is:

Information is the eliminated uncertainty. (1.1.1)

Another version of this definition treats information as a more general essence and has the following form:

Information is the eliminated uncertainty or reflected variety. (1.1.2)

For example, Rauterberg (1995) assumes that for representation of information processes in learning systems that interact with their environment, the concept of variety is more relevant than the concept of uncertainty reflected by probabilities in the formulas of Hartley and Shannon.

However, in many cases, people speak about receiving or transmitting information when the variety is undefined and there is no uncertainty. In other cases, there is variety without information.

This is illustrated by the opinions of different authors. For example, an interesting idea is suggested in the book of Knapp (1978) where variety is defined in the orthogonal way to information. In non-technical language, this means that variety, as a phenomenon, is essentially distinct from information. This approach correlates with what writes Wilson (1993): *"In the real world ... we frequently receive communications of facts, data, news, or whatever which leave us more confused than ever. Under the formal definition these communications contain no information..."*

Both definitions (1.1.1) and (1.1.2) are based on Shannon's information theory (Shannon, 1948). This theory represents statistical approach to information and is the most popular now. However, one of the followers of Shannon, the well-known French scientist Leon Brillouin (1889–1969) wrote that in this theory "the human aspect of information" is completely ignored. As a result, statistical approach has been very misleading in social sciences and humanities. So, it was not by chance that Claude Shannon called it a theory of communication but not of information. Besides, Shannon himself never defined information and wrote only about the quantity of information.

As a result, many authors tried to define information independently of Shannon. This resulted in information studies perplexity. A vivid picture of confusion about information is given in the book of Poster (1990). He begins with the statement that information *"has become a privileged term in our culture that evokes a certain feature of the new cultural conjuncture and must be treated with suspicion."* He writes about *"many forms of information: words, numbers, music, visual images"*. Describing the scientific approach, he writes that theorists like to define information in a broad sense *"simply as organization of matter and energy"*. Poster (1990) also assumes that information in the narrow sense is that part of a communication that is not "lost" in its transmission. The part that is "lost" is noise. However, there is a lot of examples when information is lost in transmission and sometimes only noise is left. For instance (cf. (Burton, 1997), one of the great mathematicians Evariste Galois (1811–1834) submitted his very important results to the French Academy of Sciences. Another outstanding mathematician Augustin-Lois Cauchy (1789–1857), a member of the Academy and professor, was appointed referee. Cauchy either forgot or lost the communication of Galois, as well as another one presented later.

In an encyclopedic dictionary of semiotics, information has been defined the following way (Moles, 1994):

"In modern semiotics the word information has two meanings. The first is the common language acceptance: a message containing novelty. In the second, technical sense, it is the measurement of the quantity of novelty conveyed by a message..."

According to Borgman (1999) information is a relation between humans and physical reality and consists of signs, which within a certain context, inform people about things in reality. In contrast to this, Parker (1974) defines information as the pattern of organization of matter and energy.

Nauta (1970) establishes a close relation between information and improbability, saying:

"Information is news: what is known already is no information. So, something is information to the extent that it is unknown, unexpected, surprising, or improbable."

Some researchers define knowledge in terms of information, while other researchers define information in terms of knowledge. For instance, Brookes (1980) suggests that "information is a small piece of knowledge" and Kogut and Zander (1992) conceive information as "knowledge which can be transmitted without loss of integrity." At the same time, Meadow and Yuan write (1997) that knowledge is the accumulation and integration of information.

If we take such an authoritative source of definitions as The American Heritage Dictionary (1996), we see the following definitions.

Information is: **1.** Knowledge derived from study, experience, or instruction. **2.** Knowledge of a specific event or situation; intelligence. **3.** A collection of facts or data: "statistical information." **4.** The act of informing or the condition of being informed; communication of knowledge: "Safety instructions are provided for the information of our passengers." **5.** (in Computer Science) A nonaccidental signal or character used as an input to a computer or communications system. **6.** A numerical measure of the uncertainty of an experimental outcome. **7.** (in Law) A formal accusation of a crime made by a public officer rather than by grand jury indictment.

Similar definitions of information are in the Roget's New Thesaurus: **1**. That which is known about a specific subject or situation: data, fact (used in plural), intelligence, knowledge, lore. **2**. That which is known; the sum of what has been perceived, discovered, or inferred: knowledge, lore, wisdom.

Information is also considered as data (both factual and numeric) that is organized and imbued with meaning or as intelligence resulting from the assembly, analysis, or summary of data into a meaningful form (McGee, 1993; Walker, 1993). According Curtis (1989), information is data processed for a purpose. According Senn (1990), information is data presented in a form that meaningful to the recipient. However, while some associate information with data, others associate it with knowledge.

As a result, the term *information* has been used interchangeably with many other words, such as content, data, meaning, interpretation, significance, intentionality, semantics, knowledge, etc. In the field of knowledge acquisition and management, information is contrasted to knowledge. Some researchers assume that if Plato took knowledge to be "justified true belief", then information is what is left of knowledge when one takes away belief, justification, and truth.

Buckland (1991) analyzes the concept of information and comes to the conclusion that the word *information* is and can be used in the same meaning as knowledge. According to Godin (2008), studies of information economy started in 1950s with the information-as-knowledge approach. However, later this was changed to the conception of information as commodity or economic activity (late 1970s and early 1980s) and then to the conception of information as technology. Information came to be defined very broadly. It included just about anything that was intangible (Godin, 2008). All this adds confusion into understanding information as a phenomenon.

Moreover, in the textbook (O'Brien, 1995) used at universities and colleges, it is written that terms *data* and *information* can be used interchangeably, but while *data are raw material resources, information are data that has been transformed into a meaningful and useful context*. In (Laudon, 1996), we find a similar notion of information, which is defined as *an organized collection of data that can be understood*.

Lerner (1999; 2004; 2007) writes that information measures uncertainty, is characterized by an equivalent reduction of uncertainty, and has different measures.

One more definition of information is presented in (Rochester, 1996). According to him, *information is an organized collection of facts and data*. Rochester develops this definition through building a hierarchy in which data are transformed into information into knowledge into wisdom. Thus, information appears as an intermediate level leading from data to knowledge.

Ignoring that an *"organized collection"* is not a sufficiently exact concept, it is possible to come to a conclusion that we have an appropriate definition of information. This definition and similar ones are used in a lot of monographs and textbooks on computer science. Disregarding slight differences, we may assume that this is the most popular definition of information. This gives an impression that we actually have a working concept.

Many will say, "If such a definition exists and people who are experts and information theory in computer science use it, then what's wrong with it? Why we need something else?"

To explain why this definition is actually incoherent, let us consider some examples where information is involved.

The first example is dealing with a text that contains a lot of highly organized data. However, this text is written in Chinese. An individual, who does not know Chinese, cannot understand this text. Consequently, it contains no information for this person because such a person cannot distinct this text from a senseless collection of hieroglyphs. Thus, we have a collection of organized data, but it contains information only for those who know Chinese. Thus, we come to a conclusion that information is something different from this collection of organized data.

It is possible to speculate that this collection of data is really information but it is accessible only by those who can understand the text. In our case, they are those who know Chinese.

Nevertheless, this is not the case. To explain this, we consider the second example. We have another text, which is a review paper in mathematics. Three people, a high level mathematician **A**, a mathematics major **B**, and a layman **C**, encounter this paper, which is in the field of expertise of **A**. After all three of them read or tried to read the paper, they come to the following conclusion. The paper contains very little information for **A** because he already knows what is written in it. The paper contains no information for **C** because he does not understand it. The paper contains a lot of information for **B** because he can understand it and knows very little about the material that is presented in it.

So, the paper contains different information for each of them. At the same time, data in the paper are not changing as well as their organization.

This vividly shows that data, even with a high organization, and information have an extremely distinct nature. Structuring and restructuring cannot eliminate these distinctions.

Although informatics, information science, and computer science are often in the spotlight, they do not provide necessary understanding of the situation and the word "information" is often used without careful consideration of the various meanings it has acquired.

For instance, *information science* is the study of the gathering, organizing, storing, retrieving, dissemination of information (Bates, 1999). In a more general interpretation, it is a study of all aspects of information: information processes, properties, functions, relations, systems, etc. For instance, Borko (1968) wrote:

"Information science is the discipline that investigates the properties and behavior of information, the forces governing the flow of information, and the means of processing information for optimum accessibility and usability. It is concerned with the body of knowledge related to the origination, collection, organization, storage, retrieval, interpretation, transmission, transformation, and utilization of information."

Giving an overview of the controversy over the concept of information, Qvortrup writes (1993):

"Thus, actually two conflicting metaphors are being used: The well-known metaphor of information as a quantity, like water in the water-pipe, is at work [see also conduit metaphor], but so is a second metaphor, that of information as a choice, a choice made by an information provider, and a forced choice made by an information receiver.

Information is the state of a system of interest (curiosity). Message is the information materialized."

Hu and Feng (2006) define *information*, carried by non-empty, well-formed, meaningful, and truthful data, as a set of states of affairs, which are part of the real world and independent of its receivers.

Bunge and Ardila (1987) distinguish between the following seven different ways in which the term *information* is used:

1. Information as meaning (semantic information).
2. Information as the structure of genetic material (genetic "information").
3. Information as a signal.
4. Information as a message carried by a pulse-coded signal.
5. Information as the quantity of information carried by a signal in a system.
6. Information as knowledge.
7. Information in a sense of communication of information (knowledge) by social behavior (e.g., speech) involving a signal.

Wersig (1997) separates six types of information understanding in information theory:

Structures of the world are information.

Knowledge developed from perception is information.

Message is information.

Meaning assigned to data is information.

Effect of a specific process, e.g., reducing uncertainty or change of knowledge, is information.

Process, commonly a process of transfer, is information.

One more example of the general confusion about information is that some, may be the majority of researchers, relate information only to society or, at least, to intelligent systems (cf., for example, (O'Brien, 1995) or (Laudon, 1996)), while others contradict information and communication, treating information as a category of solely physical systems (cf., for example, (Bougnoux, 1995)).

In addition to this, there were other problems with theoretical studies of information. For example, Shannon's information theory applies only in those contexts where its precise assumptions hold, i.e., never in reality. However, experts in information studies understood that this does not imply that an attempt to create a more general theory of information should not be pursued. On the contrary it should. The existing theories are actually too restrictive.

Many researchers consider information as an individual's brain construction (cf., for example, (Maturana and Varela, 1980; von Foerster, 1980, 1984; Flückiger, 1999)). According to Qvortrup (1993), treating information as a mental difference "doesn't necessarily imply that the difference in reality that triggered the mental difference called information is a mental construction."

Mackay (1969) suggests, "information is a *distinction* that makes a difference." A more metaphorical definition of Bateson (1972) describes *information* as "a difference that makes a difference." Clancey (1997) expanded Bateson's definition, proposing that *information* is the detection of a difference that is functionally important for an agent to adapt to a certain context. Muller (2007) assumes that "information is the facility of an object to distinguish itself." In the same venue, Markov, et al (2007) write that when a triad

(source, evidence, recipient)

exists, then the reflection of the first entity in the second one is called information. Thus, *information* is interpreted as a specific reflection.

At the same time, some researchers try to define information, using various abstract concepts. For instance, Crutchfield (1990) defines information of a source S as the equivalence class of all recordings of the symbol sequences from S. This is an interesting definition. However, it does not allow one to consider many kinds of information that is not represented by symbol sequences.

In algorithmic information theory (Kolmogorov, 1965; Chaitin, 1977), information is treated as tentatively *eliminated complexity*.

Many researchers do not discern information and a measure of information. For instance, in a well-written book of Abramson (1963), there is a section with the title "The definition of Information." However, it contains only one definition, which tells us how much information we get being told that some event occurred. Naturally, this does not explain what information is.

An interesting definition is suggested by Carl Friedrich Freiherr von Weizsäcker (1912–2007), who writes, "information is a quantitative measure of form (Gestalt)" (von Weizsäcker, 2006) and conceives information as a twofold category: (1) information is only that which is understood; (2) Information is only that which generates information (von Weizsäcker, 1974).

Losee (1997) uses the following definition of information:

Information is produced by all processes and it is the values of characteristics in the processes' output that are information.

Hobart and Schiffman (2000) suggest that the concept of information changes with time. They distinguish between classical, modern, and contemporary information ages, the meaning of information being unique to each age.

Information as a term is often closely related to such concepts as meaning, knowledge, instruction, communication, representation, and mental stimulus. For instance, information is treated as a message received and understood. In terms of data, information is defined as a collection of facts from which conclusions may be drawn. There are many other aspects of information that influence its definition. As a result, it is assumed that information is the knowledge acquired through study or experience or instruction. We see here information being a message, collection of facts and knowledge, and this is really confusing.

At the same time, challenging understanding information as a message, Stonier (1997) writes that "information is the raw material which, when information-processed, may yield a message."

In general, information has been considered as the following essences: as structures; processes (like becoming informed); changes in a knowledge system; some type of knowledge (for example, as personal beliefs or recorded knowledge); some type of data; an indication; intelligence; lore; wisdom; an advice; an accusation; signals; facts; acts; messages; as different things; as meaning; and as an effect like elimination of uncertainty (Brillouin, 1957; Ursul, 1971; Wersig and Neveling, 1976; Buckland, 1991; Wilson, 1993; etc.).

Ruben (1992) considers information on three levels or orders: the biological level, individual/psychological level, and interpersonal/ social/cultural level. Consequently, he distinguishes three rather distinct concepts of information. The first order of information, which exists on the biological level and called *Information$_e$*, is environmental data, stimuli, messages, or cues — artifacts and representations — which exist in the external environment. The second order of information, which exists on the individual/psychological level and called *Information$_i$*, is that which has been transformed and configured for use by a living system. It includes cognitive maps, cognitive schemes, semantic networks, personal constructs, images, and rules, i.e., internalized and individualized appropriations and representations. The third order of information, which exists on the social and cultural level and called *Information$_s$*, comprises the shared information/knowledge base of societies and other social system, i.e., socially constructed, negotiated, validated, sanctioned and/or privileged appropriations, representations, and artifacts.

While information has been defined in innumerable ways, there have been a lot of discussions and different approaches have been suggested trying to answer the question what information is. According to (Flückiger, 1995), in modern information theory a distinction is made between structural-attributive and functional-cybernetic types of theories. While representatives of the former approach conceive information as structure, like knowledge or data, variety, order, and so on; members of the latter understand information as functionality, functional meaning or as a property of organized systems.

Krippendorff (1994) explores different information and communication metaphors, such as information as a message transmission, the container metaphor, the metaphor of sharing common views, the argument metaphor, the canal metaphor, and the control metaphor. These metaphors, originating within different cultural environments, reflect important traits of information and information processes. However, any metaphor can be misleading and it is necessary to use it creatively, that is, to see its limits and to learn how to apply it accurately in different theoretical and practical situations.

Braman (1989) provides an important discussion of approaches to defining information for policy makers. Four major views are identified: (1) information as a resource, (2) information as a commodity, (3) information as a perception of patterns, and (4) information as a constitutive force in society. The relative benefits and problems with each of these four conceptions are discussed. The article points out that the selection of one definition or another has important consequences, and also that the tendency to neglect this problem results in conflicts rather than cooperation. Defining information is thus also a political decision.

Consequently, it is not surprising that intelligent people come to a conclusion that the main problem is that people and even experts in the field do not really know what information is (Bosma, 1985). Information is a term with too many meanings depending on the context. The costs of searching in the wrong places have been high because the superficial considering of the nature of information have left researchers in information science without a proper theoretical foundation, which is a very serious situation for an academic field.

In one of his lectures the well-known American philosopher Searle stressed that *"the notion of information is extremely misleading."* Another well-known American philosopher Hintikka writes (1984), the concept of information is multiply ambiguous. At the same time, the famous French mathematician Rene Thom (1975) calls the word *"information"* a *"semantic chameleon,"* that is something that changes itself easily to correspond to the environment. Various scientific and laymen imaginations about information stand often without the least explicit relationship to each other. Thus, we see that "too many" definitions may be as bad as "too few" ones. This has been noticed by researchers. For instance, Macgregor (2005) writes about conundrums for the informatics community caused by problems of defining information and finding the nature of information.

As Meadow assumes (1995), one of the problems all of researchers who study and try to measure the impact of information have is the multiple definitions of the very word *information.*

Van Rijsbergen and Laimas write (1996):

"Information is and always has been an elusive concept; nevertheless many philosophers, mathematicians, logicians, and computer scientists have felt that it is fundamental. Many attempts have been made to come up with some sensible and intuitively acceptable definition of information; up to now, none of these succeded."

Existing confusion with the concept of information is vividly rendered by Barwise and Seligman, who explain (1997):

"There are no completely safe ways of talking about information. The metaphor of information flowing is often misleading when applied to specific items of information, even as the general picture is usefully evocative of movement in space and time. The metaphor of information content is even worse, suggesting as it does that the information is somehow intrinsically contained in one source and so is equally informative to everyone and in every context."

One more problem is that trying to tell what information is a necessary clear distinction is not made between a definition and an attribute or feature of this concept. Normally, describing or naming one or two attributes is not considered a definition. However, there are authors who stress even a single feature of information to the point that it appears to be a definition.

Scarrott (1989) attracts attention to a dictionary definition of *information* as that which informs. This definition identifies *information* using a very vague term *to inform*, which, in turn, depends on the definition of information itself.

As a result of all inconsistencies related to the term *information*, many researchers prefer not to use the word *information*. As Crutchfield remarks (1990), "information generally is left undefined in information theory." Many experts believe that the attempt at information definition could present an act of futility (Debons and Horne, 1997).

Spang-Hanssen (2001, online) argues (cf. (Capuro and Hjorland, 2003)):

"In fact, we are not obliged to accept the word *information* as a professional term at all. It might be that this word is most useful when left without any formal definition, like e.g., the word *discussion*, or the word *difficulty*, or the word *literature*."

The term *information* is used in so many different contexts that a single precise definition encompassing all of its aspects can in principle not be formulated, argues Belkin (1978), indirectly quoting (Goffman, 1970).

An extreme point of view is to completely exclude the term information from the scientific lexicon and to abandon the term from the dictionary (Fairthorn, 1973). In the same way, in his paper "Information studies without information", Furner writes (2004) that it is easy to see that philosophers of language have modeled the phenomena fundamental to human communication in ways that do not require us to commit to a separate concept of "information." Thus, he concludes, such a concept as *information* is unnecessary for information studies. Once the concepts of interest in this area have been labeled with conventional names such as "data", "meaning", "communication", "relevance", etc., there is nothing left (so Furner argues) to which to apply the term "information."

While it is possible to partially agree with Furner in the first part of his claim. Namely, it is true that in many cases researchers use the word *information* when it is necessary to use some other word. In some cases, it may be the word *data*. In other cases, it may be the word *knowledge* or *message* and so on.

All this demonstrates that the existing variety of definitions lacks a system approach to the phenomenon, and for a long time there has been a need to elaborate a concept that reflects the main properties of information. This concept has to be the base for a new theory of information, giving an efficient tool for information processing and management. Mattessich (1993) assumes that the concept of information requires a conceptualizing process no less torturous than the one that purified the notion of energy during the last 200 years.

Meadow and Yuan (1997) argue:

"How is it possible to formulate a scientific theory of information? The first requirement is to start from a precise definition."

In a similar way, many researchers insist that it is necessary to have a definition of information. For instance, Braman (1989) stresses how important it is for information policy to define information adequately. Mingers (1996) argues, if information theory is to become a properly founded discipline, then it must make an attempt to clarify its fundamental concepts, starting with the concept of information.

The opening paragraph in the preface of the monograph of Barwise and Seligman (1997) states that there is no accepted science of information. Lenski (2004) adds to this that as a broadly accepted *science* of information is not envisaged by now, a commonly acknowledged *theory* of information processing is out of sight as well.

However, knowing about information is not only a theoretical necessity but is a practical demand. Considering the United States of America in the information age, Giuliano (1983) states that the "*informatization process is very poorly understood. One of the reasons for this is that information work is very often seen as overhead; as something that is necessary but not contributory.*"

In a similar way, Scarrott writes (1989):

"*During the last few years many of the more perceptive workers in the information systems field have become uneasily aware that, despite the triumphant progress of information technology, there is still no generally agreed answers to the simple questions — What is information? Has information natural properties? What are they? — so that their subject lacks trustworthy foundations.*"

Capuro and Hjorland (2003) also stress, "for a science like information science (IS) it is of course important how fundamental terms are defined." In addition, they assume that even "discussions about the concept of information in other disciplines are very important for IS because many theories and approaches in IS have their origins elsewhere."

This supports opinion of those researchers who argued that multifarious usage of the term *information* precludes the possibility of developing a rigorous and coherent definition (cf., for example, (Mingers, 1996)). Nevertheless, if information studies are to become a properly founded discipline, then it is necessary to clarify its most fundamental concept *information* (cf., for example, (Cornelius, 2002)).

The reason is the perplexing situation with information science. Schrader (1983) who analyzed this situation came to the following conclusion.

"*The literature of information science is characterized by conceptual chaos. This conceptual chaos issues from a variety of problems in the definitional literature of information science: uncritical citing of previous definitions; conflating of study and practice; obsessive claims to scientific status; a narrow view of technology; disregard for literature without the science or technology label; inappropriate analogies; circular definition; and, the multiplicity of vague, contradictory, and sometimes bizarre notions of the nature of the term 'information'.*"

In addition, in spite of a multitude of papers and books concerning information and a lot of studies in this area, many important properties of information are unknown. As Wilson writes (1993), "*'Information' is such a widely used word, such a commonsensical word, that it may seem surprising that it has given 'information scientists' so much trouble over the years.*" This is one of the reasons why no adequate concept (as well as understanding) of information phenomenon has been produced by information theory till the last decade of the 20th century.

Machlup and Mansfield (1983) presented key views on the interdisciplinary controversy over the concept of information in computer science, artificial intelligence, library and information science, linguistics, psychology, and physics, as well as in the social sciences.

Thus, it is not a surprise that, as Dretske (1981), Lewis (1991) and Mingers (1996) point out, few books concerning information systems actually define the concept of information clearly.

Kellogg (http://www.patrickkellogg.com/school/papers/) tells that at a talk he attended, the invited speaker started out by saying that the concept of information was universal, and that nobody could argue about its definition. "Aliens from another planet", the speaker claimed, "would agree with us instantly about what information is and what it isn't" and then for the rest of his talk, the speaker avoided giving a clear definition.

According to Sholle (1999), the information society is being sold to the average citizen as providing access to knowledge, meaningful dialogue and information essential to everyday decision-making. At the same time, even within the marketing of the information society and its benefits, *the actual nature of the information and knowledge produced and distributed by information technology remains abstract and actually undefined.* Instead, government and corporate pronouncements focus on the sheer power of the network, on the technological magic of information machines, on the overall capacity of the system, and on the abstract phenomenon of "being digital".

Many researchers assume that this diversity of information uses forms an insurmountable obstacle to creation of a unified comprehensible information theory (cf., for example, (Capurro, et al, 1999; Melik-Gaikazyan, 1997)). Capuro (Capuro, Fleissner, and Hofkirchner, 1999) even gives an informal proof of the, so-called,

Capuro trilemma that implies impossibility of a unified theory of information. According to his understanding, information may mean the same at all levels (univocity), or something similar (analogy), or something different (equivocity). In the first case, we lose all qualitative differences, as for instance, when we say that e-mail and cell reproduction are the same kind of information process. Not only the "stuff" and the structure but also the processes in cells and computer devices are rather different from each other. If we say the concept of information is being used analogically, then we have to state what the "original" meaning is. If it is the concept of information at the human level, then we are confronted with anthropomorphisms if we use it at a non-human level. We would say that "in some way" atoms "talk" to each other, etc. Finally, there is equivocity, which means that information cannot be a unifying concept any more, i.e., it cannot be the basis for the new paradigm…

It has been argued, for example, by Goffman (1970) and Gilligan (1994), that the term *information* has been used in so many different and sometimes incommensurable ways, forms and contexts that it is not even worthwhile to elaborate a single conceptualization achieving general agreement.

Shannon (1993) also was very cautious writing: "It is hardly to be expected that a single concept of information would satisfactorily account for the numerous possible applications of this general field."

Flückiger (1995) came to the conclusion that those working in areas directly related to information had apparently accepted that the problem with the definition would remain unsolved and considered Shannon's concept of information as the most appropriate.

Agre (1995) argues that the notion of information is itself a myth, mobilized to support certain institutions, such as libraries. Bowker (1994) discusses other mythologies that support the notion of information.

It may be caused by poor understanding the complex term *information* by many people, that other ways of using the word *information*, for example, as bits of information (cf., Chapter 3), have had a much stronger appeal. However, while the concept of a *bit* may allow one to measure the capacity of a floppy disc or a hard-disk, it is useless in relation to tasks such as indexing, collection management, document retrieval, bibliometrics and so on. For such purposes, the meaning of the signs must be involved, making a kind of semantic information theory a much better theoretical frame of reference compared to statistical information theory. An objective and universalistic theory of information has a much stronger appeal than theoretical views that make information, meaning and decisions context-dependent. However, the costs of searching in the wrong places have been high because the superficial considering of the nature of information leaves society without a proper theoretical foundation, which is a very serious situation for academic fields.

It is possible to compare the development of information sciences with the history of geometry. At first, different geometrical objects (lines, angles, circles, triangles etc.) have been investigated. When an adequate knowledge base of geometrical objects properties was created, a new step was taken by introducing the axiomatic theory, which is now called the Euclidean geometry. In a similar way, knowledge obtained in various directions of information theory (statistical (Shannon, 1948; Shannon and Weaver, 1949; Fisher, 1922; 1925), semantic (Bar-Hillel and Carnap, 1958), algorithmic (Solomonoff, 1964; Kolmogorov, 1965; Chaitin, 1977), qualitative (Mazur, 1984), economic (Marschak, 1971; 1980; 1980a), etc.) made it possible to make a new step — to elaborate a parametric theory called the general theory of information (GTI). As it is demonstrated in this book, all other known directions of information theory may be treated inside the general theory of information as its particular cases

The base of the general theory of information is a system of principles. There are two groups of such principles: ontological and axiological. Some of other approaches in information theory are also based on principles. For instance, Dretske (1981) considers the Xerox Principle and some others. Barwise and Seligman (1997) consider several principles of information flow.

It is necessary to remark that the general theory of information does not completely eliminate common understanding of the word information. This theory allows one to preserve common usage in a modified and refined form. For instance, when people say and write that *information is knowledge of a specific event or situation* (The American Heritage Dictionary, 1996), the general theory of information suggests that it is more adequate to say and write that *information gives knowledge of a specific event or situation.* When people say and write that *information is a collection of facts or data* (The American Heritage Dictionary, 1996), the general theory of information suggests that it is more adequate to say and write that *a collection of facts or data contains information.*

Derr (1985) give an analysis of the notion of information as used in ordinary language. This analysis shows that the general theory of information allows one to refine this usage, making it more consistent.

Thus, we can see that it is possible to divide all popular definitions of information into several classes. In one perspective, information is an objective essence, e.g., some kind of enhanced data. An alternative view emphasizes the subjective nature of information when the same carrier provides different information to different systems. Another dimension of information also has two categories: information as a thing and information as a property. According to Buckland (1991), the term *information* is used in different ways, including "information-as-knowledge", "information-as-thing" (e.g., data, signals, documents, etc.), and "information-as-process" (e.g., becoming or being informed).

The situation with information reminds us the famous ancient story of the blind men and an elephant. It had been used by many for different purposes, originating from India, having different versions, and being attributed to the Hindus, Buddhists or Jainists. Here we present a combined version of this tale, in which ten blind men are participating.

Once upon a time there was a certain raja, who called to his servant and said, "Go and gather together near my palace ten men who were born blind... and show them an elephant." The servant did as the raja commanded him. When the blind men came, the raja said to them, "Here is an elephant. Examine it and tell me what sort of thing the elephant is."

The first blind man who was tall found the head and said, "An elephant is like a big pot."

The second blind man who was small observed (by touching) the foot and declared, "An elephant is like a pillar".

The third blind man who was always methodical heard and felt the air as it was pushed by the elephant's flapping ear. Then he grasped the ear itself and felt its thin roughness. He laughed with delight, saying "This elephant is like a fan."

The fourth blind man who was very humble observed (by touching) the tail and said, "An elephant is like a frayed bit of rope."

The fifth blind man who was daring walked into the elephant's tusk. He felt the hard, smóoth ivory surface of the tusk and its pointed tip. "The elephant is hard and sharp like a spear," he concluded.

The sixth blind man who was small observed (by touching) the tuft of the tail and said, "An elephant is like a brush."

The seventh blind man who felt the trunk insisted the elephant was like a tree branch.

The eighth blind man who was always in a hurry bumped into the back and reckoned the elephant was like a mortar.

The ninth blind man was very tall. In his haste, he ran straight into the side of the elephant. He spread out his arms and felt the animal's broad, smooth side and said, "This is an animal is like a wall."

Then these nine blind men began to quarrel, shouting, "Yes it is!", "No, it is not!", "An elephant is not that!", "Yes, it's like that!" and so on.

The tenth blind man was very smart. He waited until all others made observations and told what they had found. He listened for a while how they quarreled. Then he walked all around the elephant, touching every part of it, smelling it, listening to all of its sounds. Finally he said, "I do not know what an elephant is like. That is why I am going to write an Elephant Veda, proving that it is impossible to tell what sort of thing the elephant is."

One more paradox related to information studies is that what is called information theory is not regarded as part of information science (IS) by many researchers (Hjørland, 1998). Concentrating on properties of information systems, such as information retrieval systems and libraries, behavior of their users, and functioning information carriers and representations, such as descriptors, citations, documents, titles and so on, information science paid much less attention to information itself.

Nevertheless, those information theories that have been created were found very useful in many practical and theoretical areas — from technology of communication and computation to physics, chemistry, and biology. That is why we are going to consider the essence of those information theories and explain their usefulness. However, before starting such an exposition, we informally consider the role and place of information in nature, society, and technology.

1.2. Information in Society

He that has knowledge spares his words:
and a man of understanding is of an excellent spirit.
Bible, Proverbs (Ch. XVII, v. 27)

Many books and numerous papers are written on social problems related to information. Here we give a general perspective on the situation and consider some of existing problems in this area.

Information plays more and more important and explicit role in society. Researchers understand information as a basic relationship between humans and reality (cf., for example, (Borgmann, 1999)). Information and information technology acquire a central place in public discourses. For instance, a search for the word *information* gave 3,710,000,000 results in 0.33 seconds with Google, gave 15,300,000,000 results in 0.23 seconds with Yahoo, and gave 1,220,000,000 results with AOL.

The outstanding role of information in culture is well understandable. Cultures exist and grow only due to information storage and transfer. Communication, i.e., information exchange, is one of the cornerstones of culture. Many cultural phenomena are studied from the communication perspective. The communicative approach to cultural studies assumes existence of a specific communication structure as a system of information interactions (Parahonsky, 1988). For instance, communication relations in art are based on the following communication triad (Kagan, 1984):

$$\textbf{Artist} \longrightarrow \textbf{Recipient(s)} \qquad (1.2.1)$$

This shows communication of the artist with her or his audience.

It is possible to extend this communication triad to the following composition of two triads:

$$\text{Artist} \longrightarrow \text{Piece of art} \longrightarrow \text{Recipient(s)} \quad (1.2.2)$$

The first triad reflects creation of the piece of art (painting sculpture, movie, etc.) by the artist, while the second triad shows how this piece of art influences the audience. This influence has an exclusively informational nature.

In the case, when an artist is a composer, playwright or screenwriter, the cultural communication triad (1.2.1) has to be extended to the following triad:

$$\text{Artist} \longrightarrow \text{Performer(s)} \longrightarrow \text{Recipient(s)} \quad (1.2.3)$$

In the process of creation, an artist usually communicates with herself or himself. This gives us the cultural self-communication triad:

$$\text{Artist A} \longrightarrow \text{Artist A} \quad (1.2.4)$$

Such exceptionally important processes as teaching and learning have information nature (Burgin, 2001a; Parker, 2001). Teaching always goes in communication and thus, it is based on information exchange (Burgin and Neishtadt, 1993). Learning also includes explicit and/or implicit communication, such as communication with nature in an experiment.

Actually, the whole society exists only because people and organizations are communicating. Without communication society is impossible. Even animals, birds and insects (such as bees and ants) are communicating with one another when they form a social group.

In addition, the role of information in economy has been growing very fast. Information is the key to management, research and development. In modern business, information is of primary interest. As Scarrott writes (1989), the most basic function of information is to control action in an organized system and thereby operate the organization. Thus, organizations exist only due to information exchange. Business processes, such as planning, product development, management, production, purchasing, advertising, promoting, marketing, selling, not only drive all businesses, but also generate valuable information.

There are different images of organizations: as machines, organisms, political systems, cultures, and learners (Morgan, 1998; Senge, 1990; Kirk, 1999). The *machine image* suggests that information is one of the resources that keep the wheels ticking over. The *organism image* implies that information from internal and external sources is required to keep the organization in the state of equilibrium. The *learner image* portrays an organization as a system based on obtaining information.

Thus, it is possible to consider any organization as an information processing engine. Information on many issues is represented by data in various databases: personnel databases, production databases, billing and collection databases, sales management databases, customer databases, supply chain databases, accounting databases, financial databases, and so on.

This explicates and stresses importance of information management. The work of managers in organizations, such as companies and enterprises, is very information-intensive, and the result of their work depends on information management to a great extent. As a result, information management contributes (in a positive or negative way) to the achievements of organizations, is used in a political, social and cultural contexts, has ethical dimensions and is value-laden (Kirk, 1999).

An important element in information management is information politics, which reflects rules and assumptions made about how people and organizations generate, preserve and use information. Davenport, et al, (1996) distinguish four types of information politics:

Technocratic utopianism is a heavily technological approach to information management stressing categorization and modeling of organization's full information assets.

Anarchy is characterized by the absence of any overall information management policy leaving individuals to obtain and manage their own information.

Feudalism is the management of information by individual business units groups, which define their own information needs and report only limited information to the overall corporation.

Federalism is an approach to information management based on consensus and negotiation on the organization's key information elements and reporting structures.

As Giachetti (1999) writes, global competition, shorter lead times and customer demands for increasing product variety have collectively forced manufacturing enterprises to develop rapidly, more closely collaborate with customers and suppliers, and more often introduce new products to obtain a quick return on their investment.

Information technology is regarded as means to solve basic problems in new product development, manufacture, and delivery.

The information economy is one of the key concepts to explain structural changes in the modern economy (Godin, 2008). Economical concern with information goes back to 1949. The Organization for European Economic Co-Operation (OEEC) organized a working party on scientific and technical information. The goal was to measure information activity as a tool to get science policy considerations into the organization. In these studies, information was treated as knowledge and limited to scientific and technological information. In turn, measurements and the corresponding statistics were limited to documentation in these areas.

In 1961, the Organization for Economic Cooperation and Development (OECD) was created. It continued research on information in society with the emphasis on the information economy. In this process, such fields as informatics aimed at a study of scientific information and science of science with its subfields: scientometrics, methodology of science, economy of science, politics in science and some others, emerged. The perspective on scientific and technical information was threefold: (1) information flow grows so fast that it can be called "information explosion"; (2) new technologies can tame this explosion; (3) this needs a common approach and united efforts of different countries (cf. (Godin, 2008)).

In 1969, the Information Policy Group of OECD suggested a methodological manual. It identified five specific classes of scientific and technical information activities (cf. (Godin, 2008)): (1) recording; (2) editing, revising, translating, etc. (3) distribution (including conferences); (4) collection, storage, processing; and (5) acquisition.

It is interesting to compare these information activities with information operations studied in the general theory of information

(Burgin, 1997b). There are three main classes of information operations with respect to a system R:

- *Informing* is an operation when (a portion of) information acts on R.
- *Information processing* is an operation when R acts on (a portion of) information.
- *Information utilization* is an operation when the system R acts on another system Q by means of information.

In turn, informing is also divided into three classes:

- *Information reception* is an operation when R gets information that comes from some source.
- *Information acquisition* is an operation when R performs work to get existing information.
- *Information production* is an operation when R produces new information.

Processing is divided into three classes:

- *Information transformation* is an operation when changes of information or its representation take place.
- *Information transition*, or as Bækgard (2006) calls it, *information movement*, is an operation when information is not changing but the place where the information carrier is changes.
- *Information storage* is an operation when information is not changing but time for the information carrier changes.

In turn, information transformation is also divided into three classes:

Information reconstruction is an operation when the initial information and/or its initial representation change.

Information construction is an operation when the initial information or/and its initial representation is not changing but new information and/or representation are created.

Information biconstruction is an operation when the initial information and/or its initial representation change and new information and/or representation are created.

In these terms, all information operations from the methodological manual — recording; editing, revising, and translating — are different kinds of information processing.

However, the methodological manual and the corresponding list of indicators suggested by the Information Policy Group of OECD were never used to measure scientific and technical information and related activities. Two factors explained this failure: the absence of the general conceptual framework and the fuzziness of the concept of information itself (Godin, 2008). This transparently shows how the lack of a good theory disables practical endeavors.

As importance of information was growing, researchers in the field started to consider information as a commodity in economic activity. In 1977, Porat and Rubin published a nine-volume study titled *The Information Economy*. The main assumption was that the USA and other developed countries have evolved from an economy based primarily in manufacturing and industry to one based primarily in knowledge, communication, and information. The goal was to get a total value of information in economy. Separating primary and secondary information sectors, the authors found that these sectors amounted to 46% of the US GNP and 53% of labor income in the USA. Now these numbers have become even larger.

As a result, deficiencies with information quality impose significant costs on individual companies, organizations, and the whole economy, resulting in the estimated costs only to the US economy at $600 billion per year (cf., (Hill, 2004)).

Although information still remains an economic commodity and its importance grows, the third period in information economy is characterized by the main emphasis on information technology, which occupies dominating positions in society.

That is why our time is often called *information age* (cf., for example, (Bell, 1973, 1980; Mason, 1986; Crawford and Bray-Crawford, 1995; Stephanidis, 2000)), while the modern society is called the *information society* (cf., for example, (Webster, 2002)). It is often used in conjunction with the term post-industrial society as it is usually related to a period after the industrial age. Braman (1989) treats information as a constitutive force in society. Information flow controls culture, education, technology, politics, and economy of the information society. Information is applied to a social structure of any degree of articulation and complexity. Its flow and use have an enormous power in constructing social reality.

It is impossible today to avoid constant reference to the *information age*. The phrase is so imbedded in our collective psyche that it can be used in any argument. We are "overwhelmed by the information age", or "our country needs to meet the challenges of the information age", the "information age" is good or the "information age" is causing information flood, knowledge pestilence and dire societal change.

According to Bougnoux (1993, 1995) the concepts of information and communication are inversely related. Communication is concerned with forecasting and redundancy, while information deals with the new and the unforeseen. There is no pure information or "information-in-itself" (that is, information is always related to some kind of redundancy or "noise"). To inform (others or oneself) means to select and to evaluate. This is particularly relevant in the field of journalism and mass media, but, of course, also in information science.

Information is playing an increasing role in our industrialized society. A technical overview of the flourishing electronics industry stated in 1987:

"On almost every technology front, the driving force behind new developments is the ever-rising demand for information. Huge amounts of data and information, larger than anyone ever dreamed of a generation ago, have to move faster and faster through processors and networks, then end up having to be stored" {Electronics, 1987, p.83}.

Four years later the *industrial age* had already given way to the *information age*. "In 1991 companies for the first time spent more on computing and communications gear — the capital goods of the new era — than on industrial, mining, farm, and construction machines. Infotech is now as vital, and often as intangible, as the air we breathe, which is filled with radio waves. In an automobile you notice the $675 worth of steel around you, but not the $782 worth of microelectronics" (Furth, 1994).

Information has become a highly valued economic category, on par with capital and skilled labor. The collection, structuring and processing of information consequently constitute one of the focal points of business. According to Poster (1990), the mode of information must now replace the mode of production as the key concept in a critical analysis of social formation. The three superseding historic modes are:

(1) spoken mode;

(2) printed mode;

(3) electronic mode.

A similar theoretical position was independently developed by Goguen (1997), who writes:

"An item of information is an interpretation of a configuration of signs for which members of some social group are accountable. That information is tied to a particular, concrete situation and a particular social group has some important consequences ..."

Borgmann (1999) distinguishes three types of information in society: natural information, cultural information, and technological information.

Natural information is information about reality and is conveyed by natural signs, such as a tree, river, smoke, the Moon, etc.

Cultural information is information for reality and is conveyed by conventional signs, such as a digit, letter, etc.

Technological information is information as reality. It does not give an access to reality as two other types of information. Technological information replaces reality.

Economists discuss how to build economics of information and develop theories of information economy (Arrow, Marschak and Harris, 1951; Good, 1952; McCarthy, 1956; Marschak, 1954; 1959; 1964; 1971; 1972; 1974; 1976; 1980; 1980a; Stigler, 1961; Yasin, 1970; MacQueen and Marschak, 1975; Arrow, 1979; 1984; 1984a; Hirshliefer and Rilley, 1979; McCall, 1982; Phlips, 1988; Laffont, 1989; Babcock, 1990; Bernknopf, et al, 1997; Macauley, 1997; Teisberg and Weiher, 2000; Cockshott and Michaelson, 2005).

The accelerating development of the information technology has an ongoing impact on individuals and society as a whole (Brooks, 2007; Carr, 2008).

With the advent of the Internet and development of powerful search engines, people got an unexpected access to information. For many, the Internet is becoming a universal medium, the conduit for most of the information that flows through people's eyes and ears and into their mind (Carr, 2008). This new kind of information flow brought forth a new type of data called *stream data* (cf., for example, (Burgin and Eggert, 2004)). A collection of messages is an example of a stream data.

As McLuhan (1964) pointed, media are not just passive channels of information. They supply the stuff of thought, but they also shape the process of thought. This is especially true for the Internet: working with it disables capacity for concentration and contemplation. As a result, the mind expects to take in information the way the Internet distributes it: in a swiftly moving stream of particles. People loose the ability to read and absorb a longish article on the web or in print. Even a blog post or an e-mail of more than three or four paragraphs is too much to absorb. Thus, people skim them (Carr, 2008).

New information technologies always changed people behavior, abilities and the way of thinking. Long ago Plato bemoaned the development of writing. He feared that, as people came to rely on the written word as a substitute for the knowledge inside their heads, they would cease to exercise their memory and become forgetful. They would be filled with the conceit of wisdom instead of real wisdom. As Carr writes (2008), Socrates was not wrong — the new technology did often have the effects he feared — but he was shortsighted. He couldn't foresee the many ways that writing and reading would serve to spread information, spur fresh ideas, and expand human knowledge (if not wisdom).

The arrival of the printing press, in the 15th century, set off another round of worries. Italian humanists expressed a concern that the easy availability of books would lead to intellectual laziness, making men "less studious" and weakening their minds. Others argued that cheaply printed books and broadsheets would undermine religious authority, demean the work of scholars and scribes, and spread sedition and debauchery. As New York University professor Clay Shirky notes (cf., (Carr, 2008)), "Most of the arguments made against the printing press were correct, even prescient." However, those critics were unable to imagine the diversity of blessings that the printed word would deliver. Now similar concerns are related to computers and the Internet, which have both positive and negative impact on society and individuals.

1.3. Information in Nature

Absence of evidence is not an evidence of absence
Christian De Duve

There is some controversy with respect to the rôle of information in nature. The spectrum of opinions ranges from complete rejection of existence information beyond society (cf., for example, (Machlup and Mansfield, 1983) or (Janich, 1992)) to the claim that everything in the world derives its existence from information (cf., for example, (Wheeler, 1990) or (Smolin, 1999)).

It is interesting that the area of expertise makes its imprint on the opinion of a researcher. Logicians and philosophers, who are mostly concerned with human problems, as a rule, insist that information exists purely in the human realm. Many researchers define the field of information science as limited to the human use, organization, production, and retrieval of information, excluding other information phenomenon. For instance, Barwise and Seligman write (1997) that the place of information in the natural world is far from clear.

At the same time, physicists and biologists, who know the laws of nature much better, do not agree with this opinion. Today, many scientific disciplines use the term *information* in one way or another. The term is now commonplace within a wide academic spectrum that includes astronomy, physics, biology, medicine and phisiology, psychology and the behavior sciences, economics, and political sciences. For instance, such prominent physicist as von Weizsäcker (1985) writes that information is a reflexive concept, pertaining to all sciences. He, his coauthors and followers demonstrated in the, so-called, *ur*-theory that physics reduces to information, namely, to information given by measurement outcomes (von Weizsäcker, 1958; von Weizsäcker, Scheibe and Süssmann, 1958; Castell, et al, 1975–1986; Lyre, 1998). It is done on the most basic level of elementary essences called *urs* without utilization of any specific information measure, such as Hartley measure, Shannon measure or Fisher measure.

Reading also writes (2006), "one of the main impediments to understanding the concept of information is that the term is used to describe a number of disparate things, including a property of organized matter ..." He considers energy and information as the two fundamental causal agents in the natural world.

Even more radical point of view is expressed by Wheeler (1990).

"Otherwise put, every "it" — every particle, every field of force, even the space-time continuum itself — derives its function, its meaning, its very existence (even if in some contexts indirectly) from the apparatus-elicited answers to yes-or-no questions, binary choices, *bits*. "It from bit" symbolizes the idea that every item of the physical world has at bottom — a very deep bottom, in most instances — an immaterial source and explanation; that which we call reality arises in the last analysis from the posing of yes-no questions and the registering of equipment-evoked responses; in short, that all things physical are information-theoretic in origin and that this is a *participatory universe*."

Similar ideas are analyzed by a prominent physicist Lee Smolin. In his book *The Life of the Cosmos* (1999), Smolin discusses the so-called *holographic principle*, first introduced by the physicist Gerard't Hooft and later developed by many other physicists. The holographic principle exists in two forms: strong and weak. The weak holographic principle asserts that there are no things, only processes, and that these processes are merely the exchange of data across two-dimensional screens. Data are not important by themselves. Their role is to store and provide information. According to this approach, the three-dimensional world *is* the flow of information.

Based on all these approaches, Bekenstein (2003) claims that there is a growing trend in physics to define the physical world as being made of information itself.

According to Stonier (1991), structural and kinetic information is an intrinsic component of the universe. It is independent of whether any form of intelligence can perceive it or not.

Thus, we see that the whole nature is a huge system of information processes. In the information-processing cosmos, each physical situation, regardless of location, emerges from the flow of information, generates information and gives information to the environment. According to Stonier, "information exists;" that is, information exists independently of human thinking (Stonier 1997). We can see that behind each law of nature there is a program, a functioning algorithm, while, according to the contemporary theory of algorithms (Burgin, 2005), an algorithm is a compressed information representation of processes.

The aphoristic expression "*It from Bit*" (Wheeler, 1990) allows one to see the role of information from different perspectives. One of these perspectives explains that when scientists study nature, it is only an illusion that they are dealing with physical objects. Actually researchers have only information from and about these objects. In general, all people living in the physical world know about this world and its parts and elements only because they get information form these systems. In a more detailed exposition, we have the following situation.

A person (or a group) H knows about some object A only if A has a name N_A in the mind M_H of H and there are some data D_A in M_H that represent property(ies) of A. Both N_A and D_A are conceptual representations in the sense of (Burgin and Gorsky, 1991) of A in M_H. These representations are formed by means of information that comes to M_H from receptors in the nervous system or is produced in M_H or comes to M_H from other channels, e.g., electrodes, electromagnetic fields, etc. Thus, all knowledge that people have about the world they live in is the result of information reception and production. If *It* is understood as a thing and *Bit* is understood as information, then people have things, or more exactly, their representations in the mind, only from information.

Another perspective assumes that emergence of all things is caused by information processing as there are no physical or mental processes that does not include information flow. For instance, Lyre (1998) asserts that the basic object of quantum physics, the quantum wave function, is information.

One more of the perspectives under discussion conjectures that emergence of all things *is* information processing where the whole world is something like an enormous computer that "computes" all things that exist.

In essence, every scientific discipline today uses, mostly without exact definition, the notion of information within its own context and with regard to specific phenomena. Being innate to nature, information appears in different spheres of natural sciences, such as biology, "Life, too, is digital information written in DNA" (Ridley, 2000), and in physics, "Giving us its as bits, the quantum presents us with physics as information" (Wheeler, 1990) or the whole physics is derived from Fisher information theory (Frieden, 1998).

In his book "The Touchstone of Life" (1999), Werner Loewenstein persuasively demonstrates that information is the foundation of life. To do this, he gives his own definition of information, being unable to apply the conventional definition that comes from the Hartley-Shannon's information theory. Namely, according to Loewenstein (1999), "information, in its connotation in physics, is a *measure of order* — a universal measure applicable to any structure, any system. It quantifies the instructions that are needed to produce a certain organization."

Thompson (1968) asserts that "the organization is the information". Scarrott (1989) writes that every living organism, its vital organs and its cells are organized systems bonded by information.

Gibson (1966) emphasizes the importance of understanding perception of a human being or animal, writing that "perception is not based on having sensations ... but it is surely based on detecting information."

Ruben (1992) asserts that for living beings, all interactions with the environment take place through the processing of matter-energy and/or processing of information.

Roederer (2005) considers different aspects of information in physics, biology and the brain, demonstrating that interaction is the basic process in the universe and discerning information-driven interactions and force-driven interactions.

Thus, in contrast to claims that information exists only in society, a more grounded opinion is that there is a huge diversity of information processes in nature, where information exchange goes without participation of humans. One of these processes is the replication of living creatures and their parts. In such a process, the cell is an information-processing system where DNA is a subsystem that contains the assembly instructions for building proteins. For instance, Schrodinger (1967) wrote that a gene is a very large information-containing molecule.

Further, DNA transmits genetic information (the symbolic genes which store genetic information) via specifically arranged sequences of nucleotide bases to proteins. This transmission is the basic process life on

Earth is built of. With very small variations, the genetic code, which is used to store and transmit genetic information, is similar for all life forms. In this sense, we can think of the genetic system and cellular reproduction as a symbolic code whose convention is "accepted" by the collection of all life forms.

Other codes exist in nature, such as signal transduction from the surface of cells to the genetic system, neural information processing, antigen recognition by antibodies in the immune system, etc. We can also think of animal communication mechanisms, such as the ant pheromone trails, bird signals, etc. Many assume that unlike the genetic system, most information processes in nature are of an analog rather than digital nature. This is not true because symbols can be represented not only on a piece of paper but also by signals, states of a system, chemical elements, molecules, electromagnetic waves, etc.

The ability to detect meaningful information is one of the defining characteristics of living entities, enabling cells and organisms to receive their genetic heritage, regulate their internal milieu, and respond to changes in their environment (Reading, 2006). Every organism and every cell is equipped with sensory receptors that enable it to detect and respond to meaningful information in its environment (Reading, 2006). These include surface receptors for detecting kinesthetic and chemical information, distance receptors for detecting visual, olfactory, gustatory and auditory information, and molecular receptors for detecting tissue, cellular and genetic information.

Information plays very important role in evolution. Csanyi (1989) and Kampis (1991) developed an elegant theory of evolution based on the concept of information. Burgin and Simon (2001) demonstrated that information has been and is the prevailing force of evolution in nature and society. Smith and Szathmary (1998; 1999) discuss evolutionary progress in terms of radical improvements in the representation of biological information.

The pivotal role of DNA for all living beings made it clear that life as a phenomenon is based on biological structures and information they contain. Information encoded in DNA molecules controls the creation of complex informational carriers such as protein molecules, cells, organs, and complete organisms. As a result, some researchers support the idea of looking at material objects as built of information. For instance, Rucker (1987) implies that it is now considered reasonable to say that, at the deepest, most fundamental level, our world is made of information.

Some researchers assume that information is a physical essence. For instance, Crutchfield (1990) treats information as "the primary physical entity from which probabilities can be derived." Landauer (2002) stresses, information is inevitably physical. However, it is more reasonable to suggest that people observe information only when it has a physical representation. For instance, information on the stock market or on the Olympic Games, as well as other information in social organization and communities, does not consist of physical messages but of their content.

All this brings us to the conclusion, expressed by Kaye (1995):

Information is not merely a necessary adjunct to personal, social and organizational functioning, a body of facts and knowledge to be applied to solutions of problems or to support actions. Rather it is a central and defining characteristic of all life forms, manifested in genetic transfer, in stimulus response mechanisms, in the communication of signals and messages and, in the case of humans, in the intelligent acquisition of understanding and wisdom.

1.4. Technological Aspects of Information

Programming today is a race between software engineers striving
to build bigger and better idiot-proof programs,
and the Universe trying to produce bigger and better idiots.
So far, the Universe is winning.
Rich Cook

Many books and numerous papers are written on problems of information technology. Here we give a general perspective on the situation and consider some of these problems.

Information plays various roles in technological processes. Information can serve to make decisions and to organize, perform and control a technological process (e.g., feedback information or control information). For instance, as Vegas and Basili (2005) write, "the main problem met by software developers when choosing the best suited testing technique for a software project is information." The technological process, e.g., computation, can work with information as a material when information is processed. Information can be input and/or output of a process, for example, in communication. Information is utilized by mechanisms and devices used in technological processes. As we know when a machine operates, it needs energy to enable it to produce, changes its states, e.g., preserve motion or stop moving, change objects on which it operates. At the same time, the machine needs information to determine what and how to produce, how and which states and objects to change. People who work with technology use diverse information to perform their functions and to control a diversity of machines, mechanisms and devices.

More over people began to understand that society in general and modern society in particular cannot exist without information and information technology. Information has become the leading and inherent component and fuel of every activity. Information processing is everywhere. As Ruževičius and Gedminaitė, (2007) describe, we hear words "information", "information society", "information management", "information processing", "information age", information economy", etc., so often that it seems that information becomes an object of a cult and information processing devices are sacred objects of this cult.

However, information technology emerged long ago. For instance, paper was invented almost 2000 years ago, but still remains our primary information-storage medium — apart from the human brain. What is new is the extent and role of information technology in modern culture and society.

Drathen (1990) implies that information has become a highly valued economic category, on par with capital and skilled labor. The collection, structuring and processing of information consequently constitute one of the focal points of business, and it is the task of process control engineering to provide adequate support for the production process in this area. The structuring of information is an essential precondition for the optimal conduct of technical processes.

Creation of information technology of our time is, may be, the most important revolution in human evolution. Now information processing plays more and more important role in life of people and functioning of society. Exclusively due to the accelerated development of the information technology, the industrial age gave way to the information age. It happened recently even in the USA where in 1991, companies for the first time spent more on computing and communications devices than on industrial, mining, farm, and construction machines. Information technology is now as critical for people and often as elusive as the air we breathe, which was discovered by scientists only several centuries ago. Information technology gives 30% of US economy growth since 1993. Information that is processed by computers and flows in different networks drives the economy, politics and social life.

Many information technologies have essentially changed and continue to change life of multitudes of people. The Internet brought the world into people's studies and living rooms. Now it is possible to go to this virtual world being on a plane or at a beach. New communication media connects people from the remotest places on the Earth. The Internet gave birth to e-commerce and e-publications. Cell phones and e-mail has become primary tools of communication. Information and knowledge management based on electronic devices, databases and data warehouses, revolutionizes business and industry. Embedded electronic devices saturate cars, planes, ships and houses. Computer-aided design (CAD), computer-aided manufacturing (CAM), computer-aided engineering (CAE), long-distance education, and computer-supported cooperative work (CSCW) drastically change how people create, produce, and use diverse things, how people work and how they study. New technologies change knowledge and abilities of people. In the 19th century, people knew that to talk to somebody, it was necessary to be at the same place with that person, e.g., in the same room. At the end of the 20th century, people knew that they can talk to those who were at the opposite side of the Earth, e.g., people who were living in USA, China and Australia were easily able to talk to one another without leaving their places of living. In the 19th century, people living in Europe knew that when they send a letter to somebody living in America, it was necessary to wait for a long time, weeks or even months, until the letter reached its destination. Now we know that when we send an e-mail from USA to, say, Argentina, it will take minutes to reach its destination.

Often, it is projected that the deployment of new information technologies can allow companies and corporations to circumvent or transcend geographical, cultural, temporal, and organizational barriers to the development and cooperation. Information technology has created the world where big companies can operate on a global scale, where it is possible to see what is going on the opposite side of the Earth, where people from distant countries, e.g., from USA and Australia, can communicate as those who live in the same neighborhood, and the whole Earth is becoming a small village. There are more and more countries where offering technological solutions to address economical, political, social, educational, and recreational problems is not only a commonplace but also a distinguishing feature of national cultures. That is why some prefer to use the term *Society Dependent on Information* instead of the term *Information Society* (cf., for example, (Malinowski, 2008)).

However, even now when everybody well understands the crucial role of information technology in modern society, the literature does not fully recognize the role of information and information technology in the maintenance and management of social and economic relationships within and among organizations (Baba, 1999). Anthropological discussions of reciprocity emphasize the importance of information as decision criteria in managing economic exchange but give limited attention to operations in which information is itself the object of exchange. At the same time, research in economics of information studies both sides of social information reality.

The development of information technology goes very fast. Performance of information processing machines and networks doubles every 18 months. It means that the progress in the next 18 months is the same as all previous progress, for example, the amount of data stored every 18 months is equal to the amount of all data stored before.

Machines form the physical core of technology. When machines were invented they performed work and produced material products. Now more and more machines produce information. Even the processes of doing some work, performing actions, and producing physical things more and more involve information processing.

As a result, technological aspects of information are discussed in various spheres and disciplines. Economic philosophy posits information as the source of value in a global economy. Business logic focuses on the accumulation, production and management of data, information and knowledge. Media, on the one hand, claim that availability and access to information technologies represent an increase in choice and freedom. On the other hand, they discuss negative impact of information technologies on individuals and society. For instance, information technologies make possible different kinds of cybercrimes, take away people's privacy, and put people at risk when machines fail or function in a wrong way. Discussions of the promises and dangers of the new information technologies, such as computers, Internet, digital imaging, virtual reality, etc., now saturate society. For instance, electronic communication brings both risks and rewards. One of the dangers of being able to store, retrieve and transmit information electronically is that the quantity of available information increases enormously and we come to the question of T.S. Eliot who wrote in "The Rock" (1934)

Where is the knowledge we have lost in information?

There is an information flooding and it is necessary to know how to regulate this process and have means for doing this. Indeed, as the theory of optimal control shows, provisioning, processing and managing big volumes of extra or superfluous information becomes harmful as it must engage available resources, such as time, computing and communication networks and devices, and people involved. This, in turn, can cause delays in decision and action, less efficient decision-making, control and operation. As Malinowski (2008) notes, in many situations, people spend too much effort searching for and then processing redundant information. As a result, they are left with too little time and other resources to perform their main functions.

The information overflow has one more negative consequence. Namely, the need to process more and more information, for example, to read a multitude of e-mails and answer to them, while having the same limited time available, people start doing many things in a loose manner (cf., for example, (Carr, 2008)). People lose some important skills and features. For instance, the majority of those who have to answer 100 and more e-mails every day cannot read long texts, such as books or even long articles. These people lose the ability of concentration. Their attention switches to something different in a short time.

This may be good and even beneficial to some professions, e.g., for flight dispatchers, but is unacceptable and even disastrous for other forms of activity. For instance, obtaining an important result in science or mathematics demands total concentration on the problem for a long time, sometimes for years. Immediate reactions may cause very bad consequences when deep analysis and deliberation are needed. For instance, when people plan operation of their companies or their investment strategy, immediate reactions can cause financial failures and crushes.

Dependence on information technologies and, especially, on computers poses many problems. It is necessary much better than now to understand what information is and how to manage it, doing this directly and not only through data management. It is necessary much better than now to understand what computers can do and what they cannot do. It is necessary much better than now to understand how to manage rapidly growing mountains of stored data and universes of computer networks.

An important problem of contemporary information technology is whether computers process only data or also information. A solution to this problem based on the general theory of information is given at the end of Section 2.5.

One of the most threatening features of the contemporary society is the common lack of ability to discriminate between useful, sound and trustworthy information and a worthless, often harmful and misleading stuff that has always been around but now with the advent of the Internet, e-mail systems and other communication tools, it multiplies with unimaginable rate. As Malinowski (2008) implies, a lot of information carrying media are full of information that is of low quality and quite often imprecise and even false. The usage of such information to achieve a chosen objective may bring and often brings unwelcome results.

There are other problems with information technology. For instance, software construction is chaotic and we are now only on the way to software engineering. Internet grows very fast and the problem of its scalability becomes very urgent, e.g., it is necessary to know how to deal with more than three billion pages of the Web. It is not just the huge number of pages and terabyte databases spreading on the Web that is the problem, but also the disorganized state of the Web itself.

One more problem is analyzed by Borgman (1984; 2000) and some other philosophers (cf., for example, (Veerbeck, 2002)). They found that technology in general and information technology in particular creates new patterns in the way people live their lives, alienating them from reality. On the one hand, technology decreases the effort that is needed to achieve definite goals. On the other hand, this disburdening character of technology changes the nature of people's involvement with reality. Technology can both enrich and reduce human life at the same time, and thus, the role of technology in the relations between people, as well as · between people and reality, is ambivalent. Information technology produces virtual reality as a commodity and can be used both to substitute physical reality and mediate people's involvement with physical reality and with each other. It is possible to treat virtual worlds created by computers and their networks as a mental world of artificial information processing devices. The place of the mental world is analyzed in Section 2.1 in the context of the world structure as a whole, i.e., as a component of the existential triad.

Different researchers and philosophers write about the death of privacy due to unlimited abilities of information technology and its misuse by different people and organizations (cf., for example, (Westin, 1972; Garfinkel, 2000)).

In addition, introduction of a new information technology can create problems for work groups, organization divisions and organizations because it threatens to disrupt or even destroy the boundary maintenance mechanisms used to manage social relations between groups, divisions and organizations (Baba, 1999). Often the risk arises from special characteristics of information technology, creating a number of challenges to information control and management by work groups, organization divisions and organizations. First, electronic communication may replace traditional boundary maintenance mechanisms, making these mechanisms and the corresponding means of controlling information obsolete. Second, new information technologies may remove the control of information from the conventional place, e.g., from the local scene, and place it in the hands of other parties. Third, it is possible that new information technologies represent information in an abstract form versus a physical form, whose security is easier to protect in many cases. Fourth, the new information technology can be used to force a new type of communication between groups, divisions or organizations. In essence, people have to adjust their habits, ways of working and thinking, structures of connections and develop new skills to make new information technology efficient. Otherwise, this technology does not give a sufficiently positive result for an individual, work group, organization division and the whole organization.

To conclude, it is necessary to remark that technology is created by people. So, it is natural that the first theoretical advance and main practical achievements were related to information processes in the technological domain. The first big break-through in this area was the mathematical theory of communication (or more exactly, communication in technical systems) created by Shannon (1948).

1.5. Structure of the Book

> *There's only one solution: look at the map.*
> Umberto Eco, *Foucalt's Pendulum*

> *The map is not the territory,*
> *and the name is not the thing named.*
> Alfred Korzybski

The main goal of the book is to achieve a synthesized understanding of the huge diversity of information phenomena by building a *general theory of information* that will be able to systematize and bind other existing information theories in one unified theoretical system. With this goal in mind, the book is organized as three-component system. At first (in Chapter 1), contemporary interpretations and explications of the term *information* are analyzed (Section 1.1) and the role of information in contemporary society (Section 1.2), nature (Section 1.3) and technology (Section 1.4) is discussed. Then (in Chapter 2), the general theory of information is presented. The third component of the book (Chapters 3–7) contains an exposition of popular and not so popular information theories. For these theories, only some basics that allow one to better comprehend a more detailed exposition are given. It provides an opportunity for the reader to understand how the general theory of information unifies existing knowledge in information studies.

From the unification perspective, the book has a three-fold purpose. The first aspiration is to present a new approach in information studies called the *general theory of information*. The second objective is presentation and analysis of existing directions in information theory and special information theories, such as Shannon's statistical information theory or algorithmic information theory, from the point of view of the general theory of information. The third aspiration of the book is to show possibilities opened by the new approach for information technology, information sciences, computer science, knowledge engineering, psychology, linguistics, social sciences, and education.

Here we do not try to represent special information theories and directions in a complete form or even to give all important results of these theories and directions. For some of them, such as statistical information theory or algorithmic information theory, complete representation demands several books larger than this one. Our goal is to give some introduction to these theories, explaining their basics and demonstrating how they can be presented as specifications of the general theory of information. Besides, references are given to sources where an interested reader can find more information about these theories.

Thus, we do not concentrate our exposition on demonstration of a series of sophisticated theorems, do not strive to describe one or two (even the most popular) ways of understanding and modeling the phenomenon of information and do not focus on giving an overview of the deepest contributions in the considered information theories. The goal is to present a broad picture of contemporary information studies, provide a unifying information theory and synthesize all existing approaches in amalgamated structure of ideas, constructions, methods, and applications.

Chapter 2 contains an informal exposition of the general theory of information and its applications to the theory of knowledge and psychology. The main result of the general theory of information is the development of the unified definition information that encompasses the enormous diversity of types, forms and classes of information. In

Section 2.1, we develop theoretical and methodological foundations for the general theory of information. These foundations include the general structures of the world, mathematical definition of the concept *structure*, and elements of semiotics, the discipline that studies signs, symbols and their systems.

Sections 2.2 and 2.3 contain basics of the general theory of information, giving its conceptual and methodological foundations. Developing the general theory of information, we utilize three levels of formalization: *principles*, *postulates*, and *axioms*. Principles represent informal assumptions related to studied phenomena, which, in our case, are information, information processes and systems. Postulates describe connections between the theory domain and theoretical constructions. Axioms characterize properties of theoretical constructions (e.g., mathematical structures) used in this theory. Thus, the base of the general theory of information is a system of principles and there are two groups of such principles: ontological and axiological.

Ontological principles studied in Section 2.2 reflect major features of information essence and behavior. They give an answer to the question "What is information?" In particular, the Ontological Principle O2 and its versions define information as a phenomenon that exists in nature, society, mentality of people, virtual reality, and in the artificial world of machines and mechanisms created by people.

Axiological principles studied in Section 2.3 explain how to evaluate, estimate and measure information and what measures of information are necessary. Studying information and related processes, researchers have invented a diversity of information measures, such as the information quantity, information value, information cost, information entropy, information uncertainty, average information score, information effectiveness, information completeness, information relevance, information reliability, and information authenticity. Axiological principles provide a reference frame for all of these measures and a guide for constructing new measures.

Section 2.4 demonstrates what kinds of information exist and how to discern them. For instance, distinctions between genuine information, false information, misinformation, disinformation, and pseudoinformation are considered. A variety of important properties of information are explicated and analyzed. For some of them, mathematical models are built. Relations between information, data, and knowledge are studied in Section 2.5. Relations between emotions and information are studied in Section 2.6.

Chapter 3 contains a brief exposition of several directions and applications of the statistical information theory. The main and most popular direction in the statistical information theory emerged as a communication theory in the works of Nyquist (1924), Hartley (1928), and was developed to a full extent by Shannon (1948).

However, even before communication engineers developed their measures of information, Fisher (1922) was the first to introduce such an information measure as the *variety of information*, giving birth to another direction in statistical information theory because his measure is also statistical by its nature. It is considered in Sections 3.6 and 7.3.3.

Relations between information and communication are studied in Section 3.1. Relations between information, uncertainty and entropy are considered in Section 3.2. The problem of the difference between information in conventional systems and quantum information is treated in Section 3.3. Section 3.4 demonstrates what relations exist between information and problem solving. Section 3.5 describes axiomatic foundations of the statistical theory of information. How theory of information is used in physics is explained in Section 3.6.

In contrast to statistical theories of information, semantic theories of information, which are described in Chapter 4, study meaning of information. In some cases, meaning of information is understood as the assumption that every piece of information has the characteristic that it makes a positive assertion and at the same time makes a denial of the opposite of that assertion. However, meaning is a more complicated phenomenon and to understand it in the context of information theory, we start Chapter 4 with a study of three communicational aspects, or dimensions, of information: syntax, semantics and pragmatics (cf. Section 4.1). Then we present semantic information theories, that is, theories that make emphasis on the semantics of information. In Section 4.2, we give an exposition of the first semantic information theory developed by Bar-Hillel and Carnap, as well as its later developments and improvements. In Section 4.3, we reflect on knowledge oriented information theories.

Chapter 5 provides a broad perspective on algorithmic information theory with its main peculiarities and constructions. In comparison with the majority of sources on algorithmic information theory, we present this theory on several levels. At first, the conventional level of recursive algorithms is considered in Sections 5.1 and 5.2. Many believe that the recursive information size (often called Kolmogorov complexity or descriptive complexity) of an object is a form of absolute information of the individual object. However, discovery of super-recursive algorithms and emergence the theory of super-recursive algorithms demonstrated a relative nature of the recursive information size. That is why the recursive algorithmic approach to information is upgraded to the next, super-recursive level of the algorithmic universe in Section 5.3. This is a necessary step as super-recursive algorithms are essentially more efficient in processing information than recursive algorithms and give more adequate models for the majority of information systems. Finally the highest, axiomatic level is achieved in a form of an axiomatic algorithmic information theory considered in Section 5.5. The results and constructions from Section 5.5 solve Problem 15 from the list of open problems in (Calude, 1996). Namely, Problem 15 asks to axiomatize algorithmic information theory. Section 5.4 contains a relativized version of algorithmic information theory.

It is demonstrated that more powerful algorithms, or a system with more powerful algorithms, need less information to do the same thing, i.e., to understand a text, to write a novel, to make a discovery or to organize successful functioning of a company. These results mathematically prove that the quantity of information a carrier has for a system depends on that system.

Chapter 6 contains an exposition of pragmatic theories of information. In Section 6.1, we consider the economics of information, which studies the role of information in the economic activity and has developed measures for estimation of the economic value and cost of information and information sources. In Section 6.2, we consider such important characteristics as value, cost, and quality of information.

Information quality, as researchers from MIT emphasize, is a survival issue for both public and private sectors: companies, organizations, and governments with the best information have a clear competitive edge. Section 6.3 contains elements of the qualitative information theory developed by Mazur.

Dynamic theories of information are presented in Chapter 7. In Section 7.1, theories of information flow developed by Dretske, Barwise and Seligman are described. Section 7.2 contains an exposition of the operator theory of information developed by Chechkin.

Section 7.3 exhibits information algebra and geometry. Elements of the mathematical component of the general theory of information are expounded in Subsection 7.3.1. This mathematical theory is based on functional models. It is also possible to develop a mathematical component of the general theory of information based on category theory.

Elements of the theory of abstract information algebras are presented in Subsection 7.3.2. Abstract information algebras can model operations and processes with information carriers, such as data, texts, and documents.

Information geometry is discussed in Subsection 7.3.3, demonstrating application of geometrical methods and structures to information studies and practice.

The last Chapter 8 contains some conclusions and directions for future research.

Exposition of material is aimed at different groups of readers. Those who want to know more about history of information studies and get a general perspective of the current situation in this area can skip proofs and even many results that are given in the strict mathematical form. At the same time, those who have a sufficient mathematical training and are interested in formalized information theories can skip preliminary deliberations and go directly to the sections that contain mathematical exposition. Thus, a variety of readers will be able to find interesting and useful issues in this book if each reader chooses those topics that are of interest to her or to him.

It is necessary to remark that the research in the area of information studies is extremely active and information is related to everything, in other words, there is nothing that does not involve the use of information. Consequently, it is impossible to include all ideas, issues, directions, and references that exist in this area, for which we ask the reader's forbearance.

Chapter 2

General Theory of Information

Great spirits have always found violent opposition from mediocre minds.
The latter cannot understand it when a [person]
does not thoughtlessly submit to hereditary prejudices
but honestly and courageously uses their intelligence.
Albert Einstein (1879–1955)

Shallow ideas can be assimilated;
ideas that require people to reorganize
their picture of the world provoke hostility.
James Gleick, Chaos

As we have seen, while information has become the most precious resource of society, there is no consensus on the meaning of the term "information", and many researchers have considered problems of information definition. The primary goal of the general theory of information is to obtain a definition of information that has the following properties. It has to be sufficiently wide to encompass a diversity of phenomena that exist under the common name *information*, sufficiently flexible to reflect all properties people ascribe to information and sufficiently efficient to provide a powerful tool for scientific exploration and practical usage. Basics of this theory and its applications are presented in this Chapter.

However, to have a valid and efficient theory of information, it is not enough to have a correct definition of information. We need properties of information and information processes. There are many directions in modern information theory that suggest their description of information and information processes. Some approaches to the concept of information are analyzed in Introduction.

Kinds of information and their theoretical representations form an extensive diversity of phenomena, concepts, formulas, and ideas. This inspired many researchers to argue that it is impossible to develop a unified definition of information. In particular, the most prominent researcher in the field of information science, Claude Shannon, wrote (cf., (Shannon, 1993)) that it was hardly to be expected that a single concept of information would satisfactorily account for the numerous possible applications of the general field of information theory.

A persuasive argument for impossibility to give such a unified definition is given in (Capurro, Fleissner and Hofkirchner, 1999). This result and many other arguments (cf., for example, (Melik-Gaikazyan, 1997)) undermine generality of conventional definitions of information and imply impossibility of a universal definition of information.

However, Arthur Clarke wrote (1962):

"When a distinguished but elderly scientist states that something is possible, he is almost certainly right. When he states that something is impossible, he is very probably wrong."

In the case with information, as well as in many other cases, he was completely right. It has become possible to synthesize all directions and approaches in information studies and to find a solution to the important problem of understanding what information is. This was achieved by utilization of a new definition type in the general theory of information presented in this Chapter. Namely, to overcome limitations of the conventional approaches and to solve the problem of information definition a parametric definition is used. Parametric systems (parametric curves, parametric equations, parametric functions, etc.) have been frequently used in mathematics and its applications for a long time. For instance, a parametric curve in a plane is defined by two functions $f(t)$ and $g(t)$, while a parametric curve in space has the following form: ($f(t)$, $g(t)$, $h(t)$) where parameter t takes values in some interval of real numbers.

Parameters used in mathematics and science are, as a rule, only numerical and are considered as quantities that define certain characteristics of systems. For instance, in probability theory, the normal distribution has as parameters the mean μ and the standard deviation σ. A more general parameter, functional, is utilized for constructing families of non-Diophantine arithmetics (Burgin, 1997c; 2001b).

In the case of the general theory of information, the parameter is even more general. The parametric definition of information utilizes a system parameter. Namely, an infological system (see Section 2.2) plays the role of a parameter that discerns different kinds of information, e.g., social, personal, chemical, biological, genetic, or cognitive, and combines all existing kinds and types of information in one general concept "information".

Chapter 2 contains an informal exposition of the general theory of information and its applications to the theory of knowledge and psychology. The main result of the general theory of information is the development of the unified definition information that encompasses the enormous diversity of types, forms and classes of information. In Section 2.1, we develop theoretical foundations for the general theory of information. These foundations include the general structures of the world, mathematical definition of the concept *structure*, and elements of semiotics, the discipline that studies signs, symbols and their systems.

Sections 2.2 and 2.3 contain basics of the general theory of information. Developing the general theory of information, we utilize three levels of formalization: *principles*, *postulates*, and *axioms*. Principles represent informal assumptions related to studied phenomena, which, in our case, are information, information processes and systems. Postulates describe connections between the theory domain and theoretical constructions. Axioms characterize properties of theoretical constructions (e.g., mathematical structures) used in this theory. Thus, the base of the general theory of information is a system of principles and there are two groups of such principles: ontological and axiological.

Ontological principles studied in Section 2.2 reflect major features of information essence and behavior. They give an answer to the question "What is information?" In particular, the Ontological Principle O2 and its versions define information as a phenomenon that exists in nature, society, mentality of people, virtual reality, and in the artificial world of machines and mechanisms created by people. These definitions unify all existing kinds and types of information in one general concept "information", allowing separation, study and utilization of specific types of information, such as, for example cognitive information.

Axiological principles studied in Section 2.3 explain how to evaluate, estimate and measure information and what measures of information are necessary. Studying information and related processes, researchers have invented a diversity of information measures, such as the information quantity, information value, information cost, information entropy, information uncertainty, average information score, information effectiveness, information completeness, information relevance, information reliability, and information authenticity. Axiological principles provide a reference frame for all of these measures and a guide for constructing new measures.

Information is a basic multifaceted phenomenon. As a result, it has many diverse types and classes. Knowing to what class a particular portion of information or stream of information belongs helps to better process this information. Section 2.4 demonstrates what kinds of information exist and how to discern them. For instance, distinctions between genuine information, false information, misinformation, disinformation, and pseudoinformation are considered. Classes of accessible, available and acceptable information are also studied. A variety of important properties of information are explicated and analyzed. For some of them, mathematical models are built. Relations between information, data, and knowledge are investigated in Section 2.5. It is demonstrated that the conventional approach combining these essences in the Data-Information-Knowledge Pyramid has many deficiencies and limitations. To eliminate these shortcomings, the Knowledge-Information-Matter-Energy (KIME) Square and Data-Knowledge Triad are introduced and utilized. Relations between emotions and information are studied in Section 2.6. Based on the structure of the human brain, it is demonstrated that there are three fundamental classes of information: cognitive information, direct emotional/affective information, and direct effective information Note that emotional information is a newly discovered type of information being different from the cognitive emotional/affective information usually simply called emotional information and studied by many authors (cf., for example, (Sanders, 1985; Keane and Parrish, 1992; George and McIllhagga, 2000; Bosmans and Baumgartner, 2005; Knapska, et al, 2006)).

2.1. Signs, Symbols and the World

All uses of information depend on symbols.

Arno Pensias

Before launching our inquiry into the very foundations of information phenomena, we consider (from a scientific perspective) objects to which information is related and entities that are related to information. As we have found in the previous chapter, information is related to everything, to the whole world and the whole world is related to and depends on information. So, we need better understanding of the structure of the world where we live. That is why we begin with the world as a whole.

To achieve its relevant understanding is not a simple problem. Max Born (1882–1970) wrote that the notion of reality in the physical world had become, during the last century, somewhat problematic (Born, 1953).

As we know, people live in the *physical* (*material*) *world* and many perceive that this is the only reality that exists. However, in contrast to this opinion, as Born (1953) explained, there were subjective philosophies which taught that only the mental world is real and the physical world is merely an appearance, a shadow without substance. For instance, some Eastern philosophical and religious systems, e.g., Buddhism, teach that the whole physical reality is a great illusion and the only reality is the spiritual world. As science does not have enough evidence to accept or to reject this idea, we are not going to discuss it. Nevertheless, science has enough evidence to accept existence of the *mental world*. As contemporary psychology states, each individual has a specific inner world, which forms mentality of the individual and is based on the psyche. These individual inner worlds form the lowest level of the mental world, which complements our physical world. On next level of the mental world, there are mental systems (worlds) of groups of people, communities and the whole society. It is possible to further develop this hierarchy of mental worlds but it is done elsewhere.

Some thinkers, following Descartes, consider the mental world as independent of the physical world. Others assume that mentality is completely generated by physical systems of the organism, such as the nervous system and brain as its part. However, in any case, the mental world is different from the physical world and constitutes an important part of our reality.

Moreover, our mentality influences the physical world and can change it. We can see how ideas change our planet, create many new things and destroy existing ones. Even physicists, who explored the very foundation of the physical world, developed the, so-called, observer-created reality interpretation of quantum phenomena. A prominent physicist, Wheeler, suggests that in such a way it is possible to change even the past. He stresses (Wheeler, 1977) that elementary phenomena are unreal until observed. This gives a dualistic model of reality.

However, the dualistic model is not complete. This incompleteness was prophesied in ancient Greece and proved by modern science. One of the great ideas of ancient Greece is the world of ideas (or forms), existence of which was postulated by Plato. In spite of the attractive character of this idea, the majority of scientists and philosophers believe that the world of ideas does not exist because nobody has any positive evidence in support of it. The crucial argument of physicists is that the main methods of verification in modern science are observations and experiments, and nobody has been able to find this world by means of observations and experiments. Nevertheless, there are modern thinkers who, like such outstanding scholars as philosopher Karl Popper, mathematician Kurt Gödel, and physicist Roger Penrose, continue to believe in the world of ideas, giving different interpretations of this world but suggesting no ways for their experimental validation.

However, science is developing, and this development provided recently for the discovery of the *world of structures*. On the level of ideas, this world may be associated with the Platonic world of ideas in the same way as atoms of modern physics may be related to the atoms of Democritus. The existence of the world of structures is demonstrated by means of observations and experiments. This world of structures constitutes the structural level of the world as whole. Each system, phenomenon, or process either in nature or in society has some structure. These structures exist like things, such as tables, chairs, or buildings, and form the structural level of the world. When it is necessary to investigate or to create some system or process, it is possible to do this only by means of knowledge of the corresponding structure. Structures determine the essence of things.

Let consider the development of ideas related to the *global world structure*.

In the Platonic tradition, the global world structure has the form of three interconnected worlds: *material*, *mental*, and the *world of ideas* (or *forms*).

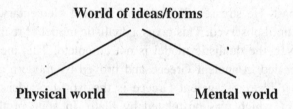

World of ideas/forms

Physical world ———— **Mental world**

Figure 2.1. The Plato triad of the world

However, independent existence of the World of Ideas has been severely criticized by many philosophers and scientists, starting with Aristotle. Many argue that taking a long hard look at what the Platonist is asking people to believe, it is necessary to have faith in another "world" stocked with something called ideas. This results in many problems. Where is this world and how do we make contact with it? How is it possible for our mind to have an interaction with the Platonic realm so that our brain state is altered by that experience? Plato and his followers have not provided convincing answers to these questions.

Another great philosopher Popper (1974; 1979) also formed his ontology of the world, which consists of three parts/worlds:

World 1: Physical objects or states.
World 2: Consciousness or psychical states.
World 3: Intellectual contents of books, documents, scientific
theories, etc.

As Popper uses the words *information* and *knowledge* interchangeably, World 3 consists of knowledge and information and we have the following triad.

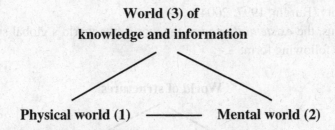

**World (3) of
knowledge and information**

Physical world (1) ——— **Mental world (2)**

Figure 2.2. The Popper triad of the world

Popper's triad is much more understandable than the Plato triad because people know what knowledge is much better than what ideas are, especially, when these are Plato ideas, or forms.

Note that the term *triad* used here is different from the notion of a *triplet*. A triad is a system that consists of three parts (elements or components), while a triplet is any three objects. Thus, any triad is a triplet, but not any triplet is a triad. In a triad, there are ties and/or relations between all three parts (objects from the triad), while for a triplet, this is not necessary.

Other authors refer World 3 in the triad of the world to signs in the sense of Charles Pierce (1839–1914), although they do not insist that it consists of objects that Pierce would classify as signs (cf., for example, (Skagestad, 1993; Capuro and Hjorland, 2003)).

Only recently, modern science made it possible to achieve a new understanding of Plato ideas, representing the global world structure as the *Existential Triad* of the world (cf. Figure 2.3). In this triad, the Physical (material) World is interpreted as the physical reality studied by natural sciences, while ideas or forms might be associated with structures, and the Mental World encompasses much more than individual conscience (Burgin, 1997; Burgin and Milov, 1999). In particular, the Mental World includes social conscience. This social conscience is projected on the collective unconscious in the sense Jung (cf. (Jung, 1969)) by the process of internalization (Atkinson, et al, 1990). In addition, the World of structures includes Popper's World 3 as knowledge, or the intellectual contents of books, documents, scientific theories, etc., is a kind of structures that are represented in people's mentality (Burgin, 1997; 2004).

Thus, the *existential triad* of the world (the world's global structure) has the following form:

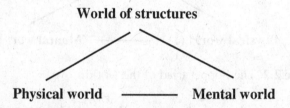

Figure 2.3. The Existential Triad of the world

In the mental world, there are real "things" and "phenomena". For example, there exist happiness and pain, smell and color, love and understanding, impressions and images (of stars, tables, chairs and etc.). In the physical world, there are the real tables and chairs, sun, stars, stones, flowers, butterflies, space and time, molecules and atoms, electrons and photons. It has been demonstrated (Burgin, 1997) that the world of structures also exists in reality. For instance, some fundamental triads (cf. Appendix A) exist in the same way as tables, chairs, trees, and mountains exist. It is possible to see such triads around us. Knowledge, *per se*, forms a component of the world of structures. It is an important peculiarity of the world (as a whole) that it exists in such a triadic form not as a static entity but as a dynamic structure.

Each of these three worlds is hierarchically organized, comprising several levels or strata. For instance, the hierarchy of the physical world goes from subatomic particles to atoms to molecules to bodies to cells to living beings and so on. On the first level of the mental world, individual mentality is situated. The second level consists of group and social mentalities, which include collective subconscience (Jung, 1969) and collective intelligence (Nguen, 2008a).

It is necessary to understand that these three worlds from the Existential Triad are not separate realities: they interact and intersect. Thus, individual mentality is based on the brain, which is a material thing. On the other hand, physicists discuss a possibility that mentality influences physical world (cf., for example, (Herbert, 1987)), while our knowledge of the physical world to a great extent depends on interaction between mental and material worlds (cf., for example, (von Baeyer, 2001)).

Even closer ties exist between structural and material worlds. Actually no material thing exists without structure. Even chaos has its chaotic structure. Structures determine what things are. For instance, it is possible to make a table from different stuff (material): wood, plastics, iron, aluminum, etc. What all these things have in common is not their material substance; it is specific peculiarities of their structure. As some physicists argue, physics studies not physical systems as they are but structures of these systems, or physical structures. In some sciences, such as chemistry, and some areas of practical activity, such as engineering, structures play a leading role. For instance, the spatial structure of atoms, chemical elements, and molecules determines many properties of these chemical systems. In engineering, structures and structural analysis even form a separate subject (cf., for example, (Martin, 1999)).

In addition, we can see existential triads in every individual and in each computer. An individual has the physical component — her body, the mental component studied by psychologists, and the structural component, which comprises all structures and systems from the other two components. A computer has the physical component — its hardware, the mental component, which consists of everything that is in computer memory, and a structural component, which comprises all structures and systems from the other two components.

While physical and mental objects have been treated for a long time and thus, do not cause difficulties in understanding, structure is a much more imprecise concept and needs additional explanation. The main question is what structure is. Is it possible to define structure in exact terms or it is as vague as ideas of Plato or forms of Aristotle?

Now the notion of a structure is used in a variety of fields. Mathematicians study mathematical structures, physicists study physical structures, chemists study chemical structures, biologists study biological structures, and so on (cf., for example, (Haken, 1988)). Actually scientists study theoretical structures, which they correspond to natural phenomena. These structures are called models of studied phenomena and scientific laws describe properties and functioning of these structures.

Structures are extremely important in computer science, programming, and computer technology (cf., for example, (Bloch, 1975; Yourdon, 1979; Burgin, 1996e)). Data structures lie at the core of information technologies. It is impossible to build a database without explicit construction of data structures (cf., for example, (Date, 1975; Tamassia, 1996; Date, et al, 2002; Burgin, 2008)). Data structures are extremely important for algorithms and programming (cf., for example, (Wirth, 1975; 1986; Burgin and Eggert, 2004)). To emphasize importance of data structures, Wirth (1975) wrote the book "Algorithms + Data Structures = Programs." Computer architecture is a structure of a computer. Researchers study knowledge structuration for the Web (cf., for example, (Scharffe and Ding, 2006)).

In the 20th century, knowledge became the prime interest in several areas: in artificial intelligence, science, business, and technology to mention but a few. For instance, knowledge is now recognized by most businesses as a very important, valuable, and sometimes even critical resource. Good knowledge management helps companies to achieve their goals. Poor knowledge management results in mistakes, miscalculations and sometimes put company's future at risk.

The new strong emphasis on knowledge explicated importance of knowledge structure. Different methodologies have been developed to reconstruct knowledge structure and utilize it for various purposes, such as learning, knowledge management, knowledge representation and knowledge processing (cf., for example, (Wu, 2004; Campbell, 2007)).

In linguistics, information structure of sentences is widely used (Halliday, 1967; Roberts, 1996; Steedman, 2000). This structure reflects the partitioning of sentences into categories, such as the background, focus, topic, comment, etc.

Structures play important roles in a lot of other areas. For instance, in computer programming there a direction called *structured programming*. It is a programming technique that enforces a logical structure on the program being written to make it more efficient, manageable, understandable and reliable (Dahl, et al, 1972).

One of the most famous discoveries of genetics was the discovery of the spatial structure of DNA. This structure determines how parts of the body will be formed, where the parts will go, how growth and aging will proceed, and what influences from the past will be imprinted through the brain onto the mind. The DNA helix is made of repeating units called nucleotides. Each nucleotide consists of three molecules. Although there is a lot of experimental evidence supporting this representation of the spatial structure of DNA as a double helix, there is still a controversy and some researchers, for example, claim that the spatial structure of DNA is a quad helix.

Structures are often treated in various psychological theories. For instance, it is assumed that the mind, where consciousness dwells, consists of myriad structures. A *psychological structure* refers to a relatively stable organization of component parts that perform one or more related psychological functions. For instance, according to psychoanalytic theories, the mind, or more general, the personality has an internal structure, which consists of three parts with separate motivations and active interactions: *Id*, which is the irrational and emotional part; the *Ego*, which is the rational part; and the *Superego*, which is the moral part.

In sociology, structure is one of the basic concepts. Although a social structure is not clearly conceptualized (cf., for example (Porpora, 1989; Abercrombie, et al, 2000), it can refer to the majority of objects studied in sociology, including individuals and groups in definite relations to each other, relatively enduring patterns of behavior and relationship within society, social stratification, and social institutions, norms, and cognitive frameworks that determine actions of actors (i.e., people, groups, and organizations) in society. Family, culture, religion, science, education, law, economy, and politics, all are social structures. Some researchers believe that social structures are naturally developed. The notion of social structure is inherent for a variety of central topics in social sciences. For instance, the structure of organization can determine organization's flexibility, capacity to change, potential for development and many other factors, and is therefore an important issue in management.

As a scientific notion, social structure has been identified as:

(a) a relationship of definite entities or groups to each other (static structure);

(b) an enduring pattern of behavior by participants in a social system in relation to each other (dynamic structure);

(c) institutionalized norms of behavior and attitude.

Sociologists identify several types of social structures:

- *normative structure* is a pattern of relations and attitudes in a social system, implying norms and modes of operation of people in varying social positions and situations;
- *ideal structure* is a pattern of relations between beliefs and views of people in varying social positions;
- *interest structure* is a pattern of relations between goals and desires of people in varying social positions and situations;
- *interaction structure* is a system of forms of communication and other interaction of people in varying social positions and situations.

Social structures are divided into two big classes: microstructures and macrostructures. A *microstructure* is a pattern of relations between the most basic elements of social life that cannot be further split on the social level and have no social structure of their own, e.g., a pattern of relations between individuals in a group composed of individuals or a structure of organization as a pattern of relations between social positions and social roles. A *macrostructure* is a second-level structures representing a pattern of relations between social systems that have their own social structure, e.g., a political structure between political parties.

Social institutions and social groups are more structures than physical systems. Social relations are structures. Social forms in the sense of Georg Zimmel (1858–1918) are patterns of structures. In such a way, structures become explicit, and sociologists study them.

Assuming this situation, Giddens (1984) developed a structuration theory of society. In it, social structures are the medium of human activities. The structuration theory is a formal social theory, which suggests a unification of the classic actor/structure dualism. The theory is a logic, conceptual and heuristic model of human behavior and action and is based on three assumptions:

- Social life is not the sum of all micro-level activity, but social activity cannot be completely explained from a macro perspective.
- The repetition of the acts of individual agents reproduce the structure.
- Social structures are neither inviolable nor permanent.

A social system can be understood by its structure, modality, and interaction. *Social structure* is constituted by rules and resources governing and available to social agents.

Giddens treats information and knowledge systems as key components of the creation and maintenance of modern institutions and identifies three types of structures in social systems: *signification*, *legitimation*, and *domination*. These structures produce basic social phenomena.

Signification produces meaning through organized webs of language (semantic codes, interpretive schemes and discursive practices).

Legitimation produces a moral order via naturalization in societal norms, values and standards.

Domination produces (and is an exercise of) power, originating from the control of resources.

An important question in cosmology is the cosmological generation of structure (Bojowald and Skirzewski, 2008). The current understanding of structure formation in the early universe is mainly built on a magnification of quantum fluctuations in an initial vacuum state during an early phase of accelerated universe expansion.

Information structures play an important role in decision theory and information economics (cf. Section 6.1). These structures represent the situation (state of the world) where customers/users are operating.

Structures occupy the central place in mathematics. At the beginning mathematics was the discipline that studied numbers and geometrical shapes (Burton, 1997). However, the development of mathematics changed its essence and mathematics has become the discipline that studies mathematical structures (Bourbaki, 1957; 1960; Burgin, 1998). For instance, for Bourbaki, pure mathematics is to be thought of as nothing other than the study of pure structure.

While formalized abstract structures are studied by pure mathematics, formal and informal concrete structures are studied by all sciences (natural, social and life sciences), applied mathematics and philosophy. For instance, physicists study physical structures. To represent physical structures in a rigorous way, they use mathematical structures. Mathematical structures help physicists to better understand physical reality, provide resources for innovative physical theorizing, and alleviate making new discoveries in physics. For instance, the mathematical structure associated with the special unitary group SU(2), i.e., the group of unitary 2×2 matrices with determinant 1, has played a pivotal role in three key episodes in 20th century physics — introduction of intrinsic spin and isospin, building gauge theory, and achieving the electro-weak unification (Anderson and Joshi, 2006). In addition the same structure plays an important role in loop quantum gravity.

Dirac (1931) wrote that a powerful method of advancing physics was one that drew on the resources of pure mathematics by attempting "to perfect and generalize the mathematical formalism that forms the existing basis of theoretical physics, and after each success in this direction, to try to interpret the new mathematical features in terms of physical entities…"

Tegmark (2007) advocates an extreme approach to physics, where the external physical reality is assumed to be a mathematical structure (the Mathematical Universe Hypothesis). He believes that this unification of physics and mathematics would allow physicists to understand reality more deeply than most dreamed possible.

In modern science and philosophy, a general approach based on revealing intrinsic structures kept invariant relative to transformations is called *structuralism* (cf., for example, (Piaget, 1971)). The main assumption of the structural approach is that different things and phenomena have similar essence if their structures are identical. To investigate a phenomenon means to find its structure.

At first, structuralism emerged in linguistics. Although Ferdinand de Saussure (1857–1913) did not use the term *structure* preferring the term *form*, his understanding in many respects coincides with the structuralistic approach. The terms *structure* and the *structural linguistics* was introduced by Rasmus Viggo Brendal (1887–1942) in (Brendal, 1939). Louis Hjelmslev (1899–1965) considered structure as an "independent essence with inner dependencies". He emphasized that the properties of each element of such integrity depend on its structure (Hjelmslev, 1958).

Afterwards, structuralism was extended to other fields. Levi-Strauss developed and successfully applied the structuralistic methods and structural analysis in anthropology (cf., for example, (Lévi-Strauss, 1963; 1987)). On this basis, he had restored the common (for all cultures) function, which intermediates the fundamental contradictions of the human existence. This function (in his opinion) has been lost by the modern European civilization.

Jacques Marie Émile Lacan (1901–1981) applied the structural analysis to psychoanalysis. Subconscious (in his opinion) is structured as a language. It is possible to describe his main conception by the triad:

real — imaginary — symbolic

Real is chaos (Lacan, 1977). Inaccessible is a name. Imaginary is an individual variation of the symbolical order. Symbolical order is objective. Lacan has been proving the thesis that an idea and existence are not identical: it is a language, which is a mediator between them.

Structuralism as a scientific methodology based on the concept of structure and mathematics as a science of structures brought the notion of structure to many other areas, including sociology, anthropology, esthetics, linguistics, and psychology. However, in spite of many achievements in a variety of areas, structuralism was not able to develop a general definition of structure as it is demonstrated in the book of Piaget (1971). Those definitions that were suggested in particular areas are more descriptions than definitions. For instance, Hjelmslev (1958) considered a structure to be an autonomous essence with inner relations. As a result, structuralism studied structures as they are interpreted in mathematics, natural sciences, linguistics, social sciences, psychology, etc. (Piaget, 1971; Caws, 1988).

Thus, there is no a generally accepted encompassing definition of a structure and the word *structure* has different meanings in different areas. Here are some examples. According to McGraw-Hill Encyclopedia of Science and Technology (2005), *structure* is an arrangement of designed components that provides strength and stiffness to a built artifact such as building, bridge, dam, automobile, airplane, or missile. According to Dictionary of Business Terms (2000), *structure* is any constructed improvement to a site, e.g., building, fence and enclosures, greenhouses, and kiosks. At the same time, *financial structure* has a very different meaning. Namely, *financial structure* is makeup of the right-hand side of a company's Balance Sheet (Dictionary of Business Terms, 2000). In archeology, *structure* refers to the basic framework or form of society. In geography and geology, *structure* is a configuration of rocks of the earth's surface. The *chemical structure* refers both to molecular geometry and to electronic structure. The structural formula of a chemical compound gives a graphical representation of the molecular structure, describing the atom arrangement. A *social structure* is a pattern of relations. In mathematical logic, a structure is a logical or mathematical object that gives semantic meaning to expressions of a logical language.

In contrast to this, achievements of mathematics allow one to elaborate a sufficiently exact and feasible definition of structure in general. Here we give formal definitions of concepts related to the World of Structures to achieve better understanding of this sphere of reality. It results in informal and formal definitions of structure in general. Cognizance of the concept of structure is important because structures are substantial for understanding the phenomenon of information.

It is necessary to remark that at this point and in similar situations later, readers who are interested only in ideas and general explanations can skip formal construction of the concept of structure. However, in this case, reader's understanding of structures either will be confined to some special area, e.g., mathematics or physics, or will stay on the level of Plato when he wrote about ideas or forms.

As we know things have structures. At the same time, Plato postulated independent existence of his world of ideas, and it is possible to consider structures as scientific counterparts of Plato ideas. Thus, it is natural to ask the question whether structures exist without matter. Here we are not going into detailed consideration of this fundamental problem. What is important for the topic of this book is that as a coin has two sides, material things also have two aspects — *substance* and *structure*.

Aristotle taught that each thing (physical object) is a *synholon*, which is a composition of matter (*hyle*) and form (*eidos*), i.e., it has *substance* and *form*. Each form is based on a *prote hyle* (a "first matter"), which is the *hypokeimenon* (foundation) of the form. Forms, according to Aristotle, determine essence of things.

Arthur Stanley Eddington (1882–1944) wrote (1922) that physical reality has *substance* and *structure*. If we say that structures exist only embodied in material things, then we have to admit that material things exist only in a structured form, i.e., matter (physical entities) cannot exist independently of structures. For instance, atomic structure influences how the atoms are bonded together, which in turn helps one to categorize materials as metals, ceramics, and polymers and permits us to draw some general conclusions concerning the mechanical properties and physical behavior of these three classes of materials. Even chaos has its structure and not a single one.

While it is possible only to name substance because any attempt to further that naming immediately involves attachment of some structure, we can develop the concept of structure to a definite extent of precision. On the intuitive level, structure means a complex of relations comparable with one another in principle. There are other definitions. For instance, Grossmann (1990) defines structure in the following way.

Definition 2.1.1. A *structure* is (a representation of) a complex entity that consists of parts in relations to each other.

This definition implies that two structures are identical if and only if: (a) their nonrelational parts are the same; (b) their relation parts are the same; and (c) corresponding parts stand in corresponding relations.

A similar definition of a mathematical structure is used in logic and mathematics (cf., for example, (Robinson, 1963; Yaglom, 1980)).

Definition 2.1.2. A *mathematical structure* is an abstract (symbolic) system

$$A = < M; R_1, R_2, \ldots, R_n >$$

where M is a set and $R_1, R_2, \ldots, R_n$ are relations in this set.

This definition is relatively simple and intuitive. However, there are some problems with this definition. First, the definition of a system in the general system theory is a particular case of Definition 2.1.2 (cf., for example, (Mesarovic and Takahara, 1975)). This makes systems particular cases of structures, while as we know, material systems are not structures but only have structures. Second, some authors (cf., for example, (Malcev, 1970)) use the name model or relational system for the concept defined in Definition 2.1.2.

Physicists understand mathematical structures in a similar way. For instance, Tegmark (2007) gives the following definition.

Definition 2.1.3. A *mathematical structure* is a system that consists of:

1. the set of entities that belong to the union $S = S_1 \cup S_2 \cup \ldots \cup S_n$ of sets $S_1, S_2, \ldots, S_n$;

2. functions (functional relations) on these sets and their Cartesian products

$$R_k: S_{i1} \times S_{i2} \times \ldots \times S_{it} \to S_n$$

In their informal definition of a mathematical structure, Bourbaki (1948) add one more condition.

Definition 2.1.4. A *mathematical structure* is a set with relations in this set and axioms that these relations satisfy.

Here our goal is to elaborate an exact formal definition of structure. To achieve this without loosing the intuitive meaning of this notion, we begin with informal definitions.

Definition 2.1.5. A *structural representation* R(**K**) of an entity (system, process) **K** is an abstract (symbolic or mental) image (representation) of **K** consisting of representations of parts (elements, components) of **K** and connections (ties, relations) between them.

In system theory, a system consists of elements and relations between elements. It is a set theoretical model. In mereology (cf., for example, (Leśniewski, 1916; Leonard and Goodman, 1940)), a system consists of parts and relations between parts.

Each system has inherent structural representations. At the same time, inherent structural representations are (better or worse) revealed by reflected structural representations presented by other systems.

There are three main types of structural representations:

1. *Material* in the Physical World.
2. *Symbolic* in the Mental World.
3. *Intrinsic* in the Structural World.

Types 2 and 3 of structural representations are abstract systems. An abstract system consists of such elements as linguistic expressions, e.g., letters, digits, and words, and relations that are subsets of Cartesian powers of sets of elements.

Definition 2.1.6. A *structure* S(**K**) of an entity (e.g., of an object, system or process) **K** is an invariant with respect to structural equivalence (isomorphism) ideal/abstract/symbolic representation of object (system) parts (elements and/or components) and connections (ties, relations) between them.

For instance, a structure of a sentence in English is a linear sequence of symbols each of which denotes a word.

Structures of chemical elements and their combinations, such as molecules, are very important in chemistry. Figure 2.4 gives some examples of chemical structures.

a) H_2O b) $H - O - H$ c) $H - \ddot{O} - H$ d) $H : \ddot{O} : H$

e)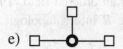

Figure 2.4. Chemical structures:

      (a) the structure of a molecule of water in the form of a chemical formula;

      (b) the structure of a molecule of water in the linear form;

      (c) the Lewis structure of a molecule of water (cf. (Lewis, 1916));

      (d) another form of the Lewis structure of a molecule of water (cf. (Lewis, 1916));

      (e) planar triangular molecule structure.

There are different ways to formalize this notion of a structure. One is based on the approach of Landry (1999). It is called the *generative definition* of structure and uses the concept of a schema from (Burgin, 2006b). The following semi-formal definition is the first step in the process of formalization.

Definition 2.1.7. A *structure* is a schema with variables of two types: object variables and connection variables.

If Q is a structure, then any system (object) R that allows one to get by abstraction the structure Q has the structure Q. It is easy to see that one and the same system may have different structures. Any complex system has, as a rule, several structures.

Example 2.1.1. Let us consider mathematical structures. Then such a mathematical system as the set of all real numbers R has several standard structures (cf., for example, (Van der Varden, 1971) and (Kelly, 1957)). One of them is the *order* of real numbers, according to which each number is either less or larger than every other number. There are several algebraic structures in this set. With respect to addition of numbers, R is an Abelian group. With respect to multiplication of numbers, the set $R \setminus \{0\}$ also is an Abelian group. With respect to addition and multiplication of numbers, R is a field. In addition, R has a measure and metric in it. With respect to the natural metric where the distance between two numbers a and b is equal to $|a - b|$, R is a metric space and a topological space determined by this metric. Besides, the order relation and the algebraic structure of a field make R into an ordered field. The algebraic structure of a group and topology make R into a topological group.

To make Definition 2.1.7 exact, we need to define schemas and variables. Both concepts are frequently used but, as a rule, in an informal way. Here we give sufficiently exact definitions for both concepts.

Definition 2.1.8. A *variable* is a triad of the form

(name, interpretation map, object domain)

Here *name* means the name of the variable, e.g., x as a name of an unknown variable, *mass*, t as a name of variable time, T as a name of variable temperature, or C as a name of variable cost; the *object domain* of the variable consists of all tentative values of this variable; and the interpretation map is the relation that connects the name of the variable with all elements (objects) from the object domain of this variable.

There are two types of variables:

Simple variables have the name that is a constant word e.g., x, a, A, G, Set, and *Cat*, or another linguistic structure, e.g., a picture.

Structured variables have the name that includes names of other variables, e.g., $V(x, y)$, $R(A, B)$, f_t, and $x + y$. Variables names of which are included in the name of a structured variable X are called components of X.

The concept of variable allows us to define the concept of constant.

Definition 2.1.9. A *constant* is a variable the object domain of which consists of a single element.

A mathematical definition of a schema is given in (Burgin, 2006b). Here we use only the concept of a static schema as we consider only static structures. Dynamic structures are considered elsewhere.

Definition 2.1.10. A *(static) schema R* is a triad (V_{NR}, c_R, V_{CR}) where V_{NR} is the class (multiset) of all object variables from R, V_{CR} is the class (multiset) of all connection variables (also called variable connection) from R and c_R is the correspondence between components of the structured variables from V_{CR} to variables from V_{NR} and V_{CR}.

There are three main types of connections:

- Connections between object variables (objects) are called *object connections*.
- Connections between object variables (objects) and connection variables (connections) are called *object-connection connections*.
- Connections between connection variables (connections) are called *connection-connection connections*.

Now we can make Definition 2.1.7 more formalized.

Definition 2.1.11 (the *generative definition*). A *structure* is a schema (V_{NR}, c_R, V_{CR}) in which the correspondence c_R connects all components of the structured variables from V_{CR} to variables from V_{NR} and V_{CR}.

Example 2.1.2. The structure of a group can be described in the following way. There are object variables G, x, y, z, x^{-1}, e, and connection variables, which denote operations in the group. There is one binary operation called multiplication $(x, y) \to xy$ (addition $(x, y) \to x + y$ in Abelian groups), one unary operation called taking the inverse (taking the opposite in Abelian groups), and one nullary operation of taking the unit element *e* of a group taking the unit element of a group (taking 0 in Abelian groups). There are following identities:

$$xx^{-1} = e,$$
$$x(yz) = (xy)z,$$

and

$$ex = xe = x.$$

Operations are relations (connections) between object variables (elements of the group), while identities are relations (connections) between object variables (elements of the group) and relation variables (operations).

Example 2.1.3. The structure of a linear over the field *R* space can be described in the following way. A set *L* is called a *linear space* or a *vector space over the field R* of real numbers and its elements are often called *vectors* if it has two operations:

addition: $L \times L \rightarrow L$ denoted by $x + y$, where *x* and *y* belong to *L*, and

scalar multiplication: $F \times L \rightarrow L$ denoted by $a \cdot x$, or ax, where *a* belongs to *R* and *x* belongs to *L*,

satisfying the following axioms

1. Addition is associative:
   For all x, y, z from *L*, we have $x + (y + z) = (x + y) + z$.
2. Addition is commutative:
   For all x, y from *L*, we have $x + y = y + x$.
3. Addition has an identity element:
   There exists an element **0** from *L*, called the *zero vector*, such that $x + 0 = x$ for all *x* from *L*.
4. Addition has inverse element:
   For any *x* from *L*, there exists an element *z* from *L*, called the *additive inverse* of *x*, such that $x + z = 0$.
5. Scalar multiplication is distributive over addition in *L*:
   For all numbers *a* from *R* and vectors *y*, *w* from *L*, we have

$$a \cdot (y + w) = a \cdot y + a \cdot w.$$

6. Scalar multiplication is distributive over addition in R:
   For all numbers a, b from R and any vector y from L, we have

   $$(a + b) \cdot y = a \cdot y + b \cdot y.$$

7. Scalar multiplication is compatible with multiplication in R:
   For all numbers a, b from R and any vector y from L, we have

   $$a \cdot (b \cdot y) = (ab) \cdot y.$$

8. The number 1 is an identity element for scalar multiplication:
   For all vectors x from L, we have $1x = x$.

Operations and axioms are represented by relations between elements of the field R and space L, while these relations are generated by corresponding schemas.

Example 2.1.4. The structure of a word can be described in the following way. There are two object variables w the object domain of which consists of all words and the object domain of which consists of all words in a given alphabet, one constant ε, which is interpreted as the empty word, one connection constant $C(\varepsilon, \varepsilon, \varepsilon)$, and three connection variables $C(w, \varepsilon, w)$, $C(\varepsilon, w, w)$ and $C(w, x, wx)$ called concatenation. The expression $C(\varepsilon, \varepsilon, \varepsilon)$ means that the concatenation of ε with ε is equal to ε, the expression $C(w, \varepsilon, w)$ means that the concatenation of w with ε is equal to w, the expression $C(\varepsilon, w, w)$ means that the concatenation of ε with w is equal to w, and the expression $C(w, x, wx)$ means that the concatenation of w with x is equal to wx.

Let us look what it involves when a system R has the structure Q. By Definition 2.1.6, Q is a schema that has object and connection variables. Then R *has the structure* Q called a *generative structure* of R when the following condition is satisfied.

Condition SC 1. It is possible to divide R into parts (or components) such that taking these parts (components) as concordant interpretations (values) of all object variables in Q and connections between these parts (components) as concordant interpretations (values) of all connection variables in Q, we obtain the system R.

Concordance here means that each variable takes only one value in this process and if v_c is a variable connection between object variables v_{O1} and v_{O2}, then the chosen interpretation (value) of v_c is a connection between the chosen interpretations (values) of v_{O1} and v_{O2}. Formally, it means that we have an isomorphism of Q and R (cf. (Burgin, 2006b)).

Note that it is possible that a system has several structures when we can separate it into several parts (or components).

Example 2.1.5. The structure of the system (*a cloud, rain, the ground*) is the fundamental triad (*Essence* 1, *connection*, *Essence* 2). Here Essence 1 is interpreted as a cloud, Essence 2 is interpreted as a ground, and connection is interpreted as the rain.

A system, as a rule, has many structures. It is possible to classify these structures.

Definition 2.1.12. A structure Q of a system R is called *primary* if different variables have different values, all values of object variables are elements of R and all values of connection variables are direct connections (ties) between elements of R.

Relations in a structure form an extensive hierarchy. We can determine this *structural hierarchy* in the following way (cf. Figure 2.5). On the first level of the structural hierarchy, we have the set $\mathfrak{I}_0$ of all object variables from this structure. On the second level of the structural hierarchy, we have the set $\mathfrak{R}_1$ of all relations between object variables from this structure. On the second level of the structural hierarchy, we have two sets: $\mathfrak{R}_{01}$ is the set of all relations between object variables from this structure and relations from $\mathfrak{R}_1$, and $\mathfrak{R}_{11}$ is the set of all relations between relations from $\mathfrak{R}_1$. On the third level of the structural hierarchy, we have five sets: $\mathfrak{R}_{(0)(01)}$ is the set of all relations between object variables from this structure and relations from $\mathfrak{R}_{01}$, $\mathfrak{R}_{(1)(01)}$ is the set of all relations between relations from $\mathfrak{R}_1$ and relations from $\mathfrak{R}_{01}$, $\mathfrak{R}_{(0)(11)}$ is the set of all relations between object variables from this structure and relations from $\mathfrak{R}_{11}$, $\mathfrak{R}_{(1)(11)}$ is the set of all relations between relations from $\mathfrak{R}_1$ and relations from $\mathfrak{R}_{11}$, and $\mathfrak{R}_{(01)(11)}$ is the set of all relations between relations from $\mathfrak{R}_{11}$ and relations from $\mathfrak{R}_{01}$. Then we can build the fourth level, the fifth level and so on. In general, structural hierarchy can grow as far as we want and even be infinite.

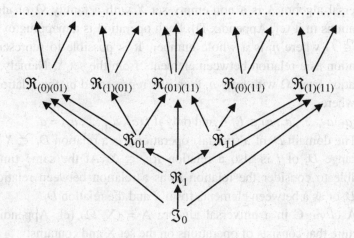

Figure 2.5. The structural hierarchy

In a similar way, we build the structural hierarchy SH(Q) for any system Q. On the first level of SH(Q), we have the set T_0 of all elements of Q. On the second level of the structural hierarchy, we have the set R_1 of all relations between elements from T_0. On the second level of SH(Q), we have two sets: R_{01} is the set of all relations between elements from T_0 and relations from R_1, and R_{11} is the set of all relations between relations from R_1. On the third level of SH(Q), we have five sets: $\mathfrak{R}_{(0)(01)}$ is the set of all relations between elements from T_0 and relations from $\mathfrak{R}_{01}$, $\mathfrak{R}_{(1)(01)}$ is the set of all relations between relations from $\mathfrak{R}_1$ and relations from $\mathfrak{R}_{01}$, $\mathfrak{R}_{(0)(11)}$ is the set of all relations between object variables from this structure and relations from $\mathfrak{R}_{11}$, $\mathfrak{R}_{(1)(11)}$ is the set of all relations between relations from $\mathfrak{R}_1$ and relations from $\mathfrak{R}_{11}$, and $\mathfrak{R}_{(01)(11)}$ is the set of all relations between relations from $\mathfrak{R}_{11}$ and relations from $\mathfrak{R}_{01}$. Then we can build the fourth level, the fifth level and so on. In general, the structural hierarchy of SH(Q) can grow as far as we want and even be infinite.

Example 2.1.6. The clone of a universal algebra (Cohn, 1965) is an example of an algebraic structure that is presented as a structural hierarchy. Indeed, it is possible to find relations between elements, between elements and relations between elements, between relations between elements, between relations between elements and relations between elements, and so on in a clone of a universal algebra. A universal algebra A is a non-empty set X with a family Ω of algebraic operations in it (cf. Appendix D). Each operation is a mapping $\omega_i: X^{ni} \to X$ ($i \in I$) where n_i is a whole number. It is possible to represent each operation as a relation between elements from the set X. Namely, if f an operation from Ω with arity n, then it is represented by the relation $R_f \subseteq X_{n+1}$ where

   $(a_1, a_2, \ldots, a_n, a) \in Rf$ if and only if $f(a_1, a_2, \ldots, a_n) = a$

   The domain D_f of a (partial) operation f is a relation $D_f \subseteq X^n$, while the range D_f of f is also a relation $Rg_f \subseteq X$, At the same time, it is possible to consider the relation R_f as a relation between relations Rg_f and D_f or as a between elements from X and the relation D_f.

   A *clone* C in a universal algebra A = (X, Ω) (cf. Appendix) is a structure that consists of operations on the set X and contains:

1. all operations from Ω;
2. *natural projections* $p_j^n: X^n \to X$ defined by the following identities $p_j^n(a_1, a_2, \ldots, a_n) = a_j$ ($j = 1, 2, 3, \ldots n; n = 1, 2, 3, \ldots$);
3. *finitary multiple composition* (also called superposition) of its elements, i.e., if $g_1, g_2, \ldots, g_m$ are operations from C with arity n and f an operation from C with arity m, then the operation h defined by the following identity

   $h(a_1, a_2, \ldots, a_n) = f(g_1(a_1, a_2, \ldots, a_n), g_2(a_1, a_2, \ldots, a_n), \ldots, g_m(a_1, a_2, \ldots, a_n))$
   also belongs to C.

   An *abstract clone* C on a set X is a structure that consists of operations on the set X and satisfies only conditions 2 and 3.

   In the theory of universal algebras, it is proved that it is possible to realize any abstract clone as the clone in a suitable universal algebra.

   Taking a clone C on a set X, we can represent each of its elements (operation) as a relation between elements from the set X. Namely, if f an operation from C with arity n, then it is represented by the relation $R_f \subseteq$ Xn + 1 where

   $(a_1, a_2, \ldots, a_n, a) \in Rf$ if and only if $f(a_1, a_2, \ldots, a_n) = a$

   The domain D_f of a (partial) operation f is a relation $D_f \subseteq X^n$, while the range D_f of f is also a relation $Rg_f \subseteq X$. Thus, the relation R_f is a relation between relations Rg_f and D_f or as a between elements from X and the relation D_f. In the same way, the operation h defined by the third condition of the clone is a relation between relations defined by operations g_1, g_2, ..., g_m. In other words, if all relations defining operations g_1, g_2, ..., g_m is on the kth level of the clone hierarchy, then the relation defining the operation h is on the $(k + 1)^{st}$ level of the clone hierarchy. For instance, if all relations R_{g1}, R_{g2}, R_{g3}, ..., R_{gn} are relations between relations between relations between elements, then R_h is a relation between relations between relations between relations between elements.

Example 2.1.7. As we have seen in the previous example, subsets of sets are defined by their characteristic functions or characteristic relations, which indicate membership (belonging) of elements to these subsets (cf. Appendix A). Thus relations between subsets and elements are relations between relations and elements.

There are many such examples in calculus and topology. The relation "to be the limit of a sequence" is a relation between an element, e.g., a number, and the relation "to be an element of a sequence". The relations between two sequences "to have the same limit" or "to be equivalent to" are relations between relations. These relations are very important in calculus and topology because they are at the core of definitions of real numbers, real hypernumbers (Burgin, 2002b) and other basic mathematical constructions.

The generative definition of a structure gives a dynamic or operational representation of this concept. At the same time, it is also possible to develop actualized or analytical definition of the structure of a system. One of them is the *extensive definition* of structure based on the cognitive operation of abstraction. Abstraction is combining some systems in class with respect to common properties, or more generally, with respect to some equivalence of systems. To define structure, we use structural similarity and structural equivalence.

Let us consider two systems A and B and assume that they can be represented as collections ObA and ObB of objects and collections Rel$A$ and Rel$B$ of relations with the following property. There is a one-to-one correspondence $C_{obA,B}$ between sets of objects ObA and ObB and this correspondence induces a one-to-one correspondence $C_{reA,B}$ between sets of relations Rel$A$ and Rel$B$. The pair $(C_{obA,B}, C_{reA,B})$ is called a *structural similarity* between systems A and B.

Note that structural similarity is not necessarily an equivalence relation because structural similarity depends on decomposition of systems. We can see this from the following example.

Example 2.1.7. Let us consider three systems *A*, *B* and *C*.

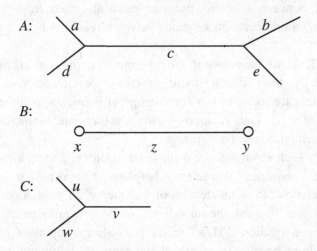

We can do the following decomposition of the system *A* into three elements: *H*, *K* and *c* where *H* consists of *a* and *d* and *K* consists of *b* and *e*. Then we have a correspondence $C_{obA,B}$ where $C_{obA,B}(H) = x$, $C_{obA,B}(K) = y$, and $C_{obA,B}(c) = z$. Relations between elements of all three objects *A*, *B* and *C* is the adjacency relation, e.g., *a* and *c* are adjacent in *A*, while *x* and *z* are adjacent in *B*. Thus, the correspondence $C_{obA,B}$ induces a one-to-one correspondence $C_{reA,B}$ between sets of relations Rel*A* and Rel*B*. Consequently, the pair $(C_{obA,B}, C_{reA,B})$ is a structural similarity between systems *A* and *B*.

We can also do the following decomposition of the system *A* into three elements: *a*, *d* and *U* where *U* consists of *b*, *c* and *e*. Then we have a correspondence $C_{obA,C}$ where $C_{obA,C}(a) = u$, $C_{obA,C}(d) = w$, and $C_{obA,B}(U) = v$. We can see that the correspondence $C_{obA,C}$ induces a one-to-one correspondence $C_{reA,C}$ between sets of relations Rel*A* and Rel*C*. Consequently, the pair $(C_{obA,C}, C_{reA,C})$ is a structural similarity between systems *A* and *C*.

Nevertheless, these decompositions do not allow one to establish structural similarity between systems *B* and *C*.

Definition 2.1.13. A collection of objects (systems) $\mathbf{K} = \{A_i; \ i \in I\}$ consists of *structurally equivalent* objects if there is a representation of each A_i as collections ObA_i of objects and collections RelA_i of relations ($i \in I$) with the following property:

(E) For any A_i and A_j from $\mathbf{K}$, there is a one-to-one correspondence $C_{obAi,Aj}$ between sets of objects ObA_i and ObA_j and this correspondence induces a one-to-one correspondence $C_{reAi,Aj}$ between sets of relations Rel A_i and Rel A_j.

In other words, the systems A_i and A_j are structurally similar with respect to these decompositions.

In particular, two systems A and B are *structurally equivalent* if there is a representation of A as collections ObA of objects and collections Rel$A$ of relations and is a representation of B as collections ObB of objects and collections Rel$B$ of relations such that there is a one-to-one correspondence $C_{obA,B}$ between sets of objects ObA and ObB and this correspondence induces a one-to-one correspondence $C_{reA,B}$ between sets of relations Rel A and Rel B.

This equivalence depends on the level of structuring. This level determines what elements and ties are distinguished. For instance, let us consider three functions: f from the set N of all natural numbers into itself, g from the set R of all real numbers into itself, and h from the set N into the set R. In mathematics, these functions are denoted by $f: N \to N$, $g: R \to R$, and $h: N \to R$. If we treat sets N and R as some mathematical objects, i.e., members of the class with the name *a mathematical object* and do not discern them, then all three functions have the same structure of the fundamental triad

$$\text{mathematical object } 1 \ \to \ \text{mathematical object } 2$$

In contrast to this, if we look upon sets N and R as different mathematical objects, then all three functions have the different structures, although all of them are isomorphic to the fundamental triad. In the first case, we do not pay attention at distinctions between sets N and R, while in the second case, such distinctions are essential. As a result, the same systems (objects) acquire different structures: in the first case, when we do not make distinctions between sets N and R, the structure is $\bullet \to \bullet$, while in the second case, when such distinction are made, the structure is $\bullet \to \circ$.

It is possible to ask whether object representation allows us to get objective structures of objects, i.e., to explicate structures that exist in reality, or the structure of a system depends only on those who build the representation. At the first glance, the process of structure explication looks very subjective as it rests on how distinguishability of elements and ties is defined. However, similar situation exists with what people see, hear and describe. For instance, in this situation when an individual sees a table, we can ask the question, *Is it an image of a real table, an image on a screen or only an illusion of the brain*?

Some philosophers, e.g., Berkeley (1685–1753) or Hume (1711–1776), as well as some religious systems, e.g., Buddhism, argue that everything what people perceive by their senses is an illusion. Now some philosophers, scientists, and science fiction writers have further developed this idea and suggested possible technical mechanisms that could be used for sustaining such an illusory world. For instance, Rucker (1987) suggests that for postmodern people, reality is a pattern of information. Tegmark (2007) suggests that the external physical reality is a mathematical structure.

In contrast to this, science developed means to discern objective reality from illusions. These means are base on repetition of observations and experiments. Although these methods are not perfect, they provide for the whole existence of science and development of the human civilization based on science as a source of the technological progress.

Lemma 2.1.1. The relation of structural equivalence is an equivalence relation in **K**.

Example 2.1.8. Graphs G_1, G_2, and G_3 (cf. Figure 2.6) are structurally equivalent.

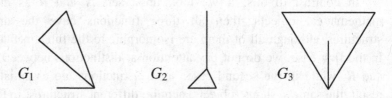

Figure 2.6. Structurally equivalent graphs

Structure in the sense of Definitions 2.1.7 and 2.1.11 is a specific representation of the corresponding object (system), namely, a structural representation. This allows us to explicate relations between generative and extensive definitions of structure.

Proposition 2.1.2. Two objects (systems) are *structurally equivalent* if they have the same generative structure.

Corollary 2.1.2. There is a one-to-one correspondence between all relations from the structural hierarchy $SH(Q)$ and some subset of named relations between elements of T.

To express that two objects (systems) A and B are structurally equivalent, it is said that they have the same *extensive structural type*, or the same *extensive structure*. This gives us the following definition.

Definition 2.1.14 (the extensive definition). An (*extensive*) *structure* of a object (system) R is the extensive structural type of R.

This definition is similar to the definition of a cardinal number in set theory. Namely, if there is a one-to-one correspondence between elements from a set A and elements from a set B, then it is said that they have the same cardinality, or the same *cardinal number* (cf. (Kuratowski and Mostowski, 1967)).

When we restrict ourselves to objects (systems) that belong to some well-defined entity, e.g., to a set, then it is possible to give the extensive definition of a structure in a different form. Let us consider a set of objects (systems) $\mathbf{H} = \{A_i; i \in I\}$.

Definition 2.1.15 (the relativistic extensive definition). A *structure* S with respect to the class $\mathbf{H}$ is a maximal subclass $\mathbf{K}$ of $\mathbf{H}$ that consists of structurally equivalent objects from $\mathbf{H}$.

This definition represents structures as classes of objects (systems) that satisfy an equivalence condition, i.e., these classes are invariant (compare with Definition 2.1.5) with respect to transformations that preserve their structures.

However, it might be more relevant to consider structural representations of systems instead of the initial systems in the Definition 2.1.14. The reason for this is that for theoretical studies, we need only symbolic structures.

Definition 2.1.16. A *structure* S with respect to the class **H** is a class **K** that consists of all equivalent structural representations of systems from **H**.

For instance, a countable Abelian group is a structural representation of the system of all integer numbers.

Definition 2.1.17 (the intensive definition). A *structure* of a system (object) *A* is a more abstract system structurally similar to *A*.

Example 2.1.9. The system *R* of all real numbers has the structure of a field. The same system has the structure of a topological space.

Example 2.1.10. The system of all integer numbers has the structure of an Abelian group. The same system has the structure of a semigroup.

After we have elaborated an exact definition of structure, it is also possible to define some concepts related to structures and used by philosophers, scientists and mathematicians in an informal way.

Explicit structures or *forms* of a physical macroscopic object K are structures of K that are directly comprehensible, e.g., geometric forms or shapes, differential forms, etc.

Symbolic form is a formalized structure. There are non-mathematical forms: formulas in chemistry, syllogisms in formal logic

Pattern is a stable structure.

Idea is an informal inexact structure that is related or belongs to the mental world.

It is necessary to remark that Bourbaki (1957; 1960) elaborated a formal definition of a general mathematical structure as a very abstract concept and classified all mathematical structures into three main groups (Bourbaki, 1948; 1960): order structures, algebraic structures and topological structures. For instance, the set of interval in a line has an order structure; arithmetic, algebraic groups, rings and fields have an algebraic structure; and the set of all real numbers and linear spaces have a topological structure. We do not give here the exact definition of Bourbaki because it demands much more mathematical knowledge than it is expected from the reader of this book. However, to explain the high level of abstractness of this concept, we would like to inform that it is based on well-known concepts of logic, such as a logical term, formal theory and strength relation between formal theories. In addition, the concept of structure uses new concepts introduced by Bourbaki, such as layer (*échelon* in French) construction schema, canonical expansion of mappings by a schema, typification of letters, and some others. This definition is abstract even for the majority of mathematicians, who prefer to use an informal notion of a mathematical structure. As Corry, 1996) writes, Bourbaki's concept of *structure* was, from a mathematical point of view, a superfluous undertaking. Even Bourbaki themselves did not use this concept in their later books of the *Eléments* after they had introduced it in *Theory of Sets* (1960).

It is also necessary to remark that all given definitions specify inner structures of objects (systems). At the same time, it is possible to give a structure characterization be specifying relations between objects with the same structure. To achieve this, we extend Definitions 2.1.13 and 2.1.14 in such a way that the class **K** of all objects (systems) from the class **H** with the same structure is enhanced with relations between these objects. This was the approach used by Ore (1935; 1936) for formalizing the concept of an algebraic structure and developed in category theory for formalizing the concept of a mathematical structure (cf. (Corry, 1996)). Both approaches are based on an idea that the structural nature of algebra in the case of Ore and mathematical structure in general in the case of categories can best be understood by entirely ignoring the existence of the elements of a given algebraic domain and concentrating on the relationships between the domain and its subdomains. The great power of category theory is in exploitation of mapping between objects with the same structure to reveal their properties (cf., for example, (Herrlich and Strecker, 1973; Goldblatt, 1979)).

Analyzing situation with structures as invariant representations of existing things (real or abstract systems), we come to the conclusion that structures, at least, structures of physical things, such as tables, chairs, and buildings, exist in the same way as these physical things exist. We do not give here explicit proofs of this deep-rooted statement because the topic of the book is information. It is possible to find details of these proofs in (Burgin, 1997).

In the context of the theory of named sets, it was discovered that everything in the world has a structure, even chaos (Burgin, 1996d). Thus, structures exist not only in languages, society or human personality, but everywhere. Consequently, it is necessary to study structures not only in linguistics, anthropology, psychology or sociology, but in all sciences: in natural sciences, such as physics, social sciences and humanities, such as sociology and psychology, and technical sciences, such as computer science.

The transition to the study of a structural reality reflects a new stage of the structuralistic doctrine. In particular, investigation of the structural level of the world and scientific research of structures as real phenomena has made possible to find the most basic structure called *fundamental triad* or *named set*. However, the study of the World of Structures is only at its beginning.

Structures are represented by signs and symbols, being themselves symbols of things, systems and processes. To understand this, let us consider theoretical understanding implied by models of the concepts *symbol* and *sign*.

If we analyze the usage of the word *symbol*, we come to the conclusion that it has three different, however, connected, meanings. In a broad sense, symbol is the same as sign. For example, the terms "symbolic system" and "sign system" are considered as synonyms, although the first term is used much more often. The basic property of the sign is that sign points to something different than itself, transcendent to it.

The second understanding of the word "symbol" identifies symbol with a physical sign, that is, some elementary entity inscribed on paper, papyrus or stone, presented on the screen of a computer monitor, and so on. Letters are signs and symbols at the same time. Decimal digits 0, 1, 2, 3, 4, 5, 6, 7, 8, and 9 are also signs and symbols.

However, we are interested in the third meaning of the word "symbol" when it is considered in a strict sense. Such understanding was developed in *semiotics* as a general theory of signs. Semiotics studies structure of signs and their communicative function. As signs exist in a huge diversity of situations, the founder of semiotics, Charles Pierce (1839–1914) and his follower Charles William Morris (1901–1979) defined semiotics very broadly. They hoped that it would influence as many disciplines as possible. For instance, Morris wrote (1938):

"The sciences must look to semiotic for the concepts and general principles relevant to their own problems of sign analysis. Semiotic is not merely a science among other sciences but an organon or instrument to all sciences."

Indeed, today semiotics is an important tool in communication research, information theory, linguistics and the fine arts, as well as in psychology, sociology and esthetics. Yet, although many other disciplines recognize the potential importance of semiotic paradigms for their fields, they have not yet found a satisfying way of integrating them into their domain.

While many use the word *symbol* in the same contexts as the word *sign*, the French linguist Ferdinand de Saussure (1857–1913), sometimes called the father of theoretical linguistics, understood the concept *sign* as a category under the concept *symbol* (Saussure, 1916). To represent the main property of signs, de Saussure introduced a structural model of sign in the form of the *dyadic sign triad* by (see Figure 2.7).

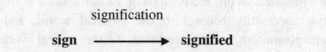

Figure 2.7. The dyadic sign triad of de Saussure

This triad is a kind of the fundamental triad.

Considering the relation between the concepts *sign* and *symbol*, Pierce inverted this relation, making *sign* the general term and *symbol* its particular case as the convention-based sign. According to Pierce, there are three kinds of signs: *icon*, *index*, and *symbol*.

The dyadic sign triad explicates important properties of sign, but not all of them. Namely, sign represents something different than itself due to the meaning. That is why Pierce extended this dyadic model by further splitting the signified into essentially different parts: the sign's object and interpretant (meaning of the sign), and thus, coming to the triadic model of a sign, the *balanced sign triad*:

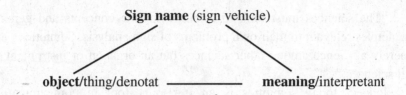

Figure 2.8. The balanced sign triad of Pierce

Thus, in the model of Pierce, a sign is understood as a relation consisting of three elements: the Name (Pierce called it Sign Vehicle), Object and Meaning of the sign. Usually, a physical representation of a sign is what people call sign in everyday life, that is, a sign is some elementary image inscribed on paper, clay tablet, piece of wood or stone, presented on the screen of a computer monitor, and so on.

The balanced sign triad is similar to the existential triad of the World. In it, the component *name* corresponds to the structural world as a syntactic system; the component *object/denotat* corresponds (but does not necessarily belong) to the physical world; and the component *meaning/interpretant* corresponds to the mental world as a semantic system. In many cases, the object is a material thing and as such, is a part of the physical world. However, object can be non-material and thus, does not belong to the physical world in some cases. For instance, the word *joy* is the name of emotion, which is nonmaterial. The word *theorem* is the name of a mathematical statement with definite properties. A statement is also nonmaterial. Nevertheless, object as a component of sign plays the same role as thing, implying that the Pierce triad is homomorphic to the existential triad.

In contrast to de Saussure and Pierce, Morris (1938) defines sign in a dynamic way. He writes, S is a sign of an object D for an interpreter I to the degree that I takes the account of D in virtue of the presence of S. Thus, the object S becomes a sign only if somebody (an interpreter) interprets S as a sign. This gives us the following diagram.

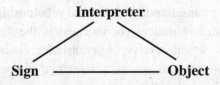

Figure 2.9. The dynamic sign triad of Morris

The dynamic sign triad of Morris is similar to the existential triad of the World. In it, the component *Sign* corresponds to the structural world as a syntactic system; the component *Object* corresponds (but does not necessarily belongs) to the physical world; and the component *Interpreter* corresponds to the mental world as the interpreter interprets sign in his/her mentality. Thus, the dynamic sign triad of Morris is homomorphic to the existential triad.

Let us consider three kinds of signs: icons, indices, and symbols. An *icon* resembles what it signifies. For instance, photographs at the level of direct resemblance or likeness are therefore heavily iconic. We all use computer icons, which helped popularize such a word processor as the Word, as well as with the pictographs, such as used on "pedestrian crossing" signs. There is no real connection between an object and an icon of it other than the likeness, so the mind is required to see the similarity and associate an object and its icon. A characteristic of the icon is that by observing it, we can derive information about the object the icon signifies. The more simplified the image, the less it is possible to learn from it. No other kind of signs gives that kind of pictorial information.

Pierce divides icons further into three kinds: images, diagrams and metaphors. *Images* have the simplest quality, the similarity of aspect. Portraits and computer icons are images. *Diagrams* represent relationships of parts rather than tangible features. Examples of diagrams are algebraic formulae. Finally, *metaphors* possess a similarity of character, representing an object by using a parallelism in some other object. Metaphors are widely used in poetry and language. Examples are the frogs of Aesop who desired a king, a computer mouse or fields in physics.

An *index* has a causal and/or sequential relationship to its signified. A key to understanding indices (or indexes) is the verb "indicate", of which "index" is a substantive. For instance, indices are directly perceivable events that can act as a reference to events that are not directly perceivable, or in other words, they are something visible that indicates something out of sight. You may not see a fire, but you do see the smoke and that indicates to you that a fire is burning. Like a pointed finger, words "this", "that", "these", and "those" are also indices. The nature of the index can be unrelated to that of the signified, but the connection here is logical and organic — the two elements are inseparable — and there is little or no participation of the mind.

Contemporary microphysics extensively utilizes indices called subatomic, or subnuclear, particles. Physicists cannot see these particles even with the best modern microscopes. However, physicists see the results of particle interactions, e.g., in the Wilson cloud chamber, and know properties of these particles.

A *symbol* represents something in a completely arbitrary relationship. The connection between signifier and signified depends entirely on the observer, or more exactly, what the observer was taught or invented. Symbols are subjective. Their relation to the signified object is dictated either by social and cultural conventions or by habit. Words are a prime example of signs. Whether as a group of sounds or a group of characters, they are only linked to their signified because we decide they are and because the connection is neither physical nor logical, words change meaning or objects change names as time goes by. Here it all happens in the mind and depends on it.

However, often, especially in science, people try to create words so that they show/explicate connections to the signified. For instance, a computer is called *computer* because it/he/she computes. A teacher is called *teacher* because she/he teaches. Some elementary particles are called neutrons because they are electrically neutral, i.e., their electrical charge is zero.

Symbols are abstract entities, and whenever we use one, we are only pointing to the idea behind that symbol. For instance, computer aliases (or shortcuts) are symbols of programs and files, namely, a shortcut is a file that opens the actual file or starts the program it refers to. If you trash the alias/shortcut, it does not affect the related file or program. All symbols work in a similar way in relation to the concept they serve. People use a great diversity of symbols. The &, %, @, © and $ symbols, logical symbols, such as ∀, ∃, ∧, ∨, and →, astrological symbols, such as ♀ and ♂, mathematical symbols, such as ∂, Δ, $\prod$, $\sum$, and ∞, road signs, V of victory, all are symbols.

Pierce divides symbols further into two kinds: a *singular symbol* denotes tangible things, while an *abstract symbol* signifies abstract notions. However, it is not always easy to make a distinction. For instance, such symbol as "elephant" signifies an abstract notion of an elephant as a specific animal. At the same time, this symbol as "an elephant" signifies the set of all elephants. Thus, it is more tangible to introduce one more class of symbols, which we call general symbols. A *general symbol* signifies both an abstract notion and a collection of things encompassed by this notion. For instance, "a lover" is a general symbol, while "love" is an abstract symbol.

One and the same word can be used as a name for different symbols and even for different types of symbols. For instance, on the social level, the word a "field" is used as an individual symbol when it denotes a specific place on the Earth. At the same time, it will be an abstract symbol used in mathematical community and denoting a specific mathematical structure, or more exactly, two kinds of structures — fields in algebra, such as the field of all real numbers, and fields in functional analysis, such as a vector field. On another, wider group level, the same word is used as a name of some system, such as a field of mathematics, field of activity or field of competence. Important examples of symbols are general concepts and formal expressions.

Antonov (1988) introduces the psychological structure of information as the triadic hierarchy: *sign — meaning — sense*. It gives us the following triad:

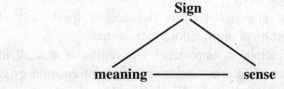

Figure 2.10. The representation triad of Antonov

This triad is also isomorphic to the Existential Triad of the world. Systems of signs are used to represent information based on syntactic properties and correspond to the Structural World. Meaning corresponds to the Physical World and sense to the Mental World.

Connections between signs and information are, for example, reflected in the cartographic triad introduced by Berlyant (1986).

sign → image → information

Figure 2.11. The cartographic triad

Thus, we can see that the structure of signs has, as rule, the form of triad. It is either a fundamental triad or is built of fundamental triads.

2.2. What Information Is: Information Ontology

Progress is impossible without change,
and those who cannot change their minds
cannot change anything.
George Bernard Shaw (1856–1950)

The main question of this section is "What is information?" To answer this question, we start with describing the basic properties of information. These properties are represented in the form of ontological principles. The first principle of the general theory of information determines a perspective for information definition, i.e., in what context information is defined.

Ontological Principle O1 (the *Locality Principle*). It is necessary to separate information in general from information (or a portion of information) for a system R.

In other words, empirically, it is possible to speak only about information (or a portion of information) for a system.

Why is this principle so important? The reason is that all major conventional theories of information assume that information exists as something absolute like time in the Newtonian dynamics. Consequently, it is assumed that this absolute information may be measured, used, and transmitted. On the abstract level, it is possible to build such a mathematical model that makes sense of absolute information, but in the practical environment, or as scientists say, empirically, it is not so.

To demonstrate this, let us consider the following situation. We have a book in Japanese and want to ask what information it contains. For a person who does not know Japanese, it contains no information. At the same time, its information for those who know Japanese may be immense.

Another situation is also possible. Let us consider a textbook, for example, in mathematics. If it is a good textbook, then it contains a lot of information for a mathematics student. However, if we show this book to a professional mathematician, she or he might say, "Oh, I know everything in this book, so it contains no information for me."

We will have the same result but for a different reason if we give this book to an art student who is bored with mathematics. Weinberger (2002) writes that the meaning of a message can only be understood relative to its receiver. Thus, the latter student will not read the book and the book will not contain information for her.

To make the situation more evident, imagine a completely deaf and blind person who comes to a movie theater without any devices to compensate his deficiencies. How much information this person will get there?

It is interesting that the Ontological Principle O1 demonstrates tendencies and changes similar to those that were prevalent in theoretical physics in the 20th century. Classical Newtonian-Laplacian physics is global, that is, all is the same whatever place in the universe we take. New physics has developed more refined methods. Relativity theory states that inertial systems that move with different speeds have different time. Quantum electrodynamics models quantum phenomena by gauge fields, which are invariant with respect to local transformations.

As Rovelli writes (1996), in physics, the move of deepening our insight into the physical world by relativizing notions previously used as absolute has been applied repeatedly and very successfully. The most popular examples are the Relativity Principle introduced by Galileo and relativity theory. By the Galileo's Principle of Relativity, the notion of the velocity of an object has been recognized as meaningless, unless it is indexed with a reference body with respect to which the object is moving. Thus, correct representation of motion demands a definite frame of reference. With special relativity, simultaneity of two distant events has been recognized as meaningless, unless referred to a specific light signal connecting these events.

Definition 2.2.1. The system R with respect to which some information is considered is called the *receiver*, *receptor* or *recipient* of this information.

Such a receiver/recipient can be a person, community, class of students, audience in a theater, animal, bird, fish, computer, network, database and so on. Necessity to have a receiver stated in the Ontological Principle O1 implies, as Buckland (1991) explains, "that the capability of being informative, the essential characteristic of information-as-thing, must also be *situational*". In this context, to be informative for people means to have an answer to somebody's question. The *informativeness*, in the sense of Buckland (1991), is a relation between the question and the thing. Thus, there is no such a thing that is inherently informative. To consider something as information for an individual or group of people is always to consider it as informative in relation to some possible questions of this individual or group. We do not always realize this, because it is mostly assumed. It is assumed, for example, that an article about the Sun may help answering questions about the Sun. It is less obvious, however, that a meteorite from outer space may answer questions about the origin of life. A good deal of scientific knowledge is needed to understand why this is the case (and a claim about the informativeness of something is knowledge-dependent and may turn out to be wrong). In a wider sense, background knowledge is always important for a person to extract information from any object (including documents and texts).

The Ontological Principle O1 well correlates with the assumption of Dretske (1981) that information is always relative to a receiver's background knowledge, as well as with the statement of Meadow and Yuan (1997) that the information in a message is a function of the recipient as well as of the message.

Some believe that dependence on prior knowledge in information extraction brings us to subjectivity in defining information and becomes the source of elusiveness of the concept of information (von Baeyer, 2004). Consequently, the first impression is that the Ontological Principle O1 supports this subjective approach to the concept of information. However, in this case, subjectivity is confused with relativity. The Ontological Principle O1 states that information has to be considered not in the absolute way, as the majority of researchers in the field are doing, but as a relative essence properties of which depend on a chosen system. Dependence on an individual is usually called subjectivity. However, subjectivity is what depends only on the opinion of an individual. At the same time, information for a person A does not necessary coincides with what A thinks about information for herself or himself. For instance, A listens to a lecture and thinks that she gets a lot of information from it. This is a subjective estimate of information in the lecture. Nevertheless, if A forgets all she had heard the next day, the lecture actually has no information for her. This is an objective estimate of information in the lecture.

Another situation is when a person B reads some mathematical paper and finds nothing interesting there. Then B thinks that she has received no information from this paper. This is a subjective estimate of information in the paper. However, if B remembers something connected to this paper, then objectively she gets information from the text. Moreover, it is possible that B finds after some time that ideas from that paper are very useful for her work. This changes the subjective estimate of the paper and B starts to think that that paper contained a lot of useful information. Moreover, as axiological principles considered in Section 2.3 show, for person B, information in that paper objectively also grows. This demonstrates that both objective and subjective estimates of information for a recipient depend not only on the recipient but also on time, interaction between the recipient and the carrier of information, work of this recipient and some other parameters.

Thus, information has objective but relativistic properties and subjective estimates of these properties. This well correlates with the situation in the classical physics where objectivity is the pivotal principle. The very fact that we treat science as a method of handling human experience inevitably involves the presence of observer in scientific theories (Lindsay, 1971). This allows us to better understand the role of the observer in interpretations of quantum theory. What seemingly began as a technical measurement problem in a specific area became gratuitously generalized into a metaphysical assertion that "observer-created" reality is all the reality that exists. The positivist idea that it is meaningless to discuss the existence of something which cannot be measured (position and velocity, within certain limits) has been developed into the idea that subatomic particles are unreal, formal structures, which only achieve actuality upon observation. In such a way, positivism became transformed into subjectivism (and even, solipsism), promoting the idea that the observer somehow creates reality by the act of observation. Heisenberg first stated that the electron does not have a well-defined position when it is not interacting. The next step in this direction is called the relational interpretation of quantum reality (Rovelli, 1996). It states that, even when interacting, the position of the electron is only determined in relation to a certain observer, or to a certain quantum reference system, or similar.

In the light of the general theory of information, we can understand the relativity principle in quantum physics interpretation in the following way. Any (material) thing exists for people only when they get information from (in a generalized sense, about) this thing. To get information from something, e.g., subatomic particle, we need an observer, i.e., recipient of information from this object. One may ask a question whether such particles existed before they were discovered. Albert Einstein (1879–1955) once asked Niels Henrik David Bohr (1885–1962) if the moon exists when no one is looking at it. Science gives a positive answer to such questions although, for example, there were no observers before these particles were discovered.

Werner Karl Heisenberg (1901–1976) writes, "The conception of objective reality of the elementary particles had thus evaporated ... into the transparent clarity of a mathematics that represents no longer the behavior of particles but rather our knowledge of this behavior" (Heisenberg, 1958). According to the general theory of information, this knowledge is formed based on information that we get from particles.

The Ontological Principles O1 and O4 provide an information theoretical explanation for this cognitive and methodological phenomenon. Indeed, for existence of something, it is necessary to consider observation in a generalized sense. Namely, we do not need any implication that the observer, or the observer system in quantum mechanics, is human or has any other peculiar property besides the possibility of interacting with the "observed" system S.

If we take the problem of subatomic particle existence for people before these particles were discovered, we see that recipients of information existed although those recipients did not know that they receive information from particles. For instance, psychological experiments show that people receive information but cannot identify it on the level of conscience (Luck, et al, 1996). Besides, there are several kinds of information in addition to cognitive information (Burgin, 2001).

To explain this phenomenon and to solve the puzzle of physical existence, let us consider the following mental experiment. A particle, say electron, passes through a Wilson cloud chamber and produces a visible track of droplets condensed on ionized molecules. A digital or film camera makes a picture of this track. Only after 30 days, a physicist looks at this picture. It is evident that the electron existed before the observer looked at the picture and found evidence of its presence. Actually, we have information that the electron existed, at least, at the moment when it interacted with the molecules in the Wilson cloud chamber.

The Ontological Principle O1 also correlates with the idea of Roederer (2002) and some other researchers that interaction plays very important role in information processes. In other words, there exists no explicit information without interaction of the carrier of information with the receiver of information. However, it is possible to speak of information not only when we have both a sender and a recipient because the recipient can extract information from a carrier when the carrier does not send it. So, the classical *communication triad* (2.2.1) is not necessary for existence of information.

$$\textbf{\textit{Information}}$$
$$\textbf{Sender} \xrightarrow{\hspace{3cm}} \textbf{Receiver/Receptor} \qquad (2.2.1)$$

The intrinsic necessary structure is the *input information triad* (2.2.2).

$$\textbf{\textit{Information}}$$
$$\textbf{Carrier} \longrightarrow \textbf{Receiver/Receptor} \qquad (2.2.2)$$

Note that in many situations, it is possible to treat a set or a sequence of carriers as one carrier. However, the structure of a carrier, e.g., whether it is integral or consists of separate parts, and the history of its interactions with the receptor can be important for some problems.

The triad (2.2.2) is complemented by the *output information triad* (2.2.3).

$$\textbf{\textit{Information}}$$
$$\textbf{Sender} \longrightarrow \textbf{Carrier} \qquad (2.2.3)$$

Together, the output and input information triads form the communication triad (2.2.1) as their sequential composition. Note that it is possible that the carrier of information in the information triads (2.2.2) and/or (2.2.3) coincides with the Sender.

Besides, even if information gives some image of a pattern from a sender, this correspondence is not necessarily one-to-one.

It is also possible to speak about some implicit (potential) information in a carrier for a given system as a receptor. An important issue is how the reciever/receptor uses information. Recently researchers take much more interest in the information user perspectives on information (Choo, 1998). It is demonstrated that though a document may be a specific information representation, the user wraps this, in a sense objective, content in an interpretative envelope and produces knowledge, thereby giving the information a subjective meaning (Stenmark, 2002). The meaning and value of any given piece of information reside in the relationship between the information and the user's knowledge. Consequently, the same information may result in different subjective meanings and values.

Being more adequate to reality than previous assumptions about the essence of information, the first ontological principle makes it possible to resolve a controversy that exists in the research community of information scientists. Some suggest that information exists only in society, while others ascribe information to any phenomenon. Utilizing the Ontological Principle O1, the general theory of information states that if we speak about information for people, then it exists only in society because now people exist only in society. However, when we consider a more general situation, then we see that information exists in everything and it is only a problem how to extract it.

Thus, the first principle explicates an important property if information, but says nothing what information is. This is done by the second principle, which has several forms.

Ontological Principle O2 (**the *General Transformation Principle***). In a broad sense, *information* for a system R is a capacity to cause changes in the system R.

Thus, we may understand information in a broad sense as a capacity (ability or potency) of things, both material and abstract, to change other things. In what follows, information in a broad sense is also called *general information*.

The concept of information in a broad sense makes information an extremely widespread and comprehensive concept. Nevertheless, this situation well correlates with the etymological roots of the term *information*. This term originated from the Latin word "*informare*", which can be translated as "*to give form to*", "*to shape*", or "*to form*".

However, as it has happened with many other words, the meaning of the word information has essentially changed. Since approximately the 16th century, the term *information* appears in ordinary French, English, Spanish and Italian in the sense we use it today: "*to instruct*", "*to furnish with knowledge*", whereas the ontological meaning of "*giving form to something*" became more and more obsolete. Although, as Capurro thinks (1978; 1991), "*information ... came to be applied, as a more or less adequate metaphor, to every kind of process through which something is being changed or in-formed*". This opinion strongly supports the Ontological Principle O2.

In addition, the Ontological Principle O2 well correlates with understanding of von Weizsäcker (2006), who writes "we rate information by the effect it has" and with opinion of Boulding (1956), who writes that messages consist of information, while the meaning of a message is the change that it produces in the image.

The Ontological Principle O2 allows for subjectivity. This is an aspect important for information theory and its applications, as Ingversen stated (1992).

Information is a general term. Like any general term, it has particular representatives. Such a representative is called a *portion of information*. For instance, information in this sentence is a portion of information. Information in this chapter is a portion of information. Information in this book is also a portion of information. Information in your head, dear reader, is also a portion of information.

To understand better the situation, let us consider some general terms. For instance, a book is a general term. This book that you are reading now is a representative of this general term. A human being or a reader is a general term. At the same time, you, dear reader, is representative for both of these general terms.

Thus, the Ontological Principle O2 implies that information exists only in form of portions of information. Informally, a portion of information is such information that can be separated from other information.

Remark 2.2.1. In some cases, we use the term "information" instead of the term "a portion of information" when it does not cause misunderstanding.

When we speak, we send sequences of information portions. When we listen, we receive sequences of information portions.

A sequence of information portions is called a *stream of information*, or *information stream*.

There are finite and infinite information streams.

Communication consists of separate but connected actions of sending and receiving information. Thus, it is possible to represent communication as two streams of information going into opposite directions. The information projection of information processing also is a stream of information. Data streams contain streams of information.

We also consider such generic terms as a piece of information and a slice of information.

Definition 2.2.2. A *piece of information* is information that comes to a system in one interaction of this system.

Thus, a piece of information is also a portion of information. However, not any portion of information is a piece of information.

Note that the concept of a piece of information is relative, i.e., what is a piece of information for one system may be not a piece of information for another system.

Definition 2.2.3. A *slice of information* is a portion of information about some object (domain, system or subject).

Information is, as a rule, about something.

Definition 2.2.4. What information is about is called an *object* of this information.

Devlin (1991) introduced another quantization of information based on the concept of *infon* as an 'item of information'. An infon has the form

$$(R, a_1, \ldots, a_n, 1)$$

or

$$(R, a_1, \ldots, a_n, 0)$$

where R is an n-place relation and $a_1, \ldots, a_n$ are objects appropriate for R.

The class of *compound infons* is constructed from the infons by means of operations of conjunction and disjunction and bounded existential and universal quantification (over parameters).

Infons (or compound infons) may be true or false about a situation. When an infon σ is true in a situation s, and, it is denoted by

$$s \mid = \sigma$$

to indicate that the infon σ is "made factual by" the situation s and to express that s supports σ. The expression $s \mid = \sigma$ is a proposition.

Operations with infons, such as $\wedge$ and $\vee$, are introduced and studied. Infons (in the sense of Devlin) are very similar to structured elements of relational databases.

In contrast to the descriptive definition of Devlin, Longo (2003) introduces infons, as units or pieces of information, in an axiomatic way. A formal model of the structure of information is presented by five axioms. According to these axioms, infons have members although it is not defined what is a member of an infon. Axioms define identity, containment, and joins of infons. Axioms of identity and containment are similar to the standard axioms used in set theory. There are different joins, which are commutative and associative operations with infons. With respect to these operations, there are identity elements and inverses of infons. Multiplicative joins correspond to adding or removing new bits to a system, while additive joins correspond to a change of state. The order or size of an infon is defined.

Infons in the sense of Devlin are special cases of the basic element called an *elementary fact* of the infological data model (Langefors, 1963; 1974; 1977; 1980; Sundgren, 1974; Tsichritzis and Lochovsky, 1982). An elementary fact has the following form:

$$(a_1, \ldots, a_n, R, t)$$

where $a_1, \ldots, a_n$ are objects, R is an n-place relation between these objects or their property and t is time. When t is equal either to 1 or to 0, an elementary fact coincides with an infon.

The Ontological Principle O2 has several consequences. First, it demonstrates that information is closely connected to transformation. Namely, it means that information and transformation are functionally similar because they both induce changes in a system. At the same time, they are different because information is potency for (or in some sense, cause of) change, while transformation is the change itself, or in other words, transformation is an operation, while information is what induces this operation.

Second, the Ontological Principle O2 explains *why* information influences society and individuals all the time, as well as why this influence grows with the development of society. Namely, reception of information by individuals and social groups induces transformation. In this sense, information is similar to energy. Moreover, according to the Ontological Principle O2, energy is a kind of information in a broad sense. This well correlates with the Carl Friedrich von Weizsäcker's idea (cf., for example, (Flükiger, 1995)) that *energy might in the end turn out to be information*. At the same time, the von Weizsäcker's conjecture explains existing correspondence between a characteristic of thermal energy such as the thermodynamic entropy given by the Boltzmann-Planck formula $S = k \cdot \ln P$ and a characteristic of information such as the information entropy given by a similar Hartley formula $I = K \cdot \ln N$, which is, in some sense, a special case of the Shannon formula $I = -\sum_{i=1}^{n} p_i \cdot \log_2 p_i$ (cf. Chapter 3).

Third, the Ontological Principle O2 makes it possible to separate different kinds of information. For instance, people, as well as any computer, have many kinds of memory. It is even supposed that each part of the brain has several types of memory agencies that work in somewhat different ways, to suit particular purposes (Minsky, 1986). It is possible to consider each of these memory agencies as a separate system and to study differences between information that changes each type of memory. This might help to understand the interplay between stability and flexibility of the mind, in general, and memory, in particular.

In addition, information that is considered in conventional theory and practice is only cognitive information. At the same time, there are two other types: direct affective/emotional information and direct effective information (cf. (Burgin, 2001) and Section 2.6). For instance, direct affective/emotional information is very important for intelligence. As Minsky (1998) said:

"Emotion is only a different way to think. It may use some of the body functions, such as when we prepare to fight (the heart is beating faster, the level of awareness is higher, muscles are tighter, etc.). Emotions have a survival value, so that we are able to behave efficiently in some situations. Therefore, truly intelligent computers will need to have emotions. This is not impossible or even difficult to achieve. Once we understand the relationship between thinking, emotion and memory, it will be easy to implement these functions into the software."

In essence, we can see that all kinds and types of information are encompassed by the Ontological Principle O2.

However, the common usage of the word information does not imply such wide generalizations as the Ontological Principle O2 does. Thus, we need a more restricted theoretical meaning because an adequate theory, whether of the information or of anything else, must be in significant accord with our common ways of thinking and talking about what the theory is about, else there is the danger that theory is not about what it purports to be about. Though, on the other hand, it is wrong to expect that any adequate and reasonably comprehensive theory will be congruent in every respect with common ways of thinking and speaking about its subject, just because those ways are not themselves usually consistent or even entirely clear. To achieve this goal, we use the concept of an *infological system* IF(R) of the system R to introduce information in the strict sense. It is done in two steps. At first, we make the concept of information relative and then we choose a specific class of infological systems to specify information in the strict sense.

As a model example of an infological system IF(R) of an intelligent system R, we take the system of knowledge of R. It is called in cybernetics the *thesaurus* Th(R) of the system R.

Another example of an infological system is *infological representation* of reality in the sense of Langefors (1963; 1974; 1977; 1980) and Sundgren (1974).

Infological system plays the role of a free parameter in the general theory of information, providing for representation in this theory different kinds and types of information. Identifying an infological system IF(R) of a system R, we can define information relative to this system. This definition is expressed by the following principle.

Ontological Principle O2g (**the *Relativized Transformation Principle***). *Information* for a system R *relative to the infological system* IF(R) is a capacity to cause changes in the system IF(R).

Now we can define information as the potency of objects (things, texts, signals, etc.) to produce changes in an infological system. In a more exact way, information for a system R is the capacity of objects to produce changes in an infological system IF(R) of R.

This definition is similar to the definition of energy as the capacity of a physical system to do work, produce heat, light, electricity, motion, chemical reactions, etc., i.e., to make physical changes (Lindsay, 1971). This is not accidental because, as we have seen, energy is a kind of information in a broad sense and energy is measured by work. Information is also related to work.

Work, in general, can be defined as something that has been accomplished through the effort, activity or agency of a person or thing. When energy acts on an object (system) changing it or its relations to other objects, e.g., changing object's location, we say that work was done on the object by the energy.

The Existential Triad allows us to consider objects of three types: physical, mental, and structural objects. Respectively, it gives us three types of work: physical work, mental work, and structural work.

Physical work is done on physical objects by energy. As we know, energy is one of the most important features of the physical world, where nothing can change or be changed without energy. As a result, the concept of energy has become pivotal to all natural science disciplines. One of the most fundamental laws of physics is the law of energy conservation in a closed physical system. In physics, students learn about the conversion of potential energy to kinetic energy when a physical body is set in motion, the transfer of energy when one physical body collides with another, and study the famous formula $E = mc^2$, which estimates how much energy is stored in a physical body with a definite mass. In biology, students learn that all organisms need energy to carry out their life functions. In chemistry, students learn that chemical reactions need energy to start and can give energy. Contemporary cars, planes and rockets that go from the Earth to the outer space use energy of chemical reactions. Now problems of energy are crucial to society because some countries have more sources of energy than others.

Mental work is done on mental objects by mental energy, also called psychic energy. Usually mental energy and psychic energy are differentiated although both are not well defined. As a result *mental energy* is considered as a mood, ability or willingness to engage in some mental work. Mental energy is often related to the activation level of the mind. *Psychic energy* has become an essential component of several psychological theories. At first, the concept of psychic energy, also called *psychological energy*, was developed in the field of psychodynamics by German scientist Ernst Wilhelm von Brücke (1819–1892). In 1874, Brucke proposed that all living organisms are energy-systems governed by the principle of the conservation of energy. However, this principle has been never proved. There are no experimental evidence that supports this principle. Nevertheless, the idea of psychic energy was adopted in later psychology.

Brucke was the supervisor of Sigmund Freud (1856–1939). Developing psychoanalysis, Freud adopted this new idea about energy and suggested that it was possible to apply both the first law of thermodynamics and the second law of thermodynamics to mental processes, describing functioning of a mental energy. In *The Ego and the Id*, Freud argued that the Id was the source of the personality's desires, and therefore of the psychic energy that powered the mind. The psychoanalytic approach assumes that the psyche of people needs some kind of energy to make it work. This energy is used in mental work, such as thinking, feeling, and remembering. It is assumed that psychic energy comes from the two main drives: Eros (or libido, the life and sexual instincts) and Thanatos (death instinct).

The theory of psychic energy was further developed by Carl Gustav Jung (1875–1961), a student of Freud. In 1928, Carl Jung published a seminal essay entitled "On Psychic Energy" (cf. (Jung, 1969a; Harding, 1973)). Later, the theory of psychodynamics and the concept of "psychic energy" was developed further by such psychologists as Alfred Adler (1870–1937) and Melanie Klein (1882–1960).

Emotional energy is an important special case of psychic energy. For instance, Collins (1993) considers emotional energy as the common denominator of rational action. This is one of the reasons why Minsky (1998) and some other researchers consider emotions as the necessary component of artificial intelligence.

In the context of the general theory of information, there is a difference between mental energy and psychic energy. *Mental energy* is information in the broad sense that has a potential to do mental work. Individual mentality of people constitutes only one level of the Mental World. The next level is the mentality of society. This implies that psychic energy is a kind of mental energy on the level of individual mentality of people like in the Physical World, there are different kinds of energy: kinetic energy, potential energy, gravitational energy, solar energy, tidal energy, wind energy, hydro energy, etc. *Psychic energy* is information that is related to the human psyche as the corresponding infological system belongs to the brain (cf. the Ontological Principle O2g).

Structural work is done on structural objects (structures) by information *per se*, or information in the strict sense (cf. the Ontological Principle O2a). Thus, information *per se* is structural energy, which has a potency to do structural work and being received and accepted performs structural changes. Computers and people are power plants for information as structural energy. As conventional power plants change energy forms, computers change symbolic information forms (information carriers) to a more efficient representation aimed at making information more appropriate and transparent for consumption, storage and processing. For instance, computers perform data mining and derive knowledge from data. A new technology called information extraction maps natural-language texts into structured relational data (Etzioni, et al, 2008). In other words, natural-language texts, which are information representations, are converted into relational structures, which can be more efficiently processed by computers than natural-language texts, fusing relevant pieces of information into a coherent summary and reducing the time required to perform complex tasks from hours to minutes. In similar but more advanced way, scientists search for scientific data by measurement and experimentation and infer knowledge about our world from these data.

In this context, all these concepts and corresponding phenomena, i.e., energy, mental energy, and information *per se* as structural energy, are particular cases of information in a broad sense.

Thus, we can see that infological system concept introduction results in an even more general than in the Ontological Principle O2 definition of information. Indeed, in a general case, we can take as an infological system IF(R) of the system R any subsystem of R. In the case when IF(R) of the system R is the whole system R, the Ontological Principle O2g becomes the same as the Ontological Principle O2. At the same time, taking different kinds of infological systems, it is possible to differentiate different kinds of information, while all these kinds are not differentiated with respect to the Ontological Principle O2.

Generality of the Ontological Principle O2g allows one to consider energy as a specific kind of information. Namely, taking physical bodies (things) as a class **M** of infological systems, we see that information with respect to systems from **M** is what changes material bodies. However, we know that it is energy, which changes material bodies. Thus, energy is a kind of information that acts directly at material bodies. Other kinds of information act indirectly. Their action is mediated by energy.

However, being more general, the relativistic definition of information makes the concept more exact, flexible and selective than the concept of information in a broad sense introduced in the Ontological Principle O2. Information in a broad sense encompasses too much for the traditional understanding of information and intuition behind this that has been formed by practice of many generations.

The relativistic definition of information can be tuned up to the diversity of existing descriptions, interpretations, definitions, understandings and ideas of information. A choice of a definite infological system IF(R) in a system R allows a researcher, philosopher or practitioner to find such an interpretation of the concept of information that the best suits the goals, problems and tasks of this researcher, philosopher or practitioner.

Selection of a definite infological system IF(R) allows one to reflect information concepts on different reality levels and domains, as well as to differentiate the natural sciences meaning of information, which is empirical, and the social sciences and humanities notion of information, which is conceptual. For instance, genetic information is often considered as something physical (a DNA molecule, a part of it or its structure), while economic information (e.g., the quotation of a negotiated price or a message from the stock market) is conceptual.

The chosen infological system IF(R) functions as a storage of information that has a definite type and serves as an indicator of information reception, transformation and action. Thus, the representation space of the infological system IF(R) is naturally called by the name *information space*. For instance, taking a database **B** as an infological system IF(R), we have such an information space as the system of all data in **B**.

Relativization of the concept of information also allows one to formalize the idea of von Weizsäcker that information is inherently a context-related, relative concept and exists only in relation to the difference of two semantic levels (cf. (Lyre, 2002; 2003)).

The symbol IF may be considered as a name (denotation) of an operator that is defined in a space of systems of some chosen kind. It may even be defined, in the space of all systems although the space of all system is a notion that, like the notion of the set of all sets, can cause contradictions (cf., for example, (Fraenkel and Bar-Hillel, 1958)). Being applied to a system R, the operator IF defines in R its infological system $IF(R)$. For instance, taking such systems as people, we can correspond the mind to each individual H as an infological system $IF(H)$ of this individual. Another option to define value of the operator IF is to correspond the system of knowledge of an individual H to this individual as her or his infological system $IF(H)$. One more option is to correspond the system of beliefs of this individual H to this individual as her or his infological system $IF(H)$.

The infological system becomes a parameter of the definition and allows one to vary the scope, meaning and features of the defined concept. As a result, this parametric definition makes it possible to overcome the limitations in the theory of information imposed by the, so-called, Capurro trilemma (Fleissner and Hofkirchner, 1995; Capurro, Fleissner and Hofkirchner, 1999). This trilemma states that information may mean either the same at all levels (*univocity*) or something similar at all levels (*analogy*) or something different at different levels (*equivocity*). In the first case of *univocity*, as Capurro suggests (Capurro, Fleissner and Hofkirchner, 1999), "we lose all qualitative differences, as for instance, when we say that e-mail and cell reproduction are the same kind of information process. Not only the 'stuff' and the structure but also the processes in cells and computer devices are rather different from each other. If we say the concept of information is being analogically used, then we have to state what the 'original' meaning is." If the concept of information is considered at the human level, then we are confronted with anthropomorphism when we use the same concept at a non-human level. By the same token, we would say that "in some way" atoms "talk" to each other, etc. Finally there is *equivocity*, which means that, for example, information in physics and information in education are wholly different concepts. In this case, information cannot be a unifying concept any more. Further reasoning brings Capurro to the conclusion that "we are faced with infinite concepts of information, something which cannot be overlooked by any kind of theory."

However, having a sufficiently variable parameter, we can generate a family of concepts that represent existing understandings and interpretations of the word *information*. The relativistic definition of information provides such a flexible parameter as the infological system. This definition possesses, at the same time, univocity, analogy and equivocity. As a result, this definition encompasses, at the same time, the broad concept of information given in the Ontological Principle O2 and a variety of more restricted concepts. Examples of such more restricted concepts are: information as *elimination of uncertainty* (statistical approach, cf. Chapter 3), as *making a distinction* (the approach suggested by Hofkirchner (1999)) or as "*any difference that makes a difference*" (the approach suggested by Bateson (1972; 2000)).

Any system has an infological system and not a single one. Consequently, in contrast to the opinion of some researchers, information is important both for the biotic and abiotic worlds. Information enters non-living physical world even without living beings.

This implies that for a complex system there are different kinds of information. Each type of the infological system determines a specific kind of information. For example, information that causes changes in the system of knowledge is called cognitive information. Existing approaches to information theory and problems with understanding information as natural, social, and technological phenomenon resulted in a current situation when researchers consider only cognitive information.

At the same time, Roederer defines information as *the agent that mediates the correspondence between features or patterns in the source system A and changes in the structure of the recipient B*. This definition strongly correlates with the definition from the Ontological Principle O2a. Taking such infological system as genetic memory, we come to the concept of biomolecular information considered by Roederer (2002).

The concept of infological system is very flexible. Indeed, it is possible to consider even dynamic infological systems. As Scarrott writes (1989), the most basic function of information is to control action in an organized system and thereby operate the organization. The concept of infological system allows one to reflect this situation, taking behavior of a system as a dynamic infological system. Indeed, the standard example of an infological system is a thesaurus, or a system of knowledge. Knowledge is often represented as texts (cf. Chapter 4). At the same time, Rikur (1995) suggests that it is natural to treat any action as a text and to apply hermeneutics to studies of actions. For instance, any action has the content and we can read actions. Thus, a system of actions *A* can be an infological system and it is possible to consider a specific kind of information with respect to this infological system *A*.

The structure of the world considered in the previous section induces similar structures in many systems that exist in this world. It is true for language, symbols, social institutions and organizations, psyche, intelligence and many others. In particular, the world structure has definite implications for infological systems.

When *R* is a material system, its infological system IF(*R*) often consists of three components:
- the *material component*, which is a system of physical objects;
- the *symbolic component* realized by the material component;
- the *system of structures*, which form the structural component of the infological system.

The symbolic component plays the role of a connecting link between the material and structural components.

For instance, the material component of the infological system of a human being is the brain or its part that is called memory. What is commonly called memory is not a single, simple system. It is an extraordinarily complex system of diverse components and processes. Memory of a person has three, or perhaps even more, distinct components (Minsky, 1986). The most important and best documented by scientific research are *sensory information storage* (SIS), *short-term memory* (STM), and *long-term memory* (LTM). Memory researchers do not employ uniform terminology. Sensory information storage is also known as *sensory register*, *sensory store*, and *eidetic memory* and *echoic memory*. Short- and long-term memories are also referred to as *primary memory* and *secondary memory*. Each component of memory differs with respect to its function, the form of information held, the length of time information is retained, and the amount of information-handling capacity. Memory researchers also posit the existence of an interpretive mechanism, as well as an overall memory monitor or control mechanism that guides interaction among various elements of the memory system.

The corresponding to the brain symbolic component is the mind. In the case of the memory as the material component, we take the symbolic representation of the data and knowledge system as the corresponding symbolic component.

Actually any entity can be used as a symbol, even a process. For instance, neurons do not have memory. However, neural networks allow one to organize some kind of memory, being capable of storing and retrieving data from this memory (Burgin, 2005). There are two formats for saving data: static and dynamic. Dynamic storage is utilized for temporary data. In this case, a part of a network is used only for preserving information. If we take a neuron, which can be in two states: firing and silent, then it is possible to interpret a silent neuron as containing the symbol "0", while a firing neuron is considered as containing the symbol "1". When a neuron can fire several kinds of output, for example any rational number, then we can store in it more than two symbols. To preserve the firing state, a neuron can be initiated to circulate in a loop, until it is stopped. Since the output of the neuron feeds back to itself, there is a self-sustaining loop that keeps the neuron firing even when the top input is no longer active. Activating the lower input suppresses the looped input, and the node stops firing. The stored binary bit is continuously accessible by looking at the output. This configuration is called a latch. Thus, a symbol, e.g., 1, is stored in form of a process in a neural network. Consequently, this process becomes a symbol itself. Existence of such memory is supported by the experimental evidence that some patterns in the brain are preserved in a dynamical fashion (Suppes and Han, 2000; Suppes, Han, Epelboim and Lu, 1999; 1999a).

The corresponding to the brain structural component includes knowledge of the individual. Structural component also includes beliefs, images, ideas, conjectures, problems, etc. However, we can take the system of knowledge, also called thesaurus, as an infological system of the brain, or of an individual. Infological elements in this case will be units of knowledge of the individual.

However, it is possible that an infological system IF(R) of a material system R has only two or even one component. For instance, we can take the system of knowledge of an individual as an infological system of this individual.

Another example of an infological system is the memory of a computer. Such a memory is a place in which data and programs are stored. Data and programs are represented by symbols in a material form: as states of electronic elements or written on paper, board or some other material objects. At the same time, data texts, symbols, knowledge, programs, algorithms and many other essences are structures (Burgin, 1997; 2004; 2005).

The computer memory is also a complex system of diverse components and processes. Memory of a computer includes such three components as the random access memory (RAM), read-only memory (ROM), and secondary storage. While RAM forgets everything whenever the computer is turned off and ROM cannot learn anything new, secondary storage devices allow the computer to record information for as long period of time as we want and change it whenever we want. Now the following devices are utilized for log-term computer memory: magnetic tapes and corresponding drives, magnetic disks and corresponding drives, and optical disks and corresponding drives.

Remark 2.2.2. In an arbitrary system R, it is possible to select different infological systems. Fixing one of these subsystems, we determine the type of information for R and changing our choice of IF(R) we change the scope of information entities (portions of information).

For instance, computers have different kinds of memory: processor registers, addressed memory, main storage, buffer storage, external storage, working storage, etc. Each of them or any combination of memory components may be considered as an infological system of a computer R. If the processor registers are treated as an infological system, then a program (even such that is kept in the main storage of this computer) does not give information for R until execution of instructions from this program begins.

Definition 2.2.4. Elements from IF(R) are called *infological elements*.

There is no exact definition of infological elements although there are various entities that are naturally considered as infological elements as they allow one to build theories of information that inherit conventional meanings of the word *information*. For instance, knowledge, data, images, ideas, fancies, abstractions, beliefs, and similar objects are standard examples of infological elements. If we consider only knowledge and data, then the infological system is the *system of knowledge* of a given system R. Such a system of knowledge is called a *thesaurus* in cybernetics.

The situation with infological elements looks similar to the situation in contemporary physics where physicists do not give a definition of matter but explain how matter is built and what elements of matter are. As such elements on the lowest level of the hierarchy, physicists take subatomic particles, physical fields, atoms, molecules and so on.

The infological system plays the role of a free parameter in the definition of information (cf. the Ontological Principle O2g). One of the ways to vary this parameter is to choose a definite kind of infological elements. Additional conditions on infological elements imply a more restricted concept of information.

To better understand how infological system can help to explicate the concept of information in the strict sense, we consider cognitive infological systems.

Definition 2.2.5. An infological system IF(R) of the system R is called *cognitive* if IF(R) contains (stores) such elements or constituents as knowledge, data, images, ideas, fancies, abstractions, beliefs, etc.

Cognitive infological system is a standard example of infological systems, while its elements, such as knowledge, data, images, ideas, fantasies, abstractions, and beliefs, are standard example of infological elements. Cognitive infological system is very important, especially, for intelligent systems. The majority of researchers believe that information is intrinsically connected to knowledge (cf. (Flückiger, 1995)).

A cognitive infological system of a system R is denoted by CIF(R) and is related to cognitive information.

Ontological Principle O2c (the *Cognitive Transformation Principle*). *Cognitive information* for a system R, is a capacity to cause changes in the cognitive infological system IFC(R) of the system R.

As the cognitive infological system contains knowledge of the system it belongs, cognitive information is the source of knowledge changes. This perfectly correlates with the approach of Dretske (1983) and Goldman (1967) who defined knowledge as information-caused belief, i.e., information produces beliefs that are called knowledge. Moreover, it is impossible to obtain knowledge without information. Dretske (1983) develops this idea, implying that information produces beliefs, which, according to our definition, are also elements of the cognitive infological system. Moreover, many researchers relate all information exclusively to knowledge. For instance, Mackay (1969) writes:

"Suppose we begin by asking ourselves what we mean by information. Roughly speaking, we say that we have gained information when we know something now that we didn't know before; when 'what we know' has changed."

Barwise and Seligman write (1997) that "information is closely tied to knowledge". Meadow and Yuan suggest (1997) that the recipient's knowledge increases or changes as a result of receipt and processing of new information. Cognitive information related to knowledge was also studied by Shreider (1967). His approach is presented in Chapter 4.

At the same time, other researchers connect cognitive information to experience. For instance, Boulding (1956) calls a collection of experiences by the name *image* and explains that messages consist of information as they are structured experiences, while the meaning of a message is the change that it produces in the image.

Cognitive information is what people, as a rule, understand and mean when they speak about information. Indeed, since approximately the 16th century, we find the word *information* in ordinary French, English, Spanish and Italian in the sense we use it today: to instruct, to furnish with knowledge (Capurro, 1991). However, scientific usage of the notion of information (cf, for example, (Loewenstein, 1999)) implies a necessity to have a more general definition. For instance, Heidegger pointed to the naturalization of the concept of information in biology in the form of genetic information (Heidegger and Fink, 1970). An example, of such a situation is when biologists discuss information in DNA in general or in the human genom, in particular.

As a result, we come to the world of structures. Change of structural features of a system, and transformation of other system characteristics through these changes, is the essence of information in the strict sense. This correlates with the von Weizsäker's remark that information being neither matter nor energy, according to (Wiener, 1961), has a similar status as the "platonic eidos" and the "Aristotelian form" (Weizsäcker, 1974).

These ideas are crystallized in the following principle.

Ontological Principle O2a (the *Special Transformation Principle*). *Information in the strict sense* or *proper information* or, simply, information for a system *R*, is a capacity to change structural infological elements from an infological system IF(*R*) of the system *R*.

To understand this principle and the definition of information it contains, we need to understand the concept of a structure. Otherwise, this definition will be incomplete, containing an undefined term. This brings us to the most fundamental question of ontology how our world is organized (built). Both problems are treated in Section 2.1.

Some researchers related information to structure of an object. For instance, information is characterized as a property of how entities are organized and arranged, but not the property of entities themselves (Reading, 2006). Other researchers have related information to *form*, while *form* is an explicit structure of an object. For instance, information is characterized as an attribute of the form (in-*form*-ation) that matter and energy have and not of the matter and energy themselves (Dretske, 2000).

However, two issues, absence of the exact concept of structure and lack of understanding that structures can objectively exist, result in contradictions and misconceptions related to information. For instance, one author writes that information is simply a construct used to explain causal interaction, and in the next sentence, the same author asserts that information is a fundamental source of change in the natural world. Constructs cannot be sources of change, they can only explain change.

The Ontological Principle O2a implies that information is not of the same kind as knowledge and data, which are structures (Burgin, 1997). Actually, if we take that *matter* is the name for all substances as opposed to *energy* and the *vacuum*, we have the relation that is represented by the following diagram called the Structure-Information-Matter-Energy (SIME) Square.

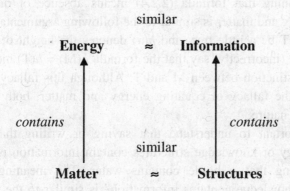

Figure 2.11. The Structure-Information-Matter-Energy (SIME) Square

In other words, we have the following principle:
Information is related to structures as energy is related to matter.

Here it is necessary to remark that many people think and write that there is no distinction between matter and energy. They base their belief on the famous formula from the relativistic physics

$$E = mc^2 \qquad (2.2.4)$$

However, this belief is only a misconception because the letter m in (2.2.4) does not stand for matter, as well as the letter E in (2.2.4) does not stand for energy. Here m means "mass" and E means the quantity of energy. "Mass" is only one of the characteristics or measures of material objects. What concerns the symbol E in (2.2.4), it is necessary to make a distinction between energy as a natural phenomenon and energy as some physical quantity. The latter is a measure of the former. It is natural to cal this quantity by the name "the quantity of energy". However, traditionally it is also called energy. There are other measures of energy. For instance, entropy is another measure of thermal energy.

Besides, what Einstein really proved was that if a body at rest emits a total energy of E remaining at rest, then the mass of this body decreases by E/c^2.

Thus, formula (2.2.4) gives a relation between two measures of two distinct natural phenomena. Namely, this formula estimates how much energy is stored in matter.

The reasoning that formula (2.2.4) means absence of distinction between energy and matter is similar to the following argumentation. Let M be a man, T be a little tree, and $h(x)$ denotes the height of x. Then somebody can (incorrectly) say that the formula $h(M) = h(T)$ means that there is no distinction between M and T. Although this fallacy is more evident than the fallacy of equating energy and matter, both fallacies have the same nature.

It is important to understand that saying or writing that matter contains energy or knowledge structures contain information is not the same as saying that a bottle contains water. The meaning of the expression "knowledge contains information" is similar to the meaning of expressions "the brain contains knowledge" or "a person has knowledge". In other words, it is possible to extract energy from matter, as well as it is possible to extract information from knowledge. In some cases, such extraction goes on automatically (on the unconscious level). It gives an illusion that information comes itself into a system without any mental work.

Mey (1986) considered matter, energy and information as a triad elements of which in their interaction constitute those features of (an) object that can be perceived by people. The Structure-Information-Matter-Energy (SIME) Square shows that it is necessary to add structure to this triad.

Some researchers, such as von Weizsäcker or Mattessich, wrote about similarities between energy and information. For instance, Mattessich (1993), assuming a holistic point of view and tracing information back to the physical level, concluded that every manifestation of active or potential energy is connected to some kind of information, just as the transmission of any kind of information requires active or potential energy.

It is possible to say that the concept of an infological system is too ambiguous and fuzzy. However, ambiguity may be a positive property if you can use it. For instance, if you can control and change ambiguity, it becomes not an ambiguity but a parameter that is utilized to tune and control the system. In a similar way, utilization of fuzzy logic instead of conventional logic has allowed researchers to build more precise and adequate models of real systems, while practioners have been able to achieve higher efficiency in production and management.

This is just the case with the infological system in the general theory of information. Thus, it is natural that considering a human being, we do not chose the same infological systems as we do for biological cells or computers. Besides, any complex system has, as a rule, several infological systems.

The main infological system of an individual is the mind with its material component — the brain. This infological system controls all human actions. As a result, behavior of people is determined by information that flows in the organism of this individual. It allows individual to adapt to environment both in nature and society.

A possibility to choose an infological system in a different way is very beneficial. It explicates existence of different types and kinds of information. Each type of information corresponds to some type of infological system. Examples of such types are considered in the Section 2.6.

In what follows, we consider only information in the strict sense.

It is possible to separate (cf., for example, (Flükiger, 1995)) three approaches to information: information as a thing, information as a structure, and information as a property. Thus, it is natural to ask which of these approaches is true. The general theory of information supports the third approach as the most grounded. Indeed, attempts to treat information in the same category as and knowledge have brought researchers to various contradictions and misconceptions resulting in the claim that information studies do not need the concept of information (Furner, 2004).

For instance, some researchers assert that information is enhanced data, i.e., a thing. At the same time, others imply that information is processed knowledge, i.e., a structure. However, both approaches remained inefficient providing no means for constructive differentiation between data, information, and knowledge.

Let us consider a situation when one system Q sends a signal to another system R. This signal carries a message, which, in turn, contains information. It brings us to the following principle.

Let I be a portion of information for a system R.

Ontological Principle O3 (**the *Embodiment Principle***). *For any portion of information I, there is always a carrier C of this portion of information for a system R.*

Really, people get information from books, magazines, TV and radio sets, computers, and from other people. To store information people use their brains, paper, tapes, and computer disks. All these entities are carriers of information.

For adherents of the materialistic approach, the Ontological Principle O3 must be changed to its stronger version.

Ontological Principle OM3 (**the *Material Embodiment Principle***). *For any portion of information I, there is some substance C that contains I.*

For example, according to Landauer (2002), information is always physical and hence ultimately quantum mechanical. However, what Landauer really claims is that nobody can have information without some physical carrier. People often confuse information with its carrier. From this confusion, such definitions as "information is a message" or "information is a collection of facts and data" or "information is knowledge derived from something" come.

However, if we identify information with its carrier, then we would inevitably come to the paradoxical conclusion suggested by Furner (2004) that information studies do not need the concept of information.

In reality, the situation is opposite. It is not the case that information laws are derived from physical laws, as Landauer (2002) suggested. In contrast to this, physical laws are derived from information laws as the title of the book "Physics from Fisher information" (Frieden, 1998) states.

In a general case, carriers of information belong to three classes: *material*, *mental*, and *structural*. For example, let us consider a book. It is a physical carrier of information. However, it contains information only because some meaningful text is printed in it. Without this text it would not be a book. The text is the structural carrier of information in the book. Besides, the text is understood if it represents some knowledge and/or other structures from the cognitive infological system. This knowledge and other corresponding structures form the mental carrier of information in the book.

Definition 2.2.6. The substance *C* that is *a carrier* of the portion of information *I* is called the *physical*, or *material*, *carrier* of *I*.

These types of information carriers (material, mental, and structural) are related to elements of the basic triplet (triad) of Jumarie (1986): the system *S* that is the *material medium* where information is physically defined; the *universe of discourse U* where information is semasiologically defined; and the *observer R* who considers *S* and *U* in his own subjective framework. The system *S* that is the material medium where information is physically defined corresponds to the physical carrier of information. The universe of discourse *U* where information is semasiologically defined corresponds to the structural carrier of information. The observer *R* who considers *S* and *U* in his own subjective framework corresponds to the mental carrier of information.

Distinctions between types of information carriers are of a great importance when something (such as a film or educational computer program) is produced for communication or/and entertainment. To achieve better information transmission, it is necessary to pay attention to how all three types of information carriers are organized and produced.

Some people think that only physical matter is what gives emergent properties. They claim that with the same physical matter and with exactly the same physical structure, e.g., all microparticles and their states and coordinates, it is possible to get the same informational content. As a consequence, they believe, it is impossible to change informational content without changing the material representation of information.

This looks so evident. However, our practice shows that this is not the case. Let us consider a textbook on physics written in Japanese. To an individual who does not know either Japanese or physics, this book will give very little (if any) information. To an individual who knows physics but does not know Japanese, this book will give more information because this person will understand formulas. To an individual who knows Japanese but does not know physics, this book will give much more information. This person will be able to learn physics using this textbook. However, to an individual who knows both Japanese and physics to a higher degree that this textbook represents, this book will also give very little (if any) information. Thus, the material representation of information in the book is not changing, while the information content is different for different people. That is, what information is in the same material representation depends on means that the receiver of this information has for information extraction.

This conclusion is supported by statements of experts in statistical information theory where information content of a message depends on knowledge of the recipient/receiver.

Existence of a definite information carrier allows one to speak about this carrier as a *representation of information*. According to the Ontological Principle O2, information is the same if it causes the same changes in a given infological system. Thus, the same information can be represented in different information carriers, e.g., by different texts, or even by information carriers of different nature, e.g., there cases when it is possible to convey the same information by an oral message, written or printed text, and picture.

However, there is a difference between information carrier and information representation. An information carrier only contains information, while an information representation contains and represents information. Thus, an information representation is always its carrier, while an information carrier is not always an information representation. An information carrier is a broader concept than a representation of information. For instance, a text written on a piece of paper is a representation of information and a carrier of this information as well. At the same time, the piece of paper with this text is only a carrier of the corresponding information. Note that one portion of information I can represent another portion of information J. Thus, I will be a carrier of J, but it will be a non-material carrier. A symbol is only partially a material carrier of information.

It is natural to consider three basic types of information representations:

- *System information representation* is a system that represents information in some activity.
- *Function information representation* is a transformation that represents information in some activity.
- *Process information representation* is a process that represents information in some activity.

Let us consider examples of information carriers and information representations.

Example 2.2.1. A file that contains some text is both a representation and carrier of information, but the computer where this file is stored is a carrier but hardly a representation of information in the file.

Example 2.2.2. A human being is a carrier but not, as a rule, representation of information she or he has.

Example 2.2.3. A book is both a carrier and representation of the information *I* it contains. However, if you put the book in a bag, the bag will become a carrier but not a representation of *I*.

Example 2.2.4. A document is a system information representation.

Example 2.2.5. A book is a system information representation.

Example 2.2.6. A web site is a system information representation.

Example 2.2.7. An e-mail or letter is a system information representation.

Example 2.2.8. Writing a response to letter or e-mail, taken as an action, is a function information representation, while taken as a process, it is a process information representation.

Building a good information representation demands specific skills. Representation of information in accessible and useful form is different from knowing the content of the information.

Ontological Principle O4 (the *Representability Principle*). *For any portion of information I, there is always a representation C of this portion of information for a system R.*

As any information representation is, in some sense, its carrier the Ontological Principle O4 implies the Ontological Principle O3.

The first three ontological principles ((O1)-(O3) or (O1)-(OM3)) imply that, in some sense, information connects the carrier C with the system R and thus, information is a component of the following fundamental triad (cf. Appendix A)

$$(C, I, R) \tag{2.2.5}$$

As a rule, there is some channel through which information comes from C to R. For instance, the carrier C of an information portion I is a piece of paper and R is a person reading the text written on C. Then the corresponding channel is the space between the paper and the eyes of the person.

People empirically observed that for information to become available, the carrier must interact with a receptor that is capable of detecting information the carrier contains. The empirical fact is represented by the following principle.

Ontological Principle O5 (**the** *Interaction Principle*). *A transaction/transition/transmission of information goes on only in some interaction of C with R.*

This principle introduces the interaction triad (2.2.6) in the theory of information.

$$C \xrightarrow{\text{Int}} R \tag{2.2.6}$$

Interaction between C and R may be direct or indirect, i.e. it is realized by means of some other objects.

The property of information explicated in the Ontological Principle O5 may look evident. However, it has important consequences. For instance, if you know that some information has passed from system to another and want to find how it happened, you have to look for a channel of transaction. Although, if it is known only that the second system possesses the same information as the first one, it not necessary that it has been a transmission. It might be possible that the same information has been created by the second system. Existence of a channel makes transmission possible but does not necessitate it.

The next principle is a specification of the Ontological Principle O5.

Ontological Principle O5a (**the *Structured Interaction Principle***).
*A system R receives information **I** only if some carrier C of information **I** transmits **I** to the system R or R extracts this information from C.*

Information transmission/extraction can be direct or go through some channel *ch*. When somebody touches a hot rod and feels that the rod is hot, it is a direct information transmission from the rod to this person. At the same time, when somebody comes close to a hot rod and feels that the rod is hot, it is a indirect information transmission from the rod to this person because such a channel as air is used.

Here, we have two ways of information transaction: transmission and extraction. Transmission of information is the passive transaction with respect to *R* when *R* receives information and active transaction with respect to *C* when *C* transmits information. Extraction of information is the active transaction with respect to *R* when *R* extracts information and passive transaction with respect to *C* when information is taken from *C*. When the carrier *C* is the system *R* itself, then we have the third type of information operations — information processing. It includes information transformation and production (Burgin, 1997b).

These two ways of information exchange reflect important regularities of education and entertainment media. At first, let us consider education where these features emerged much earlier than in entertainment.

There is an essential difference between Western and Eastern approaches to education. The main principle of the Western tradition is that a teacher comes to students to teach them. Contrary to this, the main principle of the Eastern tradition is that a student comes to Teacher to learn from him. This means that the Western approach is based on information transmission, while the Eastern approach stems from information extraction.

This is an essential difference. When students come to school without preparing to work hard to get knowledge and only wait when the teacher will put everything in their head, the results usually are not good. Such students either do not receive knowledge or receive mush less than they can and what the teacher gives them. In fact, any teacher gives his students only information, while students themselves have to accept this information and to transform it to knowledge. This transformation demands substantial work. Gifted students do this work in their brains, often subconsciously. This creates an impression that they do nothing to achieve excellent results in learning. Not so gifted students need to work hard to achieve sufficient results in learning. It is a responsibility of a teacher to teach her students how to efficiently acquire information. Many problems with education, even in developed countries, are connected to the misconception that a teacher has to give knowledge to her students. This orients the teacher to give more subject material to students without taking sufficient care of helping the students to accept information in this material and to build knowledge using accepted information.

A similar situation, for example, exists in entertainment. To make it clear, let us consider theater, movies, television, and computer games. Theater evidently represents the Eastern position. Spectators come to the theater where plays are created. Film industry creates movies not in the presence of the audience. However, traditionally people have to come to movie theaters to see movies. TV, radio, and DVDs, as it is peculiar for the Western tradition, come to each home, making entertainment consumption easier. Computer games and other computer entertainment, in the majority of cases, are also coming to the audience. Nevertheless in the latter case, this audience consists not of spectators but of participants. Modern technology provides for the film industry opportunities to come directly to the audience by means of the video. Thus, technology, mostly created in the West, supports and promotes active approach of producers at the cost of transformation of the audience (of spectators, students, etc.) into passive consumers. This is an essential trait of our society.

At the same time, the same Western technology has developed means for active participation in entertainment. Computer games represent only the first step in this direction.

The Ontological Principle O4a introduces the second communication triad (2.2.7) in the theory of information.

$$\text{channel}$$
$$C \quad \rightarrow \quad R \qquad\qquad (2.2.7)$$

Two more principles explicate dynamic properties of information processes.

Ontological Principle O6 (**the** *Actuality Principle*). *A system R accepts a portion of information I only if the transaction/transition/ transmission causes corresponding transformations.*

For instance, if after reading this paper, your knowledge remains the same, you do not accept cognitive information from this text. In a general case, when the recipient's knowledge and belief structure was not changed, there is no cognitive information reception. That is why, the concern of the people from the entertainment industry how their production influences the intended audience is of the greatest importance to the industry. General theory of information can explain many features of this impact. However, this theory does not solve all problems, and to have a complete picture, it is necessary to include sociologists, psychologists, economists, linguists, and semiologists in the study of entertainment.

If the knowledge system was changed, it was caused by the acceptance of information. Thus, the same message can give information to one receiver and give nothing to another one. Thus, Principle O6 implies that a book or report sitting on a library shelf or a record in a database contains only potential information to a potential recipient. This information becomes actual only when it is transmitted to and comprehended by the recipient.

There are clearly many different forms and levels of comprehension. At the first glance, the receipt of an encrypted message would seem to have no significance to a recipient without the key but, in fact, useful knowledge can be derived simply by knowing the origin and intended destination of the message, without understanding its content. Psychologists know that many people often judge other people by their appearance, clothes and language proficiency. Empirical studies demonstrated that students better evaluated professors who were better dressed and looked better. Even in science and mathematics, which look so objective and just, the situation is not much better. Although it is not generally admitted, scientific papers are often judged by their citations, the number of works cited, their dates, their authors, the scientific community the authors belong, the authors not included, and by the writing style. In other words, the judgment is not always based on content (Meadow and Yuan, 1997). Shallow ideas and poor results disguised by elaborate language constructions have much more chances to be published and become popular than profound ideas and groundbreaking results conveyed in simple language. A reviewer can write a negative review only because the author did not give a reference to reviewer's publications or criticized ideas or results of the reviewer.

Ontological Principle O7 (**the *Multiplicity Principle***). *One and the same carrier C can contain different portions of information for one and the same system R.*

Really, let us consider some person *A* as the system *R* and a book written in Japanese as the carrier *C*. At first, *A* does not know Japanese and *C* contains almost no information for *A*. After some time, *A* learns Japanese, reads the book *C* and finds in it a lot of valuable information for himself. Note that knowing Japanese, *A* is, in some sense, another person.

Another example is given by advertising. It changes comprehension of different things. Let us consider a situation when some person *A* comes to a movie theater to see a new film. It is, as a rule, a great difference in comprehension depending whether this person never has heard about this film or *A* has read quite a deal about good actors, talented producer, interesting plot of the film and so on.

In other words, if you want to convey some information to an audience efficiently, you have to prepare this audience to acceptation of the transferred information. This is essentially important for contemporary entertainment industry based on mass communication.

As Hjørland writes (1998), the same document (text, message or sign) has different meaning for and conveys different information to the same receiver in different situations. Information is always situational.

There are many examples when unprepared community did not accept even the highest achievements of human intellect and creativity. Thus, it is known that when many outstanding works of art were created and many great discoveries in science were made, society did not understand what was done and rejected in some cases the highest achievements of geniuses. Only consequent generations understood the greatness of what had been done before. As examples, we can take the great Austrian composer Wolfgang Amadeus Mozart (1756–1791), who died in poverty, the great mathematicians Evariste Galois (1811–1832) and Niels Hendrik Abel (1802–1829), who wrote outstanding works but were neglected and died at young age because of this.

One more example of misunderstanding gives the life of the great English physicist Paul Adrien Maurice Dirac (1902–1984). He was well known and respected by physical community when he theoretically discovered a positive "electron", which was later called positron. However, other physicists did not understand Dirac's achievement and even mocked at him.

The great German mathematician Gauss made one of the most outstanding discoveries of the 19th century, the discovery of the non-Euclidean geometry. Nevertheless, he did not want to publish his discovery because correctly considered the contemporary mathematical community unprepared to the comprehension of this discovery.

The last three principles reflect only the situations when transformation of an infological system takes place. However, it is important to know and predict properties of these transformations, for example, to evaluate the extent or measure of transformations. These aspects of the general theory of information are treated in the next section.

2.3. How to Measure Information: Information Axiology

> *Don't worry about people stealing your ideas.*
> *If your ideas are any good, you'll have to ram them down people's throats.*
> Howard Aiken, IBM engineer

The general theory of information lays emphasis on making a distinction between information and measures of information. Often people do not do this and call, for example, Shannon's entropy (cf. Chapter 3) or the difference between two values of algorithmic complexity (cf. Chapter 4) by the name *information*. Similar situation exists with energy in physics. On the one hand, energy is defined as a scalar physical quantity that is a property of theoretical representations of physical objects and systems, i.e., a potency to perform work. On the other hand, energy is often defined as the ability to do work, i.e., as a natural property, which in some cases can be measured. At the same time, a physical quantity is, as rule, a theoretical representation, often a measure, of some natural property or phenomenon.

Thus, it is very important to distinguish a phenomenon and its measured features, especially, in the case of information because there are so many different measures of the same portion of information (cf., for example, (Sharma and Taneja, 1974; Kahre, 2002)). For instance, we have such measures as the Shannon entropy, Rényi entropy, and Fisher variety of information in statistical information theory, as well as absolute and conditional information size in algorithmic information theory. All these measures can be used to measure the same portion of information, e.g., information in some text.

This peculiarity of information is similar to the situation with physical objects and their measures. For instance, taking a table, we can measure its length, width, height, mass, weight and so on. As it has been demonstrated in the previous section and in Chapter 1, information is a natural property of systems and processes, and we want and need to know how to measure this property.

As Hjørland writes (1998), many researchers in information science share the hope or the ideal that it should be possible to retrieve not only documents but directly to get information from these documents and even to measure information retrieved. Measuring information would help to improve the design of information systems and services. The goal of information systems and services is to inform actual and potential users, improving their possibility to find relevant answers to questions they have.

Studying information and related processes, researchers have invented a diversity of information measures, such as the quantity, value, cost, entropy, uncertainty, average information score, effectiveness, completeness, relevance, reliability, and authenticity. Wand and Wang (1996) propose to assess information in four dimensions: correctness, unambiguity, meaningfulness, and completeness. Researches studied information quality (IQ) as a compound measure of information. Information measures are related to different dimensions of information. Accuracy and completeness assess the correspondence between information representation and the real world and thus, are related to semantic dimension of information (cf. Section 4.1). Usefulness and usability assess how information is used and thus, are related to pragmatic dimension of information (cf. Section 4.1). Below examples of mathematical expressions for information measures introduced by different authors are given:

$$I = \int p'^{\,2}(x) \,/\, p(x) \,\mathrm{d}x \qquad \text{(Fisher, 1925)}$$

$$S(D) = -K\,\mathrm{Tr}\,\eta(D) \qquad \text{(von Neumann, 1927; 1932)}$$

$$I = \log_2 n \qquad \text{(Hartley, 1928)}$$

$$J(P \,\|\, Q) = \sum_{i=1}^{n} (p_i - q_i)\log(p_i \,/\, q_i) \qquad \text{(Jeffreys, 1946)}$$

$$H(X) = -\sum_{i=1}^{n} p_i \log_2 p_i \qquad \text{(Shannon, 1948)}$$

$$H(X \,\|\, Y) = -\sum_{y \in Y} p(y) \sum_{x \in X} p(x \,|\, y)\log_2 p(x \,|\, y) \qquad \text{(Shannon, 1948)}$$

$$I(X; Y) = H(X) - H(X \,|\, Y) \qquad \text{(Shannon, 1948)}$$

$$\mathrm{cont}(p) = m(\neg\, p) \qquad \text{(Bar-Hillel and Carnap, 1952)}$$

$$\inf(p) = \log\left(1 \,/\, (1 - \mathrm{cont}(p))\right) \qquad \text{(Bar-Hillel and Carnap, 1952)}$$

$$D(P \,\|\, Q) = \sum_{i=1}^{n} p_i \log_2(p_i \,/\, q_i) \qquad \text{(Kullback and Leibler, 1951)}$$

$$I(a) = \mathrm{E}\{\log_2(\pi(\Theta \,|\, X) \,/\, (\pi(\Theta)) \qquad \text{(Lindley, 1956)}$$

$$Q(I) = p_0(R, g) - p_1(R, g) \qquad \text{(Harkevitch, 1960)}$$

$$H_\alpha(p_1, p_2, \ldots, p_n) = (1/(1 - \alpha))\ln\left(\sum_{i=1}^{n} p_i^{\,\alpha}\right) \qquad \text{(Rényi, 1960)}$$

$$D_\alpha(P \,\|\, Q) = (1/(1 - \alpha)) \sum_{t=1}^{n} \ln(p_i^{\,\alpha} q_i^{\,1-\alpha}) \qquad \text{(Rényi, 1961)}$$

$$H(P \,\|\, Q) = \sum_{t=1}^{n} p_i \log q_i \qquad \text{(Kerridge, 1961; Bongard, 1963)}$$

$$C_A(x) = \min \{l(p);\ A(p) = x\}$$
(Solomonoff 1964; Kolmogorov 1965; Chaitin 1966)

$$C(x) = \min \{l(p);\ U(p) = x\}$$
(Solomonoff 1964; Kolmogorov 1965; Chaitin 1966)

$$I_T(y{:}\ x) = C_T(x) - C_T(x\,|\,y) \quad \text{(Kolmogorov 1965; Chaitin 1966)}$$

$$I_n = \sum_{i=1}^{n} p_{1i} \log_2 q_{1i} - \sum_{i=1}^{n} p_{0i} \log_2 q_{0i} \quad \text{(Bongard, 1967)}$$

$$I(P \,\|\, Q) = \sum_{i=1}^{n} [((p_i \log p_i + q_i \log q_i)/2) - ((p_i + q_i)/2)\log((p_i + q_i)/2)]$$
(Sibson, 1969)

$$I_n(P^n, Q^n) = A \sum_{i=1}^{n} p_i \log p_i + B \sum_{i=1}^{n} p_i \log q_i$$
(Sharma and Taneja, 1974)

$$K(x) = \min\{l(p);\ UPFTM(p) = x\}$$
(Gács,1974; Levin,1974; Chaitin,1975)

$$v(I_x) = U_1 - U_0 = \mathrm{E}\{\max_a u(a, x)\} - \max_a \mathrm{E}\{u(a, x)\}$$
(Marschak, 1980)

$$C_P(x) = \min \{l(p);\ U_p(p) = x\} \qquad \text{(Burgin, 1982)}$$

$$U(r) = \int_0^1 (r_i - r_{i+1})\log_2 i \qquad \text{(Higashi and Klir, 1983)}$$

$$U(r) = \sum_{i=1}^{n} (r_i - r_{i+1})\log_2 i \qquad \text{(Higashi and Klir, 1983)}$$

$$C_{IT}(x) = \min \{l(p);\ UITM(p) = x\} \qquad \text{(Burgin, 1990)}$$

$$R(Z;Y) = I(Z;Y)\,/\,H(Z) \qquad \text{(Kononenko and Bratko, 1991)}$$

$$I_1(X; y_j) = \sum_{i=1}^{n} p(x_i\,/\,y_{ij}) \log_2 [p(x_i\,/\,y_{ij})\,/\,p(x_i)]$$
(DeWeese and Meister, 1999)

$$I_2(X; y_j) = -\sum_{i=1}^{n} p(x_i)\log_2 p(x_i) + \sum_{i=1}^{n} p(x_i / y_{ij})\log_2 p(x_i / y_{ij})$$

(DeWeese and Meister, 1999)

$$I_{tot}(\overrightarrow{p}) = \text{Tr} \, (\rho - \mathbf{1}/n) \qquad \text{(Brukner and Zeilinger, 1999)}$$

$$\left| I(d, K_I, K_F) \right| = \left| (K_A, K_R) \right| = \left| K_A \right| + \left| K_R \right| \qquad \text{(Mizzaro, 2001)}$$

$$E_c(Z; Y, W) = I(Z;Y) / I(Z;W) \qquad \text{(Hill, 2004)}$$

$$I_{IT}(y: x) = C_{IT}(x) - C_{IT}(x \mid y) \qquad \text{(Burgin, 2005)}$$

$$I_P(y: x) = C_P(x) - C_P(x \mid y) \qquad \text{(Burgin, 2005)}$$

In his paper "Some varieties of information" (1984), Hintikka introduces eight measures of information: Inf_{add} (A/B), Cont_{add} (A/B), Inf_{cond} (A/B), Cont_{cond} (A/B), Inf_{trans} (A/B), Cont_{trans} (A/B), Inf(A), Cont(A). He calls these measures by the name *information* and concludes that it is not at all clear which of the multiple senses of information we are presupposing on different occasions on which we are thinking and talking about information.

Sharma and Taneja (1974) introduced an infinite family of information measures in the form

$$I_n (P^n, Q^n) = A \sum_{i=1}^{n} p_i \log p_i + B\sum_{i=1}^{n} p_i \log q_i$$

Here A and B are arbitrary fixed real numbers. Thus, there is a continuum of such measures.

Kahre (2002) also shows a necessity in an infinite set of information measures to evaluate various properties of information.

All this diversity of information measures confirms that it is necessary not only to invent new measures of information but also to elaborate methodology and theoretical foundations of information measurement.

To know how to measure information in a scientific way, we need not only some empirically invented measures, even if these measures are as successful and popular as the quantity of information and information entropy suggested by Shannon, but also a theory for information measurement. To build such a theory, called *information axiology*, we use general axiology.

In the traditional context, *axiology* is a branch of philosophy concerned with developing a theory of values (cf., for example, (Kessler, 1998)). Ethics, which treats moral values, and aesthetics, that treats artistic values, are two main subdivisions of axiology.

At the same time, in modern methodology of science, axiology has a more general meaning. Namely, *scientific axiology* is an area concerned with developing structures and methods for evaluation, estimation and measurement. Correspondingly, axiology in methodology of science is a field concerned with developing methodology and philosophy for scientific evaluation, estimation and measurement (cf., for example, (Burgin and Kuznetsov, 1991; 1994)).

Information axiology in the general theory of information has similar goals in developing theoretical tools and principles for information evaluation, estimation and measurement. Axiological principles explain different approaches to the problem of information measurement, find what measures of information are necessary, investigate what are their general properties and instruct how to build such measures.

Besides, physicists understood necessity of a physical measurement theory much earlier as physics has a longer history than information science. As a result, there is a developed physical measurement theories, which continues to grow. Theory of measurement has been also developed in psychology, sociology, economics and some other disciplines. Thus, it is also reasonable to use achievements of theory developed in various disciplines for constructing an efficient and flexible information measurement theory.

To elaborate principles of information measurement, we use information properties described in the ontological principles. According to Ontological Principle O2 and its versions considered in the previous section, information reception causes changes either in the whole system R that receives information or in some infological system $IF(R)$ of this system. Consequently, it is natural to assume that it is possible to measure information by estimating changes caused by the reception of information. This is reflected in the first axiological principle.

Axiological Principle A1. *A measure of a portion I of information for a system R is some measure of changes caused by I in R (or more exactly, in the chosen infological system* IF(*R*))*.*

The Axiological Principle A1 well correlates with a suggestion of von Weizsäcker, who writes (2006), "we rate information by the effect it has", as well as with other approaches where information is related to change (cf., for example, (MacKay, 1956; 1961; 1969; Boulding, 1956; Shreider, 1968; Otten, 1975; Pratt, 1977; Mizzaro, 2001)).

Other axiological principles describe types of information measures that might be useful for different purposes, instruct what kinds of information measures we need, and explain why we need such measures.

Time is a very important resource. Thus, the first criterion for measure classification, as well as an important parameter for measuring procedures, is the time of changes.

Axiological Principle A2. *With respect to time orientation, there are three temporal types of measures of information:* 1) *potential or perspective*; 2) *existential or synchronous*; 3) *actual or retrospective.*

Actual or retrospective information measures are the most constructive as they are related to the past. Here we give some examples of such measures.

Yasin (1970), Parker, et al (2006) and some other researchers consider such measures of information as utility, value, range, cost, accessibility, significance, usability, timeliness, and reliability. Here are definitions of some of these measures. Let us consider some of them.

Significance of an information portion *I* is defined with respect to a chosen problem *P* and reflects to what extent information *I* is necessary for solving *P*.

Utility of an information portion *I* is defined with respect to a chosen problem *P* and reflects how *I* is used for solving *P*.

Usability of an information portion *I* is the frequency its utilization for solving problems from some class **P**. Usability depends both on the class **P** of problems and on how often these problems are solved.

The *range* of an information portion *I* is defined with respect to a chosen problem *P* and reflects how *I* is used for solving *P*.

Usually the same definitions are used to define significance, utility, usability, and range of data and/or knowledge. This gives an impression that it is necessary to measure more tangible essences, such as data and knowledge, and not information. However, both data and knowledge are important, useful and reliable only if information contained in these data and knowledge is important, useful and reliable.

For instance, in information retrieval systems, such measures are used: completeness and precision of a search (cf., for example, (Lancaster, 1968)). *Completeness ratio* of a search is determined by the formula

$$Cm(s) = (R_s/R_A) \cdot 100\% \qquad (2.3.1)$$

Here R_s is the number of relevant results of the search s, e.g., of relevant found documents, and R_A is the number of relevant objects in the domain of the search, e.g., if the search is done in a document database X, then R_A is the number of relevant documents in X.

Precision ratio of a search is determined by the formula

$$Pr(s) = (IR_s/A_s) \cdot 100\% \qquad (2.3.2)$$

Here IR_s is the number of irrelevant results of the search s and A_s is the number of all results of the search s.

Completeness and precision ratios characterize operating efficiency of the information retrieval system.

Both formulas (2.3.1) and (2.3.2) look simple. However, the problem is how to evaluate relevance of the search results. The most exact way to do this is to compare information of each search result with information needs of the user.

Properties of all these examples are synthesized in the following definition.

Definition 2.3.1. A *retrospective measure* of an information portion I determines (reflects) the extent of changes caused by I in the system R from the time when information I came to R (was accepted by R) till the time of the measurement.

Although this concept looks very simple for understanding, evaluation and manipulation, there is a problem how to separate changes caused by I from other changes in the system R. For instance, there were many studies of how movies and TV influence those who watch them. For instance, there is evidence that violence demonstrated on screen of TV and in various movies contributes to violence in society. However, it is difficult to separate influence of movies and TV programs from other impacts: from the family, friends, neighbors, etc.

As an example of such a measure, we may take the measure that is used to estimate the computer memory content as well as the extent of a free memory. Namely, information in computers is represented as strings of binary symbols and the measure of such a string is the number n of these symbols. The unit is called a bit.

When such a string is written into the computer memory, it means that some information is stored in the memory. Changes in the memory content might be measured in a different way. The simplest is to measure the work that has been performed when the string has been written. The simplest way to do this is to count how many elementary actions of writing unit symbols have been performed. However, this number is just the number of bits in this string. So, conventional measure of the size of memory and its information content correlates with the axiological principles of general theory of information.

At the same time, according to measures used in the algorithmic information theory, the algorithmic information size of such string may be much less that n (cf., Chapter 5).

Synchronous measures of information estimate changes in the present and thus, they depend on how we define present time. People have an illusion that everybody knows what present time is. However, if we ask whether present time is this second, this minute, this hour, this day, this month, this year or this century, the would depend on the situation and perspective. For instance, when a professor comes to a lecture and begins, saying, "Now we have a lecture on complexity", the word *now* means, say, one or two hours. However, when a journalist writes, "Now we live in information society", the word *now* means some number of years.

Because we do not want to go here into philosophical discussions about the concept of present, we simply fix some positive real number t_p. Then taking a temporal point x, we assume that everything (events, processes, episodes, circumstances, etc.) that goes within the temporal interval $(x - t_p, x + t_p)$ is present with respect to the temporal point x. This gives us a symmetric model of present with t_p as its parameter.

Definition 2.3.2. A synchronous measure of an information portion I determines (reflects) the extent of changes in R that go on in the present with respect to the time (temporal point) of receiving I.

There are different frames of reference for synchronous information. For instance, it is possible to define synchronous information measures with respect to time of accepting or comprehending the information portion I.

The trickiest thing is to predict future. Nevertheless, considering possibilities instead of future prediction, we come to potential or perspective measures of information.

Definition 2.3.3. A *perspective measure* of an information portion I determines (reflects) the extent of changes that I can cause in R.

An example of a perspective measure of information is Kolmogorov complexity, or recursive information size, $C_R(x)$ (cf. Chapter 5).

Let us consider the following example. Some student A (considered as a system R) studies a textbook C. After two semesters she acquires no new knowledge from C and finishes to use it. At this time an existential measure of information contained in C for A is equal to zero. The actual measure of information in C for A is very big if A is a good student and C is a good textbook. But the potential measure of information in C for A may be also bigger than zero if in future A returns to C and finds material that she did not understand before.

Von Weizsäcker (2006), who does not make a strict distinction between information and a measure of information, separates *potential*, or *virtual*, information, equating it with positive entropy, and *actual* information.

Hjørland (1998) discusses how such a measure as the value of information changes with time. Any given tool (including a sign, concept, theory, finding or document), he writes, has given potentialities in relation to new social practices and paradigms. Fr instance, users of libraries may discover that hitherto neglected documents may be very useful to develop a new social practice or to defend an existing practice. In a similar way, they may find that well-known documents may be re-interpreted and used in new ways. Especially the kinds of documents known as "classics" have such abilities to be reused and re-interpreted by each generation. But even the most humble documents may turn out to be of value. "The dust of one age is the gold of another" (Hesselager, 1980). This does not, of course, imply relativism regarding the relevance of documents: mostly tools quickly become obsolete and "only of historical interest". There may of course also be a discrepancy between the potential value of some documents and the cost of storing, organizing and using them.

All these considerations show that some information measures have three modalities with respect to time. For instance, there are three modalities of information value: *potential value of information*, *synchronic value of information*, and *retrospective value of information*.

The second criterion for measure classification is derived from the system separation triad, which posit a system into its environment and specifies links between the system and environment:

$$(R, l, W)$$

Here, R is a system, W is the environment of this system and l are the links between R and W. This triad shows what is included in the system for which information is considered, what belongs to the environment of this system, and how the system is connected to the environment.

Axiological Principle A3. *With respect to space orientation, there are three structural types of measures of information: external, intermediate, and internal.*

Note that, as a rule, space in Principle A3 means a system space, which often does not coincide with the physical space (cf. Section 7.3.1).

Definition 2.3.4. An *internal* information measure of an information portion I reflects the extent of inner changes caused by I.

Examples are given by the change of the length (the extent) of a thesaurus or knowledge system. In other words, an internal information measure evaluates information by its traces in the knowledge system. This is close to the approach of Mizzaro (2001). Only Mizzaro identifies information with the traces, i.e., changes in the knowledge system, while in the general theory of information, information is the cause of these traces/changes. The situation is similar to physics where energy is estimated by the work it produces but is not identified with this work.

Definition 2.3.5. An *intermediate* information measure of an information portion I reflects the extent of changes caused by I in the links between R and W.

Examples are given by the change of the probability $p(R, g)$ of achievement of a particular goal g by the system R. This information measure $Q(I) = p_0(R, g) - p_1(R, g)$ suggested by Harkevitch (1960) is called the quality of information.

Definition 2.3.6. An *external* information measure of an information portion I reflects the extent of outer changes caused by I, i.e. the extent of changes in W.

Examples are given by the change of the dynamics (functioning, behavior) of the system R or by the complexity of changing R.

Axiological Principle A4. *With respect to the way in which the measure is determined and evaluated, there are three constructive types of measures of information: abstract, grounded, and experimental.*

There are no strict boundaries between these types of information measures.

Definition 2.3.7. An *abstract information measure* is determined theoretically under general assumptions.

The change of the length (the extent) of a thesaurus, measures cont(p) and inf(p) defined by Bar-Hillel and Carnap (1952), and measures built by Sharma and Taneja, 1974) are examples of abstract information measures.

Definition 2.3.8. A *grounded information measure* is determined theoretically subject to realistic (observable, experienced, etc.) conditions.

Shannon entropy (cf. Chapter 3) is also a grounded information measure if we understand probabilities used in the formula for entropy as relative frequencies obtained in a large number of experiments and observations.

Kolmogorov complexity (information size, cf. Chapter 5) is also a grounded information measure because it is non-computable.

Definition 2.3.9. Values of an *experimental information measure* are obtained by experiments and constructive (physical) measurement operations.

The number of books in a library is an experimental information measure as it is possible to count these books. Another example of an experimental information measure is the number of bits (or bytes or kilobytes) in a computer file. Contemporary operating systems perform such measurements almost instantly.

Note that the difference between grounded and experimental information measures depends on means used for measurement. For instance, it is possible to consider computations of an inductive Turing machine as a sequence of experiments (Burgin, 2005). At the same time, it is proved (Burgin, 1982) that inductive Turing machines compute Kolmogorov complexity (information size). Thus, Kolmogorov complexity (information size) becomes an experimental information measure. In other words, when measurement is performed by recursive algorithms, such as Turing machines, Kolmogorov complexity (information size) is a grounded measure, while utilization of inductive Turing machines makes Kolmogorov complexity (information size) an experimental information measure.

The axiological principles A2–A4 have the following consequences.

A unique measure of information exists only for oversimplified system. Any complex system R with a developed infological system $IF(R)$ has many parameters that may be changed. As a result, complex systems demand many different measures of information in order to reflect the full variety of system properties, as well as of conditions in which these systems function. Thus, the problem of finding one universal measure for information is unrealistic. For instance, uncertainty elimination measured by the Shannon's quantity of information is only one of possible changes.

Another important property of information is a possibility to obtain a better solution of a problem (which is more complete, more adequate, demands less resources, for example, time, for achievement a goal) given adequate information. Changes of this possibility reflect the utility of information. Different kinds of such measures of information are introduced in the theory of information utility (Harkevitch, 1960) and in the algorithmic approach in the theory of information (Kolmogorov, 1965; Chaitin, 1977; Burgin, 1982).

Axiological Principle A5. *With respect to determination orientation, there are three constructive types of measures of information: absolute, fixed relative, and variable relative.*

This distinction is important for measurement interpretations.

Definition 2.3.10. An *absolute information measure* gives a (quantitative) value of an information portion *per se*.

For instance, Hartley (1928) introduced the quantity of information in the message *m* that informs about the outcome of an experiment *H* or about some event *E* in the following form

$$I(n) = \log_2 n$$

when the number of possible outcomes of this experiment is equal to *n* or on the number of possible alternatives to this event is equal to *n*.

This is an absolute measure of information.

Definition 2.3.11. A *fixed relative information measure* shows a (quantitative) relation between an information portion and some chosen (quantitative) value.

For instance, the measures $\inf(p)$ introduced by Bar-Hillel and Carnap (1952) is equal to $\log (1/(1 - \text{cont}(p))) = \log 1 - \log (1 - \text{cont}(p))$ shows a relation between the value of $\text{cont}(p)$ and number 1. This relation is presented in the arithmetical form. Taking some real number *k*, it is possible to consider a similar measure $\inf_k(p) = \log (k/(1 - \text{cont}(p)))$. This measure shows a relation between the value of $\text{cont}(p)$ and number *k*. In it, the number *k* gives some initial level of information. For instance, when the size of a computer file is measure in bytes, such an initial (zero) level is equal to seven bits.

Definition 2.3.12. A *variable relative information measure* shows a (quantitative) relation between a given information portion and another varying (quantitative) value.

Information transmission efficiency introduced in the Ontological Principle O6 is a variable relative information measure.

The average information score used to construct the *relative information score*,

$$R(Z;Y) = I(Z;Y) / H(Z)$$

is normalised as a percentage of total information required (Kononenko and Bratko 1991). Thus, it is a variable relative information measure.

Another example of a variable relative information measure is the overall *effectiveness* of the classification process (Hill, 2004):

$$E_C(Z; Y, W) = I(Z;Y) / I(Z;W)$$

The Ontological Principle O5 states that transmission of information goes on only in some interaction of the carrier C with the system R. This directly implies the following result.

Proposition 2.3.1. The measure of a portion of information I transmitted from C to a system R depends on interaction between C and R.

Stone (1997) gives an interesting example of this property. Distortions of human voice, on one hand, are tolerable in an extremely wide spectrum, but on the other hand, even small amounts of distortion create changes in interactive styles.

We can measure information by measuring processes related to this information. This allows us to build such an important measure as information transmission efficiency.

Namely, *information transmission efficiency reflects a relation (like ratio, difference etc.) between measures of information that is accepted by the system R in the process of transmission and information that is presented by C*, i.e., *tentative information in C, in the same process.*

It is known that the receiver accepts not all information that is transmitted by a sender. Besides, there are different distortions of transmitted information. For example, there is a myth that the intended understanding may be transmitted whole from a sender to a receiver. In almost every process of information transmission, the characteristic attitudes of the receiver "interfere" in the process of comprehension. People make things meaningful for themselves by fitting them into their preconceptions. Ideas come to us raw, and we "dress" and "cook" them, i.e., we process and refine raw ideas. The standard term for this process is *selective perception*. We see what we wish to see, and we twist messages around to suit ourselves. All this is demonstrated explicitly in the well-known "Mr. Biggott" studies (Tudor, 1974). An audience was shown a series of anti-prejudice cartoons featuring the highly prejudiced Mr. Biggott. Then people from the audience were subjected to detailed interviews. The main result was that about two thirds of the sample clearly misunderstood the anti-prejudice intention of the cartoons. The major factors accounting for this selective perception, according to the researchers, were the predispositions of the audience. Those who were already prejudiced saw the cartoons as supporting their position. Even those from them who understood the intentions of the cartoons found ways of evading the anti-prejudice "effect". Only those with a predisposition toward the message interpreted the films in line with the intended meanings of the communicators. This is best understood in psychology and cognitive science experiments. Our predisposition to search for the support of what we already believe and to neglect negative evidence has evolutionary roots.

2.4.　Types and Properties of Information: Information Typology

By three methods we may learn wisdom:
First, by reflection, which is noblest;
Second, by imitation, which is easiest;
and third by experience, which is the bitterest.
Confucius

Trust but verify.
Ronald Reagan

As we already know, there are many different types and kinds of information. To organize this huge diversity into a system, it is worthwhile to classify information with respect to various criteria. One of these criteria is the *nature of the carrier* of information. According to this criterion, we have the following types: *digital* information, *printed* information, *written* information, *symbolic* information, *molecular* information, *quantum* information, and so on. For instance, digital information is represented by digits, printed information is contained in printed texts, and quantum information is contained in quantum systems.

Another criterion is the system that receives information. According to this criterion, we have the following types: *visual* information, *auditory* information, *olfactory* information, *cognitive* information, *genetic* information, and so on. For instance, according to neuropsychological data, 80% of all information that people get through their senses is visual information, 10% of all information is auditory information, and only 10% of information that people get through other senses.

One more criterion is what information is about. According to this criterion, we have the following types: *weather* information, *car* information, *emotional* information (in the sense of (Sanders, 1985; Keane and Parrish, 1992; George and McIllhagga, 2000; Bosmans and Baumgartner, 2005; Knapska, et al, 2006)), *author* information, political information, *health care* information, *quality* information, *geological* information, *economical* information, *stock market* information, and so on.

One more criterion is the system where information functions. According to this criterion, we have the following types: *biological* information, *genetic* information, *social* information, *physiological* information, *ethic* information, *esthetic* information, and so on. Physical information refers generally to the information that is contained in a physical system or to the information about physical systems.

It is necessary to make a distinction between cognitive emotional information, which is cognitive information about emotions, and direct affective, or direct emotive, information, which influences the emotional state of the receiver (cf. Section 2.6).

Relations between information and its receptor/receiver determine three important types: accessible, available and acceptable information.

Definition 2.4.1. If a system Q has access to some information, then this information is called *accessible* for Q.

We know that it is possible to have access to some information to a different extent. For instance, imagine that you need two books A and B. The first one, A, is in your library at your home. You can go to the shelf where the book is and take it any time you want. At the same time, the second book, B, is only in the library of another city. You can get it, but only by the interlibrary exchange. Thus, both books are accessible, but the first one is much easier to retrieve. This shows that *accessibility* is a property of information, which can be estimated (measured) and used in the information quality assessment (cf. Section 6.2).

In addition, it is possible to distinguish conditionally accessible information.

Definition 2.4.2. If a system Q has access to the information carrier, then this information is called *conditionally accessible* for Q.

To see the difference between accessible information and conditionally accessible information, imagine a book in English. For a person D who knows English and has the book, information in this book is accessible. At the same time, for a person D who does not know English and has the book, information in this book is only conditionally accessible. There are different conditions for accessibility. One condition is that D learns English. Another condition is that D finds an interpreter. One more condition assures that that the book is translated from English to the language D knows.

However, even if a person has access to some information, it does not mean that this person can get this information. Imagine you come to a library that has one million books. You know that information you need is in some of these books but you do not know in which one. If you do not have contemporary search tools to get this information and can only read books to find it, then it will not be available to you. You cannot read all million books.

Definition 2.4.3. If a system Q can get some information, then this information is called *available* to Q.

For instance, information about Lebesgue integration is available only to those who know mathematics. Some laws of physics, e.g., Heisenberg's uncertainty principle, state that there is information about physical reality unavailable to anybody. Some mathematical results, e.g., Gödel's incompleteness theorems, claim that there is information about mathematical structures unavailable to anybody. In computer science, it is proved (cf., for example, (Sipser, 1997; Burgin, 2005)) that for a universal Turing machine, information whether this machine halts given arbitrary input is unavailable.

As information may be available to a different extent, *availability* is a graded property of information, which can be estimated (measured) and used in the information quality assessment (cf. Section 6.2).

Definition 2.4.4. If a system Q can accept some information, then this information is called *acceptable* to Q.

As we know, when a person can get some information, it does not mean that this person accepts this information. Imagine you read about some unusual event in a newspaper, but you do not believe that it is possible. Then information about this event is available to you, but you cannot accept it because you do not believe that it is possible. There are many historical examples of such situations.

For millennia, mathematicians tried to directly prove that it is possible to deduce the fifth postulate of the Euclidean geometry from the first four postulates. Being unable to achieve this, mathematicians were becoming frustrated and tried some indirect methods. Girolamo Saccheri (1667–1733) tried to prove a contradiction by assuming that the first four postulates were valid, while the fifth postulate was not true (Burton, 1997). To do this, he developed an essential part of what is now called a non-Euclidean geometry. Thus, he was able to become the creator of the first non-Euclidean geometry. However, Saccheri was so sure that the only possible geometry is the Euclidean geometry that as some point he claimed a contradiction and stopped. Actually his contradiction was only applicable in Euclidean geometry. Of course Saccheri did not realize this at the time and he died thinking he had proved Euclid's fifth postulate from the first four. Thus, information about non-Euclidean geometries was available but not acceptable to Saccheri. As a result, he missed an opportunity to obtain one of the most outstanding results in the whole mathematics.

A more tragic situation due to biased comprehension involved such outstanding mathematicians as Niels Henrik Abel (1802–1829) and Carl Friedrich Gauss (1777–1855). As history tells us (Bell, 1965), there was a famous long-standing problem of solvability in radicals of an arbitrary fifth-degree algebraic equation. Abel solved this problem proving impossibility of solving that problem in a general case. In spite of being very poor, Abel himself paid for printing a memoir with his solution. This was an outstanding mathematical achievement. That is why Abel sent his memoir to Gauss, the best mathematician of his time. Gauss duly received the work of Abel and without deigning to read it he tossed it aside with the disgusted exclamation "Here is another of those monstrosities!"

Moreover, people often do not want to hear truth because truth is unacceptable to them. For instance, the Catholic Church suppressed knowledge that the Earth rotates around the Sun because people who were in control (the Pope and others) believed that this knowledge contradicts to what was written in the Bible.

Relations between these three types of information show that any available information is also accessible. However, not any accessible information is available and not any acceptable information is available or accessible. For instance, there are many statements that a person can accept but they are inaccessible for this person. A simple example gives theory of algorithms. It is known that given a word x and a Turing machine T, it is impossible, in general, to find whether T accepts x or not using only recursive algorithms (cf., for example, (Burgin, 2005)). Thus information about acceptance of x by T is acceptable to any computer scientist because this information is neutral. At the same time, this information is in principle inaccessible for computer scientists and other researchers who use only recursive algorithms.

There are different levels at which information may be acceptable. For instance, information about yesterday's temperature is acceptable as knowledge, while information about tomorrow's temperature is acceptable only as a hypothesis. Thus, *acceptability* is a graded, fuzzy or linguistic property (Zadeh, 1973) of information, which can be estimated (measured) and used in the information quality assessment (cf. Section 6.2).

What is accepted must have some meaning. As a result, some researchers connect information to meaning, claiming that all information has to be meaningful. For instance, researchers actively discuss the problem whether information is *any* meaningful data or it is only meaningful data that have additional properties to be information (cf., for example, (Menant, 2002) and (Roederer, 2002)). This discussion has three aspects. The first one is *methodological*. It deals with the scientific base of the discussion. The second aspect is *ontological* and is related to the essence and nature of information. The third aspect is *theoretical* and is aimed at elucidation of the scope of this theory.

Taking the general theory of information as the scientific base and applying ontological and axiological principles introduced in Sections 2.2 and 2.3, we can reason that meaning is only one of many properties of information.

Stonier (1997) comes to the same conclusion on the empirical grounds, stressing that it is important to distinguish between information and meaning. To explain this, he gives the following example. On his opinion, the letters of a written alphabet or the nucleotides of a strand of DNA are information (or contain information according to the general theory of information). In this context, two identical sequences of nucleotides contain twice as much information as one of these sequences. However this information may yield a message with a meaning if and only if it has been processed. If the nucleotides in the second sequence are identical with the nucleotides in the first one, then when both sequences are processed, its message is merely redundant. The message may acquire a meaning if and only if it has been information-processed by a recipient.

Types of information can be differentiated by their actions on the receiver R. It gives us pure and extended information.

Definition 2.4.5. If (a portion of) information I directly changes only the infological system $IF(R)$ (and nothing more), then I is called (a portion of) *pure information* for the system R.

However, it is possible that information for the system R changes not only the infological system $IF(R)$ but also other parts of the system R.

Definition 2.4.6. If (a portion of) information I changes the infological system $IF(R)$ and some other parts of R, then I is called *extended information* or more strictly, a *portion of extended information*.

It is useful to make a distinction between pure and extended information when we analyze information impact. This distinction allows us to get more exact knowledge about the system that receives information.

Note that the distinction between pure and extended information is relative, i.e., it depends on the system R. Pure information for one system can be extended information for another system. For instance, when something falls on the man A, it gives him extended information.

However, if a different person B watches this situation, it gives her only pure information.

Another example is when a boxer is knocked, then he receives extended information because it adds to the boxer's knowledge (changes his infological system $\text{IF}(R)$) that he has received a knock and at the same time, it changes his physical state. However, when a processor receives an instruction to perform some operation, e.g., addition of two numbers, then this instruction, as well as the set of signals that codify this instruction, has extended information for the computer because it changes not only the memory of the computer (its infological system $\text{IF}(R)$) but also other parts. At the same time, when computer receives some data to store in its memory, this gives only pure information to the computer.

However, sometimes it is difficult to make an exact distinction between pure and extended information. For instance, a woman watched TV and what she saw impressed her so much she changed her life. In particular, being fat she stopped eating too much and became slender and active. Thus, at first, TV information changed her cognitive infological system and then changed this woman even physically. Consequently, directly received information is pure for the woman because it had changed only her knowledge. However, in a long run, by Definition 2.4.6, the received information is extended.

Even a text may contain extended information. Let us consider a letter L that contains information about some tragic event. Thus, we can imagine the following situation. A person A receives this letter, reads it and then has a stroke because the news from the letter have been so horrible for him. Thus, we can state that the letter L contained some portion of extended information for A.

Taking such a criterion as truthfulness, we come to three basic types of information:

- *true* or *genuine information*,
- *pseudoinformation*,
- *false information* or *misinformation*.

However, it is possible to separate these types of information in a more general situation, utilizing the concept of *information measure*.

There are different ways to do this. Analyzing different publications, we separate two classes of approaches: relativistic definitions and universal definitions. The latter approach is subdivided into object-dependent, reference-dependent and attitude-dependent classes. At first, we consider the relativistic approach to this problem.

Let us consider an information measure m that takes values in the set R of all real numbers or in the interval $[-n, n]$.

Definition 2.4.7. A portion of information I is called *positive information relative to the measure m* if $m(I) > 0$.

Definition 2.4.8. A portion of information I is called *negative information relative to the measure m* if $m(I) < 0$.

The concepts of positive and negative information show the advantage of the relativistic approach. Really, a portion of information I can be negative information with respect to one measure, decreasing the infological system $IF(R)$ of a system R according to this measure, and it can be positive information, increasing $IF(R)$ with respect to another measure.

For instance, a portion of information I comes to a database D with corrections for some data. Treating D as an infological system, we want to evaluate changes in D and to find to what class this information belongs. Changing incorrect data in D increases correctness of data in D, implying that I is positive information with respect to correctness. However, new (correct) data have shorter description. As a consequence, with respect to such measure m as the volume of data, I is negative information because it decreases the volume of data in D, i.e., the number of bytes in D diminishes.

Another example is related to a description d of a person A in a database D. Then information I comes to us, telling that d is not a description of A, but a description of another person K. Then with respect to such measure m as the volume of data, nothing changes in D and thus, I is pseudoinformation as the number of bytes in D stays the same. But because the previous attribution of d to A was incorrect and the new attribution of d to K is correct, I is true or genuine information with respect to such measure as correctness.

A specific class of misinformation is neginformation.

Definition 2.4.9. A portion of information *I* is called *neginformation relative to the measure m* if it is intended negative information relative to the measure *m*.

Neginformation may be useful. For instance, when we want to solve a problem or to perform a computation and information *I* comes reducing complexity of solution (computation), then it is useful information, which is neginformation by Definition 2.4.9.

Definition 2.4.10. A portion of information *I* is called *neutral information relative to the measure m* if $m(I) = 0$.

A portion of information *I* can be negative information with respect to one measure, it positive information with respect to another measure, and neutral information with respect to the third measure.

Thus, we have three basic types of information:

- *positive information,*
- *neutral information,*
- *negative information.*

One more class is quasiinformation.

Definition 2.4.11. If a portion of information *I* (in the general sense) does not change the chosen infological system IF(*R*), then *I* is called *quasiinformation* for a system *R* relative to the system IF(*R*).

This definition implies the following property of quasiinformation.

Proposition 2.4.1. a) A portion of quasiinformation *I* is neutral information with respect to any information measure.

b) If all changes of an infological system IF(*R*) are measurable, then a portion of information *I* is quasiinformation if and only if it neutral information with respect to any information measure.

In what follows, we do not consider quasiinformation, i.e., general information, such that does not include pure information. However, this concept is very important for politics, mass media, entertainment industry, and some other institutions, which are reproached for disseminating quasiinformation.

To have genuine information in the conventional sense, we take such measure *m* as correctness of knowledge or such measure as validity of knowledge or of knowledge acquisition. Let us specify the relativistic approach in the case of cognitive information, taking a measure *m* that reflects such property as truthfulness. There are many other measures related to the same infological systems, e.g., to the system of knowledge, but not related to truthfulness. For instance, we can take such an information measure m_v that measures the changes in the volume of a system of knowledge T in bytes, kilobytes, and gigabytes for a computer thesaurus or in number of books and journals for a library thesaurus. This measure has nothing to with the truthfulness of information that is contained in computer files or in books.

Thus, to get more exact representation of the convenient meaning of the term *true information*, we consider only cognitive information and assume that true cognitive information gives true knowledge, or more exactly, make knowledge truer than before. To measure these changes, we utilize a measure *tr* of truthfulness of knowledge in the system T. Such a measure, allows one to develop the comprehensive approach to true information.

There are different ways to introduce truthfulness of knowledge. All of them belong to one of three types: *object-dependent*, *reference-dependent* and *attitude-dependent*. For the first type, we have the following definitions.

A popular approach to truthfulness involves a system R, or as it is now fashionable to call it, an intelligent agent A, that (who) has knowledge. Then, the truthfulness $tr(R, D)$ of knowledge the system R has about a domain (object) D is a function of two variables where the range of the variable R is a set of different systems with knowledge and the range of the variable D is a set of different object domains. It is supposed that each element or system of knowledge refers to some object domain because knowledge is always knowledge about something (cf. (Burgin, 2004) and Section 2.5). In addition, the function $tr(R, D)$ satisfies conditions that differentiate knowledge from similar structures, such as beliefs, descriptions or fantasies.

Let us consider such an infological system as *thesaurus* or *knowledge system T* and fix some numerical measure m of information, which reflects changes of T. Taking T as the infological system restricts our study to cognitive information, for which distinctions between genuine information, misinformation, disinformation, and pseudoinformation are more natural. However, it is possible to find an information measure m that is not related to the knowledge system and to define genuine information, misinformation, disinformation, and pseudoinformation relative to this measure m.

Truthfulness, or correctness, of knowledge shows absence of distortions in knowledge. This property is related to *accuracy of knowledge*, which reflects how close is given knowledge to the absolutely exact knowledge. However, truthfulness and accuracy of knowledge are different properties. For instance, statements "π is approximately equal to 3.14" and "π is approximately equal to 3.14159" are both true, i.e., their truthfulness is equal to 1. At the same time, their accuracy is different. The second statement is more accurate than the first one.

However, it is necessary to understand that the truth of knowledge and the validity of it its acquisition are not always the same. For instance, the truth of knowledge represented by propositions and the validity of reasoning are distinct properties, while there are relations between them (cf., for example, (Suber, 2000)). This relationship is not entirely straightforward. It is not true that truth and validity, in this sense, are utterly independent because the impossibility of "case zero" (a valid argument with true premises and false conclusion) shows that one combination of truth-values is an absolute bar to validity. According to the classical logic, when an argument has true premises and a false conclusion, it *must* be invalid. In fact, this is how we define invalidity. However, in real life, people are able to take a true statement and to infer something false. An example of such a situation gives the Cold War Joke, which is considered in this section.

To formalize the concept of knowledge truthfulness, we use the model developed in (Burgin, 2004) and Section 2.5. According to this model, *general knowledge K* about an object F has the structure represented by the diagram (2.4.1) and high level of validation.

$$
\begin{array}{ccc}
 & g & \\
W & \longrightarrow & L \\
\uparrow t & & \uparrow p \\
U & \longrightarrow & C \\
 & f &
\end{array}
\qquad (2.4.1)
$$

This diagram has the following components:

1) a class U containing an object F;
2) an intrinsic property of objects from U represented by an abstract property $T = (U, t, W)$ with the scale W (cf. Appendix C);
3) a class C of names, which includes a name «F» of the object F;
4) an ascribed property names from C represented by an abstract property $P = (C, p, L)$ with the scale L (cf. Appendix C);
5) the correspondence f assigns names from C to objects from U where in general case, an object has a system of names or more generally, conceptual image (Burgin and Gorsky, 1991) assigned to it;
6) the correspondence g assigns values of the property P to values of the property T. In other words, the correspondence g corresponds values of the intrinsic property to values of the ascribed property. For instance, when we consider a property of people such as height (the intrinsic property), in weighting any thing, we can get only an approximate value of the real height, or height with some precision (the ascribed property).

In more detail, the structure of knowledge is discussed and described in Section 2.5.

To evaluate the degree to which a given knowledge is true, we use the diagram (2.4.1), validating correspondences g, f, and p. There are different systems of knowledge validation. Science, for which the main validation technique is experiment, is a mechanism developed for cognition of nature and validation of obtained knowledge. Mathematics, which is based on logic with its deduction and induction, is a formal device constructed for empowering cognition and knowledge validation. Religion with its postulates and creeds is a specific mechanism used to validate knowledge by its compliance with religious postulates and creeds. History, which is based on historical documents and archeological discoveries, is also used to obtain and validate knowledge on the past of society. Each system of validation induces a corresponding validation function $tr(T, D)$ that gives a quantitative or a qualitative estimate of the truthfulness of knowledge T about some domain D. Note that the domain D can consists of a single object F.

If a validation function $tr(T, F)$ is defined for separate objects, then it is possible to take some unification of all values $tr(T, F)$ ranging over all objects F in the domain D to which knowledge T can be related. This allows us to obtain a truthfulness function $tr(T, F)$ for systems of knowledge. In general, such unification of values $tr(T, F)$ is performed by some integral operation in the sense of (Burgin and Karasik, 1976; Burgin, 1982a).

Definition 2.4.12. a) An *integral operation* W on the set R of real numbers is a mapping that corresponds a number from R to a subset of R, and for any $x \in R$, we have $W(\{x\}) = x$.

b) A *finite integral operation* W on the set R of real numbers is a mapping that corresponds a number from R to a finite subset of R, and for any $x \in R$, we have $W(\{x\}) = x$.

As a rule, integral operations are partial. That is, they attach numbers only to some subsets of R. At the same time, it is possible to define integral operations in arbitrary sets.

Examples of integral operation include: summation, multiplication, taking the minimum or maximum, determining the infimum or supremum, evaluating integrals, taking the first element from a given subset, taking the sum of the first and second elements from a given subset, and so on.

Examples of finite integral operation include summation, multiplication, taking minimum, determining maximum, calculating the average or finite weighted average for finite sets, taking the first element from a given finite subset, and so on.

The following integral operations are the most relevant to the problem of information truthfulness estimation are:

1) taking the average value;
2) taking the minimal value;
3) taking the maximal value.

Let us consider some measures of correctness.

Example 2.4.1. One of the most popular measure $tr(T, D)$ of truthfulness is correlation between the experimental data related to an object domain D and data related to D that are stored in the knowledge system T.

Correlation r is a bivariate measure of association (strength) of the relationship between two sets of corresponded numerical data. It often varies from 0, which indicates no relationship or random relationship, to 1, which indicates a strong relationship or from −1, which indicates a strong negative relationship to 1, which indicates a strong relationship. Correlation r is usually presented in terms of its square (r^2), interpreted as percent of variance explained. For instance, if r^2 is 0.1, then the independent variable is said to explain 10% of the variance in the dependent data. However, as experts think, such criteria are in some ways arbitrary and must not be observed too strictly.

The *correlation coefficient* often called Pearson product-moment correlation coefficient is a measure of linear correlation and is given by the formula:

$$r = \frac{\sum_{i=1}^{n}(x_i - \overline{x})(y_i - \overline{y})}{\sqrt{\sum_{i=1}^{n}(x_i - \overline{x})^2 \sum_{i=1}^{n}(y_i - \overline{y})^2}}$$

In the context of knowledge validation, numbers x_i are numerical data about objects from the domain D obtained by measurement and experiments, while numbers y_i are numerical data about objects from the domain D obtained from the knowledge K.

In general, the correlation coefficient is related to two random variables X and Y and is defined in the context of mathematical statistics as

$$r_{X,Y} = \frac{\text{Cov}(X, Y)}{\sigma_X \sigma_Y} = \frac{E((X - \mu_X)(Y - \mu_Y))}{\sigma_X \sigma_Y}$$

Here μ_X (μ_Y) is the mean and σ_X (σ_Y) is the standard deviation of the random variable X (Y).

Beside correlation coefficient r, which is the most common type of correlation measure, other types of correlation measures are used to handle different characteristics of data. For instance, measures of association are used for nominal and ordinal data.

One more measure of correlation is given by the formula

$$\sum_{i=1}^{n}(x_i - \overline{x})(y_i - \overline{y})$$

This formula has the following interpretation:

$\sum_{i=1}^{n} (x_i - \bar{x})(y_i - \bar{y}) > 0$ means positive correlation;

$\sum_{i=1}^{n} (x_i - \bar{x})(y_i - \bar{y}) < 0$ means negative correlation;

$\sum_{i=1}^{n} (x_i - \bar{x})(y_i - \bar{y}) = 0$ indicates absence of correlation.

Example 2.4.2. It is possible to take validity of stored in the system of knowledge T about the object domain D as a measure $tr(T, D)$. There are different types of validity. For instance, researchers have introduced four types of validity for experimental knowledge: conclusion validity, internal validity, construct validity, and external validity. They build on one another, and each type addresses a specific methodological question.

However, we need to estimate not only knowledge but also other elements of cognitive infological systems, such as beliefs.

Belief relations in a system are estimated by belief and plausibility measures (cf. (Klir and Wang, 1993)) or by extended belief measures that are described below. It makes possible to take some belief or plausibility measure as a measure $tr(T, D)$ of truthfulness of knowledge or adequacy of beliefs in the system T.

A belief and plausibility measures are kinds of fuzzy measures introduced by Sugeno (1974).

Let $\mathbf{P}(X)$ be the set of all subsets of a set X and $\mathbf{A}$ be a Borel field of sets from $\mathbf{P}(X)$, specifically, $\mathbf{A} \subseteq \mathbf{P}(X)$.

Definition 2.4.13. A *fuzzy measure* on $\mathbf{A}$ in X (in the sense of Sugeno) is a function g: $\mathbf{A} \rightarrow [0,1]$ that assigns a number in the unit interval $[0,1]$ to each set from A so that the following conditions are valid:

(FM1) $X \in \mathbf{A}$, $g(\varnothing) = 0$, and $g(X) = 1$, i.e., the function g is normed.

(FM2) the function g is monotone, i.e., for any A and B from $\mathbf{A}$, the inclusion $A \subseteq B$ implies $g(A) \leq g(B)$.

(FM3) For any non-decreasing sequence $A_1 \subseteq A_2 \subseteq \ldots \subseteq A_n \subseteq A_{n+1} \subseteq \ldots$ of sets from $\mathbf{A}$, the following equality is valid

$$g(\cup_{n=1}^{\infty} A_n) = \lim_{n \to \infty} g(A_n)$$

(FM4) For any non-increasing sequence $A_1 \supseteq A_2 \supseteq \ldots \supseteq A_n \supseteq A_{n+1} \supseteq \ldots$ of sets from $\mathbf{A}$, the following equality is valid

$$g(\cap_{n=1}^{\infty} A_n) = \lim_{n \to \infty} g(A_n)$$

If $A \in \mathbf{A}$, then the value g(A) is called the fuzzy measure of the set A.

Then this definition was improved further through elimination of the condition (FM4) (Sugeno, 1977; Zimmermann, 2001).

Many measures with infinite universe studied by different researchers, such as probability measures, belief functions, plausibility measures, and so on, are fuzzy measures in the sense of Sugeno.

A more general definition of a fuzzy measure is used in (Klir and Wang, 1993; Burgin, 2005d).

Let $\mathbf{P}(X)$ be the set of all subsets of a set X and $\mathbf{B}$ be an algebra of sets from $\mathbf{P}(X)$, in particular, $\mathbf{B} \subseteq \mathbf{P}(X)$.

Definition 2.4.14. A *fuzzy measure* on $\mathbf{B}$ in X is a function g: $\mathbf{B} \rightarrow R^+$ that assigns to each set from $\mathbf{B}$ a positive real number, is monotone (FM2):

(FM1) g($\varnothing$) = 0.

(FM2) For any A and B from $\mathbf{B}$, the inclusion $A \subseteq B$ implies g(A) $\leq$ g(B).

In what follows, we call g simply a fuzzy measure and call $\mathbf{B}$ the *algebra of fuzzy measurable sets* with respect to the fuzzy measure g.

Popular examples of fuzzy measures are possibility, belief and plausibility measures.

Possibility theory is based on possibility measures. Let us consider some set X and its power set $\mathbf{P}(X)$.

Definition 2.4.15 (Zadeh, 1978; Zimmermann, 2001). A *possibility measure* in X is a partial function *Pos*: $\mathbf{P}(X) \rightarrow [0,1]$ that is defined on a subset $\mathbf{A}$ from $\mathbf{P}(X)$ and satisfies the following axioms:

(Po1) $\varnothing, X \in \mathbf{A}$, *Pos*($\varnothing$) = 0, and *Pos*($X$) = 1.

(Po2) For any A and B from $\mathbf{A}$, the inclusion $A \subseteq B$ implies *Pos*(A) $\leq$ *Pos*(B).

(Po3) For any system $\{A_i; i \in I\}$ of sets from $\mathbf{A}$,

$$Pos(\bigcup_{i \in I} A_i) = \sup_{i \in I} Pos(A_i)$$

Possibility can be also described by a more general class of measures (cf. (Oussalah, 2000; Zadeh, 1978)).

Definition 2.4.16. A *quantitative possibility measure* in X is a function P: $\mathbf{P}(X) \rightarrow [0,1]$ that satisfies the following axioms:

(Po1) $\varnothing$, $X \in \mathbf{A}$, $P(\varnothing) = 0$, and $P(X) = 1$;

(Po2a) For any A and B from $\mathbf{A}$, $P(A \cup B) = \max \{P(A), P(B)\}$.

A quantitative possibility measure is a fuzzy measure.

Dual to a quantitative possibility measure is a necessity measure (Oussalah, 2000).

Definition 2.4.17. A *quantitative necessity measure* in X is a function N: $\mathbf{P}(X) \to [0,1]$ that satisfies the following axioms:

(Ne1) $N(\varnothing) = 0$, and $N(X) = 1$.

(Ne2) For any A and B from $\mathbf{P}(X)$, $N(A \cap B) = \min \{N(A), N(B)\}$.

A quantitative necessity measure is a fuzzy measure.

Possibility and necessity measures are important in support logic programming (Baldwin, 1986), which uses fuzzy measures for reasoning under uncertainty and approximate reasoning in expert systems, based on the logic programming style.

In addition to ordinary measures, fuzzy measures encompass many kinds of measures introduced and studied by different researchers. For instance, beliefs play an important role in people's behavior. To study beliefs by methods of fuzzy set theory, the concept of a belief measure was introduced and studied (Shafer, 1976).

Definition 2.4.18. A *belief measure* in X is a partial function *Bel*: $\mathbf{P}(X) \to [0,1]$ that is defined on a subset $\mathbf{A}$ from $\mathbf{P}(X)$ and satisfies the following axioms:

(Be1) $\varnothing$, $X \in \mathbf{A}$, $Bel(\varnothing) = 0$, and $Bel(X) = 1$.

(Be2) For any system $\{A_i; i = 1, 2, \ldots , n \}$ of sets from $\mathbf{A}$ and any n from N,

$$Bel(A_1 \cup \ldots \cup A_i) \geq$$

$$\sum_{i=1}^{n} Bel(A_i) - \sum_{i < j} Bel(A_i \cap A_j) + \ldots + (-1)^{n+1} Bel(A_1 \cap \ldots \cap A_i)$$

Axiom (Be2) implies that belief measure is a super-additive fuzzy measure as for $n = 2$ and arbitrary subsets A and B from X, we have

$$Bel(A \cup B) \geq Bel(A) + Bel(B) - Bel(A \cap B)$$

Axiom (Be2) also implies the following property

$$0 \leq Bel(A) + Bel(\overline{A}) \leq 1$$

where $\overline{A}$ is a complement of A.

For each set $A \in \mathbf{P}(X)$, the number $Bel(A)$ is interpreted as the degree of belief (based on available evidence) that a given element x of X belongs to the set A. Another interpretation treats subsets of X as answers to a particular question. It is assumed that some of the answers are correct, but we do not know with full certainty which ones they are. Then the number $Bel(A)$ estimates our belief that the answer A is correct.

One more class is plausibility measures, which are related to belief measures.

Definition 2.4.19. A *plausibility measure* in X is a partial function $Pl: \mathbf{P}(X) \to [0,1]$ that is defined on a subset A from $\mathbf{P}(X)$ and satisfies the following axioms (Shafer, 1976):

(Pl1) $\varnothing, X \in A$, $Pl(\varnothing) = 0$, and $Pl(X) = 1$.

(Pl2) For any system $\{A_i; i = 1, 2, ..., n\}$ of sets from A and any n from N,

$$Pl(A_1 \cap ... \cap A_i) \leq$$

$$\sum_{i=1}^{n} Pl(A_i) - \sum_{i<j} Pl(A_i \cup A_j) + ... + (-1)^{n+1} Pl(A_1 \cup ... \cup A_i)$$

Belief measures and plausibility measures are *dual measures* as for any belief measure $Bel(A)$, $Pl(\overline{A}) = 1 - Bel(A)$ is a plausibility measure and for any plausibility measure $Pl(A)$, $Bel(\overline{A}) = 1 - Pl(A)$ is a belief measure.

When Axiom (Be2) for belief measures is replaced with a stronger axiom

$$Bel(A \cup B) = Bel(A) + Bel(B) \text{ whenever } A \cap B = \varnothing$$

we obtain a special type of belief measures, the *classical probability measures* (sometimes also referred to as *Bayesian belief measures*).

In a similar way, some special kinds of probability measures are constructed (Dempster, 1967).

It is possible to consider dynamical systems in spaces with a belief, plausibility or possibility measure. These systems allow one to model mental processes, cognition, and information processing in intelligent systems. For example, it is possible to consider data- and knowledge bases as dynamical systems with a belief measure and study their behavior. Such a belief measure can reflect user beliefs in correctness and validity of data, as well as user beliefs in truth and groundedness of knowledge systems.

A belief system, either of an individual or of a community, contains not only beliefs, but also disbelief, i.e., belief that something is not true. To represent this peculiarity, we use extended belief measures.

Definition 2.4.20. An *extended belief measure* is a function Bel: $P(X) \to [-1,1]$ that satisfies the same axioms as belief measures.

Remark 2.4.1. It is possible to represent an extended belief measure by an intuitionistic fuzzy set in the sense of (Atanasov, 1999).

The attitude-dependent approach to truthfulness of knowledge is based on a function

$$at\colon \mathbf{K} \to L$$

where $\mathbf{K}$ is a collection of knowledge items or knowledge systems, and L is a partially ordered set. This function evaluates truthfulness of knowledge items (systems) based on attitudes that reflect confidence that a knowledge item (system) is true. When an individual or a group evaluates knowledge, their estimates are assigned to different knowledge items (systems). For instance, a knowledge item k_1 is presumed to be absolutely true, a knowledge item k_2 is estimated as true only in some situations, while a knowledge item k_3 is treated as rarely true. It is possible to give numerical values of truthfulness. For instance, it is 30% true that it will be raining tomorrow (item k_1) and 70% true that it will not be raining tomorrow (item k_2). These estimates reflect individual or group confidence, which may be based on some grounded procedures, e.g., on measurement, or represent only dispositions, preferences, and beliefs, e.g., gut feeling.

For instance, Bertrand Russell (1926) uses an attitude-dependent approach in his definition of knowledge. Consideration of beliefs forms the starting point for Russell's definition of knowledge.

According to the *attitude-dependent approach*, we have the following definitions.

Definition 2.4.21. *General knowledge T* about an object *F* for a system *R* is the entity that has the structure represented by the diagram (2.4.1) that is estimated (believed) by the system *R* to represent with high extent of confidence true relations.

In the *reference-dependent approach*, we measure truthfulness of knowledge not with respect to an object domain, but with respect to another thesaurus. For instance, one person A gives some information I to another person B. Then it is viable to measure truthfulness of I with respect to A or more exactly, to the system of knowledge T_A of A.

To achieve this goal, we use some measure $cr(T, D)$ of correlation or consistency between knowledge systems T and D. Let T_A be the system of knowledge of A and T_B be the system of knowledge of B.

Definition 2.4.22. The function $cr(I, T_A) = cr(I(T_B), T_A) - cr(T_B, T_A)$ is called a measure of *relative truthfulness* of information.

Object-depended truthfulness also works when the real situation is known. However, it is impossible to compare knowledge directly with a real system. So, in reality, we always compare different systems of knowledge. It means that the object-dependent approach can be reduced to the reference-dependent approach. However, one system of knowledge can be closer to reality than another system of knowledge. In this case, we can assume that the corresponding truthfulness is object related (at least, to some extent). For instance, in the object-dependent approach, it is possible to compare theoretical knowledge to experimental data.

This assumption is used as the main principle of science: it is presupposed that correct experiment gives true knowledge about reality and to find correctness of a model or theory, we need to compare the model or theory with experimental data.

Such an approach to knowledge is developed in the externalist theories of knowledge (Pollock and Cruz, 1999).

Example 2.4.3. Let us consider the system T_A of knowledge of a person A. The system T_{SA} is generated from all (some) statements of A. Then the value of the function $tr(T_{SA}, T_A)$ reflects sincerity of A, while the value of the function $tr(T_{SA}, T_A)$ reflects sincerity of information I given by A.

Example 2.4.4. System related truthfulness is useful for estimating statements of witnesses. In this case, the measure of inconsistency $incons(T_A, T_B)$ between T_A and T_B is equal to the largest number of contradicting pairs (p_A, p_B) of simple statements p_A from T_A and p_B from T_B when p_A and p_B are related to the same object or event. The measure of inconsistency $incons(T_A, T_B)$ between T_A and T_B determines several measures of consistency $cons(T_A, T_B)$ between T_A and T_B. One of such measures of consistency $cons(T_A, T_B)$ between T_A and T_B is defined by the formula

$$cons(T_A, T_B) = 1/(1 + incons(T_A, T_B)) \qquad (2.4.2)$$

It is possible to normalize the measure of inconsistency $incons(T_A, T_B)$, defining $incons_N(T_A, T_B)$ as the ratio of the number of contradicting pairs (p_A, p_B) and number of all pairs (p_A, p_B) of simple statements p_A from T_A and p_B from T_B when p_A and p_B are related to the same object or event. This measure generates the corresponding normalized consistency measure $cons_N(T_A, T_B)$ by the formula (2.4.2). Another normalized consistency measure $cons_N(T_A, T_B)$ is defined by the formula (2.4.3).

$$cons(T_A, T_B) = 1 - incons(T_A, T_B) \qquad (2.4.3)$$

Relations of a portion of information I to its object O_I reflect important properties of this information. One of the most important relations of this kind is truth, i.e., correct representation of O_I by I. We can say that information is *true*, or *factual*, if it allows one to build true knowledge. However, it is necessary to understand that in the context of the general theory of information, truth is only one of many properties of information. Thus, we consider here a more general situation and distinguish true information as a specific kind of genuine information (cf. Definition 2.4.7).

It is reasonable to call cognitive information *true* when it gives knowledge with a high level of validity. However, genuine information is not always true information. For instance, the Hartley measure of uncertainty (entropy) of an experiment E that can have k outcomes is equal to $\log_2 k$. Then it is possible to measure information I in a message M that tells us that the experiment E can have h outcomes as $m(I) = \log_2 k - \log_2 h$. The message and consequently, information I can be false, i.e., it is not true that the experiment E can have h outcomes, but if h is less than k, information I is genuine with respect to the chosen measure.

Treating information as the quality of a message that is sent from the sender to the receiver, we can see that information does not have to be accurate. It may convey a truth or a lie, or just be something intangible. Even a disruptive noise used to inhibit the flow of communication and create misunderstanding would in this view be a form of information. That is why, the problem of measuring information accuracy is so important for information and library science (cf., for example, (Fallis, 2004)).

The measure $tr(T, D)$ of correctness validity allows us to define truthfulness for information. If we have a thesaurus T and a unit of information I, then the truthfulness of I about an object domain D is given by the formula

$$tr(I, D) = tr(I(T), D) - tr(T, D)$$

Here $I(T)$ is the thesaurus T after it receives/processes information I. The difference shows the impact of I on T.

Definition 2.4.23. The function $tr(I, D)$ is called a measure of *domain related truthfulness* or simply, *correctness* of information.

Truthfulness, or correctness, of information shows what changes it makes with truthfulness of knowledge. This property is related to accuracy of information, which measures how changes the distance of the initial knowledge to the absolutely exact knowledge.

Consequently, we come to three main types of information about some object (domain):

- *objectively true information,*
- *objectively neutral information,*
- *objectively false information.*

Taking some object domain D and tr as the measure m in Definitions 2.4.7, 2.4.8 and 2.4.10, we obtain unconditional concepts of true and false information.

Definition 2.4.24. A portion of information I is called *objectively true information* about D if $tr(I, D) > 0$.

Some researchers, such as Dretske (1981; 2000), Barwise and Perry (1983), Barwise and Seligman (1997) and Graham (1999), assert that truth, or factuality, is the intrinsic property of information. For instance, Dretske (1981) writes, "… *false* information and *mis*-information are not kinds of information — any more than decoy ducks and rubber ducks are kinds of ducks". Weiss (1996) directly asserts, "Information is truth".

However, many others, such as Fox (1983), Devlin (1991), Colburn (2000), Fetzer (2004), and Dodig-Crnkovic (2005), are proponents of semantic information neutrality and argue that meaningful and well-formed data already qualify as information (or according to the general theory of information, carry information), no matter whether they represent or convey a truth or a falsehood or indeed have no alethic, i.e., related to truth, value at all.

To adequately discuss a possibility of false information existence from a methodological point of view, it is necessary to take into account three important issues: multifaceted approach to reality, historical context, and personal context. Thus, we come to the following conclusion.

First, there is a *structural issue* in this problem. Namely, the dichotomous approach, which is based on classical two-valued logic, rigidly divides any set into two parts, in our case, true and false information. As a result, the dichotomous approach gives a very approximate image of reality. Much better approximation is achieved through the multifaceted approach based on multivalued logics, fuzzy reasoning, and linguistic variables.

Second, there is a *temporal issue* in this problem. Namely, the problem of false information has to be treated in the historical or, more exactly, temporal context, i.e., we must consider time as an essential parameter of the concept. Indeed, what is treated as true knowledge in one period of time can be discarded as false knowledge in another period of time.

Third, there is a *personal issue* in this problem, i.e., distinction between genuine and false information often depends on the person who estimates this knowledge. For instance, for those who do not know about non-Diophantine arithmetics (Burgin, 1997d; 2001b), $2 + 2$ is always equal to 4. At the same time, for those who know about non-Diophantine arithmetics (Burgin, 1997d; 2001b), it becomes possible that $2 + 2$ is not equal to 4.

In light of the first issue of our discussion about false information, we can see that in cognitive processes, the dichotomous approach, which separates all objects into two groups, A and *not A*, is not efficient. Thus, if we take the term "false information", then given a statement, it is not always possible to tell if it contains genuine or false information. To show this, let us consider following statements:

1. "π is equal to 3".
2. "π is equal to 3.1".
3. "π is equal to 3.14".
4. "π is equal to 3.1415926535".
5. "π is equal to $(4/3)^2$".

According to the definition of π and our contemporary knowledge that states that π is a transcendent number, all these statements contain false information. In practice, they are all true but with different exactness. For example, the statement (4) is truer than the statement (1). Nevertheless, in the ancient Orient, the value of π was frequently taken as 3 and people were satisfied with this value (Eves, 1983). Archimedes found that π is equal to 3.14. For centuries, students and engineers have used 3.14 as the value for π and had good practical results. Now calculators and computers allow us to operate with much better approximations of π, but nobody can give the exact decimal value of this number.

Importance of the temporal issue is demonstrated by the following example from the history of science that helps to better understand the situation with false information. Famous Greek philosophers Leucippus (fl. 445 B.C.E.) and Democritus (460–360 B.C.E.) suggested that all material bodies consist of small particles, which were called atoms. "In reality", said Democritus, "there are only atoms and the void".

We can ask the question whether this idea about atoms contains genuine or false information. From the point of view of those scientists who lived after Democritus but before the fifteenth century, it contained false information. This was grounded by the fact that those scientists were not able to look sufficiently deep into the matter to find atoms.

However, the development of scientific instruments and experimental methods made it possible to discover micro-particles such that have been and are called atoms. Consequently, now it is a fact, which is accepted by everybody, that all material bodies consist of atoms. As a result, now people assume that the idea of Leucippus and Democritus contains genuine or true information.

This shows how people's comprehension of what is genuine information and what is understood as false information changes with time. Lakatos (1976) and Kline (1980) give interesting examples of similar situations in the history of mathematics, while Cartwright (1983) discusses analogous situations in the history of physics.

All these examples demonstrate that it is necessary to consider false information as we use negative numbers, as well as not to discard pseudoinformation as we do not reject utility of such number as 0. History of mathematics demonstrates that understanding that 0 is a number and a very important number demanded a lot of hard intellectual efforts from European mathematicians when Arab mathematicians brought to them knowledge about 0 from India.

Going to the third point of the discussion about false information related to the personal issue, let us consider other examples from the history of science as here we are studying information by scientific methods.

In his lectures on optics, Isaac Newton (1642–1727) developed a corpuscular theory of light. According to this theory, light consists of small moving particles. Approximately at the same time, Christian Huygens (1629–1695) and Robert Hooke (1635–1703) built a wave theory of light. According to their theory, light is a wave phenomenon. Thus, somebody would like to ask the question who of them, i.e., was it Newton or Huygens and Hooke, gave genuine information and who gave false information. For a long time, both theories were competing. As a result, the answer to our question depended whether the respondent was an adherent of the Newton's theory or of the theory of Huygens and Hooke. However, for the majority of people who lived at that time both theories provided pseudoinformation because those people did not understand physics.

A modern physicist believes that both theories contain genuine information. So, distinction between genuine and false information in some bulk of knowledge depends on the person who estimates this knowledge.

Existence of false information is recognized by the vast majority of people, but some theoreticians insist that that false information is not information. There is persuasive evidence supporting the opinion that false information exists. For instance, readers find a lot of false or inaccurate information in newspapers, books and magazines. Recent studies found a considerable amount of inaccurate information on the Internet (Hernon, 1995; Connell and Triple, 1999; Bruce, 2000; Berland, et al, 2001).

The new and truly wonderful medium, the Internet, unfortunately has one glaring downside. Namely, along with all the valid information it provides, the Internet also contains much misleading information (disinformation), false information, and outright hype. This is as true in the fields of science and science criticism as in any other. Certainly many so-called "discussion groups" and informal "book review" sites are good examples of the blind leading the blind (Fallis, 2004).

On an Internet blog, the author of this book has once encountered an assertion of one of the bloggers that there is a critique of a book *A* in a paper *D*. Finding the paper *D*, the author was very surprised because the paper *D* had been published in 2003, while the book *A* was published only in 2005.

Another example of the situation when "the blind leads the blind" we can take the critique of the professor *P* aimed at the book *B*. Indeed, this critique was irrelevant and contained essential logical and factual mistakes. However, being asked if he read the book he criticized, *P* answered that he did not need to do that because he saw an article by the same author and understood nothing. Naturally, in this situation, his critique was an example not only of incompetent but also indecent behavior because using Internet and other contemporary means of communication, professor *P* transmitted this false information to those who read his writings on this topic.

Ketelaar (1997) writes:

"Why do we demand more of the quality of food or a car than we demand of that other essential: information? Reliability and authenticity determine the credibility and the usefulness of information. These concepts, developed in different cultures and at different times, are essential for our information society in its dependence on trust in information. In the creation and distribution of digital information, conditions should be met to ensure the reliability and authenticity of the information."

Thus, we see that the problem of false information is an important part of information studies and we need more developed scientific methods to treat these problems in an adequate manner.

There is also such type as biased information. This type naturally exists only for information in society. For instance, Mowshowitz and Kumar (2009) write:

"Search engine consolidation raises the specter of biased information and free speech abridgement ..."

One of the main problems in information management is to discern genuine information, biased information, and misinformation.

To have genuine information relevant to usual understanding, we take such measure as correctness of knowledge or such measure as validity of knowledge. Thus, we can call cognitive information *false* when it decreases validity of knowledge it gives. Definition 2.4.8 and 2.4.23 show that we have false information when its acceptance makes our knowledge less correct. For instance, let us consider people who lived in ancient Greece and accepted ideas of Leucippus and Democritus that all material bodies consist of atoms. Then they read Aristotle's physics that eliminated the idea of atoms. Because we now know that the idea of atoms is true, Aristotle's physics decreased true knowledge about such an aspect of the world as the existence of atoms and thus, gave false information about atoms.

We see that false information is also information because it has a definite impact on the infological system. Only this impact is negative.

It is necessary to understand that the concept of false information is relative, depending on the chosen measure. Let us consider the following situation. A message *M* comes, telling that something is completely incorrect. Thus, it will be negative information with respect to a semantic measure of information (cf. Chapter 4). At the same time, if all letters in the message *M* were transmitted correctly, it will contain genuine information with respect to the (statistical) Shannon's measure of information (cf. Chapter 3).

It is interesting that there is no direct correlation between false information and meaningless information. Bloch in his book "Apology of History" (1961) gives examples when false information was meaningful for people, while genuine information was meaningless for them.

There is a difference between quasiinformation and pseudoinformation. Quasiinformation does not change a chosen infological system at all and thus, any its measure is equal to 0. At the same time, for pseudoinformation, it is sufficient to have only one measure equal to 0. Consequently, any quasiinformation is pseudoinformation with respect to the same infological system.

Moreover, pseudoinformation with respect to one measure can be true information with respect to another measure and false information with respect to the third measure. For instance, let us consider some statement X made by a person A. It can be true with respect to what A thinks. Thus, information in X is genuine with respect to what A thinks (i.e., according to the measure m_2 that estimates correlation between the statement X and beliefs of A). The statement X can be false with respect to the real situation. Thus, information in X is false with respect to the real situation (i.e., according to the measure m_3 that estimates correlation between the statement X and reality). At the same time, information in X can be pseudoinformation from the point of view of the person D who does not understand it (i.e., according to the measure m_1 that estimates correlation between the statement X and knowledge of D).

Definition 2.4.25. A portion of information I is called *objectively false information* or *misinformation* (*misleading information*) about D if $tr(I, D) < 0$.

Remark 2.4.3. It is possible to use not only zero, but some other number $a \neq 0$ as a threshold that separates correct and false information.

Considering different types of information that are produced and utilized by people, it is promising to take into account intentions. This separates disinformation from other types of information.

Definition 2.4.26. A portion of information I is called *disinformation* about D if it is intended misinformation about D.

Definition 2.4.27. A portion of information I is called *objective disinformation* if it is intended misinformation relative to the measure $tr(I, D)$ where D is some object domain.

There are different levels of disinformation: personal, group, and social.

Definition 2.4.28. A portion of information I is called *personal* (*group, social*) *disinformation* if it is intended misinformation relative to the measure $cor(I, T_A)$ ($cor(I, T_G)$ or $cor(I, T_S)$, correspondingly) where A is some person (G is some group and S is some society, correspondingly) and T_A is the system of knowledge of A (G or S, correspondingly).

This analysis shows that not all kinds of misinformation are disinformation. In addition, it gives us a criterion to discern these kinds of information. In some cases, such distinction is very important, for example, in scientific research or in politics.

Definition 2.4.29. A portion of information I is called *objectively neutral information* about D if $tr(I, D) = 0$.

For instance, information that is not related to the domain D is objectively neutral information about D. The same is true for information that does not change knowledge about D.

Let us take a real number k. Usually, it belongs to the interval $[0, 1]$.

Definition 2.4.30. a) A portion of information I is called *true to the degree k for T_A information* about D if $tr(I, T_A) > k$.

b) A portion of information I is called *false* or *misleading to the degree k for T_A information* about D if $tr(I, T_A) < k$.

Informally it means that a portion of information I is true to the degree k for T_A the system if it allows one to achieve sufficiently high, i.e., higher than k, level of truthfulness in T_A. Note that if the level of truthfulness in T_A is already higher than k, and I does not change this level, then I is true to the degree k for T_A.

In a similar way, a portion of information I is false to the degree k for T_A the system if it does not allow one to achieve sufficiently high, i.e., higher than or equal to k, level of truthfulness in T_A. Note that if the level of truthfulness in T_A is less than k, and I does not change this level, then I is false to the degree k for T_A.

The concept of relative truthfulness strongly depends on the knowledge system T_A. For instance, let us assume, for simplicity, that knowledge can be either true or false without any intermediate values of truthfulness, i.e., the function $tr(I, T_A)$ takes only two values 0 or 1. Then for a person A who knows that the length of the diagonal of the square with the side a is equal to $a\sqrt{2}$, information I that the length of the diagonal d of the square with the side 1 cm is equal to $\sqrt{2}$ cm is true to the degree 1 information about d. Indeed, A can deduce from his knowledge that the length of d is equal to $\sqrt{2}$ cm.

At the same time, if a person X believes that the length of the diagonal of the square is equal to the length of its side, the same information I is false to the degree 1 information about d because X will deduce in his system of knowledge that the length of d is equal to 1 cm.

The degree of truthfulness allows one to measure, for example, relative truthfulness of information in a message. Let us consider a message m that tells us about n objects, e.g., n events or n people, and information about 1/3 of these objects is true, while all other information in m is false. Then we may assume that information in m is true to the degree 1/3 and false to the degree 2/3. It is possible to make these estimates of truthfulness more exact by giving weights to the objects under consideration. For instance, a message p that tells us about two objects, U and K, one object U has the weight (e.g., importance) 9, while another objects K has the weight (e.g., importance) 1. Then if the message p tells us truth about K and lies about U, this message is true only to the degree 0.1 and false to the degree 0.9.

Another characteristic of information is *correctness*.

Correctness of knowledge is defined with respect to some rules or conditions. Let us consider a system of conditions Q.

Definition 2.4.31. Knowledge K is correct with respect to Q if K satisfies all conditions from Q.

Example 2.4.5. Let us consider such procedural/operational knowledge as programs (Burgin and Kuznetsov, 1994). This knowledge informs computer what to do or how to perform computations. Correctness becomes a critical issue in software and hardware production and utilization as a result of the increased social and individual reliance upon computers and computer networks. A study of the National Institute of Standards found that only software defects resulted in \$59.5 billion annual losses (Tassey, 2002; Thibodeau, 2002). It is possible to find a detailed analysis of the concept *software correctness* and the most comprehensive development of this concept in (Burgin and Debnath, 2006; 2007). There are different forms of software correctness, such as functional, descriptive, procedural, temporal, and resource correctness.

Example 2.4.6. Let us consider such propositional knowledge, i.e., knowledge expressed by propositions in some logical language. This is, may be, the most popular form of knowledge representation (cf. (Bar-Hillel and Carnap, 1952; 1958; Halpern and Moses, 1985)). Then a knowledge system is usually represented by a propositional calculus. Traditionally it is assumed that such knowledge is correct if the calculus is consistent.

Applying such an information measure as truthfulness, we come to three basic types of information:

- *correct information,*
- *incorrect information,*
- *unverified information.*

Let us take a number k from the interval [0, 1].

Definition 2.4.32. a) A portion of information I is called *true* or *genuine information* about D with respect to a measure *cor* if $cor(I, T_A) > 0$.

b) A portion of information I is called *true* or *genuine to the degree k information* about D with respect to a measure *cor* if $cor(I, T_A) > k$.

This definition looks natural and adequately works in many situations. However, there are some problems with it. Imagine that information that gives correct knowledge about some domain (object) D comes to A but it does not change the knowledge system T_A because correct knowledge about D already exists in T_A. In this case, $cor(I, T_A) = cor(I(T_B), T_A) - cor(T_B, T_A) = 0$.

This implies that it is necessary to distinguish relative, i.e., relative to a knowledge system T_A, information correctness and absolute correctness. To define absolute correctness, we take a knowledge system T_{0D} that has no *a priori* knowledge about the domain (object) D.

Definition 2.4.33. A portion of information I is called *purely true information* about D with respect to a measure *cor* if $cor(I, T_{0D}) > 0$.

It is necessary to understand that it is not a simple task to find such a knowledge system T_{0D} that has no *a priori* knowledge about the domain (object) D. Besides, truthfulness depends on other properties of the knowledge system T_{0D}, e.g., on algorithms that are used for conversion of received information into knowledge. It is possible to have true information that gives a wrong impression resulting in acquisition of incorrect knowledge. For instance, during the Cold War, witty people told the following joke.

Russia challenged the United States to a foot race. Each country sent their fastest athlete to a neutral track for the race. The American athlete won. The next day, the major Soviet newspaper "Pravda", which means *truth* in Russian, published an article with the following title:

Russia takes second in international track event, while the United States comes in next to last.

We see that what is said in the article of "Pravda" is true, but people's *a priori* knowledge makes them to assume a big race with many participants and in such a way, they have a wrong impression and false knowledge if they did not know how many participants was in the race.

It is possible to treat truthfulness and correctness as linguistic variables in the sense of (Zadeh, 1973). For instance, we can separate such classes as highly correct/true information, sufficiently correct/true information, weakly correct/true information, weakly false information, and highly false information.

It is possible to consider other characteristics and features of information, such as usefulness, clarity, compliance, and value (cf. Section 6.2). For instance, Mazur (1970) introduces and studies a diversity of similar information types, such as *simulating information, disimulating information, confusing information, transinformation, comparative transinformation, compensation transinformation, exclusive transinformation, pseudoinformation, simulating pseudoinformation, disimulating pseudoinformation, useful information,* and *parasitic information.*

Different authors explicated and discussed various attributes of information. For instance, Cleveland (1982) considers the following properties of information:

1. Information, as a rule, is *expandable*, meaning that information expands as it is used and there are no obvious limits to this expansion.
2. Information often is *compressible*, meaning that information can be concentrated, integrated, summarized for easier handling.

However, one of the discoveries of the algorithmic information theory (cf. Chapter 5) was that there are symbolic sequences information in which cannot be compressed. They are called random sequences.

3. Information is *substitutable*, meaning that information can replace, in some sense, capital, labor, or physical materials. More exactly, information allows one to use less money, labor, or physical materials than it is possible to do without this information.

However, if somebody has enough information for building a car but has neither necessary parts nor instruments nor money to buy them, this person would not be able to really build a car.

4. Information is *transportable*, meaning that it is possible to transmit information at the speed of light, i.e., faster than any material object. Moreover, quantum mechanics tells us that information can travel even faster than light. It happens, for example in the phenomenon called quantum entanglement, which describes effectively direct or causal relationships between apparently distinct or spatially separated particles.

This property of information shows that in spite of many claims about physical nature of information, pure information is not physical.

5. Information is *diffusive*, meaning that information tends to leak (the more it leaks the more we have) and is aggressive in striving to break out of the unnatural bonds of secrecy.

This property of information is a part (component) of information expandibility (property 1).

6. Information is *shareable*, meaning that information does not decrease when it is given to others.

Property 6 is often used to show that information differs from material resources to a great extent. However, this implication is based on an illusion due to confusion between information, information carrier and measure of this information. Often experts say that if a person *A* shares information *I* with a person *C*, then both have the same information. However, it only means that both individuals have the same data or knowledge, i.e., the same carriers of information, but the measure of information can decrease. For instance, taking an information measure, such as the value of information, we see that the value of information about some invention decreases when the inventor does not have a patent, while another person or company gets this information. At the same time, if we measure information, or more exactly, its symbolic representation, in bits (or bytes), we observe that the quantity of bites has become twice as big as *A* had before sharing information with *B*. The reason is that they have two similar texts instead of one. The same happens when you rewrite a file from one disc to another. It means that some measures can increase, while others decrease. Thus, information does not stay the same in the process of sharing.

Braman (1989) distinguishes four approaches to information:

1) *information as a resource*, coming in pieces unrelated to knowledge or information flows into which it might be organized;

2) *information as a commodity* is obtained using information production chains, which create and add economic value to information;

3) *information as a perception* of patterns has past and future, is affected by motive and other environmental factors, and itself has its effects;

4) *information as a constitutive force* in society, essentially affecting its evironment.

It is important to understand that these approaches do not give definitions of information but only identify information roles in society.

Banathy (1995) considers three important types of information. With respect to a system *R*, it is possible to consider referential, non-referential, and state-referential information.

1. *Referential information* has meaning in the system *R*.
2. *Non-referential information* has meaning outside the system *R*, e.g., information that reflects mere observation of *R*.
3. *State-referential information* reflects an external model of the system *R*, e.g., information that represents *R* as a state transition system.

Joshi and Bayrak (2004) assume that the following properties are inherent to the nature of information and do apply to all types of information.

1. *Temporality*, meaning that there are temporal limits for information validity or information includes information about temporal characteristics of its object domain.

2. *Semantics* or *meaning* of information is always a language issue.

However, as the general theory of information shows, not all types of information have both properties. For instance, when information is represented by a logical proposition in the classical logic or by a mathematical theorem about numbers or algebraic groups, it has no temporal properties at all. Besides, information, as we know, can be conveyed by natural signs that do not belong to any language. For instance, smoke gives information about fire.

Goguen (1997) considers the following properties of information:

1. Information is *situated*, meaning that information can only be fully understood in relation to the particular, concrete situation in which it actually occurs.

2. Information is *local*, meaning that interpretations of information are constructed in some particular context, including a particular time, place and group.

3. Information is *emergent*, meaning that information cannot be understood at the level of the individual, that is, at the cognitive level of individual psychology, because it arises through ongoing interactions among members of a group. In terms of the general theory of information, this means that information reception is a process that depends on the interaction of the accepting system and the carrier of information.

4. Information is *contingent*, meaning that the interpretation of information depends on the current situation, which may include the current interpretation of prior events. In particular, interpretations are subject to negotiation, and relevant rules are interpreted locally, and can even be modified locally.

5. Information is *embodied*, meaning that information is tied to bodies in particular physical situations, so that the particular way that bodies are embedded in a situation may be essential to some interpretations.

6. Information is *vague*, meaning that in practice, information is only elaborated to the degree that it is useful to do so; the rest is left grounded in tacit knowledge.

7. Information is *open*, meaning that information (for both participants of communication and analysts) cannot in general be given a final and complete form, but must remain open to revision in the light of further analyses and further events.

Analyzing these properties from the perspective of the general theory of information, we see that in general, information has such properties. At the same time, the general theory of information allows us to make these properties more precise and to deduce them from the principles of this theory.

Information is *situated*, meaning that information can only be fully understood in relation to its context.

Information is *local*, meaning that according to the Ontological Principle O1, empirical level relates information to a definite system.

Information is *emergent*, meaning that information reveals in the process of interaction with the information carrier and subsequent processing.

Information is *contingent*, meaning that information extraction depends on many parameters, which are changing and change extracted information.

Information is *embodied*, meaning that information has physical carriers.

Information is *vague*, meaning that in practice, information extraction is not predetermined and can vary from situation to situation.

Information is *open*, meaning that extraction of information contained in the carrier is an ongoing process.

These properties of information have important consequences. One of such consequences is that openness of information implies that traditional models of information processing systems, such as Turing machines or other recursive or subrecursive algorithms, are very limited because they imply that after obtaining a result the system stops functioning (cf., for example, (Rogers, 1987; Sipser, 1997; Hopcroft, et al, 2001)). In contrast to this, new models, such as inductive and limit Turing machines, and some other superrecursive algorithms (cf., for example, (Burgin, 2005)), information produced does not have a final and complete form, but remains open to revision in the light of further analyses and further events.

Meadow and Yuan (1997) also give a list of various attributes of information.

1. *Commodification*: When information is considered as a commodity, its price or cost is an important factor.

2. *Value*: Information price is not the same as information value because the latter may vary considerably from person to person or organization to organization. However, information value may depend on information price.

3. *Reliability of content*, which includes *accuracy*, *veracity*, *credibility*, *correctness*, and *validity*. All these properties are components of information quality considered in Section 6.2.

4. *Reliability of source*: The attributes of content can be applied to the author or corporate source of information in the same way as to its content. This may actually be a rating of the previous content reliability of information from this source, or of the circumstances under which a particular message originated. This property is also a factor of information value and a component of information quality considered in Section 6.2.

5. *Time*: There are different types of time related to information. It can be the date/time of creation of the message as an information carrier (representation) or the elapsed time since its creation. It can be time when the message was received. Also, the time range of validity of the content, e.g., is it true at moment of transmission, for all time, etc.

6. *Generality* shows whether the information content is applicable to a broad subject or to highly specific one.

7. *Novelty* shows the extent to which the content is original, or its relation to other works, e.g., whether it repeats, duplicates, adds to or contradicts the previous work.

8. *Subject domain* reflects what this information is about. It is expressed as subject matter or it names or implies persons, places, institutions, devices, etc.

9. *Specificity or depth* refers to the depth of coverage or the degree of detail of the information in a message.

10. *Clarity* and *comprehensibility* show easiness of understanding and often vary with the individual user of information, e.g., the reader.

11. *Amount* of information has many different meanings and measures. One measure is the number of characters, pages or other physical characteristics of a text. Another measure of information amount is the Hartley-Shannon entropy (cf. Chapter 3). One more

estimate for information amount is the recipient's sense of number of new facts learned although there is not yet a formal measure for this.

12. *Instructional value* shows the extent to which information instructs about its subject rather than simply reports on it. Such a value would also vary considerably from person to person or group to group.

Thus, we see an extremely active research in the information studies domain, which has allowed researchers to find many properties of information and information processes.

2.5. Information, Data, and Knowledge

> *Where is the Life we have lost in living?*
> *Where is the wisdom we have lost in knowledge?*
> *Where is the knowledge we have lost in information?*
> T.S. Eliot

> *We're drowning in information and starving for knowledge.*
> Rutherford D. Rogers

As a whole, the triad *Data — Information — Knowledge* attracted attention of researchers only in the 20th century. In contrast to this, the best minds in society were concerned with the problem of knowledge from ancient times. Concepts of information and data in contemporary understanding were introduced much later.

In Upanishad (written from the end of the second millennium B.C.E. to the middle of the second millennium C.E.), which is one of the principal classical texts in Indian culture, two kinds of knowledge, higher knowledge and lower knowledge, were discerned. Later in Nyaya school of Hindu philosophy, four types of knowledge acquisition were considered: *perception* when senses make contact with an object, *inference*, *analogy*, and *verbal testimony* of reliable persons. Inference was used in three forms: *a priory* inference, *a posteriory* inference, and inference by common sense.

In philosophy of the great Chinese philosopher Confucius (551–479 B.C.E.), knowledge was thoroughly considered. Confucius discerned two kinds of knowledge: one was *innate*, while the other came from learning.

According to him, knowledge consisted of two components: *knowledge of facts* (statics) and *skills of reasoning* (dynamics). Contemporary methodology of science classifies the first type as a part of the logic-linguistic subsystem and the second type as a part of the procedural subsystem of a developed knowledge system (Burgin and Kuznetsov, 1994).

For Confucius, to know was to know people. He was not interested in knowledge about nature, studied by modern science. As philosophy of Confucius had the main impact on Chinese society for many centuries, his approach resulted in a much lower level of Chinese science and technology in comparison with Europe and America. It was so for a long time, but now China is catching up.

However, the most profound analysis of the problem of knowledge was done by Greek philosophers. As many of his works, e.g., *Meno*, *Phaedo*, *Phaedo*, *Symposium*, *Republic*, and *Timaeus*, show, Plato (427–347 B.C.E.) treated knowledge as a *justified true belief* (Cornford, 2003). However, in one of the Plato's dialogues, *Theœtetus*, Sokrates and Theaetetus discuss the nature of knowledge and challenge this approach. They consider three definitions of knowledge: *knowledge as* nothing but *perception*, *knowledge as true judgment*, and, finally, *knowledge as a true judgment with justification*. Analyzing these definitions, Sokrates demonstrates that each of these definitions is unsatisfactory. Bertrand Russel (1872–1970), Edmund Gettier in (Gettier, 1963), Elliot Sober in (Sober, 1991) and some other thinkers gave examples demonstrating that the definition of knowledge as a justified true belief is not adequate.

Aristotle (384–322 B.C.E.) was also interested in problems of knowledge, categorizing knowledge based on knowledge objects (domains) and the relative certainty with which one could know those objects. He assumed that certain domains (such as in mathematics or logic) permit one to have a knowledge that is true all the time. However, his examples of absolute knowledge, such as *two plus two is always equal to four* or *all swans are white*, failed when new discoveries were made. For instance, the statement *two plus two always equals four* was disproved when non-Diophantine arithmetics were discovered (Burgin, 1977). The statement *"all swans are white"* was disproved when Europeans came to Australia and found black swans.

According to Aristotle, absolute knowledge is characterized by certainty and precise explanations. Other domains, such as human behavior, do not permit precise knowledge. The corresponding vague knowledge is characterized by probability and imprecise explanations. Knowledge that would fall into this category would include ethics, psychology, or politics. Unlike Plato and Socrates, Aristotle did not demand certainty in everything. One cannot expect the same level of certainty in politics or ethics that one can demand in geometry or logic. In his work *Ethics*, Aristotle defines the difference between knowledge in different areas in the following way:

"we must be satisfied to indicate the truth with a rough and general sketch: when the subject and the basis of a discussion consist of matters which hold good only as a general rule, but not always, the conclusions reached must be of the same order. . . . For a well-schooled man is one who searches for that degree of precision in each kind of study which the nature of the subject at hand admits: it is obviously just as foolish to accept arguments of probability from a mathematician as to demand strict demonstrations from an orator."

Aristotle was also interested in how people got knowledge. He identified three sources of knowledge: *sensation* as the passive capacity for the soul to be changed through the contact of the associated body with external objects, *thought* as the more active process of engaging in the manipulation of forms without any contact with external objects at all, and *desire* as the origin of movement toward some goal.

Many other philosophers, such as René Descartes (1596–1650), George Berkeley (1685–1753), David Hume (1711–1776), Immanuel Kant (1724–1804), Georg Wilhelm Friedrich Hegel (1770–1831), and Ludwig Josef Johann Wittgenstein (1889–1951) studied problems of knowledge. For instance, Descartes defined knowledge in terms of doubt and distinguishes rigorous knowledge (*scientia*) and lesser grades of conviction (*persuasio*). He assumed that there was conviction when there remains some reason which might lead us to doubt, but knowledge was conviction based on a reason so strong that it could never be shaken by any stronger reason.

Knowledge has been always important in society. However, now importance of knowledge grows very fast as society becomes more and more advanced. Thus, in the 20th century, with the advent of computers, knowledge has become a concern of science. As a result, now knowledge is studied in such areas as artificial intelligence (AI), computer science, data- and knowledge bases, global networks (e.g., Internet), information science, knowledge engineering, and knowledge management. Philosophers also continue their studies of knowledge (Chisholm, 1989).

There is a great deal of different books and papers treating various problems and studying different issues of knowledge (cf., for example, (Pollock and Cruz, 1999)). A lot of ideas, models and several theories have been suggested in this area. All research of knowledge can be separated into the following directions:

- *Structural analysis* of knowledge that strives to understand how knowledge is built;
- *Axiological analysis* of knowledge that aims at explanation of those features that are primary for knowledge;
- *Functional analysis* of knowledge that tries· to find how knowledge functions, is produced, and acquired.

Structural analysis of knowledge is the main tool for the system theory of knowledge, knowledge bases, and artificial intelligence.

Axiological analysis of knowledge is the core instrument for philosophy of knowledge, psychology, and social sciences.

Functional analysis of knowledge is the key device for epistemology, knowledge engineering, and cognitology.

Information attracted interests of researchers much later than knowledge. Only in 20s of the 20th century, the first publications appeared that later formed the core of two principle directions of information theory. In 1922, Fisher introduced an important information measure now called Fisher information and used in a variety of areas — from physics to information geometry and from biology to sociology. Two other papers (Nyquist, 1924) and (Hartley, 1928), which started the development of the most popular direction in information theory, appeared a little bit later in the communication theory.

Interest to the third component, data, from the triad *Data — Information — Knowledge* emerged in the research community only after computers became the main information processing devices. As a result, researchers started to study the whole triad only in 80s of the 20th century. It is possible to approach relations between data, information and knowledge from many directions. Now the most popular approach to the triad *Data — Information — Knowledge* represents hierarchical relations between them in a form of a pyramid with data at the bottom,

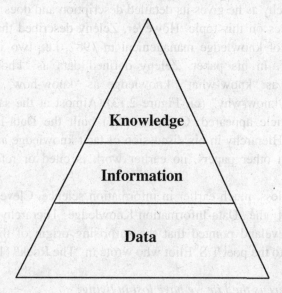

Figure 2.12. The Data-Information-Knowledge Pyramid

knowledge at the top and information in the middle (cf., for example, (Landauer, 1998; Boisot and Canals, 2004; Sharma, 2005; Rowley, 2007)).

In knowledge management literature, this hierarchy is mostly referred to as the *Knowledge Hierarchy* or the *Knowledge Pyramid*, while in the information science domain, the same hierarchy is called the *Information Hierarchy* or the *Information Pyramid* for obvious reasons. Often the choice between using the label "Information" or "Knowledge" is based on what the particular profession believes to fit best into the scope of this profession. A more exact name for this pyramid is the *Data-Information-Knowledge-Wisdom Hierarchy* (Sharma, 2005). Here we use the term *Data-Information-Knowledge Hierarchy* or *Data-Information-Knowledge Pyramid*.

As Sharma writes (2005), there are two separate threads that lead to the origin of the Data-Information-Knowledge Hierarchy. In knowledge management, Ackoff is often cited as the originator of the Data-Information-Knowledge Hierarchy. His 1988 Presidential Address to ISGSR (Ackoff, 1989) is considered by many to be the earliest mention of the hierarchy as he gives its detailed description and does not cite any earlier sources on this topic. However, Zeleny described this hierarchy in the field of knowledge management in 1987, i.e., two years earlier than Ackoff. In his paper, Zeleny defined data as "know-nothing", information as "know-what", knowledge as "know-how", and added wisdom as "know-why" (cf. Figure 2.13). Almost at the same time as Zeleny's article appeared, Cooley (1987) built the Data-Information-Knowledge Hierarchy in his discussion of tacit knowledge and common sense. As in other papers, no earlier work is cited or referred to by Cooley.

Nevertheless, much earlier in information science, Cleveland (1982) wrote about the Data-Information-Knowledge Hierarchy in detail. Besides, Cleveland pointed that the surprising origin of the hierarchy itself is due to the poet T.S. Eliot who wrote in "The Rock" (1934):

> *Where is the Life we have lost in living?*
> *Where is the wisdom we have lost in knowledge?*
> *Where is the knowledge we have lost in information?*

This is the first vague mention of the Information-Knowledge Hierarchy, which was expanded by Cleveland and others by adding the layer of "Data" or in terms of Cleveland, of "facts and ideas". Cleveland (1982) concedes that information scientists are "still struggling with the definitions of basic terms" of the hierarchy. He uses Elliot's metaphor as a starting point to explain the basic terms. Cleveland also agrees that there are many ways in which the elements of the hierarchy may be defined, yet universal agreement on them need not be a goal in itself. At the same time, many researchers assume that the difference between data, information, and knowledge has pivotal importance in our information age. For instance, Landauer (1988) writes, "the repeated failure of natural language models to succeed beyond the most superficial level is due to their not taking into account these different levels (of the Data-Information-Knowledge Pyramid) and the fundamentally different processing required within and between them."

The most popular in computer science approach to the relation between data and information implies that *information is an organized collection of facts and data*. Data are transformed into information into knowledge into wisdom. Thus, information appears as an intermediate level of similar phenomena leading from data to knowledge.

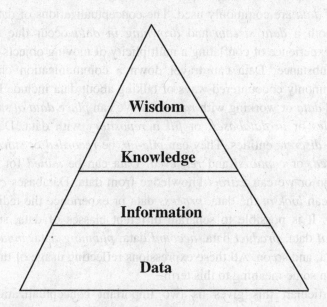

Figure 2.13. The extended Data-Information-Knowledge Pyramid

Some researchers include the one more level, "understanding", in the pyramid. As Ackoff writes (1989), descending from wisdom there are understanding, knowledge, information, and, at the bottom, data. Each of these includes the categories that fall below it. However, the majority of researchers do not include understanding in their studies of the Data-Information-Knowledge Pyramid.

The Data-Information-Knowledge Pyramid is a very simple schema. That is why it is so appealing to people. However, as we know from physics and other sciences, reality does not always comply with the simplest schemes. So, to find relevance of this pyramid, we need to understand all its levels.

Data has experienced a variety of definitions, largely depending on the context of its use. With the advent of information technology the word *data* became very popular and is used in a diversity of ways. For instance, information science defines data as *unprocessed information*, while in other domains data are treated as a representation of objective facts. In computer science, expressions such as *a data stream* and *packets of data* are commonly used. The conceptualizations of data as a flow in both a *data stream* and *drowning in data* occur due to our common experience of conflating a multiplicity of moving objects with a flowing substance. Data can travel down a communication channel. Other commonly encountered ways of talking about data include having *sources of data* or working with *raw data*. We can *place data in storage*, e.g., *in files* or *in databases*, or *fill a repository* with data. Data are viewed as *discrete* entities. They can *pile-up,* be *recorded* or *stored* and *manipulated*, or *captured* and *retrieved*. Data can be *mined* for useful information or we can *extract* knowledge from data. Databases *contain* data. We can *look at* the data, *process* data or experience the tedium of data-*entry*. It is possible to separate different classes of data, such as *operational* data, *product* data, *account* data, *planning* data, *input* data, *output* data, and so on. All these expressions reflecting usage of the term *data* assign some meaning to this term.

In particular, this gives us two important conceptualizations of data: *data* are a *resource*, and *data* are *manipulable objects*. In turn, this may implicitly imply that data are solid, physical, things with an objective existence that allow manipulation and transformation, such as rearrangement of data, conversion to a different form or sending data from one system to another.

Ackoff defines *data* as symbols that represent properties of objects, events and their environments. Data are products of observation by people or automatic instrument systems.

According to Hu and Feng (2006), *data* is a set of values recorded in an information system, which are collected from the real world, generated from some pre-defined procedures, indicating the nature of stored values, or regarding usage of stored values themselves.

Data are also considered as discernible differences between states of some systems in the world (Lloyd, 2000, 2002), i.e., *light* versus *dark*, *present* versus *absent*, *hot* versus *cold*, 1 versus 0, + versus −, etc. In many cases, binary digits, or bits, represent such differences and thus, carry information about these differences. However, according to Boisot (2002), information itself is a relation between these discernible states and an observer. A given state may be informative for someone. Information is then what an observer will extract from data as a function of his/her expectations or prior knowledge (Boisot, 1998).

Many understand data as discrete atomistic tiny packets with no inherent structure or necessary relationship between them. However, as we show in this section, this is not true. Besides, there are different kinds and types of data. In addition, data as an abundant resource can pile-up to such an extent that many people find themselves drowning in data. As a substance, data can be measured. The most popular way to measure data is in bits where bit is derived from the term a *binary digit*, 0 or 1. The reason for this is very simple. Namely, the vast majority of information processing and storage systems, such as computers, calculators, embedded devices, CD, DVD, flash memory storage devices, magneto-optical drives, and other electronic storage devices, represent data in the form of sequences of binary digits.

The next level in the Data-Information-Knowledge Pyramid is information. However, as definitions of information are considered in Sections 1.1 and 2.2, let us look how knowledge is defined and treated in contemporary studies. Logical analysis shows that knowledge is difficult to define. Taking knowledge as the third and upper component of the Data-Information-Knowledge Pyramid, we see that in spite of the long history of knowledge studies, there is no consensus on what knowledge is. As we have seen, over the millennia, the philosophers of each age and epoch have added their own ideas on the essence and nature of knowledge to the list. Science has extended this list as well. As a result, there is a lot of confusion in this area. As Land, et al (2007) write, knowledge itself is understood to be a slippery concept, which has many definitions.

In any case, our civilization is based on knowledge and information processing. That is why it is so important to know what knowledge is. For instance, the principal problem for computer science as well as for computer technology is to process not only data but also knowledge. Knowledge processing and management make problem solving much more efficient and are crucial (if not vital) for big companies and institutions (Ueno, 1987; Osuga, 1989; Dalkir, 2005). To achieve this goal, it is necessary to distinct knowledge and knowledge representation, to know regularities of knowledge structure, functioning and representation, and to develop software (and in some cases, hardware) that is based on theses regularities. Many intelligent systems search concept spaces that are explicitly or implicitly predefined by the choice of knowledge representation. In effect, the knowledge representation serves as a strong bias.

At the same time, there are essentially diverse descriptions of knowledge. For instance, in the Webster's Revised Unabridged Dictionary (1998), knowledge is defined as:

1. A dynamic process:
    (a) the act or state of knowing;
    (b) clear perception of fact, truth, or duty;
    (c) certain apprehension;
    (d) familiar cognizance;
    (e) learning;
    (f) cognition.

2. An object:
    (a) that which is or may be known;
    (b) the object of an act of knowing;
    (c) a cognition.
3. An object:
    (a) that which is gained and preserved by knowing;
    (b) instruction;
    (c) acquaintance;
    (d) enlightenment;
    (e) scholarship;
    (f) erudition.

4. A property:
  (a) that familiarity which is gained by actual experience;
  (b) practical skill; as, a *knowledge* of life.
5. As a domain:
  (a) scope of information;
  (b) cognizance;
  (c) notice; as, it has not come to my *knowledge*.

In the Oxford English Dictionary, knowledge is defined as:

(i)   expertise, and skills acquired by a person through experience or 'education; the theoretical or practical understanding of a subject,

(ii)   what is known in a particular field or in total; facts and information,

(iii)   awareness or familiarity gained by experience of a fact or situation.

In monographs on knowledge engineering (Osuga, S. et al, 1990), we find the following definitions:

1.   *Knowledge* is a result of cognition.
2.   *Knowledge* is a formalized information, to which references are made or which is utilized in logical inference.

This gives some general ideas about knowledge but is not constructive enough even to distinguish knowledge from knowledge representation and from information. The following example demonstrates differences between knowledge and knowledge representation. Some event may be described in several articles written in different languages, for example, in English, Spanish, and Chinese, but by the same author. These articles convey the same semantic information and contain the same knowledge about the event. However, representation of this knowledge is different.

Nonaka and Takeuchi (1995) assume that knowledge is always about action — the knowledge must be used to some end. In contrast to this, information is conceived of as a process, whereas knowledge is perceived as a state (Machlup, 1983). According to Sholle (1999), *information* and *knowledge* are distinguished along three axes: multiplicity, temporality, and spatiality.

If we take formal definitions of knowledge, we see that they determine only some specific knowledge representation. For instance, in logic knowledge is represented by logical propositions and predicates. On the one hand, informal definitions of knowledge provide little opportunities for computer processing of knowledge because computers can process only formalized information. On the other hand, there is a great variety of formalized knowledge representation schemes and techniques: semantic and functional networks, frames, productions, formal scenarios, relational and logical structures. However, without explicit knowledge about knowledge structures *per se*, these means of representation are used inefficiently.

Now let us look in more detail where and how information fits into a unified picture with knowledge and data. The conventional systemic approach demands not to consider components from the Data Information-Knowledge Pyramid but to define these concepts in a related way:

Information is structuring of data (or structured data)

Knowledge is structuring of information (or structured information)

However, another approach (cf., for example, (Meadow and Yuan, 1997)) suggests a different picture:

Data usually means a set of symbols with little or no meaning to a recipient.

Information is a set of symbols that does have meaning or significance to their recipient.

Knowledge is the accumulation and integration of information received and processed by a recipient.

From the perspective of knowledge management, information is used to designate isolated pieces of meaningful data. These data integrated within a context constitute knowledge (Gundry 2001; Probst, Raub, and Romhard, 1999).

As a result, it is often assumed that data themselves are of no value until they are transformed into a relevant form. This implies that the difference between data and information is functional, not structural.

Stenmark (2002) collected definitions of the components from the Data-Information-Knowledge Pyramid from seven sources.

Wiig (1993) considers information as facts organized to describe a situation or condition, while knowledge consists of truths and beliefs, perspectives and concepts, judgments and expectations, methodologies and *"know how"*.

Nonaka and Takeuchi (1995) consider information as a flow of meaningful messages, while knowledge consists of commitments and beliefs created from these messages.

According to Spek and Spijkervet (1997) data are not yet interpreted symbols, information consists of data with meaning, while knowledge is the ability to assign meaning.

According to Davenport (1997) data are simple observations, information consists of data with relevance and purpose, while knowledge is valuable information from the human mind.

According to Davenport and Prusak (1998) data are discrete facts, information consists of messages meant to change the receiver's perception, while knowledge is Experiences, values, insights, and contextual information.

Quigley and Debons (1999) treat data as a text that does not answer questions to a particular problem. They look upon information as a text that answers the questions who, when, what or where. In addition, they deal with knowledge as a text that answers the questions why and how.

Choo, et al (2000) understand data as facts and messages, assume that information is data vested with meaning, and perceive knowledge as justified, true beliefs.

Dalkir (2005) treats data as "content that is directly observable or verifiable", information as "content that represents analyzed data" or as "analyzed data — facts that have been organized in order to impart meaning", while knowledge is defined in his book as subjective and valuable information.

One more view of the difference between data and information is that data is *potential information* (Meadow, 1996). A message as a set of data may potentially be information but the potential is not always realized.

This distinction is similar to the distinction between potential energy and kinetic energy. In mechanics, kinetic energy is that associated with movement. Potential energy is associated with the position of a thing, e.g., a weight raised to a height. If the weight falls, its energy becomes kinetic. Impact is what happens if the weight, with its kinetic energy, hits something.

Meadow and Yuan suggest (1997) that in the information world, impact is what happens after a recipient receives and in some manner acts upon information. This perfectly correlates with Ontological Principles O1 and O2.

The most popular in computer science approach to information is expressed by Rochester (1996) who defines information as an *organized collection of facts and data*. Rochester develops this definition through building a hierarchy in which data are transformed into information into knowledge into wisdom. Thus, information appears as an intermediate level of similar phenomena leading from data to knowledge.

An interesting approach to understanding data, information and knowledge in a unified context of semiotics is developed by Lenski (2004). He suggests that d*ata* denote the syntactical dimension of a sign, *knowledge* denotes the semantic dimension of a sign, and *information* denotes the pragmatic dimension of a sign.

Data, in the sense of Lenski, are a sytem that is organized by structural and grammatical rules of sign. In contrast to this, knowledge results from neglecting the amount of individual contribution to the semantic abstraction process. However, such an abstraction process may be only subjectively acknowledged resulting in personal knowledge. Comprising pragmatical dimension, information is bound to a (cognitive) system that processes the possible contributions provided by signs that constitute data for a possible action. Moreover, information inherits the same interpretation relation as knowledge with the difference that the latter is abstracted from any reference to the actual performance whereas information, in contrast, emphasizes reaction and performance. As a result, knowledge and information are closely tied together.

To elaborate his own definition of information, Lenski (2004) uses seven principles as a system of prerequisites for any subsequent theory of information that claims to capture the essentials of a publication-related concept of information. These principles were introduced by other researchers as determining features of information.

Principle 1. According to Bateson (1980), information is a difference that makes a difference.

Principle 2. According to Losee (1997), information is the values of characteristics in the processes' output.

Principle 3. According to Belkin and Robertson (1976), information is that which is capable of transforming structure.

Principle 4. According to Brookes (1977), information is that which modifies ... a knowledge structure.

Principle 5. According to Brookes (1980), knowledge is a linked structure of concepts.

Principle 6. According to Brookes (1980), information is a small part of such a structure.

Principle 7. According to Mason (1978), information can be viewed as a collection of symbols.

After formulating these principles, Lenski gives his own interpretation, explicating their relation to the "fundamental equation" (2.5.1) of Brookes (1980).

$$K(S) + \Delta I = K(S + \Delta S) \qquad (2.5.1)$$

This equation reflects the situation when a portion of information ΔI acts on a knowledge structure $K(S)$, transforming it into the structure $K(S + \Delta S)$ with ΔS as the effect of this change.

In this context, principles of Lenski acquire the following meaning.

Principle 1 reflects the overall characterization of information along with its functional behavior and refers to the Δ-operator in the "fundamental equation".

Principle 2 expresses a process involved with results ΔI that are constituents of information.

Principle 3 specifies the concept of *difference* as a transformation process resulting in $K[S + \Delta S]$.

Principle 4 shows on what information acts, namely, on the knowledge (or structure) system $K[S]$

Principle 5 explains what is knowledge (or knowledge structure).

Principle 6 relates information to knowledge structures.

Principle 7 specifies carriers of information.

To achieve his goal, Lenski (2004) formulates one more principle.

Principle 8. The emergence of information is problem-driven.

Based on these principles, Lenski (2004) presents a working definition for a publication-related concept of *information*.

Definition 2.4.10. *Information* is the result of a problem-driven differentiation process in a structured knowledge base.

Thus, very often, it is assumed that being different, knowledge and information nevertheless have the same nature. For instance, the sociologist Merton writes (1968), that knowledge implies a body of facts or ideas, whereas information carries no such implication of systematically connected facts or ideas.

In many books and papers, the terms *knowledge* and *information* are used interchangeably, even though the two entities, being intertwined and interrelated concepts, are far from identical. Moreover, some researchers define information in terms of data and knowledge. At the same time, other researchers define knowledge and/or data in terms of information. For instance, Kogut and Zander (1992) conceive information as "knowledge which can be transmitted without loss of integrity", while Meadow and Yuan write (1997) that knowledge is the accumulation and integration of information. MacKay (1969) also assumes that knowledge itself turns out to be a special kind of information. In a similar way, Davenport (1997) treats information as data with relevance and purpose, while Tuomi (1999) argues that data emerge as a result of adding value to information.

In a similar way, commonplace usage of words *data* and *information* blurs differences between these concepts. For instance, Machlup and Mansfield write (1980):

"Data are the things given to the analyst, investigator, or problem-solver; they may be numbers, words, sentences, records, assumptions — just anything given, no matter in what form and of what origin... Many writers prefer to see data themselves as a type of information, while others want information to be a type of data."

All these and many other inconsistencies related to the Data-Information-Knowledge Pyramid cause grounded critic of this approach to understanding data, knowledge, and information. For instance,

Capurro and Hjorland write (2003) that the semantic concept of information, located between data and knowledge is not consistent with the view that equates information management with information technology. Boisot and Canals (2004) criticize distinctions that have been drawn between data, information, and knowledge by those who analyzed the Data-Information-Knowledge Pyramid. Fricke (2008) also offers valid arguments that the hierarchy is unsound and methodologically undesirable.

In addition, researchers criticized various implications of the Data-Information-Knowledge Pyramid. For instance, Tuomi (1999) argues that in contrast to the conventional estimation that knowledge is more valuable than information, while information is superior to data, these relations have to be reversed. Thus, data emerge as a result of adding value to information, which in turn is knowledge that has been structured and verbalized. Moreover, there are no "raw" data as any observable, measurable and collectible fact has been affected by the very knowledge that made this fact observable, measurable and collectible. According to Tuomi, knowledge, embedded in minds of people, is a prerequisite for getting information. Some researchers treat knowledge in the mind as information, which, in turn, is explicit and appropriate for processing when it is codified into data. Since only data can effectively be processed by computers, Tuomi (1999) explains, data is from the technological perspective the most valuable of the three components of the Data-Information-Knowledge Pyramid, which consequently should be turned upside-down.

The general theory of information suggests different (in comparison with the Data-Information-Knowledge Pyramid and Lenski's approach) schema called the Knowledge-Information-Matter-Energy (KIME) Square. Ontological Principle O2a implies that information is not of the same kind as knowledge and data, which are structures (Burgin, 1997). Taking *matter* as the name for all substances as opposed to *energy* and the *vacuum*, we have the relation that is represented by the diagram in Figure 2.14.

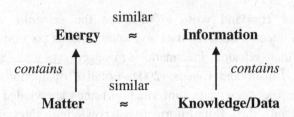

Figure 2.14. The Knowledge-Information-Matter-Energy (KIME) Square

The KIME Square visualizes and embodies the following principle:

> *Information is related to knowledge and data as*
> *energy is related to matter.*

This schema means that information has essentially different nature than knowledge and data, which are of the same kind. Knowledge and data are structures, while information is only represented and can be carried by structures. Note that the KIME Square is a particular case of the SIME Square (cf. Figure 2.11).

Existence of information in nature gives direct evidence to the essential difference between information and knowledge. For instance, stones do not have knowledge. They only contain information. People who know enough about stones, e.g., geologists, can take a stone, extract information it contains, e.g., make measurements, and build knowledge about this stone, e.g., to know its weight, form, color, stone type, hardness, reflectivity, etc.

This approach to knowledge is to a great extent similar to the theory of MacKay (1969), who asserted that knowledge must be understood as a coherent representation. Here a representation is any structure (a pattern, picture or model) whether abstract or concrete which, by virtue of its features, could symbolize some other structure. Such representation structures contain information elements. Information elements in the sense of MacKay are not shapeless isolated things, but are imbedded in a structure, which helps the receiver to infer the meaning of the information intended by the sender. MacKay illustrates this thought with the following example. A single word rarely makes sense. Usually it takes a group of words or even several sentences acting together to clarify the role of a single word within a sentence. The receiver will find it all the easier to integrate a piece of information in what she knows, the more familiar she is with all words in the group or, at least, with some parts of it. In particular, when people receive the same message several times from different sources, usually it will each time become more familiar and seem more plausible to them. This effect is by no means limited to verbal communication. For instance, the result of a scientific experiment gains in evidence the more often it is reproduced under similar conditions.

However, the Knowledge-Information-Matter-Energy representation of relations between knowledge and other basic phenomena (the KIME Square) is different from the approach of MacKay in several essential issues. First, MacKay does not specify representation of what knowledge is, while the Knowledge-Information-Matter-Energy model does it as it is demonstrated in the second part of this section. Second, MacKay does not specify what kind of representation knowledge is, while the Knowledge-Information-Matter-Energy model does it, assuming that knowledge is a structure or a system of structures. Third, the Knowledge-Information-Matter-Energy model provides an explication of the elementary knowledge structure presented in the diagrams (2.4.1) and (2.5.7).

The KIME Square is in a good accordance with the distinction some researchers in information sciences have continually made between information and knowledge. For instance, Machlup (1983) and Sholle (1999) distinguish information and knowledge along three axes:

1. *Multiplicity*: Information is piecemeal, fragmented, particular, while knowledge is structured, coherent and universal.
2. *Time*: Information is timely, transitory, even ephemeral, while knowledge is enduring and temporally expansive.
3. *Space*: Information is a flow across spaces, while knowledge is a stock, specifically located, yet spatially expansive.

These distinctions show that information is conceived of as a process, whereas knowledge is a specific kind of substance. This understanding well correlates with the information concept in the general theory of information (cf. Ontological Principle 2c), which is incorporated in the KIME Square. Indeed, cognitive information is conceived as a kind of energy in the World of Structures, which under definite conditions induces structural work on knowledge structures (cf. Section 2.2). At the same time, knowledge is a substance in the World of Structures.

Some may be concerned that data and knowledge occupy the same place in the Knowledge-Information-Matter-Energy Square (cf. Figure 2.14). It is possible to explain similarities and distinctions between data and knowledge with the following two metaphors.

Metaphor A. *Data and knowledge are like molecules, but data are like molecules of water, which has two atoms, while knowledge is like molecules of DNA, which unites billions of atoms.*

Metaphor B. *Data and knowledge are like living beings, but data are like bacteria, while knowledge is like a human being.*

The new understanding of relations between information and knowledge suggests that a specific fundamental triad called the *Data-Knowledge Triad* provides a more relevant theoretical explication for Elliot's metaphor than the Data-Information-knowledge Pyramid. Namely, the Data-Knowledge Triad has the following form:

$$\textbf{Data} \xrightarrow{\;\textbf{information}\;} \textbf{Knowledge} \qquad (2.5.2)$$

Combining data, information and knowledge into a conceptual structure, the triadic representation (2.5.2) has the following interpretation, which reveals data-knowledge dynamics:

Data, under the influence (action) of additional information, become knowledge

That is, information is the active essence that transforms data into knowledge. It is similar to the situation in the physical world, where energy is used to perform work, which changes material things, their positions and dynamics.

This relation between information and knowledge is well understood in the area of information management. For instance, Kirk (1999) writes, "the effectiveness of information management can be measured by the extent of knowledge creation or innovation in organizations".

The triadic representation also well correlates with opinions of Bar-Hillel (1964), Heyderhoff and Hildebrand (1973), Brookes (1980), and Mizzaro (1996; 1998; 2001), who describe an information process as a process of knowledge acquisition often understood as reduction of uncertainty. The information triad (2.5.1) also well correlates with the approach of Boisot (1998) where information is what an observer extracts from data as a function of his/her expectations or *prior knowledge*. Boisot illustrates the point by means of the diagram given in Figure 2.15.

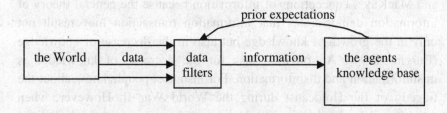

Figure 2.15. Transformation of data into knowledge

The diagram (2.5.1) indicates that it is data, or more exactly, data representations, that are physical, but not information. Data are rooted in the world's physical properties. Knowledge, by contrast, is rooted in the estimates and expectations of individuals. Information is what a knowing individual is able to extract from data, given the state of his/her knowledge. Physical extraction essentially depends on used types of knowledge, i.e., whether it is represented by (contained in) algorithms and logic, models and language, operations and procedures, goals and problems. The situation is similar to the situation with (physical) energy when energy is extracted from different substances and processes: from petroleum, natural gas, coal, wood, sunlight, wind, ocean tides, etc. This extraction essentially depends on used devices, technologies and techniques. For instance, solar cells are used to convert sunlight into electricity. Wind and water turbines rotate magnets and in such a way create electric current. Petroleum-powered engines make cars ride and planes fly.

The KIME Square shows essential distinction between knowledge and information in general, as well as between knowledge and cognitive information, in particular. This distinction has important implications for education. For instance, transaction of information (for example, in a teaching process) does not give knowledge itself. It only causes such changes that may result in the growth of knowledge.

This is in a good accordance with the approaches of Dretske (1981) and MacKay (1969), who declare that information increases knowledge and knowledge is considered as a completed act of information.

However, the general theory of information differs from Dretske's and MacKay's conceptions of information because the general theory of information demonstrates that information transaction may result not only in the growth of knowledge but also in the decrease of knowledge (Burgin, 1994). An obvious case for the decrease of knowledge is misinformation and disinformation. For instance, people know about the tragedy of the Holocaust during the World War II. However, when articles and books denying the Holocaust appear, some people believe this and loose their knowledge about the Holocaust. Disinformation, or false information, is also used to corrupt opponent's knowledge in information warfare.

Moreover, even genuine information can decrease knowledge. For instance, some outstanding thinkers in ancient time, e.g., Greek philosophers Leucippus (fifth century B.C.E.) and Democritus (circa 460–370 B.C.E.), knew that all physical things consisted of atoms. Having no proofs of this feature of nature and lacking any other information supporting it, people lost this knowledge. Although some sources preserved these ideas, they were considered as false beliefs. Nevertheless later when physics and chemistry matured, they found experimental evidence for atomic structure of physical object and knowledge about these structures was not only restored but also essentially expanded.

In a similar way, knowledge about America was lost in Europe and America was rediscovered several times. A thousand years ago, nearly half a millennium before Columbus, the Norse extended their explorations from Iceland and Greenland to the shores of Northeastern North America, and, possibly, beyond. According to the Olaf saga, the glory of having discovered America belongs to Bjarni, son of Herjulf, who was believed to have discovered Vinland, Markland, and Helluland as early as 985 or 986 on a voyage from Iceland to Greenland (Reeves, et al, 1906). There are also stories of other pre-Columbian discoveries of America by Phoenician, Irish, and Welsh, but all accounts of such discoveries rest on vague or unreliable testimony.

Sometimes people loose some knowledge and achieve other knowledge. For instance, for centuries mathematicians knew that there was only one geometry. The famous German philosopher Kant (1724–1804) wrote that knowledge about the Euclidean geometry is given to people *a priory*, i.e., without special learning. However, when mathematicians accepted the discovery of non-Euclidean geometries, they lost the knowledge about uniqueness of geometry. This knowledge was substituted by a more exact knowledge about a diversity of geometries.

Some may argue that it was only a belief in uniqueness but not a real knowledge. However, in the 18th century, for example, the statement

There is only one geometry   (2.5.3)

was sufficiently validated (for that time). In addition, it was well correlated with reality as the Euclidean geometry was successfully applied in physics. Thus, there all grounds to assume that the statement (2.5.3) represented knowledge of the 18th century in Europe. Likewise, now the vast majority of people know that *two plus two is equal to four* although existence of non-Euclidean geometries shows that there are situations when *two plus two is not equal to four*.

As we have seen, usually people assume that information creates knowledge. However, utilizing information, it is possible to create data from knowledge. This shows that the triad (2.5.4) inverse to the triad (2.5.2) is also meaningful and reflects a definite type of information processes.

$$\textbf{Knowledge} \longrightarrow \textbf{Data} \qquad (2.5.4)$$

At the same time, data generate not only knowledge, but also beliefs and fantasies. It gives us two more information diagrams.

$$\textbf{Data} \xrightarrow{\textbf{information}} \textbf{Beliefs} \qquad (2.5.5)$$

and

$$\textbf{Data} \xrightarrow{\textbf{information}} \textbf{Fantasy} \qquad (2.5.6)$$

To explain why it is more efficient, i.e., more adequate to reality and more productive as a cognitive hypothesis, to consider information as the knowledge content than to treat information in the same category as data and knowledge, let us consider people's beliefs. Can we say that belief is structuring of information or structured information? No. However, we understand that people's beliefs are formed by the impact some information has on people's mind. People process and refine information and form beliefs, knowledge, ideas, hypotheses, etc. Beliefs are structured in the same way as knowledge, but in contrast to knowledge, they are not sufficiently justified. Thus, the concept of information in the sense of Stonier (1991; 1992) does not allow fitting beliefs into the general schema, while the concept of information in the sense of the general theory of information naturally integrates beliefs into the system *data, information,* and *knowledge.*

There are other examples that give evidence to support the statement that diagrams (2.5.2) and (2.5.4) correctly explains relation between information, data and knowledge. In spite of this, people are very conservative in their beliefs and do not want to change what they "know" to a more grounded knowledge mostly because it demands an intellectual effort (action) and even people are systems that comply with the principle of minimal action.

It would be useful to better clarify our understanding of data. At first glance, data look like some material things in contrast to knowledge, which is structural in essence. However, when we start analyze what data are, we see that this is not so. Indeed, numbers are a kind of data. Let us take a number, say 10. In mathematics, it is called a natural number. From mathematics, we also know that natural numbers are equivalent collections of sets that have the same number of elements (Bourbaki, 1960). This is a structure. Another way to define natural numbers is to axiomatically define the set N of all natural structures and to call its elements by the name natural number (Kuratowski and Mostowski, 1967). Abstract sets and their elements definitely are structures. One more way to represent natural numbers is to use abstract properties (Burgin, 1989). This approach develops the idea of Cantor, who defined the cardinal number of a set A as a property of A that remains after abstraction from qualities of elements from A and their order (cf., for example, (Kuratowski and Mostowski, 1967)). Properties, abstract and real, are also structures. Thus, in any mathematical sense, a natural number, in particular, number 10, is a structure. Moreover, the number 10 has different concrete representations in mathematical structures. In the decimal numerical system, it is represented as 10. In the binary numerical system, it is represented as 1010. In the ternary numerical system, it is represented as 101. In English, it is represented by the word *ten*, and so on. In the material world, it is possible to represent number 10 as a written word on paper, as a written word on a board, as a said word, as a geometrical shape on the computer or TV screen, as signals, as a state of some system and so on. All these representations form a structure that we call *number* 10.

We come to similar situations analyzing letters and words, which are also data. Let us consider the letter a. It also has different physical representations: a, a, a, *a*, *a*, **a**, *a*, A, *A*, **A**, a, *a*, ɑ, *a*, a, A, **A**, A, **A**, *A*, and many others. In addition, it is possible to represent the letter a as a written symbol on paper, as a said word, as a written symbol on a board, as a geometrical shape on the computer or TV screen, as signals, as a state of some system and so on. Thus, the letter a, as well as any other letter or word, is a structure. It is known that words are represented by written and printed symbols, pixels on the screen, electrical charges and brain-waves (Suppes and Han, 2000).

All these considerations show that it is necessary to make a distinction between data, knowledge, beliefs, ideas and their representations. For knowledge, it is a well-known fact and knowledge representation is an active research area (cf., for example, (Ueno, et al, 1987)). Change of representation is called codification. Knowledge codification serves the pivotal role of allowing what is collectively known to be shared and used (Dalkir, 2005). At the same time, differences between data and data representations are often ignored.

In comparison with knowledge, information is an active structure. As it has been observed by some researchers (cf., for example, (Hodgson and Knudsen, 2007)), information causes some action.

Distinction between knowledge and cognitive information implies that transaction of information (for example, in a teaching process) does not give knowledge itself. It only causes changes that may result in the growth of knowledge. In other words, it is possible to transmit only information from one system to another, allowing a corresponding infological system to transform data into knowledge. In microphysics, the main objects are subatomic particles and quantum fields of interaction. In this context, knowledge and data play role of particles, while information realizes interaction.

In a mathematical model of information, a cognitive infological system is modeled by a mathematical structure, for example, a space of knowledge, beliefs, and fantasies (cf. Section 7.3.1). Information, or more exactly, a unit of information, is an operator acting on this structure. A global unit of information is also an operator. Its place is in a higher level of hierarchy as it acts on the space of all (or some) cognitive infological systems.

Knowledge constitutes a substantial part of the cognitive infological system. Some researchers even equate cognition with knowledge acquisition and consider the system of knowledge often called thesaurus as the whole cognitive infological system. In a general case, knowledge, as a whole, constitutes a huge system, which is organized hierarchically and has many levels. It is possible to separate three main levels: microlevel, macrolevel, and megalevel (Burgin, 1997). On the megalevel, we consider the whole system of knowledge as it exists in society and its commensurable subsystems, such as mathematics, physics, biology, advanced mathematical and physical theories. On the macrolevel, we have such systems of knowledge as formal theories and abstract models. Scientific and mathematical theories form a transition from the macrolevel to the megalevel. When a mathematical or scientific theory appears, it is, as a rule, small and is situated on the macrolevel of knowledge. This was the situation with the calculus at the end of 17th century, with non-Euclidean geometries in the middle of the 19th century or with non-Diophantine arithmetics now. At the same time, mature theories, such as geometry, algebra, genetics or quantum physics, are on the megalevel. There are different models of knowledge on the megalevel. The majority of them give a structural representation of a scientific theory as one of the most advanced systems of knowledge.

The most developed and relevant model of knowledge megasystems is the structure-nominative model developed by Burgin and Kuznetsov (1991; 1992; 1993; 1994). This model encompasses all other existing models of theoretical and practical knowledge. The most popular models, such as positivists logical representation of knowledge where scientific theory is represented by logical propositions and/or predicates (cf. (Popper, 1965; 1979))) or model-oriented representation of knowledge, which assumes that a theory is a collection of models (cf. (Suppes, 1967; Suppe, 1979; Sneed, 1979; van Frassen, 2000)), work well on the macrolevel but are essentially incomplete on the megalevel.

The most popular approaches to represent knowledge on the macrolevel is to use logical languages and calculi. For instance, Halpern and Moses (1985) consider knowledge of an agent (a person or an artificial system) as a set of propositions. The conceptual world of such an agent consists of formulas of some logical language concerning the real world and agent's knowledge. Systems of propositions (propositional calculi) describe actual and possible worlds in theories of Bar-Hillel and Carnap (1952; 1958), Hintikka (1970; 1971; 1973; 1973a) and other researchers (cf. Section 4.2).

Propositional knowledge in general and agent's knowledge, in particular, are connected to the world through the possible-world semantics where a logical universe W is considered, states of which are fixed assignments of truth values to primitive propositions from a propositional language L. It is possible to think of these states of the universe W as the worlds that the agent thinks or cognitive system assumes are possible. Taken another way, states are the propositional assignments that do not contradict agent's knowledge. A more realistic approach treats systems of propositions as descriptions of situations in the world W.

The microlevel contains "bricks" and "blocks" of knowledge that are used for construction of other knowledge systems. For instance, such knowledge macrosystems as formal theories in logic are constructed using knowledge microsystems or elements: propositions and predicates. Their "bricks" or elementary logical units are atomic formulas, i.e., simple logical functions and propositions, such as "*Knowledge is power*" or "*Information can give knowledge*", while composite propositions, logical functions and predicates, such as "$(2 + 2 = 4)\vee (2 + 2 = 2)$" or " If X is a metric space, then X is a topological space", are "blocks" or compound logical units.

Studies in the area of artificial intelligence (AI) attracted attention of many researchers to knowledge representation and modeling. Different models of knowledge on the macrolevel have been introduced. The most known models are semantic networks (cf., for example, (Ueno, et al, 1987; Burgin and Gladun, 1989)), systems of frames (Minsky, 1974), scripts or formal scenarios (cf. (Schank and Abelson, 1975)), productions (cf., for example, (Ueno, et al, 1987)), relational and logical structures (cf., for example, (Thayse, et al, 1988)).

Here we consider only the microlevel of knowledge, aiming at construction of a mathematical model of knowledge units, finding elementary knowledge units, and study of their integration into complex knowledge systems. This study helps to separate of data and knowledge, as well as to improve efficiency of information processing by computers.

To study knowledge as an essence, we have to begin with an important observation that there is no knowledge *per se* but we always have knowledge about something (about weather, the Sun, the size of a parameter, occurrence of an event, and so forth). In other words, knowledge always involves some real or abstract object. Plato was may be the first to formulate this explicitly in his dialogue Republic. Such an object may be a galaxy, planet, point, person bird, letter, novel, love, surprise, picture, song, sound, light, square, triangle, etc.

However, to discern an object, as well as to be able to speak and to write about it, we have to name it. A name may be a label, number, idea, text, and even a physical object of a relevant nature. For instance, a name may be a state of a cell in computer memory or a sound or a sequence of sounds when you pronounce somebody's name. In the context of named set theory (cf. Appendix A), any object can be used as a name for another object.

Besides, the simplest knowledge about an object is some property of this object. As Aristotle wrote, we can know about things nothing but their properties. The simplest property is existence of the object in question. However, speaking about properties, we have to differentiate *intrinsic* and *ascribed properties* of objects. In this, we are following the long-standing tradition that objects have intrinsic properties. Intrinsic properties of real, e.g., physical, objects are reflected by ascribed properties. Intrinsic properties of abstract, e.g., mathematical, objects are given in the form of assumptions, axioms, postulates, etc. Ascribed properties are obtained by measurement, calculation or inference.

Intrinsic and ascribed properties are similar to *primary* and *secondary qualities* discerned by the English philosopher John Locke (1632–1704). He assumed that primary qualities are objective features of the world, such as shape and size. By contrast, secondary qualities depend on mind. Examples of secondary qualities are color, taste, sound, and smell.

To formalize the concept of knowledge, we consider a set U of some objects. Note that it is possible that the set U consists of a single object. We call this set U the knowledge domain (universe). Then knowledge about an object F from U involves:

(1) the class U or some its subclass that contains the object F;

(2) an intrinsic property represented by an abstract property $T = (U, t, W)$ with the scale W, which is defined for objects from U;

(3) some class C, which includes a name «F» of the object F; and

(4) an ascribed property represented by an abstract property $P = (C, p, L)$ with the scale L, which is defined for names from C. This property P is ascribed to objects from U, although not directly, but through their names. Thus, we come to the following definition.

Definition 2.5.1. *An elementary unit K of general knowledge* about an object F is defined by the diagram (2.5.7):

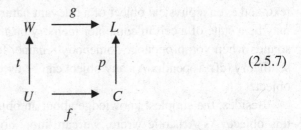

$$(2.5.7)$$

In the diagram (2.5.7), the correspondence f relates each object H from U to its name (or to its system of names or, more generally, to its *conceptual image* (Burgin and Gorsky, 1991)) «H» = $f(H)$ from C and the correspondence g assigns values of the property T to values of the property P. In other words, g relates values of the intrinsic property to values of the ascribed property. For instance, when we consider such property of material things as weight (the intrinsic property), in weighting any thing, we can get only an approximate value of the real weight, or weight with some precision (the ascribed property).

Relation f may have the form of some algorithms/procedures of object recognition, construction or acquisition. Relation g may have the form of some algorithms/procedures of measurement, evaluation or prediction.

Note that the object F (or H) can be a big system that consists of other objects. For instance, it can be a galaxy or the whole physical universe.

Remark 2.5.1. It is possible that objects from U are characterized by a system of properties. However, this does not demand to change our representation of an elementary unit of property because due to the construction of composition of properties and the result of Theorem A.1 (cf. Appendix A), the system of properties is equivalent to one property.

Some suggest that knowledge does not exist outside some knowledge system. Elementary units of knowledge form such minimal knowledge systems.

Knowledge may range from general to specific (Grant, 1996). General knowledge is broad, often publicly available, and independent of particular events. Specific knowledge, in contrast, is context-specific.

General knowledge, its context commonly shared, can be more easily and meaningfully codified and exchanged, especially among different knowledge or practice communities. Codifying specific knowledge so as to be meaningful across an organization or group requires its context to be described along with the *focal knowledge*. This, in turn, requires explicitly defining contextual categories and relationships that are meaningful across knowledge communities. Taking Definition 2.5.1 as a base, we define specific knowledge in the following way.

Definition 2.5.2. *An elementary unit K of specific knowledge about an object F is represented by the diagram (2.5.8):*

$$
\begin{array}{ccc}
D_W & \xrightarrow{\ g_W\ } & B_L \\[4pt]
\Big\uparrow{\scriptstyle t_U} & & \Big\uparrow{\scriptstyle p_C} \\[4pt]
D_U & \xrightarrow[\ f_U\ ]{} & B_C
\end{array}
\qquad (2.5.8)
$$

Here D_U is a subset of U that contains F; D_W is the set of values of the property T on objects from D_U, i.e., $D_W = \{t(u); u \in D_U\}$; B_C is a subset of C that consists of the names of objects from D_U, i.e., $B_C = \{f(u); u \in D_U\}$; B_L is the set of values of the property P on the names of objects from D_U; f_U, t_U, p_C, and g_W are corresponding restrictions of relations f_U, t_U, p_C, and g_W.

As a result, we obtain a commutative cube (2.5.9), in which all mappings r_W, r_L, r_U, and r_C are inclusions.

$$(2.5.9)$$

Any system of knowledge is built from such elementary units by means of relations, which glue these "knowledge bricks" together. However, it is possible that some larger blocks are constructed from elementary units and then systems of knowledge are built from such blocks.

Often elementary units of knowledge are expressed by logical propositions of the type:

An object F has the property P

or

The value of a property P for an object F is equal to a.

These propositions correspond to the forms of elementary data in the operator information theory of Chechkin (1991) (cf. Section 7.2).

The elementary unit of knowledge (2.5.7) has specific components. The first one is the *attributive* or *estimate component* of knowledge. It reflects relation of the intrinsic property T to the ascribed property P and is represented by the diagram (2.5.10).

$$W \xrightarrow{\;g\;} L \qquad\qquad (2.5.10)$$

The second is the *naming component of knowledge*. It reflects the process of naming of an object when this object is separated, discovered or constructed. It is represented by the diagram (2.5.11).

$$U \xrightarrow{} C \qquad\qquad (2.5.11)$$
$$f$$

The third is the *object component of knowledge*. It is the intrinsic property of objects from the knowledge domain U. It is reflected by the diagram (2.5.12).

$$U \xrightarrow{} W \qquad\qquad (2.5.12)$$
$$t$$

The object component may be material, for example, U consists of all elementary particles or of computers, or may be ideal. For instance, U consists of some texts or of real numbers.

The fourth is the *information component* or *representation component of knowledge*. It is the ascribed property of objects from the knowledge domain U. The information component of knowledge renders information about the knowledge domain U. It is reflected by the diagram (2.5.13).

$$C \xrightarrow{\ p\ } L \qquad\qquad (2.5.13)$$

Any knowledge system has two parts: cognitive and substantial. The elementary unit of knowledge the object component (U, t, W) as its *substantial part* and all other elements, that is, the information component (C, p, L) and two relations f and g, as its *cognitive part*. Cognitive parts of knowledge form knowledge systems *per se* as abstract structures, while adding to substantial parts forms extended knowledge systems.

It is necessary to emphasize that here we consider only the simplest units of knowledge. More extended systems are formed by means of specific relations and estimates that glue these small "bricks" and larger "blocks" of knowledge together.

Giving an exact definition of knowledge by building its mathematical model allows us to study data in a more exact form than before, discern data from knowledge, and specify several types of data.

The first type is *raw* or *uninterpreted data*. They are represented by the diagram (2.5.14).

$$B \xrightarrow{\ q\ } M \qquad\qquad (2.5.14)$$

In this diagram, B consists of names of some objects, M consists of some semiotic objects such as symbols, words or texts, and q is a relation between B and M.

Raw data resemble the information component of knowledge. The difference is that raw data are not related to any definite property.

Having raw data, a person or a computer system can transform them into knowledge by means of other knowledge that this person or computer system has.

Remark 2.5.2. All other types of data may be also treated as incomplete knowledge.

The second type is *formally interpreted data*. They are related to the abstract property P, forming a named subset of the information component of a knowledge unit K, and are represented by the diagram (2.5.15).

$$B_C \xrightarrow{\ p_C\ } B_L \qquad\qquad (2.5.15)$$

Example 2.5.1. Automatic instrument systems perform measurement, which is a process of data acquisition, producing formally interpreted data. An unmanned weather station, for example, may record daily maximum and minimum temperatures. Such recordings are formally interpreted data because numbers (the set B_L) are corresponded to names "maximal temperature" and "minimal temperature" (the set B_C).

The third type is *attributed data*. They are related not to the abstract property P but to values of an intrinsic property T. Attributed data are represented by the diagram (2.5.16).

$$D_W \xrightarrow{\ q\ } B_L \qquad\qquad (2.5.16)$$

Two other types of data are more enhanced and are closer to knowledge.

The fourth type is *object interpreted data*. They are represented by the diagram (2.5.17).

$$
\begin{array}{c}
B_L \\
q \ \ \uparrow p \\
D_U \xrightarrow{\ \ \ } B_C
\end{array}
\qquad\qquad (2.5.17)
$$

Taking Example 2.5.1 and adding that maximal and minimal temperatures are temperatures of air (or water), we obtain object interpreted data as the diagram (2.5.15) is extended to the diagram (2.5.17).

The fifth type is *object attributed data*. They are represented by the diagram (2.5.18).

$$D_W \xrightarrow{\quad q \quad} B_L$$
$$\uparrow p \qquad\qquad (2.5.18)$$
$$B_C$$

Thus, we have demonstrated that knowledge on the microlevel has the structure presented in the diagrams (2.5.7) and (2.5.8). However, this does not mean that other essences cannot have this structure. As a matter of fact, cognitive essences related to knowledge have on the microlevel the same structure. Namely, knowledge is closely related to beliefs and fantasy. To distinct knowledge from beliefs and fantasy, we use measures of truthfulness. There are two types of truthfulness criteria: *object-dependent* and *attitude-dependent* (cf. Section 2.4). According to the first approach, we have the following definitions.

Definition 2.5.3. *An elementary unit K of specific knowledge* (of *general knowledge*) about an object F is the entity that has the structure represented by the diagram (2.5.8) (by the diagram (2.5.7)) that is justified as knowledge.

There are different systems of justification: science, for which the main validation technique is experiment; mathematics, which is based on logic with its deduction and induction; religion with its postulates and creeds; history, which is based on historical documents and archeological discoveries.

Contemporary theory of knowledge (Pollock and Cruz, 1999) discerns several types of justification: *doxatic justification*, *internal justification*, and *external justification*. Doxatic approach assumes that the justifiability of a belief is a function exclusively of what beliefs one holds. Internal justification takes into account only internal states of an individual. External justification includes not only internal states of an individual but also what is going on in the objective reality.

Knowledge has been always connected to truth. Namely, the condition, that a person knows something, *P*, always implied that *P* is true (cf., for example, (Pollock and Cruz, 1999)). However, history of mankind shows that knowledge is temporal, that is, what is known at some time can be disproved later. For instance, for a long time people knew that the Earth is flat. However, several centuries ago it was demonstrated that this is not true. For a long time philosophers and scientists knew that the universe always existed. However, in the 20th century it was demonstrated that this is not true. Modern cosmology assumes that the universe erupted from an enormously energetic, singular event metaphorically called *"big bang"*, which spewed forth all the space and all of matter. As time passed, the universe expanded and cooled.

A more recent example tells us that before 1994, physicists knew that the gravitational constant G_N had a value between $6.6709 \cdot 10^{-11}$ and $6.6743 \cdot 10^{-11}$ $m^3 kg^{-1} s^{-2}$ (Montanet, et al, 1994) Since then three new measurements of G_N have been performed (Kiernan, 1995) and we now have four numbers as tentative values of the gravitational constant G_N, which do not agree with each other.

Thus, we see that it is possible to speak about false knowledge and knowledge of some time, say knowledge of the 19th century. It means that there is *subjective knowledge*, which can be defined as beliefs with the high estimation that they are true, and *objective knowledge*, which can be defined as beliefs that correspond to what is really true

In his description of the World 3 (cf. Section 2.1), Popper (1965; 1979) asserted that this world contains all the knowledge and theories of the world, both true and false. Thus, Popper assumed existence of false knowledge.

Definition 2.5.4. *An elementary unit B* of *specific belief* (of *general belief*) about an object *F* is the entity that has the structure represented by the diagram (2.5.8) (by the diagram (2.5.7)) that is sufficiently justified.

As Bem (1970) writes, "beliefs and attitudes play an important role in human affairs. And when public policy is being formulated, beliefs *about* beliefs and attitudes play an even more crucial role." As a result, beliefs are thoroughly studied in psychology and logic. Belief systems are formalized by logical structures that introduces structures in belief spaces and calculi, as well as by belief measures that evaluate attitudes to cognitive structures and are built in the context of fuzzy set theory. There are developed methods of logics of beliefs (cf., for example, (Munindar and Nicholas, 1993) or (Baldoni, et al, 1998)) and belief functions (Shafer, 1976). Logical methods, theory of possibility, fuzzy set theory, and probabilistic technique form a good base for building cognitive infological systems in computers.

Definition 2.5.5. An *elementary unit M* of *specific fantasy* (of *general fantasy*) about an object *F* is the entity that has the structure represented by the diagram (2.5.8) (by the diagram (2.5.7)) that is not validated.

For instance, looking at the Moon in the sky, we know that we see the Moon. We can believe that we will be able to see the Moon tomorrow at the same time and we can fantasize how we will walk on the Moon next year.

According to the attitude-dependent approach, we have the following definition.

Definition 2.5.6. An *elementary unit K* of *specific knowledge* (of *general knowledge*) about an object *F* for a system *R* is the entity that has the structure represented by the diagram (2.5.8) (by the diagram (2.5.7)) that is estimated (believed) by the system *R* to represent with high extent of confidence true relations.

In a similar way, we define beliefs and fantasies.

Definition 2.5.7. An *elementary unit B* of *specific belief* (of *general belief*) about an object *F* for a system *R* is the entity that has the structure represented by the diagram (2.5.8) (by the diagram (2.5.7)) that is estimated (believed) by the system *R* to represent with moderate extent of confidence true relations.

Definition 2.5.8. An *elementary unit M* of *specific fantasy* (of *general fantasy*) about an object *F* for a system *R* is the entity that has the structure represented by the diagram (2.5.8) (by the diagram (2.5.7)) that is estimated (believed) by the system *R* to represent with low extent of confidence true relations.

If confidence depends on some validation system, then there is a correlation between the first and the second stratification of cognitive structures into three groups — knowledge, beliefs, and fantasy.

Separation of knowledge from beliefs and fantasies leads to the important question whether knowledge gives only true/correct representation of object properties. Many think that knowledge has to be always true. What is not true is called misconception. However, as we have mentioned, history of science and mathematics shows that what is considered knowledge at one time may be a misconception at another time. For instance, the Euclidean geometry was believed for 2200 years to be unique (both as an absolute truth and a necessary mode of human perception). People were not even able to imagine anything different in the field of geometry. The famous German philosopher Emmanuil Kant claimed that (Euclidean) geometry is given to people *a priory*, i.e., without special learning. In spite of this, almost unexpectedly some people began to understand that geometry is not unique. For example, in his book *Theorie der Parallellinien* (1766), Johann Heinrich Lambert (1728–1777) deduced a large number of non-Euclidean geometry results, assuming a possibility of such geometries. Two amateur researchers, Ferdinand Karl Schweikart (1728–1777) and his nephew Franz Adolph Taurinus (1794-1874), came to the conclusion that non-Euclidean geometries existed. In 1818, Schweikart wrote to Gauss about his investigations in this area, calling new geometry by the name *astral geometry*. Taurinus published two books on this topic in 1825 and 1826. However, only outstanding mathematicians were able to actually build non-Euclidean geometries.

Trying to improve the axiomatic system suggested for geometry by Euclid, three great mathematicians of the 19th century (C.F. Gauss (1777–1855), N.I. Lobachewsky (1792–1856), and Ja. Bolyai (1802–1860)) discovered a lot of other geometries (cf., for example, (Burton, 1997)). At first, even the best mathematicians opposed this discovery and severely attacked Lobachewsky and Bolyai who published their results. Forecasting such antagonistic attitude, the first mathematician of his times Gauss was afraid to publish his results on non-Euclidean geometry, understanding that this discovery would shock layman and mathematician alike, causing vulgar attacks and damaging his reputation. Nevertheless, progress of mathematics brought understanding and then recognition of non-Euclidean geometry. This discovery is now considered as one of the highest achievements of the human genius. It changed to a great extent understanding of mathematics and improved comprehension of the whole world.

In the 20th century, a similar situation existed in arithmetic. For thousands of years, much longer than that for the Euclidean geometry, only one arithmetic has existed. Mathematical establishment has treated arithmetic as primordial entity inherent to human intelligence. For instance, German mathematician Leopold Kronecker (1825–1891) wrote: *"God made the integers, all the rest is the work of man"*. By British mathematician Henri John Stephen Smith (1826-1883), arithmetics (the Diophantine one) is one of the oldest branches, perhaps the very oldest branch, of human knowledge. His older contemporary Carl Gustav Jacobi (1805-1851) said: *"God ever arithmetizes"*.

But in spite of such a high estimation of the Diophantine arithmetic, its uniqueness and indisputable authority has been recently challenged. A family of non-Diophantine arithmetics was discovered (Burgin, 1997). Like geometries of Lobatchewsky, these arithmetics depend on a special parameter, although this parameter is not a numerical but a functional one.

These examples show that it is necessary to distinguish exact knowledge, approximate knowledge, and false knowledge or misconception. To separate them, we consider the object and representation components of knowledge (cf. diagrams (2.5.12) and (2.5.13)).

Definition 2.5.9. An elementary unit *K* of knowledge (2.5.7) or (2.5.8) is:

a) *exact* if the properties of and relations between ascribed properties are the same, or more exactly, isomorphic to the properties of and relations between intrinsic properties;

b) *approximate* if properties of and relations between ascribed properties are similar, i.e., are not essentially different from (are sufficiently close to) the properties of and relations between intrinsic properties;

c) *false* if the properties of and relations between intrinsic properties are essentially different from the properties of and relations between ascribed properties;

To ascribe exact meaning to the expressions *similar* and *essentially different* in this definition, it is possible to use the concept of tolerance (Zeeman and Buneman, 1968) and different measures and metrics in systems of cognitive structures.

There are three kinds of features (and their estimates) for cognitive structures: *groundedness*, *confidence*, and *relevance*. For each of them, a corresponding measure exists that allows one to find numerical values representing differences between separate cognitive structures and their classes with respect to the corresponding feature.

It is necessary to remark that the approach to knowledge evaluation considered here is relevant to all types of knowledge. According to systemic typology developed in the structure-nominative model of developed knowledge systems (Burgin, and Kuznetsov, 1993; 1994), there are four main types of knowledge: *logic-linguistic knowledge* that uses linguistic and logical means to reflect relations and connections existing in the object domain; *model knowledge* is knowledge in a form of a model (for instance, knowledge of a place or a person); *procedural/operational knowledge* describes, or prescribes, how to do something; and *problem knowledge* exposes absence of knowledge about something. This typology stays true for microknowledge considered in this section as properties may be composite and have a sophisticated structure.

2.6. Emotions and Information

> *No aspect of our mental life is more important*
> *to the quality and meaning of our existence than emotions.*
> Ronald de Sousa

Many have asserted that information is definitely connected to meaning. As von Weizsäcker writes (2006), "information is only what can be understood". If we understand *understanding* as a cognitive process, then this assumption restricts information to cognitive information. Although the general theory of information provides a very broad understanding of information, all given examples of information describe only cognitive information, which is information in the conventional sense. In this section, we demonstrate that there are other types of information and they are not connected to understanding, and thus, are different from cognitive information. At first, we consider information related to emotions.

Emotions play an extremely important role in human life. They are prime determinants of person's sense of subjective well being and play a central role in many human activities. Emotions give color, meaning and motivation to the whole of the vast range of actions of which people are capable, including intellectual activity. Emotions give values and are values themselves.

As Darwin (1896) writes, "a definite description of the countenance under any emotion or frame of mind, with a statement of the circumstances under which it occurred, would possess much value". Economists discovered that even economical activity could not be correctly understood and studied with out taking emotions into account. In the contemporary situation, financial experts are concerned that large number of investors basing their decisions more on their affect than the value of the stock might have adverse, if not disastrous consequences for the stock market, the economy, and possibly world peace (Shiller, 2000).

The word *emotion* is derived from the Latin word *emovere*, where "*e*" means *out* and "*movere*" means *move*. This implies that emotions cause changes, namely, they change, sometimes drastically, the state, behavior and thinking of an individual or a group of people. However, it is not an easy task to define emotion in exact scientific terms.

There are different understandings of what *emotion* is. As for many other basic psychological concepts, *emotion* is hard to conceptualize. A multitude of philosophers and researchers has tried to elaborate an adequate definition of emotion. Emotions have been treated as either physiological processes, or perceptions of physiological processes, or neuropsychological states, or adaptive dispositions, or evaluative judgments, or computational states, dynamical processes or even social facts. Some researchers suggest that *emotion* is a "cluster" concept (Sloman, 2002).

Emotions have been studied from ancient times. At first, philosophers contributed to the understanding of emotions. Plato (427–347) and Aristotle (384-322), as well as ancient Greek stoics suggested their ideas on emotions. Sophisticated theories of emotions can be found in works of such outstanding philosophers as René Descartes (1596-1650), Baruch (or Benedictus) Spinoza (1632–1677), and David Hume (1711–1776).

Later biologists and then psychologists started to study emotions. Charles Robert Darwin (1809–1882) gave an acceptable definition of *emotion as a frame of mind* (cf. (Darwin, 1896)). However, emotions are also connected to the physical state of the body. Thus, it is possible to assume, taking this as a working definition, that *emotion* is a person's internal state of being, which is based in and/or tied to physical state and sensory feelings and which regulates person's response to an object or a situation. In some theories, emotions are considered as a type of affective states, which also include other phenomena, such as moods, dispositions, motivational states (e.g., hunger, curiosity or fear), and feelings of pleasure or pain (cf., for example, (Schrer, 2005)). In what follows, we understand emotions in very general sense that may encompass affective states.

Scientific understanding of emotions involves several levels of explanation. The most important of them are the *functional* or *computational* level, *algorithmic* level, and *implementational* level (cf., for example, (Marr, 1982)). At the *functional level*, it is identified what emotions are *for*. This issue is appropriate even if one believes, as some traditionally have, that emotions actually represent the breakdown of smoothly adaptive functions such as thought, perception, and rational planning. In this case, the emotions may be understood precisely in terms of their failure to promote the smooth working of the cognitive and conative functions, that is, efficient interaction of perception and thought (cognitive and functions) with interpersonal relations, communication and discourse (conative functions). The second, *algorithmic level* represents procedures or algorithms of neurophysiological processes related to emotions. The third, *implementational level* describes the actual neurophysiological processes related to emotions in animals or humans.

In psychological literature, *emotion* is described as a state usually caused by an event of importance to the subject. It typically includes (Oakley and Jenkins, 1996) the following components:

(1) a conscious mental state with a recognizable quality of feeling and often directed towards some object;

(2) a body perturbation of some type;

(3) recognizable expressions of the face, tone of voice, and gesture;

(4) readiness for certain kinds of actions.

Thus, it is possible to describe emotions as internal phenomena that can, but do not always, make themselves observable through expression and behavior (Niedenthal, et al, 2006). Some researchers imply that emotions always are conscious (cf., for example, (Clore, 1994)), while others convincingly observe that one can be jealous or infatuated without being conscious or aware of the jealousy or infatuation (Sloman, 2001).

Although feelings and emotions are closely related, many researchers distinguish these phenomena, where the term *feeling* refers to the subjective experience of emotions. Some emotions can occur unconsciously, while feelings are always conscious. Often vague knowledge or belief is also interpreted as a feeling. For instance, a person may say, "I have a feeling that it will rain tomorrow."

Emotions are even closer connected to moods. A mood is usually understood as a relatively long lasting, affective or emotional state. There is also another understanding in which mood is treated as a disposition to do something. Moods differ from simple emotions in that they are less specific, often less intense, less likely to be triggered by a particular stimulus or event, and longer lasting. Moods generally have either a positive or negative valence. A mood activates specific emotions. For instance, being irritable, people are often seeking an opportunity to become angry. Most academics agree that emotions and moods are related but distinct phenomena (Beedie, et al, 2005).

However, here we do not make strict distinctions between moods and emotions, considering them as affective states. As in the case with emotions, there is no generally accepted and sufficiently exact definition of an affective state. Researchers comprise a variety of different phenomena under the name *affective state*, such as emotions, moods, motivations, desires, pleasures, pains, attitudes, preferences, and even values (cf., for example, (Scheutz and Sloman, 2001)).

However, it is necessary to make a distinction between an emotion and the cause of this emotion. For instance, when the emotion of joy is caused by thoughts about some happy event, it is possible to ask whether these thoughts are part of that emotion (Problem 1). Moreover, there is no consent between researchers if it possible to have emotions without thoughts that cause these emotions (Problem 2).

The first problem is solved in a negative way as there are emotions that can occur unconsciously. A solution to the second problem is discussed below, and existence of (direct) affective information demonstrates that in general, emotions are independent of thoughts.

The problem of emotion classification was also intensively debated. According to some researchers, emotions are distinctive discrete states. Other researchers assume that emotions smoothly vary along one or several dimensions. A popular approach is to divide emotions into basic and complex categories. Complex emotions are built from basic emotions like primary colors blend together to form all other colors. However, researchers differ on which emotions are basic and what the number of basic emotions is.

In this context, René Descartes (1596–1650) introduced six basic emotions: *love*, *hate*, *astonishment*, *desire*, *joy*, and *sorrow*, while Immanuel Kant (1724–1804) proposed five basic emotions: *love*, *hope*, *modesty*, *joy*, and *sorrow*. Later William James (1842–1910) narrowed the category down to four basic emotions: *love*, *fear*, *grief*, and *rage*. All other emotions, he argued, were variations on those four emotions.

In contemporary psychology, there are also diverse ideas with respect what emotions are basic (cf., for example, (Ortony and Turner, 1990)). For instance, Plutchik considered eight basic emotions: *acceptance*, *anger*, *anticipation*, *disgust*, *joy*, *fear*, *sadness*, and *surprise*, while Weiner and Graham considered only two basic emotions: *happiness* and *sadness*. Izard considered ten basic emotions: *anger*, *contempt*, *disgust*, *distress*, *fear*, *guilt*, *interest*, *joy*, *shame*, and *surprise*, while Watson considered only three basic emotions: *fear*, *love*, and *rage*.

Another division of emotions involves three main groups: primary emotions, secondary emotions, and background emotions. *Primary emotions*, such as anger, fear, disgust and sadness, emerge in the stimulus-response chain of events as a complement to behavioral responses. Another complement is thinking. Thus, we can observe three types of responses to a stimulus: cognitive response (thinking), affective response (emotions), and effective response (action). LeDoux (1996;

2000) contends that basic (primary) emotions constitute one of the prime factors of adaptive behaviors related to survival.

In contrast to primary emotions, *secondary emotions* begin with thinking and cognition, following the pathway created by memory and learning. That is, images and other mental structures are associated with secondary emotions, and events triggering these structures also set off the associated emotions. It is conjectured that intuition is partially due to secondary emotions.

Background emotions are affective states that exist when there are no such strong emotions as anger, fear, interest, satisfaction or sadness. These emotions monitor the conditions of the body and brain as they function in the normal mode.

In general, all emotions often but not always are expressed through both physiological changes and stereotyped motor responses, especially responses of the facial muscles. These responses accompany subjective experiences that are not easily described on the whole. There is a disagreement whether emotion expressions are similar in different human cultures (Niedenthal, et al, 2006). Some researchers assume that people from different cultures have different ways of expressing their emotions. Some cultures tend to allow their members to express their emotions freely, while others motivate their members to hold their emotions back. This difference is very prominent in Eastern and Western cultures. Eastern cultures are seen as more community oriented, therefore individuals show less emotion in order to keep the well-being of the group intact. Members of Western cultures express their emotions much more in order to help themselves out, even if it involves negative emotions towards others.

Moreover, it is possible that emotions recognized in one culture are not specified in another culture. Here are some examples of emotions that are not clearly identified in Western countries. There is an emotion that is called *abhiman* in India. In English, it can be approximately described as a feeling of prideful loving anger. There is an emotion that is called *fureal* in Japan. In English, it can be approximately described as connectedness to someone else.

At the same time, other scientists insist on similarity in emotion display, e.g., facial expressions, across cultures supporting universality of emotions. Experimental results demonstrated universality in not only choice of the primary expression, but even the second most expressed emotion in each face, effectively finding cross-cultural agreement in emotion blends (Ekman, 1987).

In addition, researchers found that expression of the emotions is closely tied to the autonomic nervous system, and it therefore involves the activity of certain brain structures (e.g., brainstem nuclei, hypothalamus, and amygdala), as well as autonomic nervous system components (e.g., preganglionic neurons in the spinal cord, autonomic ganglia, and peripheral effectors). The brain system that directly coordinates emotional responses has been called the *limbic system*. At the level of the cerebral cortex, the two hemispheres apparently differ in their governance of the emotions, with the right hemisphere more critically involved than the left hemisphere.

Emotions and feelings impact reasoning and behavior. For instance, psychologists demonstrated how emotions influence decision-making and actions (Ketelaar and Clore, 1997; Ketelaar and Goodie, 1998). Besides, there is experimental evidence that the amygdala, which is part of the system of feelings and emotions, enhances and modulates response of cortical parts.

As both emotions and information are extremely important for people, it is natural to ask a question what are relations between information and emotions. One answer is evident. Information can be about anything. Thus, it is possible to consider information about emotions. Such information has been called *emotional information* although it is more relevant to call it *cognitive emotional information*. This type of information has been extensively studied in the area called *emotional intelligence*.

Emotional intelligence is understood as the ability to perceive emotion, integrate emotion to facilitate thought, understand emotions, and to regulate emotions to promote personal growth (Salovey and Mayer, 1990). It is usually assumed that people perceive emotions observing and analyzing external expression of emotions.

Emotional intelligence manifests itself in certain adaptive behaviors and according to contemporary understanding, has four components:

- Perceiving emotions is the ability to detect, discern and accept emotions in social environment, e.g., in faces, pictures, voices, texts, and other social symbols, including the ability to identify one's own emotions. Perceiving emotions is a basic aspect of emotional intelligence, as it makes all other processing of emotional information possible.
- Understanding emotions is the ability to comprehend "emotion language", to know how emotions influence people, and to discern complicated relationships among emotions. For instance, understanding emotions encompasses the ability to be sensitive to slight variations between emotions, and the ability to recognize and describe how emotions evolve over time.
- Using emotions is the ability to harness emotions to facilitate various cognitive and physical activities, such as thinking, problem solving, fighting, running, jumping, and swimming. The emotionally intelligent person can fully take advantage of his or her changing moods to best perform the task at hand. Such people instigate and develop definite emotions in themselves and others to facilitate a chosen activity. For instance, anger, as a rule, helps fighting. Fear prompted into opponents also helps fighting with them.
- Managing emotions is the ability of people to regulate emotions both in themselves and in others. Therefore, the emotionally intelligent person can take advantage of emotions, even negative ones, and manage them to achieve intended goals.

The emotional intelligence approach treats emotions as useful sources of cognitive information that help one to make sense of and navigate the social environment and individual behavior (Goleman, 1998). Studies show that individuals vary in their ability to process and utilize information about emotions.

Observations of these phenomena changed understanding of thinking as a psychological process and resulted in new directions in psychology directed at emotional intelligence and emotional information.

The most distant roots of emotional intelligence can be traced back to Darwin, who stressed the importance of emotional expression for survival and adaptation (Darwin, 1872). In 1930s, Robert Ladd

Thorndike (1910–1990) wrote about social intelligence, which included knowledge about emotional factors of people's behavior (Thorndike, 1936; Thorndike and Stein, 1937). David Wechsler (1896–1981) described the influence of non-intellectual factors on intelligent behavior, and further argued that conventional models of intelligence would not be complete until these models can adequately represent such factors (Wechsler, 1940). Howard Gardner (1975; 2006) built the theory of multiple intelligences, which included both *interpersonal intelligence* defined as the capacity to understand the intentions, motivations and desires of other people and *intrapersonal intelligence* defined as the capacity to understand oneself, to appreciate one's feelings, fears and motivations. In Gardner's view, traditional types of intelligence, such as measured by IQ (the Intelligence Quotient), fail to fully explain cognitive ability. In such a way, psychologists started to understand that traditional definitions of intelligence were lacking in ability to fully explain performance outcomes and it was necessary to include other psychological traits, such as emotions, in intelligence definitions. From the general theory of information perspective, emotions can react to the cognitive information that comes to the level of subconscious and is not registered on the level of unconsciousness. This and conscious reasoning about registered emotional display is the main source of emotional intelligence.

The first use of the term *emotional intelligence* is attributed to Leuner (1966). Then Payne (1983/1986) and Greenspan (1989) also used the concept of emotional intelligence. They were followed by Salovey and Mayer (1990) and Goleman (1995).

When people understood importance of emotional intelligence, they came to the natural conclusion that to achieve appropriate level of artificial intelligence, it is necessary to develop emotions or, at least, their imitation for computers. This idea resulted in the development of affective computing — a branch of artificial intelligence aimed at the design of computer systems and devices that can recognize and interpret human emotions, processing their symbolic representations (Picard, 1997; 1998). Besides researchers in artificial intelligence (AI) have come to the conclusion that machines cannot reach human intelligence without emotions (cf., for example, (Simon, 1967; Pfeifer, 1988; Minsky, 1998;

Sloman, 2001; Scheutz and Sloman, 2001; Scheutz, 2002; Scheutz, 2002a)). Consequently, engineers and software developers have started designing computer systems and devices that can exhibit innate emotional capabilities and/or are able of efficiently and convincingly simulating emotions.

Moreover, as Sloman (2001) writes

"The study of emotion in cognitive science and AI has recently become very fashionable, with a rapidly growing number of workshops, conferences and publications on the topic, some reporting attempts to produce emotional behavior in robots or software agents, some concerned with detecting and responding to emotions in human users of computing systems, and some aiming to model and explain human emotions."

As we can see, emotional intelligence engages thinking about emotions and thus, is associated exceptionally with cognitive information. The problem is whether information can influence emotions only through cognitive processes or there is specific emotional information, which directly acts on emotions without involving thinking and deliberation.

To consider some phenomenon as an information process, we have, according to the general theory of information, to find the corresponding infological system. As we have seen, a model example of such a system is the system of knowledge related to cognitive information. However, this is only one type of infological systems. There are also other important types. In spite of this, when people speak or think about information, they implicitly have in mind only cognitive information. That is why people relate information to eliminated uncertainty in knowledge, to news, to novelty, etc. This is also the reason why even experts do not make clear distinctions between information and knowledge or information and data. It is not a simple task to separate substantial essences, such as knowledge and data, from cognitive information, which transforms knowledge and data into one another (cf. Section 2.5).

However, regardless of the usual identification of information with the cognitive information, the general theory of information gives means to discover other kinds and types of information. They exist and in many cases are even more important than cognitive information. There is empirical evidence that combining cognitive information with other kinds may give an unexpected result. For instance, news on TV have been traditionally considered only as a kind of cognitive information source, namely, giving knowledge about current events to people. Nonetheless, when an attempt was made to enrich news programs with emotional content (in terms of the general theory of information, it means adding direct emotional/affective information), the success of the enterprise was immensely great. We can now observe this in the case with CNN and other news broadcasting programs. Reinventing TV news, CNN transformed television from our living-room program to a multichannel window into the world. In some way, this has made the audience not simply a viewer and listener but a virtual participant of the events. With this emotions have joined cognitive information in the news and the channel became one of the most successful entertainment media.

Another example is given by the entertainment-education soap operas, which were created by Miguel Sabido (Nariman, 1993). Entertainment-education soap operas pursue the two-fold aim. Without sacrificing commercial appeal, they are designed according to elements of communication and behavioral theories in order to reinforce specific values, attitudes, and behaviors oriented at the advancement of viewers. As it is known, purely informational or educational campaigns that are directed to this goal in many countries represent an enormous financial drain of resources and time. Only synthesis of entertainment and education provides for efficient delivery of educational messages. Consequently, learning becomes entertainment and entertainment incorporates education. It is not a simple task to accomplish this because learning is a kind of cognition aimed at knowledge acquisition, while entertainment mostly appeals to emotions and feelings of people. It is necessary to remark that such a synthesis is a real art because in pursuing this double-folded goal without necessary skills and talent, it is possible to achieve neither entertainment nor education of the audience.

Thus, we see that informing may include enjoyment and amusement as well as enjoyment may encompass persuasion and instruction. Consequently, we come to the conclusion that such processes that amuse and/or persuade are not alien to information processes.

Nevertheless, in the face of all empirical evidence, it seems (at least, at the first glance) that information has very little to do (if anything) with emotions (pleasure, amusement, and gratification) and intentions (interest and attention) because information is usually considered as something opposite to emotions and intentions. However, this is a fallacy based on the conventional model of information. In contrast to this, taking means from the general theory of information, we see that it is possible to find corresponding infological systems in an individual, demonstrating that emotions and intentions are also caused by specific kinds of information, which differ from the conventional type of information. For instance, this understanding is reflected in the statement that three main goals of communication are to inform, to entertain, and to persuade.

People, as other complex systems, have several infological systems. To find an infological system directly related to emotions and feelings, the *theory of the triune brain and behavior* developed by Paul MacLean (1913–2007) is utilized.

The main conception of his approach is existence of three levels of perception and action that are controlled by three corresponding centers of perception in the human brain. These three centers together form the triune brain and have the structure of a triad (MacLean, 1973; 1982).

The idea that the brain is organized in functional levels can be dated to Herbert Spancer's mid-19th century speculations that each step of evolution added a new level of brain and a new level of behavioral complexity (cf. (Kolb and Whishaw, 1990)). The brain may be anatomically structured into three principal parts: hindbrain (including the spinal cord), midbrain, and forebrain. According to the theory of MacLean, the neural basis, or framework, of the brain consists of three parts: the spinal cord, hindbrain, and midbrain. In addition to it, centuries of evolution have endowed people with three distinct cerebral systems. The oldest of these is called the *reptilian brain* or R-*complex*. It programs behavior that is primarily related to instinctual actions based on ancestral learning and memories. The reptilian brain is fundamental in acts such as primary motor functions, primitive sensations, dominance, establishing territory, hunting, breeding, and mating.

Through evolution, people have developed a second cerebral system, the *limbic system*, which MacLean refers to as the *paleomammalian brain* and which contains hippocampus, amygdala, hypothalamus, pituitary gland, and thalamus. This system is situated around the R-complex, is shared by humans with other mammals, and plays an important role in human emotional behavior.

The most recent addition to the cerebral hierarchy is called the *neomammalian brain*, or the *neocortex*. It constitutes 85% of the whole human brain mass and receives its information from the external environment through the eyes, ears, and other organs of senses. This brain component (neocortex) contains cerebrum, corpus callosum, and cerebral cortex. The cerebrum and cerebral cortex are divided into two hemispheres, while the corpus callosum connects these hemispheres. The neocortex deals with information in a logical and algorithmic way. It governs people creative and intellectual functions like social interaction and advance planning. The left hemisphere works with symbolic information, applying step-by-step reasoning, while the right hemisphere handles images processed by massively parallel (gestalt) algorithms.

In the development of neurophysiology and neuropsychology, MacLean's theory was used as a base for the Whole Brain model, developed by Herrmann (1988). The main idea of this development is a synthesis of the triune model with the two-hemisphere approach to the brain functioning. As it is known (cf., for example, (Kolb and Whishaw, 1990)), the left hemisphere performs logical and analytical functions, operating with symbolic information representation. It is also supposed that the left hemisphere works in the sequential mode on the level of consciousness. The right hemisphere processes images and realizes intuitive, creative and synthesizing functions of the brain. It is also supposed that the right hemisphere works in the parallel (concurrent) mode on the level of consciousness. In a similar way, Herrmann differentiates functioning of the left parts and right parts of both the cerebral and limbic regions of the brain.

The theory of the triune brain (reptilian, old mammalian, and new mammalian) is used as a metaphor and a model of the interplay between instinct, emotion, and rationality in humans. Cory (1999) applied this

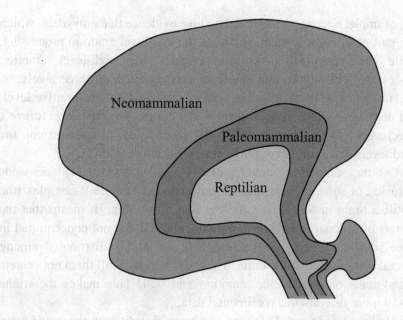

Figure 2.16. The triune brain

model to economic and political structures. In Cory's schema, the reptilian brain mediates the claims of self-interest, whereas the old mammalian brain mediates the claims of empathy. If selfish interests of an individual are denied for too long, there is discontent due to a feeling of being unjustly treated. If empathic interests are denied for too long, there is discontent due to guilt. In either case, the center of intelligence at the prefrontal cortex plays the role of a mediator. Its executive function is required to restore balance, generating the reciprocity required for effective social and economic structures.

While the triune brain has become a well-known model in contemporary psychology, MacLean's theory caused several objections on the ground of the development and structure of the triune brain system. First, there is evidence that the, so-called, paleomammalian and neomammalian brains appeared, although in an undeveloped form, on much earlier stages of evolution than it is assumed by MacLean. Second, there are experimental data that in the neocortex, such regions exist that are homological to the, so-called, paleomammalian and reptilian brains.

For example, neuropsyhological data give evidence that amygdala, which is a part of a limbic system, performs the low-level emotion processing, while the ventromedial cortex performs the high-level emotion processing. This shows that emotions exist, at least, on three levels: on the subconscious level of limbic system, on the conscious intuitive level, and on the conscious rational level in the cortex. The first level utilize s direct affective information, while the second and, to some extent, the third levels make use of cognitive emotional information.

At the same time, development of the system of will demands inclusion of some regions that are not included into the R-complex (the reptilian brain in MacLean's theory) into this system. It means that the centers of rational intelligence, emotion and will are not concentrated in three separate regions of the brain but are highly distributed among several components of the brain. Thus, it is better to call them not centers but systems of intelligence, emotion and will. This makes the triune model more relevant to experimental data.

However, even psychologists who have objections to the triune brain model admit that it is a useful, although oversimplified, metaphor. Indeed, the structure presented as the triune brain is based on a sound idea of three functional subsystems of the brain: the system of reasoning, system of emotions and feelings, and system of will and instincts. In what follows, we use the term *triune brain* for the system that comprises these three functional subsystems.

The triune brain system is a neuropsychological construction that has to be related to a model of personality to achieve better understanding of information processes in the human psyche. Psychologists have elaborated many different models of personality (cf., for example, (Monte, 1980; Ryckman, 1993; Kshirsagar and Magnenat-Thalmann, 2002; McCrae and Allik, 2002)). Some models have been suggested by sociologists. Here we use the extensive model of personality developed by Burgin (1997a; 1998). This model has a hierarchical structure and incorporates other systemic models of personality, both psychological and sociological.

According to the extensive model, the material component of personality consists of ten layers (levels) and has the structure represented in Figure 2.17.

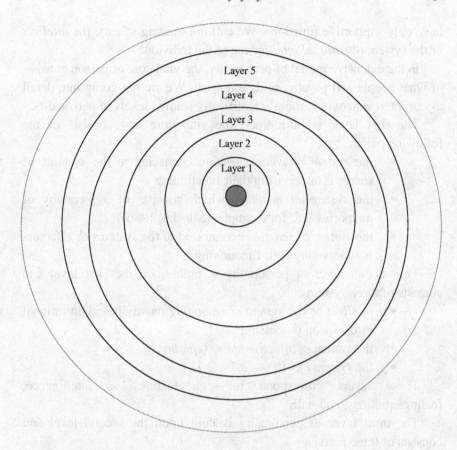

Figure 2.17. The structure of the extensive model of personality

Each layer consists of three parts. Each part constitutes a definite system and represents some essential feature of a person. On the psychological level of personality such part has three components: a system called the *structural brain scheme* (actual or virtual), a *process* that is realized in this system and through which the corresponding *feature* emerges, and what is produced by this system. For instance, intelligence of a person is a complex feature (trait) distributed in psyche and generated by collective functioning of different schemes in the nervous system of a human being. However, one of the schemes contributing to intelligence plays the leading role, while other systems have only supportive functions. We call this leading scheme the *intellect* or the system of *rational intelligence* of the individual.

In the extensive model of personality, the vital core of person consists of what people call psyche, soul, and spirit. We are not going into detail as our interest here is in more scientifically studied levels of personality.

The first layer is built upon the vital core and consists of the following parts:

- the *system of comprehension* connected to the systems of senses is supervising their functioning
- the *system of memory*, which consists of a hierarchy of memories (cf., for example, (Minsky, 1986))
- the *system of reactions* connected to the systems of effectors is supervising their functioning

The second layer of personality is built upon the first level and consists of three systems:

- *intellect* or the *system of reasoning* mostly based on rational information processing
- the *system of affective states* (*emotions*)
- the *system of will*

These layers correspond to such features as intelligence, feeling/emotions, and will.

The third layer of personality is built upon the second level and consists of three parts:

- the *system of individual gifts* (*talents*)
- *character*
- *temperament*

The fourth layer of personality is built upon the third level and contains three systems:

- the *system of values*
- the *system of norms*
- the *system of ideals*

The fifth layer of personality is built upon the fourth level and contains three systems:

- the *system of knowledge*
- the *system of skills*
- *experience*

The sixth layer of personality is built upon the fifth level and contains three systems:

- the *system of inclinations*
- the *system of interests*
- the *system of needs*

The seventh layer of personality is built upon the sixth level and contains three spheres:

- the *world outlook*
- the *ethic-moral sphere*
- the *esthetic sphere*

The eighth layer of personality is built upon the seventh level and contains three dynamic systems:

- the *system of attitudes*
- the *system of dispositions*
- the *system of precepts*

The ninth layer of personality is built upon the eighth level and contains three systems:

- *activity*
- *communication*
- *behavior*

The tenth layer of personality is built upon the ninth level and contains three complexes:

- *perception of this person by others*
- *attitudes of others to this person*
- *reactions of others to this person*

Our interest in information and information processing brings us to the second layer, components of which are represented by corresponding structures in the brain. Intelligence, as a component of the second layer, is not simply a static system. It is also a *process* in the nervous system that expresses itself in an efficient behavior of an individual and the *feature* of personality that is the result of this process. Thus, we have three constituents of intelligence: *personality trait* (*feature*), *psychic process*, and *behavior*. The same is true for two other components of the second layer, emotions and will. In other words, emotions and will include processes in human psyche, traits of personality, and behavior that is the result of these processes.

Comparing the second layer of personality to the triune brain, we come to the conclusion that during many years of evolution, the reptilian brain has expanded to a more complex system, acquiring additional functions for the advanced regulation of the human behavior. That is why we call this system in the brain, which includes the reptilian brain, by the name the *System of Will and Instinct* (SWI).

Thus, we come to the following structure (cf. Figure 2.18), which represents three basic systems of the brain:

- the *Center* (or *System*) *of Rational Intelligence* (also called *System of Reasoning*);
- the *System* (*Center*) *of Emotions* (or more generally, *of Affective States*);
- the *System* (*Center*) *of Will and Instinct.*

All three centers of the brain are schemes in the sense of the schema theory, which is developed as a specific direction of the brain theory (Anderson, 1977; Arbib, 1992; Armbruster, 1996; Burgin, 2006). According to this theory, brain schemes interact in a way of concurrent competition and coordination. All these interactions are based on physical processes but have an inherent informational essence related to a specific type of information. Information processes in the brain are more exactly reflected by the theory of the *triadic mental information* than by the conventional information theory that deals only with cognitive information.

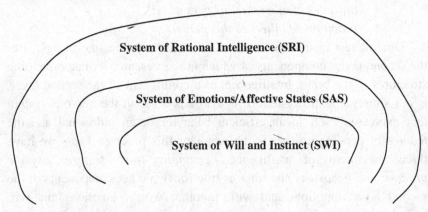

Figure 2.18. Three basic systems of the brain

If we take the more standard structuring of the brain, we also find these three systems. In the conventional setting (cf., for example, (DeArmond, et al, 1989; Russell, 1992)), the brain includes three components: the forebrain, midbrain, and hindbrain.

The *forebrain* is the largest division of the brain involved in a wide range of activities that make people human. The forebrain has a developed inner structure. It includes the *cerebrum*, which consists of two *cerebral hemispheres*. The cerebrum is the nucleus of the System (Center) of Rational Intelligence.

Under the cerebrum, is the *diencephalon*, which contains the *thalamus* and *hypothalamus*. The thalamus is the main relay center between the *medulla* and the *cerebrum*. The hypothalamus is an important control center for sex drive, pleasure, pain, hunger, thirst, blood pressure, body temperature, and other visceral functions. The forebrain also contains the limbic system, which is directly linked to the experience of emotion. The *limbic system* is the nucleus of the System (Center) of Emotions (or more generally, of Affective States).

The *midbrain* is the smallest division and it makes connections with the other two divisions — forebrain and hindbrain — and alerts the forebrain to incoming sensations.

The *hindbrain* is involved in sleeping, waking, body movements and the control of vital reflexes such as heart rate, blood pressure. The structures of the hindbrain include the pons, medulla and cerebellum. The hindbrain is the nucleus of the System (Center) of Will and Instinct.

Note that the standard structure of the brain also contains several fundamental triads. Taking the structure of the brain in general, we have:

Prosencephalon (**Forebrain**) contains:
- Diencephalon
- Telencephalon

Mesencephalon (**Midbrain**)

Rhombencephalon (**Hindbrain**) contains:
- Metencephalon
- Myelencephalon

It gives us three fundamental triads:

(forebrain, midbrain, hindbrain),

(Prosencephalon, Diencephalon, Telencephalon)

(Rhombencephalon, Metencephalon, Myelencephalon)

The *prosencephalon* consists of the *telencephalon*, *striatum*, and *diencephalon*, giving one more fundamental triad:

(telencephalon, striatum, diencephalon)

The midbrain (mesencephalon) consists of the *cerebrum*, *thalamus*, and *hypothalamus*, giving one more fundamental triad:

(cerebrum, thalamus, hypothalamus)

The hindbrain is made of the *cerebellum*, *pons*, and *medulla*, giving one more fundamental triad:

(cerebellum, pons, medulla)

The brainstem consists of the *midbrain*, *medulla oblongata*, and the *pons*, giving one more fundamental triad:

(midbrain, medulla oblongata, pons)

One more fundamental triad is formed by the two hemispheres of the brain and the corpus callosum, which is a bundle of axons and connects these two hemispheres:

(left hemisphere, corpus callosum, right hemisphere)

The pons relays sensory information between the cerebellum and cerebrum. It give us one more fundamental triad:

(cerebellum, pons, cerebrum)

The rhombencephalon is comprised of the *metencephalon*, the *myelencephalon*, and the *reticular formation*, giving one more fundamental triad:

(metencephalon, reticular formation, myelencephalon)

The cerebellum may be divided into the anterior, posterior and flocculonodular lobes, giving one more fundamental triad:

(anterior lobe, flocculonodular lobe, posterior lobe)

It is also possible to consider the following fundamental triads:

(cerebrum, thalamus, medulla)

and

(cerebrum, hypothalamus, limbic system)

The System of Rational Intelligence realizes rational thinking. It includes both symbol and image processing, which go on in different hemispheres of the brain. The System of Emotions governs sensibility and the emotional sphere of personality. The System of Will and Instincts directs behavior and thinking. Two other systems influence behavior only through the will. For instance, a person can know that it is necessary to help others, especially, those who are in need and deserve helping. But this person does nothing without a will to help. In a similar way, we know situations when an individual loves somebody but neither tells this nor explicitly shows this due to an absence of a sufficient will.

It is necessary to remark that discussing the will of a person, we distinguish *conscious will, unconscious will,* and *instinct.* All of them are controlled by the Center of Will and Instincts (SWI). In addition, it is necessary to make distinctions between thoughts about intentions to do something and the actual will to do this. Thoughts are generated in the Center of Rational Intelligence, while the will dwells in the Center of Will and Instincts. In other words, thoughts and words about wills, wishes, and intentions may be deceptive if they are not based on a will.

Will is a direct internal injunction, as well as any kind of motivation. That is, the forces that act on or within an organism to initiate and direct behavior, has to be transformed into a will in order to cause the corresponding action. The will is considered as a process that deliberates on what is to be done (Spence, 2000).

Computer science metaphors demonstrated their value in the study of the brain. For instance, neural networks provide a mathematical model (although a very approximate one) of brain structures. It is also possible to picture the triadic structure of the brain in terms of computers.

Will/instincts encompass instructions and programs (in general, software) of the psychic behavior/activity/functioning.

Emotions are states (affective states) of the psyche as a device (or the hardware) of the psychic behavior/activity/functioning.

Knowledge in all its forms (as the logic-linguistical, model, axiological and problem-heuristic knowledge) is the processed material (or infware, cf. (Burgin, 2005)) for the device (or hardware) of the psychic behavior/activity/functioning.

Taking each of these three centers, SRI, SAS, and SWI, as a specific infological system, we discover three types of information. One is the conventional information that acts on the center of reasoning and of other higher brain functions (SRI), which is situated in the neocortex. This information gives knowledge, changes beliefs and generates ideas. Thus, it is natural to call it *cognitive* information. Information of the second type acts on the system of emotions (SAS), which includes the paleomammalian brain. It is natural to call this information by the name *direct emotional information*, or *direct affective information* or *emotive information*. Information of the third type acts on the System of Will and Instinct (SWI), which contains the reptilian brain. It is natural to call this 'information by the name *direct regulative* or *direct effective information*.

Thus, triadic mental information has three dimensions:

Cognitive information changes the content of the SRI, which includes the knowledge system (thesaurus) and neocortex (neomammalian brain) as its carrier.

Direct emotional/affective information changes the content of the SAS, which includes the paleomammalian brain (limbic system).

Direct regulative/effective information changes the content of the SWI, which includes the reptilian brain or R-complex.

However, in general, emotions form only one part of affective states, which also include moods, feelings, etc. That is why in general, direct affective information is more general than direct emotional information. However, as there is no consensus on the differences between emotions and affective states, these two names of information (direct emotional and direct affective) are used interchangeably.

In a similar way, direct regulative information is more general than direct effective information. However, these two names of information are also used interchangeably.

Actions of these three types of information are represented in Figure 2.19.

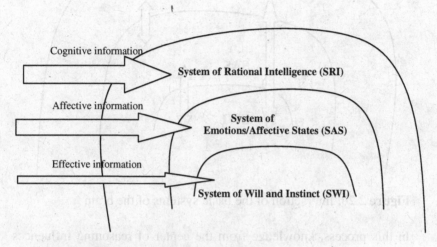

Figure 2.19. Action of information on the triune brain

It is necessary to remark that it is possible to separate other types and kinds of information, taking other subsystems of the brain and/or of the whole nervous system as infological systems. For instance, people have five senses: vision, hearing, smell, touch, and taste. A specific system in the nervous system including the brain corresponds to each of the senses. For instance, the visual cortex in the brain controls vision. Consequently, we can consider visual information. In the same way, it is possible to discern sound, smell, tactile, and taste information. Empirical considerations have already demonstrated distinction between these types of information. At the same time, the general theory of information explains and theoretically grounds these distinctions.

All three centers constantly interact (cf. Figure 2.20), exchanging information. Rephrasing the expression of Marshall McLuhan (1964), we can say that "*all types of information co-exist in a state of active interplay*".

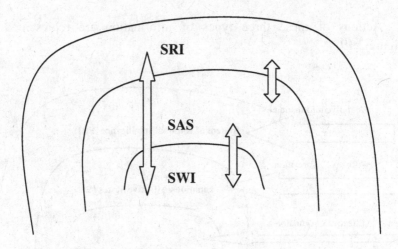

Figure 2.20. Interaction of the basic systems of the brain

In this process, knowledge from the center of reasoning influences emotions and behavior. The cognitive control of emotions is the basic postulate of the cognitive therapy (Beck, 1995). The scheme of such cognitive control, as a part of self-regulation, is given in the Figure 2.21.

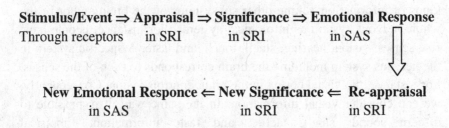

Stimulus/Event ⇒ Appraisal ⇒ Significance ⇒ Emotional Response
Through receptors   in SRI   in SRI   in SAS

New Emotional Responce ⇐ New Significance ⇐ Re-appraisal
     in SAS   in SRI   in SRI

Figure 2.21. Interaction of the basic systems of the brain

Will controls emotions and reasoning. For instance, recollections of past events can provoke grief, amusement or happiness of an individual. Knowledge of civil law restricts behavior of citizens. As psychologists have discovered, positive emotions help in cognition, learning and remembering. Will changes direction of the cognitive processes of a scientist. Conscious or unconscious will cause attraction or distraction of attention of students and so on.

Interactions between the basic brain systems imply dependencies between thinking, emotions, and actions of people. Emphasizing some of these relations, psychologists build their theories and psychotherapists develop their therapeutic approach. Recently, the so-called "cognitive revolution" has taken hold in the United States and many other countries around the world. It influenced both psychology, resulting in the emergence of cognitive psychology (Neisser, 1976), and psychotherapy, inspiring creation of cognitive therapy (Beck, 1995). In psychology, the word *cognitive* often means thinking in many contexts of contemporary life (Freeman and DeWolf, 1992). The cognitive therapeutic approach begins by using the amazingly powerful reasoning abilities of the human brain. This is important because our emotions and our actions are not separate from our thoughts. They are all interrelated. Thinking (SRI) is the gateway to our emotions (SAS) — and our emotions are the gateway to our actions through motivation and will (SWI). This is only another way of saying that information from the System of Rational Intelligence (SRI) goes to the System of Emotions (SAS) — and from it to the System of Will and Instincts (SWI) that controls our actions. Consequently, the cognitive psychotherapeutic approach, which has been successfully utilized for treating many mental disorders, gives additional supportive evidence for the theory of the triune brain and behavior, as well as for the theory of the triadic mental information. The latter explains that while going from the System of Rational Intelligence to the System of Emotions to the System of Will and Instincts information is transformed from cognitive information, to direct emotional/affective information to direct effective/regulative information (cf. Figure 2.22).

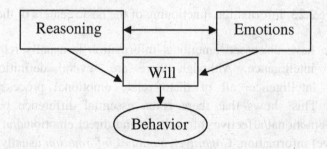

Figure 2.22. Interaction between components of personality

However, functioning of all three basic systems in the brain is not always perfectly coordinated. All people encounter situations in which feelings contradict will, emotions overflow logical reasoning or desire to do something interferes with rational arguments. Discrepancy between different centers of the brain is sometimes very hard for a person, but such a discrepancy exists. For example, there is an experimental evidence that amigdala operates independently of cognitive awareness. This partial independence of the centers enabled MacLean to call each of them a brain and to consider actual brain of human being as a triune brain.

Existing information transactions and interactions in the triune brain are represented by Figure 2.23.

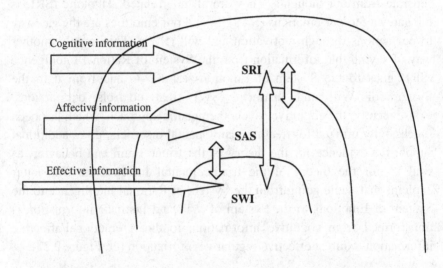

Figure 2.23. Information functioning of the basic centers of the brain

As we have discussed emotional information is usually related to emotional intelligence. Although there are several definitions of emotional intelligence, all of them relate emotional processing to cognition. This shows that there is an essential difference between cognitive emotional/affective information and direct emotional/affective (or emotive) information. *Cognitive emotional information* usually called simply emotional information is cognitive information about emotions (of us or others), while direct emotional/affective information is information that directly influences emotions by its impact on the System of Affective States. Cognitive emotional information also influences emotions but only indirectly. At first, emotional information is transformed into affective information. Then this affective information goes to the System of Affective States, where emotions are regulated and controlled. Experimental studies confirm this conjecture. For instance, Knapska, et al (2006) built an experimental rat model of between-subject transfer of emotional information and its effects on activation of the amygdala. Experiments demonstrated that undefined by the authors emotional information transferred by a cohabitant rat can be evaluated and measured and that it evokes very strong and information-specific activation of the amygdala.

As it is explained above, terms *emotional information*, *affective information* and *effective information* are also used, having different meaning. Namely, many researchers use terms *emotional information* and/or *affective information* meaning cognitive information about or related to emotions (cf., for example, (Sanders, 1985; Severino, et al, 1989; Keane and Parrish, 1992; George and McIllhagga, 2000; Drikker, 2002; Ochsner, et al, 2002; Bosmans and Baumgartner, 2005; Knapska, et al, 2006)). It would be natural to call this kind of information by the name *cognitive emotional information* or *cognitive affective information*. This type of information is very important as it has two dimensions — cognitive and emotional. At the same time, there is information that directly influences the emotional system of the brain. This information was discovered in the general theory of information (Burgin, 2001) and is called *direct emotional information* or *direct affective information* to discern it from cognitive emotional information. It is possible that the cognitive content (measure) of a message is zero, but this message has a big emotional content (measure). Experienced politicians, lawyers, and artists use this peculiarity on the empirical level. The property to attract people on the unconscious level is called charisma.

In a similar way, the term *effective information* sometimes means cognitive information that makes people to behave in a definite way, giving instructions and orders (cf., for example, (Lewis and Newton, 2006; North, 2008)). It would be natural to call this kind of information by the name *cognitive effective information*. Computer programs written in a procedural programming language contain cognitive effective information for computers. In army, officers give a lot of cognitive effective information when they address their subordinates. This type of information is very important as it has two dimensions — cognitive and regulative. At the same time, there is information that directly influences the system of will and instincts in the brain. This information was discovered in the general theory of information (Burgin, 2001) and is called *direct regulative information* or *direct effective information* to discern it from cognitive effective information. It is possible that the cognitive content (measure) of a message is zero, but this message has a big regulative content (measure). Many people, such as experienced politicians, lawyers, experts in marketing, sales persons and psychiatrists, use this phenomenon on the empirical level. Parables often contain implicit effective information.

Now let us consider if there is experimental evidence supporting existence of not only cognitive emotional information but also of direct emotional information. There is strong psychological evidence that supports the conjecture that cognitive emotional information is different from direct emotional information.

Loewenstein, et al (1991) explain that processes of decision making include *anticipatory emotions* and *anticipated emotions*. Anticipatory emotions are defined as immediate visceral reactions, e.g., fear, anxiety or dread to risk and uncertainties, while anticipated emotions are typically not experienced in the immediate present but are expected to be experienced in the future (disappointment or regret). Both types of emotions serve as additional sources of information for decision-making.

For instance, research shows that happy decision-makers are reluctant to gamble. The fact that a person is happy would make him or her decide against gambling, since he or she would not want to undermine his or her happy feeling. This can be looked upon as "mood maintenance". A happy person does not think, "I am happy. So, I will not risk." Contrary to this, happiness comes as direct emotional information to the decision center of the brain and in such a way influences made decisions.

According to the information hypothesis, feelings during the decision process affects people's choices, in cases where feelings are experienced as reactions to the imminent decision. If feelings are attributed to an irrelevant source to the decision at hand, their impact is reduced or eliminated.

Zajonc (1980) argues that emotions are meant to help people take or avoid taking a stand, versus rational reasoning that helps people make a true/false decision. However, following ones emotions may be helpful in some cases and misleading and even dangerous in other situations.

It is possible that a person thinks about something without reception any external information and this thinking involves emotion even without thinking about emotions. For instance, the emotion of joy can be caused by thoughts about some event even without thinking about emotions connected to or caused by this event.

Some theorists argue that some emotions can be caused without any thoughts or any cognitive activity at all (cf., for example, (LeDoux, 1996)). They point to very immediate reactions, e.g., bodily responses, as well as to the conjectured emotions of infants and animals as justification here. When emotions can be caused without any thoughts, then the cause is affective information, which goes directly to the Center of affective states (SAS). This information is defferent from cognitive information because reception of cognitive information results in cognitive activity, while reception of direct emotional information results only in emotions.

Additional evidence that supports independent existence of direct emotional information is given by the Damasio's theory of the role of emotions in decision-making (Damasio, 1994; 1999). This theory is based on the Hypothesis of Somatic Marker. This hypothesis explains how decision processes are organized in the brain and the role of emotions in these processes.

The main argument of Damasio is that a purely rational solution to many problems that people encounter would require a very big amount of time because such a decision has to be based on dealing with all possibilities involved, predicting outcomes, and calculating costs and benefits. In turn, this will demand memory capacity and time that are not available in the majority of decisions made by people. To reduce time and memory demands, the brain involves emotions in decision-making.

Damasio describes the process of decision-making in the following way. At first, the prefrontal cortex (i.e., the main part of the Center of Rational Intelligence (SRI)) creates images of different scenarios that may occur as a consequence of each possible decision. These images contain not only purely descriptive elements of the situation representation, which carry cognitive information about possible outcomes. In addition, these images contain direct emotional information, which corresponds to the emotional reaction to possible outcomes. An individual does not think, for example, that meeting a friend would make her happy, but the image of that friend evokes positive emotions in the Center of affective states (SAS). Thus, the corresponding image contains affective information. This information is called a somatic marker of the object, event or process represented by the image. Somatic markers give a very compact representation of possible decisions and the Center of Rational Intelligence (SRI) can easily process cognitive information carried by somatic markers. When somatic markers come to the Center of Rational Intelligence, the person has the so-called gut feeling that some decisions are better than others.

As Simon (1998) writes, it should be stressed that "somatic marking" of the possibilities presented by the images does not always go in a conscious way. Thus, the brain can acquire both cognitive information and affective information unconsciously.

However, after the Center of Rational Intelligence receives somatic markers, and the person has the so-called gut feeling, some of the possibilities are rationally re-evaluated.

There are more arguments supporting separate existence of direct emotional information. For instance, chemicals and electrical signals can change emotions without any preliminary change of the cognitive or psychomotoric systems. It means that such chemicals and electrical signals carry direct emotional information for the brain.

Analyzing three key types of emotions, we see that characteristics of primary emotions show that they come out as a result of receiving direct emotional/affective information. Direct affective information triggers (induces) primary emotions.

Characteristics of secondary emotions show that they come forward as a result of receiving cognitive emotional/affective information. Cognitive emotional information triggers (induces) secondary emotions.

Background emotions are present when both primary and secondary emotions go away.

It is necessary to remark that one and the same carrier (e.g., text or picture) can contain different types of information (e.g., cognitive and emotional information). It is similar to a situation when some text, even such simple as one sentence, contains information about different subjects and objects. For instance, let us take the sentence: "Tom quarreled with Jane when he was at the party, which was organized by the AMS". This sentence informs us about two persons, Tom and Jane, one organization — the American Mathematical Society (AMS), and an event, the party, which was organized by the AMS.

News on TV can include not only cognitive but also emotional information. For example, when people see such catastrophes as earthquakes, they have different emotions and feelings, such as pity and horror. At the same time, a historical melodrama can give definite cognitive information about historical events and do not only induce emotions.

The separation of three types of information also makes it possible to achieve deeper understanding of the role of communication when we say that three main goals of communication are to inform, to entertain, and to persuade. More correct form is to state that communication gives knowledge, entertains, and regulates behavior.

The discovery of two additional types of information also shows that entertainment needs all three types of information, although the emphasis is on the emotional information. In other words, direct affective (emotive) information is the most important for entertainment. The development of better entertainment products, such as films, TV programs, etc., demands better knowledge about direct affective information. However, people know about feelings and emotions but do not know about this type of information. Moreover, as it is stated in (Minsky, 1986), *"many people think emotion is harder to explain than intellect"*. Development of the theory of affective information, as a part of the theory of the triadic mental information, might help to explain the emotional sphere of a personality.

Besides, good entertainment attracts and keeps attention, and attention is related to the system of will. Really, because of its finite computational resources, the human brain must process information selectively in a variety of domains. For instance, we may limit brain processing according to some restrictions. As such restrictions, we usually have a subset of the many possible objects that could be perceived; a subset of the many possible decisions that could be made; a subset of the many possible memories that could be accessed; or a subset of the many possible actions that could be performed. Consequently, without direct effective/regulative information, entertainment cannot achieve its goal. Direct regulative/effective information and regularities of the functioning of the Center of Will and Instinct are the keys to attention.

Direct emotional information has a great impact on processes in the memory and thus, influences education. For instance, it would be hardly possible to find an example of the artwork that has produced a strong impression and at the same time not printed in memory by any details. The modern psychophysiology connects the memorizing process to the emotional support.

A skilled teacher knows that a theme will be acquired by students only by keeping their interest. The important specification is that the condition of interest, without which the perception can not be effective, is the optimum balance between new information and known information.

According to the information theory of requirements and behavior developed by the physiologist Simonov (1986; 1998), an emotion is a psychological reaction to the information or other deficiency and evaluation of the probability of its satisfaction. The low probability of the goal's achievement leads to negative emotions, such as fear, anxiety, anger, grief, etc., while the high probability results in positive emotions, such as delight, pleasure, happiness, etc.

The perception efficiency can be determined by the measure of information deficiency. The optimum deficiency, i.e., the relation between the unknown and known information, derivates the effective emotional response, what promotes the assimilation, the fastening of the new information, the development of real potential of representations and feelings. The too small deficiency of the information does not result in emotional excitation and do not promote the perception. The monotony is associated with boredom and exhaustion (Drikker, 2002).

However the surplus of the information is also unproductive. Certainly, the operative memory always processes the information, but for the further active usage the information should be rationally readdressed in places of a constant keeping, and the neuron ways of its access to it are precisely registered. It is well known, the destruction of the file structure on the disk is fatal: though the record is not destroyed physically, but the information is practically inaccessible. The superfluous current of the information creates the similar situation. The certain part of the information is saved in ROM memory and it can be even shown in some circumstances, but it is impossible to reliably make such information actual. Besides the large volume of the information causes overstrain of the mentality, and the protective reaction, shunting the impressions flow, disconnects the attention.

Discovery of new kinds of information, as well as achievement of the new level of understanding the essence of information, allow us to solve the important long-standing problem of contemporary information technology whether computers work with and process only data or also information and/or knowledge. The general theory of information explains that processing data, computers, as a rule, also process information contained in these data.

Computers work with data. Sometimes process raw data, sometimes they compute with interpreted data, and sometimes handle attributed data. Data are information representations. Computers change representation of information and often this also changes information in those representations. Usually new forms of information are more consumable, usable and accessible. For instance, data mining and knowledge acquisition change representation of information, making it more compressed, transparent or/and learnable.

It is similar to the situation with energy. It is usually assumed that power stations produce energy. It is not a correct understanding. Actually they change carriers, or representation, of energy. Usually energy contained in oil or gasoline or uranium or in solar rays or in water current is changed to energy contained in electric current or electric potential. This new form of energy is much more consumable, usable and accessible.

In addition, it is possible to better understand the role of information in computation and communication. On the one hand, computation is information processing when representation of information changes and often this also changes information in those representations. Communication is information exchange by sending and receiving information carriers. On the other hand, all these processes are powered by energy and controlled by information. For instance, when a computer is switched on, the action is not only giving energy to the computer but also giving information that it must start functioning to the computer.

Chapter 3

Statistical Information Theory

Statistics: The only science that enables
different experts using the same figures
to draw different conclusions.
Evan Esar (1899–1995)

The main and most popular direction in the statistical information theory emerged as a communication theory in the works of Harry Nyquist (1924), Ralph Vinton Lyon Hartley (1928), and was developed to a full extent by Claude Shannon (1948). After some period when it was completely ignored, Shannon's paper *The Mathematical Theory of Communication*, which appeared in 1948, made information a prominent notion. Although the word *information* does not figure in the title, it was later called the information theory and even now many researchers either do not know or ignore other directions in information theory.

In the 1950s and 1960s, Shannon's theory spread its influence to a variety of disciplines. However, researchers in these disciplines were not satisfied with the statistical approach and elaborated definitions of information specific for their disciplines (cf., for example, (Gibson, 1966; Neisser, 1967; Seiffert 1968; Attneave, 1974)).

The approach of Hartley and Shannon formed the mainstream of the statistical information theory and to a great extent, of the whole information theory. It became so popular that many are not aware of existence of other directions. The bias towards this approach is very strong even now. For instance, in his review of a paper with an alternative approach, one expert in the Shannon's information theory wrote that he worked all his life in this theory and did not see any necessity in anything else.

There are many excellent books on the Shannon's information theory, for example, such a canonical text as (Cover and Thomas, 1991). That is why we give only a short exposition of this magnificent theory, demonstrating that it is an important special case of the general theory of information.

However, even before communication engineers developed their measures of information, Ronald Fisher (1922) introduced variety of information as the first measure of information. This measure is also statistical by its nature. It is considered in Sections 3.2 and 3.6.

3.1. Information and Communication

> *The worst wheel of the cart creaks most of all.*
> Proverb

Information is intrinsically connected to communication. For instance, Shannon's statistical theory of information started as a mathematical theory of communication (Shannon, 1948) and only later was interpreted as information theory and called by this name. This is not by chance because information is the main object of communication.

Definition 3.1.1. *Communication is a process of information exchange*, when information is transmitted and received.

Usually two types of communication are considered: technical communication, which is the base of information technology and the object of the Shannon's information theory, and interpersonal communication, which is the foundation of society functioning and the object of many studies (cf., for example, (Brooks, 1967)). At the same time, there is also human-computer communication, which is of a different kind because it includes both human beings and technical systems. Consequently, we have three types:
- Communication between technical systems.
- Communications between people (living beings).
- Communication between people and technical systems.

Communication process as a total event has been a subject of many studies. Many models, or structural descriptions, of communication have been suggested to aid understanding of its general organization. This makes it possible to classify and to describe the parts of the process and to indicate how they fit together and interact in the process. Models provide clues that permit predictions of behavior and thus, stimulate further research and development.

There are different models of communication. Static and dynamic models form the main two classes. *Static models* represent a system in which the communication goes on. *Dynamic models* feature functioning of such a system.

In turn, dynamic models also contain two classes: function models and process models. *Function models* describe the communication process as a system of functions. *Process models* include a representation of communication events and actions, as well as relations between them.

The simplest static model of a communication event consists of three elements: a *sender*, *receiver*, and *channel* (cf. Figure 3.1).

channel

Sender ————————————▶ **Receiver**

Figure 3.1. The static communication triad

Here, the sender and/or receiver are systems that can contain other systems as their components. For instance, a receiver can be an individual, group of people, cell phone, radio set, all TV sets that show programs of the same channel or the whole society. The same is true for a receiver. Elements in the triad from Figure 3.1 have diverse interpretations. Below are two examples.

| **All people working with the Internet at 12 p.m.** | the Internet ————▶ | **All people working with the Internet at 12 p.m.** |

Figure 3.2. The Internet communication triad

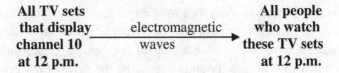

Figure 3.3. The TV communication triad

The simplest function model of communication consists of three elements: *sender*, *receiver*, and *message* (cf. Figure 3.4), as well as three functions: *message production*, *message transmission*, and *message reception/comprehension*. The first function is defined on states of the sender (e.g., mental states when the sender is an individual) and gives a message as its output. The third function is defined on messages and gives states of the receiver (e.g., mental states when the receiver is an individual) as its output.

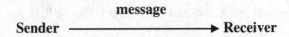

Figure 3.4. The functional communication triad

However, there is one more component of communication that influences the whole process. When the Sender creates the message, it is based on Sender's knowledge, current situation and some other factors. At the same time, the Receiver reconstructing meaning of the message also uses her/his knowledge. Meaning can also depend on a situation and so on. All these factors form the context of communication. This context is related to the message and change of the context can essentially transform the message in its semantic and pragmatic dimensions. The context of a message influences even syntax because knowledge of the language in which the message is written is important to understanding its syntax. Taking all these factors into account, we obtain the extended functional communication triad.

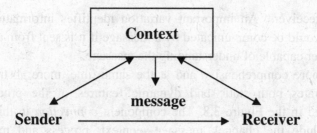

Figure 3.5. The extended functional communication triad

A simple process model of communication is given in Figure 3.6.

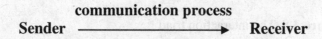

Figure 3.6. The process communication triad

In a natural way, it is useful to extend the simple process model of communication by including the context of communication. Usually the context for a process is more encompassing than the context for a function, for example, it may contain procedural and script components.

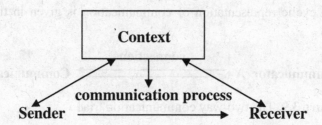

Figure 3.7. The extended process communication triad

All considered models assume there is a definite sender and at least one receiver. Models of interpersonal communication assume existence of a common language understood by the sender and, at least, by one of the receivers. An important variation identifies information as that which would be communicated by a message if it is sent from a sender to a receiver capable of understanding the message.

A more comprehensive and at the same time, more abstract model incorporates both static and dynamic features of the process. It is presented in the Figure 3.8. The component *connection* in this diagram can include the channel, message, context, process and many other things.

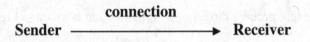

Figure 3.8. The interaction triad

All structures in Figures 3.1–3.4, 3.6, and 3.8 are special cases of fundamental triad.

It is necessary to remark that the structures presented above are models of a single communication act. They are "elementary particles" of a communication process when the sender and the receiver exchange their roles. A model of communication as whole process is a complex that is built of such elementary models.

The cyclic representation of communication is given in the Figure 3.9.

connections

Communicator A ⇄ **Communicator B**

Figure 3.9. The two-way communication triad

There are also other models of communication. One of the most popular is Shannon's model of communication (Shannon, 1948). It represents communication in technical systems (see Figure 3.10). Note that the component noise source in this model is a part of the communication context represented in Figure 3.5.

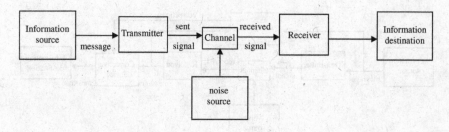

Figure 3.10. Shannon's model of communication

Shannon (1948) describes the components of the communication system as follows:

1. The information source (or message source) produces messages or parts of messages intended for a particular destination.

2. On the basis of the message, the transmitter produces a sequence of signals such that they can be transmitted over a channel.

3. The channel is merely the medium used to transmit the signal from transmitter to receiver. During transmission the signals may be perturbed and distorted by a so-called noise source.

4. The receiver usually performs the reverse operation to the transmitter, reconstructing if possible the original message from the signals.

5. The destination is the addressee of the message and can be either a person or a thing. It requires a priori knowledge about the information source that enables it to understand the message transmitted. In particular, the destination must know the set of signs available to the information source.

In the same paper, Shannon (1948) suggested a more advanced schema of communication (see Figure 3.11). Note that both components the *observer* and *noise source* in this model are parts of the communication context represented in Figure 3.5.

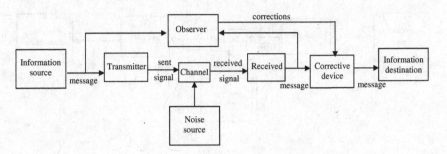

Figure 3.11. Shannon's model of communication with correction

It is also possible that an observer corrects signals in the Channel. It may be the same observer or a different one. This situation corresponds to one more model of communication.

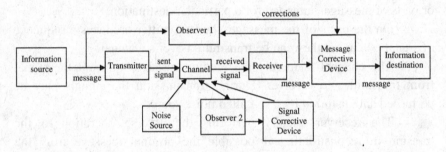

Figure 3.12. A model of communication with double correction

This model shows that it is possible to improve signals and messages in the channel. Spelling checker in an e-mail system or in a word processor is an example of such a correction device. The sender of the e-mail plays the role of the Observer in the first case. The user of the word processor plays the role of the Observer in the second case.

An Observer who/that regulates corrections in the channel can be treated as an intermediary in information transmission. This situation is analyzed in the mathematical theory of information developed by Kahre (2002). This theory is based on the following postulate as an additional axiom of the probability theory.

Law of Diminishing Information. Compared to direct reception, an intermediary can only decrease the amount of information for an ideal receiver.

Kahre (2002) gives the following illustration for this by considering an Englishman trying in vain to read a Chinese newspaper. When he gets the help of an interpreter, he finds out that after all there is a lot of information in the paper. This shows that the Englishman is not the ideal receiver (knowing all languages) because an interpreter as an intermediary could improve his understanding. This argument is turned around and used to define an ideal receiver.

Definition 3.1.2. A receiver is *ideal* if no intermediary can improve its performance.

Thus, we can easily see that the Law of Diminishing Information is a simple rephrasing of the definition of an ideal receiver. At the same time, the concept of an ideal receiver is too abstract and cannot exist in real life because to be ideal, a receiver has to possess all possible knowledge because for any piece of knowledge K, it is possible to build a message m such that to comprehend this message it is necessary to knowledge K. Thus, an ideal receiver must have all knowledge that exists now and has to be able to potentially acquire infinite knowledge. Hence, all real receivers are not ideal and we cannot consider information that an ideal receiver can get from a message because it will be an infinite amount of information.

For a real receiver, the observer from the communication models in Figures 3.11 and 3.12 can increase information in the sent message by making corrections and other changes. Noise in the channel, as a rule, decreases information, while a correction device, especially, an intelligent correction system, is able to increase information in the sent message. For instance, an educated interpreter can increase information when she or he interprets. Note that it is possible to consider the observer from communication models in Figures 3.11 and 3.12 as an intermediary channel. This shows that the Law of Diminishing Information is not true for real receivers.

The idea of possible corrections influenced mathematical linguistics where the most popular model is a generative grammar (cf., for example,

(Hopcroft, et al, 2001)). A generative grammar represents language generation or learning as a one way process, that is, everything that is generated (learned) is included in the language. However, recently new types of grammars were introduced in (Burgin, 2005c). They are called grammars with prohibition (Burgin, 2005c) or corrections grammars (Carlucci, Case, and Jain, 2007). The advantage of these grammars is that they allow one to correct what has been already generated or learned.

Program checkers (Blum and Kannan, 1995) give one more example of information correction in the communication channel.

Communication between people has additional features in comparison with communication in technical systems. In the model presented in Figure 3.13, which is an extension of the model of Miller (1966), the Referent, also called Context includes everything to what the message of the parties involved in communication (here the Source Encoder and Receiver/Decoder) is related (referred), e.g., things, situations, processes, systems, ideas, actions, etc.

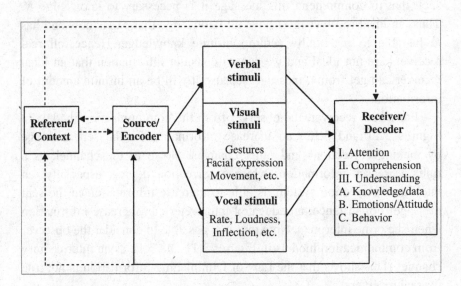

Figure 3.13. An extended Miller's model of speech communication (Miller, 1966; Keltner, 1970)

Visual and vocal stimuli form a group of *explicit physical stimuli*. In general, it is also possible to consider mental stimuli, e.g., hypnotic impact of one person (the Source) on the other (the Receiver).

In this model it is assumed that the process of message transmission consists of three stages: the Source/Encoder gains Receiver's attention, sends the message (e.g., speaks), and then the Receiver/Decoder receives and either accepts or rejects the message. The process of information transmission includes two more stages: at the beginning, the Source/Encoder encodes intended information into a message, while at the end, the Receiver/Decoder extracts information from the message by decoding it and understanding the result of decoding. The result of comprehension/understanding consists of three components: cognitive emotional/attitudinal, and physical. The cognitive component includes data and knowledge formed by the Receiver as a result of message comprehension. The emotional/attitudinal component includes emotions involved and attitudes to the Encoder and Referent formed by the Receiver as a result of message comprehension. The physical component can include a message sent to the Encoder, change of the behavior of the Receiver, and message to somebody else formed by the Receiver as a result of message comprehension.

A specific kind of communication emerges in human-computer interaction (HCI). Now computers are everywhere and human-computer interaction has become an important area of activity and research. It is interesting that, as a rule, human-computer interaction is only human-computer communication because other types of interaction, e.g., physical action, such as repair of a computer or its installation or when a frustrated user kicks its computer, are not taken into account. Thus, interaction is reduced to its special type — communication. Human-computer communication is a process of information transmission and reception. It implies new relations between communicating systems and demands new models. One of such models is developed in (Burgin, Liu, and Karplus, 2001) and called the M-model of communication.

At the beginning, human-computer communication correlated with a general schema of communication presented in Figures 3.1–3.4. However, the development of computer interface changed this situation, and the human-computer interaction became a mediated communication. Moreover, in many cases, communication in society is mediated by technical devices such as telephones, facsimile machines, satellites, computers, and so on. It is possible to treat these technical devices as some kinds of advanced channels, while the human-computer communication possesses a mediator of a different type called the *communication space*. This gives us the following schema.

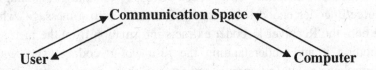

Figure 3.14. The two-way mediated human-computer communication triad

Communication space is not simply a channel because both participants of communication, a user and a computer, are represented in the corresponding communication space by special objects, which interact with one another. We call these objects *communication representatives* of the user and the computer, respectively. Communication functions are delegated to these representatives.

Now for the majority of software, communication space is the screen of the monitor of a computer. Only in some cases, this space is a virtual reality. In future, this will be the main type of communication spaces.

When human-computer communication is realized by Windows, then on the screen, as a communication space, the user is represented by the cursor and special signs (such as >) that denote commands together with the fields for these commands. The computer, or more exactly, the operating system Windows, is represented by different icons and special signs that denote computer messages to the user together with the fields for these messages.

The M-model of human-computer interaction/communication brings us to a new model of communication in general represented in Figure 3.15.

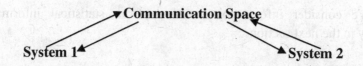

Figure 3.15. The two-way mediated communication triad

This model resembles the model from Figure 3.1. However, there is an essential difference between a channel and communication space. In a channel, only messages are represented as sets of signals that go from one place to another, while in a communication space, the participants of communication (communicating systems) have their representatives (agents). For instance, now virtual versions of real-life people created by computer programs and called avatars, are being used to draw crowds — and their money — to commercial Web sites. Moreover, software (sometimes intelligent) agents form a basis for many kinds of advanced software systems that incorporate varying methodologies, diverse sources of domain knowledge, and a variety of data types. The intelligent agent approach has been applied extensively in business applications, and more recently in medical decision support systems. All software agents are communication representatives that act/function in the communication space.

Statistical information theory in the form of Hartley and Shannon was originally developed to deal with the efficiency of information transmission in electronic channels. It is based on the definition of an information quantity that can be measured. The basic problem that needed to be solved according to Shannon was the reproduction at one point of a message produced at another point. He deliberately excluded from his investigation the question of the meaning of a message, arguing that (Shannon, 1969):

"Frequently the messages have meaning; that is they refer to or are correlated according to some system with physical or conceptual entities. These semantic aspects of communication are irrelevant to the engineering problem. The significant aspect is that the actual message is one selected from a set of possible messages. The system must be designed to operate for each possible selection, not just the one which will actually be chosen since this is unknown at the time of design."

We consider information measures from statistical information theory in the next section.

3.2. Information and Entropy

The good news: Computers allow us to work 100% *faster.*
The bad news: They generate 300% *more work.*
Unknown

The term entropy and its denotation by the letter S were introduced by Rudolf Julius Emanuel Clausius (1822–1888) in 1864 in his work on the foundations of classical thermodynamics, where he formulated the second law of thermodynamics in the form that the entropy of a closed system cannot decrease. Later the concept of thermodynamic entropy was clarified and grounded by Ludwig Eduard Boltzmann (1844–1906). The famous formula for thermodynamic entropy is

$$S = k \cdot \ln W \qquad (3.2.1)$$

where W is the thermodynamic probability that the system is in the state with the entropy S.

It is interesting that similar expression was introduced in the statistical information theory. The main question asked by the statistical information theory is "How much information do we get when we receive a message m that an experiment H had an outcome D or that some event E happened?" Harry Nyquist (1889–1976), Ralph Vinton Lyon Hartley (1888–1970) and Claude Elwood Shannon (1916–2001) studied a specific kind of communication, namely, communication in technical systems, assuming that the only purpose of technical communication is to reproduce the input data pattern in the receiver. That is why the event of the most interest to them was which symbol came in the process of message transmission and whether this symbol was not distorted in the process of transmission. Thus, the goal of statistical information theory is to measure *transmission information*.

The answer of Hartley (1928) to this question is based on the number n of possible outcomes of the experiment H or on the number n of possible alternatives to the event E. Namely, the quantity of information in the message m is given by the simple formula

$$I(n) = \log_2 n \qquad (3.2.2)$$

When all possible outcomes of this experiment or all possible alternatives to the event E have the same probability, then the probability p of E (of D) is equal to $1/n$ (cf. Appendix E). Thus, as $\log_2 n = -\log_2 p$, it is possible to rewrite the formula (3.2.2) in the following form:

$$I(n) = -p \cdot \log_2 p \qquad (3.2.3)$$

However, equal probabilities of events are a relatively rare event and it is more realistic to assume that the probabilities of experiment outcomes or of alternatives to the event, in general, are different. Thus, the answer of Shannon (1948) about information in the message m is subtler and more realistic because it takes into account the individual probability of each outcome or of each event alternative. Thus, it is assumed that initially n events $E_1, E_2, E_3, \ldots, E_n$, one of which is E (or n outcomes $D_1, D_2, D_3, \ldots, D_n$ of the experiment H, one of which is D), were possible and had probabilities $p_1, p_2, p_3, \ldots, p_n$, correspondingly. Then the *entropy* of the message m is equal to

$$H(m) = H(p_1, p_2, \ldots, p_n) = -\sum_{i=1}^{n} p_i \cdot \log_2 p_i \qquad (3.2.4)$$

Probabilities used in the formula (3.2.4) are obtained from experiments and observation, using relative frequencies. For instance, let us consider such events as encountering a given letter, say "a", in an English text. In this case, it is possible to analyze many English texts to find relative frequency of occurrence for this letter "a" in different texts, that is, to calculate the number m of occurrences of "a" in a text T and to divide it by the number n of letters in this text T. The ratio m/n is the relative frequency of occurrence of "a" the text T. For instance, the relative frequency of occurrence for this letter "e" in the sentence "Information is related to structures as energy is related to matter" is equal to 8/57 because the sentence has 57 letters and "e" is used 8 times. Usually the average of such relative frequencies is used as the probability of occurrences and thus, of receiving this letter. If we know probabilities $p_1, p_2, \ldots, p_{26}$ for all letters $a, b, c, \ldots, z$ from the English alphabet, we can easily find information in one English letter calculating the information entropy $H(p_1, p_2, \ldots, p_{26})$. Note that to correctly estimate information in one letter in a text, it is necessary to know the language in which the text is written because the same letter may have different probabilities of occurrences in different languages. For instance, the letter "*a*" has the probability (average letter frequency) 0.0817 of occurrences in English texts, the probability (average letter frequency) 0.1212 of occurrences in Spanish texts and the probability (average letter frequency) 0.0763 of occurrences in French texts.

Accurate average letter frequencies can only be obtained by analyzing a large amount of representative texts. However, these are only approximate average frequencies because no exact letter frequency distribution underlies a given language, since all writers write slightly differently. More recent analyses show that letter frequencies, like word frequencies, tend to vary, both by writer and by subject. For instance, one cannot write an essay about x-rays without frequently using the letter "x". In addition, different authors have habits, which can be reflected in their use of letters.

The base of the logarithm used in formulas (3.2.1)–(3.2.4) is arbitrary, since it affects the result by only a multiplicative constant, which determines the unit of information that is implied. If the log is taken base 2 as in formulas (3.2.2)–(3.2.4), the unit of information is *binary digit* or *bit*. If we use a natural logarithm instead as in the formula (3.2.1), we come to the unit called *nat*.

Consistent with the choice of the base 2, which represents a choice from two options, information is usually measured in bits, bytes, kilobytes, megabytes, terabytes, etc. One byte contains eight bits. In the metric (decimal) system, one kilobyte contains thousand bytes, one megabyte contains million bytes and so on. At the same time, in the binary system, one kilobyte contains $2^{10} = 1024$ bytes, one megabyte contains $2^{20} = 1048576$ bytes and so on.

There are two meanings of the word *bit*. The most popular one is a *binary digit*, 0 or 1. Such *bits* are the individual pieces of data in computers. As we know, computers use binary strings (words) of symbols 0 and 1 as processed data and information representation.

At the same time, bits are used to denote measurement units of uncertainty represented by entropy H and information I. Indeed, when the binary system of encoding messages with only two symbols, typically "0" and "1", is used, a message is encoded as a string of binary digits. The most elementary choice one can make is between two items: "0" and "1", "heads" or "tails", "true" or "false", etc. Shannon defined the bit as such an elementary choice, or unit of information content, from which all selection operations are built. In one more interpretation, bit, or binary digit, and is equivalent to the choice between two equally likely alternatives. For instance, if we know that a coin is to be tossed, but are unable to see it as it falls, a message telling whether the coin came up heads or tails gives us one bit of information.

The quantity of information in a message entirely depends on the symbolic language used and the length of the text, and not at all on the meaning of this message. For instance, the words "information" and "abdaracabra" written in the Roman alphabet with 26 symbols are one of 26^{11} (=3,670,344,486,987,776 $\approx 3.7 \times 10^{15}$) possible formal words of size 11. Therefore, both contain the same information quantity (assuming that each symbol has the same likelihood), even though the first has an English meaning and the second does not.

The term *bit* was first published by Shannon in his main 1948 paper. He attributed its origin to John Wilder Tukey (1915–2000) who contracted the expression *binary digit* to one word *bit*. However, Vannevar Bush (1890–1974) had written in 1936 of *bits of information* that could be stored on the punch cards in the mechanical devices of that time.

Formula (3.2.4) reflects uncertainty we have with respect to the situation when we know only probabilities but not the actual result (event), and the measure of this uncertainty is also the measure of information that eliminates this uncertainty. In this formula, entropy and later, information are defined as quantities that depend on symbol manipulation alone

For instance, suppose we have a device M that can produce one of three symbols, A, B, or C on each step. As we wait for the next symbol to appear, we are *uncertain* as to which symbol M will produce. Once a symbol appears and we recognize it, our uncertainty *decreases*, and we assume that we have received some *information*. That is, information is a decrease in our uncertainty. The more symbols can M produce the higher will be the uncertainty of what happens. This is reflected by the formula (3.2.2) of Hartley (1928). Namely, the uncertainty is equal to $\log_2 3$ in the case of three symbols.

At the same time, the formula (3.2.4) of Shannon (1948) shows that uncertainty also depends on the probabilities with which the device M produces these symbols. For instance, when the letter A is produced with the probability $\frac{1}{2}$, while letters B and C are produced with probabilities $\frac{1}{4}$, then by the formula (3.2.3), the uncertainty is equal to

$$- (\tfrac{1}{2} \log_2 \tfrac{1}{2} + \tfrac{1}{4} \log_2 \tfrac{1}{4} + \tfrac{1}{4} \log_2 \tfrac{1}{4}) = \tfrac{1}{2} \log_2 2 + \tfrac{1}{4} \log_2 4 + \tfrac{1}{4} \log_2 4$$
$$= \tfrac{1}{2} + \tfrac{1}{2} + \tfrac{1}{2} = 1\tfrac{1}{2}$$

The number $\log_2 3 \approx 1.585$ is larger than $1.5 = 1\frac{1}{2}$. Thus, this simple example demonstrates differences between Shannon's and Hartley's measures of uncertainty of information, namely, in the contemplated case, Shannon's measure is smaller than Hartley's measure.

With the formula (3.2.4), Shannon created a direct equivalent to entropy as defined in statistical mechanics. The denotation H in the formula (3.2.4) and the term entropy were taken from Boltzmann's H-Theorem in thermodynamics. According to letters in the American Journal of Physics, Boltzmann originally used the letter E and then later switched to H, presumably meaning the Greek letter E called "eta" (Brush, 1967; Hjalmars, 1977). The letter "eta" (E) corresponds to H in English. Although it is not known why this was done, perhaps the goal was to avoid confusing entropy with E standing for energy.

Many years later, Shannon told such a story (cf. (Tribus and McIrvine, 1971)):

After deriving the formula (3.2.4), "my greatest concern was what to call it. I thought of calling it 'information', but the word was overly used, so I decided to call it 'uncertainty'. When I discussed it with John von Neumann, he had a better idea. Von Neumann told me, "You should call it entropy, for two reasons. In the first place your uncertainty function has been used in statistical mechanics under that name, so it already has a name. In the second place, and more important, nobody knows what entropy really is, so in a debate you will always have an advantage."

When $p_1 = p_2 = p_3 = \ldots = p_n = p$, the formula (3.2.4) of Shannon becomes the formula (3.2.2) of Hartley because $\sum_{i=1}^{n} p_i = 1$ and $p = 1/n$

implies $-\log_2 p = \log_2 n$. However, the original formula (3.2.2) of Hartley is, in some sense, more general than the formula (3.2.4) because it does not use the concept of probability, depending only on the number n of possible outcomes of the experiment H or on the number n of possible alternatives to the event E.

Besides, it is possible to deduce the formula (3.2.4) from the formula (3.2.2) (Brillouin, 1956).

Defining information entropy, it is possible to consider the experiment H, event E and a message as random variables. In this case, entropy $H(X)$ gives a measure of uncertainty of the random variable X.

It is also possible to treat the system of probabilities $p_1, p_2, ..., p_n$ as a probability distribution $P(X)$ on the set of events $E_1, E_2, ..., E_n$, i.e., $P(X)$ is a function where variable X takes values in the set $\mathbf{E} = \{E_1, E_2, ..., E_n\}$ and $P(E_i) = p_i$. This allows one to consider $H(p_1, p_2, ..., p_n)$ as the entropy of the random variable X and represent it in the form

$$H(X) = H(P(X)) = \sum_{X \in \mathbf{E}} P(X) \, \log_2 P(X) \qquad (3.2.5)$$

Formulas (3.2.4) and (3.2.5) give non-conditional, or in some sense, absolute statistical information entropy, which measures uncertainty of an event (experiment outcome). Conditional entropy determines uncertainty of an event E, when we get some information about the event E in the message m that informs about an event C. Information about C changes the probability distribution $p_1, p_2, p_3, ..., p_n$ to another probability distribution $p_{C1}, p_{C2}, p_{C3}, ..., p_{Cn}$. These probabilities are used to define conditional entropy by the following formula

$$H(p_1, p_2, p_3, ..., p_n \, | \, C) = H(p_{C1}, p_{C2}, p_{C3}, ..., p_{Cn}) = -\sum_{i=1}^{n} p_{Ci} \cdot \log_2 p_{Ci}$$

$$(3.2.6)$$

It is natural to consider the event C as a member of a system of events with the probability distribution $P(Y)$ of a random variable Y that takes values in this system of events. Then we have

$$H(X \, | \, Y) = H(p_1, p_2, p_3, ..., p_n \, | \, C) \qquad (3.2.7)$$

Shannon took information to be a decrease in uncertainty, which is measured by information entropy. This allows one to express information using formulas (3.2.6) and (3.2.7). Namely, *the quantity of information* about the outcome of the experiment H or about the event E in the message m that informs about an event C is determined by the following formula

$$I(X; Y) = H(X) - H(X \mid Y)$$

Here are some basic properties of statistical information measures (cf. (Aczél and Daróczy, 1975)):

$$H(X, Y) = H(X) + H(Y)$$
$$I(X; Y) = H(X) + H(Y) - H(X, Y)$$
$$I(X; Y) = H(Y) - H(Y \mid X)$$
$$H(X) \geq H(X \mid Y)$$
$$H(X) - H(Y) = H(X \mid Y) - H(Y \mid X)$$

Statistical information theory has been successfully applied to many important problems in communication practice, allowing one to theoretically estimate different characteristics of communication systems, such as channel capacity and amount of information for discrete and continuous, noisy and noiseless systems.

One of the basic results of the statistical information theory is the noisy-channel coding theorem, also called the fundamental theorem of information theory, or just Shannon's theorem because it was first presented with an outline of the proof in the classical paper (Shannon, 1948). A rigorous proof was given later. Here we consider this theorem only as an example of proved in statistical information theory mathematical results with important practical applications. There are many books and papers with dozens similar results about communication channels.

Let us take the channel capacity defined as

$$C = \max \{I(X, Y); p_X\},$$

and the binary entropy function

$$H(p_b) = -[p_b \cdot \log_2 p_b + (1 - p_b) \log_2 (1 - p_b)]$$

Theorem 3.2.1 (Shannon, 1948). For every discrete memoryless channel, the channel capacity C has the following properties:

1. For any $\varepsilon > 0$ and $R < C$, for large enough N, there exists a code of length N and a rate larger than or equal to R and a decoding algorithm, such that the maximal probability of block error is $\leq \varepsilon$.

2. If a probability of bit error p_b is acceptable, rates up to $R(p_b)$ are achievable, where

$$R(p_b) = C/(1 - H(p_b))$$

3. For any p_b, rates greater than $R(p_b)$ are not achievable.

This theorem shows that however contaminated with noise interference a communication channel may be, it is possible to communicate digital data (information) nearly error-free up to a given maximum rate through the channel. The theorem also describes the maximum possible efficiency of error-correcting methods versus levels of noise interference and data corruption. However, it does not describe how to construct the error-correcting method, only telling us how good the best possible method can be. This theorem has wide-ranging applications in both communications and data storage applications.

There are many myths and misconceptions related to the Shannon theory of information. For instance, in one good book on information theory, it is written that far from selecting the binary code arbitrarily, or believing it to be the simplest possible scheme, Shannon proved that it is, in fact, the most economical symbol system available. To get a flavor of Shannon's proof, the author of the book suggests considering a sailor who wants to signal a number between 0 and 127 by means of flags. If the sailor decides to fly just a single flag to do the job, he needs to have 128 different flags in his locker. A less expensive strategy would be to fly three flags to spell out the number in the decimal system. To do this, he would need to own just twenty-one flags — ten for the units, ten for the tens and one for hundreds. The conclusion of the author of the book is that the cheapest technique is based on the binary coding, with only fourteen flags — seven zeroes and seven ones — he can compose any number up to 127.

However, if the sailor decides to fly flags based on the ternary coding and represent any number up to 127, he would need to own just the same number of fourteen flags. So, in this example, the ternary coding will be as cheap as the binary one. Nevertheless, to represent any number from 0 to 160, the sailor would need the same fourteen flags in the ternary system, but sixteen flags in the binary system. The advantage of the ternary system grows with the growth of the quantity of numbers that the sailor has to represent. For example, to compose any number from 0 up to 2100, the sailor would need twenty-four flags in the binary system and only twenty-one flags in the ternary system. To do the same thing for all numbers from 0 up to 17000, the sailor would need twenty-eight flags in the binary system and only twenty-four flags in the ternary system.

In spite of all its success, Shannon's theory of the amount of information was criticized by many later authors as merely a theory of syntactical information, because it excluded the semantic and pragmatic levels. The price to pay for the ability to build such an elegant and advanced mathematical theory is that the Hartley-Shannon approach does not deal at all with many important aspects of information. That is why statistical information theory is rightly seen as lacking indications for a conceptual clarification of information. Besides, any attempt to develop statistical information theory into a universal theory of information would necessarily reach an impasse. To this day, though, there is no measure in information theory which is as well-supported and as generally accepted as Shannon's quantity of information.

There were many generalizations of the Shannon's approach to information measurement aimed at improving some capabilities of this theory. Here we consider some of them.

Rényi (1960) introduced a new kind of entropy by the following formula

$$H_\alpha(p_1, p_2, ..., p_n) = (1/(1 - \alpha)) \ln \left(\sum_{i=1}^{n} p_i^\alpha \right) \qquad (3.2.8)$$

Now this information measure is called the Rényi entropy.

In the limit, when $\alpha \to 1$, the Rényi entropy is equal to a version of the Shannon entropy with the natural logarithm *ln* instead of *log₂*:

$$\lim_{\alpha \to 1} H_\alpha(p_1, p_2, ..., p_n) = - \sum_{i=1}^{n} p_i \ln p_i$$

The Rényi *entropy* was used to express quantum mechanical uncertainty relations, such as the Heisenberg uncertainty principle (Bialynicki-Birula, 2006).

Klir and his collaborators (Higashi and Klir, 1983; Klir and Mariano, 1987; Klir, 1991) developed a generalized information theory. The main idea of this generalization is based on the following consideration.

Information is related to uncertainty, namely, it is possible to interpret information entropy (quantity of information) as the measure of eliminated uncertainty. Thus, it would be natural to use possibility measures instead of probability defining possibilistic entropy. As a generalization of the Hartley measure of information (3.2.2), Higashi and Klir (1983) introduced the U-uncertainty measure of information

$$U(r) = \int_0^1 (r_i - r_{i+1}) \log_2 I \qquad (3.2.9)$$

Here a possibility distribution r is expressed in the form of a normalized fuzzy set A and $|A_\alpha|$ is the cardinality of the α–cut of the fuzzy set A.

$$U(r) = \sum_{i=1}^n (r_i - r_{i+1}) \log_2 i \qquad (3.2.10)$$

Patni and Jain (1977) introduced and studied a non-additive measures of information associated with a pair of finite probability distributions defined on sets with the same number of elements.

Lerner (1999; 2003) constructed a very different generalization of the Shannon's approach to information measurement. One of the main ideas of his approach is to use the entropy functional called *information path integral* as an integral measure of information processes instead of the Shannon's entropy.

To conclude, it is necessary to comment that the Hartley-Shannon direction has become very popular and dominates the field of information studies to such an extent that many people are not aware that this approach was not the first mathematical direction in information theory and it is not generally the best to use. Many other information theories exist and are extensively used to solve real-life problems. One of such theories was developed by Ronald Aylmer Fisher (1890–1962) who was an English statistician, evolutionary biologist, and geneticist. Based on mathematical statistics, he introduced (1922) an original information measure now called Fisher information and used it to build the classical measurement theory (Fisher, 1950).

In a particular case, which is important for physics (Frieden, 1998), Fisher information is defined by the following formula

$$\mathbf{I} = \int p'^2(x) / p(x) \, dx$$

Here $p(x)$ denotes the probability density function for the noise value x, and $p'(x) = \mathrm{d}p(x)/\mathrm{d}x$.

According to Fisher's theory, the quality of any measurement may be specified by this form of information.

By its origin, Fisher information $\mathbf{I}$ is the quality metric for the estimation/measurement procedure. Consequently, the value of $\mathbf{I}$ measures an adequate change in knowledge about the measured/estimated parameter, and is information in the sense of the general theory of information. Thus, the approach based on Fisher information is also included in the general theory of information.

3.3. Quantum Information

> *One who knows that enough is enough will always have enough.*
>
> Lao-Tzu

Quantum information is information that is stored and processed by quantum systems, such as subatomic particles and atoms. Now research community has a lot of interest in quantum information due to ideas of building quantum computers, which theoretically can be not only smaller but also more efficient than conventional computers. Quantum information is studied mostly by the statistical approach. However, there are essential differences between the classical statistical information theory and quantum information theory. In quantum information theory statistical operators are used instead of random variables from the classical theory, matrices play the role of numbers, and units of information are qubits rather than bits.

It is interesting that John von Neumann (1903–1957) introduced quantum entropy (1927) as a measure for statistical operators before Shannon introduced information entropy. The reason was that he did this extending the thermodynamic entropy to the quantum realm without any connections to information.

Taking a statistical operator D and using a thought experiment on the ground of phenomenological thermodynamics, he came to the formula

$$S(D) = - K \sum\nolimits_{i=1}^{k} \lambda_i \log_2 \lambda_i \qquad (3.3.1)$$

where λ_i are eigenvalues of the operator D and K is an arbitrary constant.

Formula (3.3.1) is usually converted to a more compressed form

$$S(D) = - K \operatorname{Tr} \eta(D) \qquad (3.3.2)$$

where $\operatorname{Tr} A$ is the trace of an operator (matrix) A and η is the continuous real function defined by the following formula

$$\eta(t) = t \log_2 t$$

Quantum entropy is one of the basic concepts of the mathematical formalism of quantum mechanics developed by von Neumann (1932). However, quantum entropy was not related to information for a long time. Only the advent of quantum computation (in its theoretical form) connected quantum entropy and information. Although von Neumann apparently did not see an intimate connection between his entropy formula and the similar Shannon information entropy, many years later an information theoretical reinterpretation of quantum entropy has become common. For instance, it is demonstrated that the von Neumann entropy describes the capacity of a quantum communication channel much in the same way as information entropy describes the capacity of a quantum communication channel (cf. Section 3.2).

There were many generalizations of the von Neumann's approach to quantum information measurement (cf., for example, (Hu and Ye, 2006)).

Classical information theory developed by Nyquist, Hartley, Shannon and their followers usually abstracts completely from the physical nature of the carriers of information. This is sensible because information can be converted easily and essentially without loss between different carriers, such as magnetized patches on a disk, currents in a wire, electromagnetic waves, and printed symbols. However, this convertibility no longer holds for microscopic particles that are described by quantum theory. As a result, the quantum level of information processing demands new tools and new understanding.

A natural way to understand the underlying unity between physical (i.e., thermodynamic or quantum) entropy and information entropy is as follows. Information entropy describes the part of the (classical) physical information contained in a system of interest (whether it is an entire physical system or just a subsystem delineated by a set of possible messages) whose identity (as opposed to amount) is unknown (from the point of view of a particular person or system). This informal characterization corresponds to both von Neumann's definition of the quantum entropy and Shannon's definition of the information entropy.

The basic questions of quantum information theory are taken from classical information theory, e.g., how much quantum information is carried by any given system or transmission channel, how much is stored in a storage device, how can such information be coded and decoded efficiently, etc. These and other questions from classical information theory are posed in this new context, but for the moment, many of them are still awaiting rigorous answers. Trying to find these answers would help deciding the feasibility of quantum computers, and other practical applications, as well as it will open new perspectives on the foundations of quantum theory. As Duwell (2008) writes, many physicists and philosophers believe that conceptual problems with quantum mechanics can be resolved by appeal to quantum information. However, this proposal is difficult to realize and even to evaluate because there is a widespread disagreement about the nature of quantum information. Different concepts of quantum information have been proposed. They depend on the quantum theory used for this purpose (cf., for example, (Freedman, et al, 2002)).

It is usually assumed that according to contemporary quantum physics, the information contained in quantum physical systems consists of two parts: *classical information* and *quantum information*. Quantum information specifies the complete quantum state vector (or equivalently, wavefunction) of a system, while classical information reflects a definite (pure) quantum state from a predetermined set of distinguishable (orthogonal) quantum states. Such a set of pure quantum states forms a basis for the vector space of all the possible pure quantum states.

Thus, the measure of classical information in a quantum system gives the maximum amount of information that can actually be measured and extracted from that quantum system for use by external classical (decoherent) systems, since only basis states are operationally distinguishable from each other. The impossibility of differentiating between non-orthogonal states is a fundamental principle of quantum mechanics, equivalent to Heisenberg's uncertainty principle.

An amount of (classical) physical information is quantified by the rules of statistical information theory. As we have seen, for a system S with n distinguishable states (orthogonal quantum states) that have equal probabilities of occurrence, the amount (quantity) of information $I(q)$ contained in the system's state q is equal to log n. The logarithm is selected for this definition since it has the advantage that this measure of information content is additive when combining independent, unrelated subsystems. For instance, if subsystem A has n distinguishable states (i.e., with the information content $I(q_A) = \log n$) and an independent subsystem B has m distinguishable states (i.e., with the information content $I(q_B) = \log m$), then the combined system C has nm distinguishable states and an information content $I(q_C) = \log nm = \log n + \log m = I(q_A) + I(q_B)$. People expect information to be additive from their everyday associations with the meaning of the word, e.g., they often think that two pages of a book can contain twice as much information as one page.

While in the classical case, the least unit of information is bit, for an isolated quantum system, the fundamental unit of information is the *quantum bit* or *qubit*, the name attributed to William Wooters and Benjamin Schumacher (cf. (von Bayer, 2004)). Qubits are represented by states of quantum two-level systems such as the spin of an electron or the polarization of a photon.

A state of a physical system is a complete description of parameters/characteristics of the experiment. These characteristics are formalized by mathematical means most commonly formulated in terms of such a mathematical field as linear algebra (cf. Appendix D) where a given quantum system is associated with a Hilbert space H as its phase (state) space. Then a pure state of this system is represented by a ray in this Hilbert space. Alternatively, it is possible to represent pure states by normalized or unit vectors, i.e., vectors that have the norm equal to 1, in H. These vectors form the unit sphere in the Hilbert space H. The portion of information that is contained in a quantum physical system is generally considered to specify that system's state.

Calculations in quantum physics make frequent use of linear operators that represent observables and specific physical notation, which is essentially different from the mathematical notation employed in the field of linear algebras and Hilbert spaces. For instance, a vector x corresponding to a pure quantum state has two names and is called either a *ket* and denoted by $|x\rangle$ or a *bra* and denoted by $\langle x|$. Inner products of vectors are called *brakets* and denoted by $\langle x|y\rangle$.

When a basis of vectors (kets or bras) is chosen, then an arbitrary state of a quantum system is represented by a linear combination of vectors from the basis. In the case of quantum information, the pair of qubit states $|1\rangle$ and $|0\rangle$ is chosen as the basis. Then an arbitrary qubit has the following mathematical representation:

$$|x\rangle = a|0\rangle + b|1\rangle$$

where a and b are the complex amplitudes of qubit states $|1\rangle$ and $|0\rangle$.

Although this is just a single unit of information, the continuous amplitudes a and b can carry an infinite amount of information, similar to analogue information carriers such as the continuous voltage stored on capacitors (cf. (Borel, 1927; 1952; Burgin, 2005; 2008)). However, this infinite information is treated as inaccessible because measurements give only one of two states $|1\rangle$ or $|0\rangle$. Even ignoring information carried by the coefficients a and b, it is possible to presume that a state of a system that stores qubit potentially contains two classical states 1 and 0. This feature of qubits (at least, theoretically) allows quantum information processing systems to achieve much higher efficiency than contemporary computers, which are based on classical information processing and work with separate digits. A collection of qubits has the potential for storing exponentially more information than a comparable collection of classical information carriers. For instance ten qubits can contain $2^{10} =$ 1024 bits of information as each qubit contains (in some sense) two tentative states. Quantum computers tend to process this information in a parallel way, achieving exponential speed up as 10 goes in the exponent of 2^{10}. Of course, there is a problem of extracting this information by a measurement.

Besides, analogue systems, which are working with information represented in a continuous form (cf. (Burgin, 2005)), are known to suffer from the cumulative build-up of noise, as opposed to digital standards like transistor–transistor logic that latch to their high or low levels through constant measurement and feedback. Quantum bits are similarly vulnerable to analogue noise, but they can offer much more in return, providing massively parallel information processing.

Thus, properties of quantum information make it look advantageous for computation. Computers process information and computation is a kind of information processing. Thus, quantum computation involves quantum information.

The main idea of quantum computing, which is attributed to Feynman (1982; 1986) and Beniof (1980; 1982), is to perform computation on the level of atoms and subatomic particles, utilizing a theoretical ability to manufacture, manipulate, and measure quantum states of physical systems. As there are different quantum process and various models are used in quantum physics, several approaches to quantum computation have been developed: a quantum Turing machine (Deutsch, 1985), quantum circuits (Feynman, 1986; Deutsch, 1989), modular functors that represent topological quantum computation (Freedman, 2001; Freedman, et al, 2002), and quantum adiabatic computation (Farhi, et al, 2000; Kieu, 2002; 2003).

According to the standard quantum mechanics approach, quantum computers run on quantum information in much the same way as a classical computer runs on classical information (cf., for example, (Deutsch, 1985)). The elementary storage units are two-state quantum systems or qubits instead of classical bits. Quantum computing units perform unitary transformations on a few (typically, one or two) qubits. Programs (or algorithms) for quantum computers are then sequences of such operations. In order to read the result of a quantum computation, one has to perform a measurement. Like all quantum measurements, it does not always give the same value, but results in a statistical distribution. However, for a good quantum algorithm, the probability of obtaining the correct result is significantly larger than what is possible to get by chance. Just as for classical stochastic algorithms this is then sufficient to reach any desired degree of certainty by repetitions of the computation.

As a result, quantum computers (if built) may perform certain tasks much better than classical ones. At first, Deutsch (1985) built a theoretical model of a general-purpose quantum computer and demonstrated that quantum computers theoretically allow massive parallelism in computations. The idea was in using a quantum superposition of the pure states $|x\rangle$ as the replacement of the union of a quantity of classical registers, each in one of the initial states $|x\rangle$. In this context, parallel mappings of register states are realized by unitary operators of quantum physics. There were other achievements in this area, but what really boosted the subject was the result of Shor (1994) that one of the notoriously difficult and practically relevant classical computation problems, namely, the factoring of large numbers, could be speeded up so as to do it rapidly on a quantum computer. This is achieved by an efficient use of highly entangled states to do a massively parallel computation. The results of the computation are then organized in interference of qubit patterns. With Shor's breakthrough, quantum computing transformed from a pure academic discipline into a national and world interest. In this context, exceptionally powerful algorithms were found in three areas: searching a database, abelian group theory, and simulating physical systems (cf. (Freedman, et al, 2002)).

The most attractive candidates for quantum information processors currently come from the area of atomic physics and quantum optics. Here, individual atoms and photons are manipulated in a controlled environment with well-understood couplings, offering environmental isolation that is unsurpassed in other physical systems.

The main hardware requirements for quantum information processors are:

1. The quantum system (that is, a collection of qubits) must be initialized in a well-defined state.
2. Arbitrary unitary operators must be available and controlled to launch the initial state to an arbitrary entangled state.
3. Measurements of the qubits must be performed with high quantum efficiency.

Quantum information theory has many potential applications (quantum computing, quantum cryptography, quantum teleportation) that are currently being actively explored by both theoreticians and experimentalists (cf., for example, (Nielsen and Chuang, 2000)). However, some researchers criticize the quantum approach to computation from a theoretical perspective and doubt that quantum computers will work. Indeed, a few potentially large impediments still remain that prevent researchers from building a real quantum computer that can rival today's modern digital computers. Among these obstacles error correction, decoherence, and hardware architecture are probably the most formidable. Thus, at present, quantum information processing and quantum communication technology remain in their pioneering stage.

3.4. Information and Problem Solving

> *With so much information now online,*
> *it is exceptionally easy to simply dive in and drown.*
> Alfred Glossbrenner

Let us consider some problems where considerations based on information entropy allow one to find the best solution.

Problem 1. Andrew knows a natural number n such that $0 < n \leq 10$. How many questions does Barbara need to ask to learn n if to any question, Andrew answers only *yes* or *no*?

Solution. Let us assume that Barbara needs to ask m questions to learn n. As $0 < n \leq 10$, the result of the questioning has 10 possible outcomes. Here the experiment is asking questions. It is natural to suppose that all outcomes have the same probability. Thus, the probability p of each outcome is 1/10. Consequently, the Hartley-Shannon entropy, i.e., information in one of these outcomes, is

$$H(Q) = -\log_2 (1/10) = \log_2 (10) \approx 3.32 \text{ bits}$$

At the same time, one question can provide only one bit of information as there are exactly two possible answers and it is possible to suppose that both have the same probability. Thus the number of questions must satisfy the following system of equalities and inequalities

$$m \geq H(E_k) \geq H(E_k) = H(Q) = \log_2 (1/10) \approx 3.32$$

Consequently, we have

$$m \geq 4$$

When we have the lower boundary for m, it is easy to find a minimal solution for this problem.

Algorithm for a minimal solution is given in a Figure 3.16.

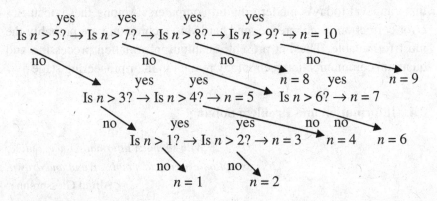

Figure 3.16. Algorithm for a minimal solution of Problem 1

This algorithm shows that the largest number of questions is equal to four although, in some cases, Barbara will need only three questions to find the number. Information estimates prove that it is impossible to improve this algorithm, obtaining the result using only three questions or less. In general, four questions are sufficient to find a number even in a larger interval from 1 to 16 because $\log_2 16 = 4$.

Problem 2. There are 25 coins. 24 have the same standard weight, while one coin is lighter than a good coin. There is also a balance that allows one to find which of two weights is heavier. How many times does one need to weigh coins to identify the lighter one?

Solution. Let assume that one needs to weigh coins m times to identify the lighter one. The result of weighing the coins has 25 possible outcomes as any of these coins may be lighter than a good coin. Thus, the probability p of each outcome is 1/25. Consequently, the Hartley-Shannon entropy, i.e., information in one of these outcomes, is

$$H(W) = - \log_2 (1/25) = \log_2 (25)$$

where W is the complete process of weighing.

However, in contrast to the previous problem, each separate weighting may have three outcomes: the right pan goes down, the left pan goes down, and both pans come to the equilibrium. Thus, each weighting gives $\log_2 3$ bits of information. When the set W consists of m weighting, we have

$$m \log_2 3 \geq \log_2 25$$

Consequently,

$$m \geq (\log_2 25/\log_2 3) \approx 2.93$$

i.e., $m \geq 3$ as m is a whole number. This shows that we need to look for an algorithm with $m = 3$.

Algorithm of a minimal solution for finding the lighter coin is given is given below:

Step 1: Divide 25 coins into three groups: 9, 8 and 8 coins, and go to the next step.

Step 2: Compare two groups of 8 coins, using the balance. This action (experiment) may give two results. In one case, the weights of both groups are equal. It means that the lighter coin is in the group of 9 coins because there is only one lighter coin. Then go to Step 3. In the other case, the balance shows that the weight of one group of 8 coins is less than the weight of the other group of 8 coins. It means that the lighter coin is in the lighter group of 8 coins because there is only one lighter coin. Then go to Step 8.

Step 3: Divide the group of 9 coins into three groups: 3, 3 and 3 coins, and go to the next step.

Step 4: Compare two selected groups of 3 coins, using the balance. This action (experiment) may give two results. In one case, the weights of both groups are equal. It means that the lighter coin is in the third group of 3 coins because there is only one lighter coin. Then go to Step 5. In the other case, the balance shows that the weight of one group of 3 coins is less than the weight of the other group of 3 coins. It means that the lighter coin is in the lighter group of 3 coins because there is only one lighter coin. Then go to Step 7.

Step 5: Divide the third group of 3 coins into three equal parts and go to the next step.

Step 6: Compare two selected coins, using the balance. This action (experiment) may give two results. In one case, the weights of both coins are equal. It means that the third selected coin is lighter than a good coin and we have the answer to the problem. In the other case, the balance shows that the weight of one coin is less than the weight of the other coin. It means that the lighter of these two coins is not a good coin and we have the answer to the problem.

Step 7: Divide the lighter group of 3 coins into three equal parts and go to Step 6.

Step 8: Divide the lighter group of 8 coins into three groups: 3, 3 and 2 coins, and go to the next step.

Step 9: Compare two selected groups of 3 coins, using the balance. This action (experiment) may give two results. In one case, the weights of both groups are equal. It means that the lighter coin is in the third group of 2 coins because there is only one lighter coin. Then go to Step 10. In the other case, the balance shows that the weight of one group of 3 coins is less than the weight of the other group of 3 coins. It means that the lighter coin is in the lighter group of 3 coins because there is only one lighter coin. Then go to Step 5.

Step 10: Compare two selected coins, using the balance. This action (experiment) may give only one result the balance shows that the weight of one coin is less than the weight of the other coin because we know that there is one lighter coin. It means that the lighter of these two coins is not a good coin and we have the answer to the problem.

This algorithm shows how to find the answer to the problem performing three or less weighings. Information estimates prove that it is impossible to improve this algorithm, obtaining the result using only two weighings or less.

Problems 1, 2 and their versions have many practical applications in such areas as information search, knowledge acquisition, ordering, control, to mention but a few.

3.5. Axiomatization of Information Measurement

> *Consequently the student who is devoid of talent*
> *will derive no more profit from this work*
> *than barren soil from a treatise on agriculture.*
> Quintilian (35–95)

Information entropy (quantity of information) has an axiomatic description. Namely, there is a set of axioms, which characterize some natural properties of this information measure, and these axioms uniquely determine Shannon's entropy.

There are different systems of axioms used for characterization of the information entropy. We follow here the approach of Shannon (1948).

Building his axiomatic system based on natural properties of information measure, Shannon starts with the problem of information transmission. He analyses how to define a quantity that will measure, in some sense, how much information is "produced" by a discrete information source, or as Shannon explains, at what rate information is produced. With such a goal, the problem is specified by taking into account only information that tells the receiver whether some event has happened. To estimate this information, it is assumed that the receiver regards a set $\{E_1, E_2, ..., E_n\}$ of possible events with known probabilities of occurrence $p_1, p_2, ..., p_n$. Then it is possible to assume that the measure for the information entropy $H(p_1, p_2, ..., p_n)$ should have the following properties:

Axiom A1 (*Continuity*). The function $H(p_1, p_2, ..., p_n)$ is continuous in all arguments p_i.

Axiom A2 (*Monotonicity*). If all p_i are equal ($p_i = 1/n$), then $H(1/n, 1/n, ..., 1/n\}$ must increase monotonically with n.

Axiom A3 (*Branching property*). If a choice is broken down into two successive choices, the original function $H(p_1, p_2, ..., p_n)$ should be the weighted sum of the two new values of this function and the weights are equal to the probabilities of branching.

Axiom A4 (*Normalization*). $H(1/2, 1/2\} = 1$.

Axiom A4 means that making a choice from equally probably events/outcomes gives one bit of information.

Axiom A5 (*Symmetry*). The function $H(p_1, p_2, ..., p_n)$ is symmetric for any permutation of the arguments p_i.

Axiom A5 means that the function $H(p_1, p_2, ..., p_n)$ depends only on probabilities $p_1, p_2, ..., p_n$ but is independent of their order.

The axiomatic characterization is obtained in the following theorem.

Theorem 3.5.1. The only function that satisfies Axioms A1–A5 is the information entropy $H(p_1, p_2, ..., p_n)$ determined by the formula (3.2.4).

Proof. Let us consider the function $F(n) = H(1/n, 1/n, ..., 1/n)$ and r^m events with equal probabilities of occurrence. It is possible to decompose choices from all of these events into a series of m choices from r potential events with equal probabilities of occurrence. In this situation, Axiom A3 gives us the following equality:

$$F(r^m) = mF(r)$$

Let us fix some number t. Then for any natural number n, there is a number m such that

$$r^m \leq t^n < r^{m+1} \qquad (3.5.1)$$

Indeed, if m is the largest number such that $r^m \leq t^n$, then $t^n < r^{m+1}$.

As before, Axiom A3 gives us the following equality:

$$F(t^n) = nF(t)$$

Taking logarithms from all terms in the formula (3.5.1), we have

$$m \log_2 r \leq n \log_2 t < (m + 1) \log_2 r \qquad (3.5.2)$$

Dividing all terms in the formula (3.5.2) by $n \log_2 r$, we have

$$m/n \leq (\log_2 t / \log_2 r) < m/n + 1/n \qquad (3.5.3)$$

This gives us

$$\left| (\log_2 t / \log_2 r) - m/n \right| < 1/n \qquad (3.5.4)$$

At the same time, Axiom A2 and inequalities (3.5.1), give us

$$F(r^m) \leq F(t^n) \leq F(r^{m+1})$$

By properties of logarithms, we have

$$mF(r) \leq nF(t) \leq (m+1)F(r) \qquad (3.5.5)$$

Dividing all terms in the formula (3.5.5) by $nF(t)$, we have

$$m/n \leq (F(t)/F(r)) < m/n + 1/n \qquad (3.5.6)$$

This gives us

$$\left| (F(t)/F(r)) - m/n \right| < 1/n \qquad (3.5.7)$$

Properties of absolute value give us the following inequality

$$\left| (F(t)/F(r)) - (\log_2 t / \log_2 r) \right| < 2/n \qquad (3.5.8)$$

Indeed, we have

$$\left| (F(t)/F(r)) - (\log_2 t / \log_2 r) \right| = \left| (F(t)/F(r)) - m/n + m/n - (\log_2 t / \log_2 r) \right| \leq$$
$$\left| (F(t)/F(r)) - m/n \right| + \left| m/n - \log_2 t / \log_2 r \right| = 1/n + 1/n = 2/n$$

As the left side in the inequality (3.5.8) is constant, while the right side in the inequality (3.5.8) is converging to zero when n goes to infinity, we can conclude that

$$F(t)/F(r) = (\log_2 t / \log_2 r)$$

Consequently, as the number t is fixed, we have

$$F(r) = (F(t)/\log_2 t) \log_2 r = K \log_2 r \qquad (3.5.9)$$

As the equality (3.5.9) is true for any r and $H(1/2, 1/2) = 1$ by Axiom A4, we have

$$F(r) = \log_2 r$$

Let us consider n events $E_1, E_2, \ldots, E_n$ with equal probabilities and divide them into k groups with n_i elements in the group with number i. Then theory of probability implies that the probability that the event belongs to the group with number i is equal to

$$p_i = n_i / (\sum_{i=1}^{k} n_i)$$

Now let us contemplate function $H(p_1, p_2, \ldots, p_n)$ with rational arguments $p_1, p_2, \ldots, p_n$. Then taking the least common denominator of all $p_1, p_2, \ldots, p_n$ as the number n, we can divide events $E_1, E_2, \ldots, E_n$ with equal probabilities them into k groups with n_i elements in the group so that we will have

$$p_i = n_i / (\sum_{i=1}^{k} n_i) = n_i / n$$

It is possible to break a choice of one event from the set $\{E_1, E_2, ...,$ $E_n\}$ into two steps: at first, to choose a group where the chosen event is and then to choose the event from the already chosen group. By Axiom A3, this gives us the following equality

$$H(1/n, 1/n, ...,1/n) = H(p_1, p_2, ..., p_n) + \sum_{i=1}^{k} p_i H(1/n_i, 1/n_i, ..., 1/n_i)$$

$$(3.5.10)$$

It is proved that

$$H(1/n, 1/n, ..., 1/n) = \log_2 n$$

and thus,

$$H(1/n_i, 1/n_i, ..., 1/n_i) = \log_2 n_i$$

for all $i = 1, 2, 3, ..., n$.

This and the equality (3.5.10), give us

$$\log_2 n = H(p_1, p_2, ..., p_n) + \sum_{i=1}^{k} p_i \log_2 n_i \qquad (3.5.11)$$

Consequently, we have

$$H(p_1, p_2, ..., p_n) = \log_2 n - \sum_{i=1}^{k} p_i \log_2 n_i$$

$$= - \sum_{i=1}^{k} p_i \log_2 n_i/n = - \sum_{i=1}^{k} p_i \log_2 p_i$$

This proves formula (3.2.4) for all rational numbers. As by Axiom A1, $H(p_1, p_2, ..., p_n)$ is a continuous function and the set of all rational numbers is dense in the set of all real numbers, we have

$$H(p_1, p_2, ..., p_n) = - \sum_{i=1}^{k} p_i \log_2 p_i$$

for all real numbers.

Theorem is proved.

If we take Axiom A4 away, we have a weaker result.

Theorem 3.5.2 (Shannon, 1948). The only function that satisfies Axioms A1–A3 and A5 is the information entropy $H(p_1, p_2, ..., p_n)$ determined by the formula

$$H(p_1, p_2, ..., p_n) = K \sum_{i=1}^{k} p_i \log_2 p_i \qquad (3.5.12)$$

where K is an arbitrary constant.

This is the theorem obtained by Shannon (1948) although its proof was later improved by other researchers.

There were different generalizations of this result.

Sharma and Taneja (1974) introduced an infinite family of information measures. They consider a generalized information entropy $H(p_1, p_2, ..., p_n; q_1, q_2, ..., q_n)$ that is a function of n pairs $(p_i; q_i)$ of probabilities $(i = 1, 2, 3, ..., n)$ and give its axiomatic characterization, proving the following result.

Theorem 3.5.3 (Sharma and Taneja, 1974). The only function that satisfies Axioms GA1–GA3 and GA5 is the generalized information entropy $H(p_1, p_2, ..., p_n; q_1, q_2, ..., q_n)$ determined by the formula

$$H(p_1, p_2, ..., p_n; q_1, q_2, ..., q_n) = A \sum_{i=1}^{k} p_i \log_2 p_i + B \sum_{i=1}^{k} p_i \log_2 q_i$$

where K is an arbitrary constant.

Axiom GA1 (*Continuity*). The function $H(p_1, p_2, ..., p_n; q_1, q_2, ..., q_n)$ is continuous in all arguments p_i and q_i.

Axiom GA2 (*Additivity*).

$$H(rp_1, (r-1)p_1, rp_2, (r-1)p_2, ..., rp_n, (r-1)p_n;$$
$$(tq_1, (t-1)q_1, tq_2, (t-1)q_2, ..., tq_n, (t-1)q_n)$$
$$= H(p_1, p_2, ..., p_n; q_1, q_2, ..., q_n) + H(r, r-1; t, t-1)$$

where $r, t \in [0,1]$.

Axiom GA3 (*Branching property*). If a choice is broken down into two successive choices, the original function $H(p_1, p_2, ..., p_n; q_1, q_2, ..., q_n)$ should be the weighted sum of the two new values of this function and the weights are equal to the probabilities of branching.

Axiom GA5 (*Symmetry*). The function $H(p_1, p_2, ..., p_n; q_1, q_2, ..., q_n)$ is symmetric for any permutation of the arguments p_i and the corresponding permutation of the arguments q_i.

As Kulblback's information measure (Kulblback, 1959) and Kerridge's inaccuracy (Kerridge, 1961) are special cases of the generalized information entropy $H(p_1, p_2, ..., p_n; q_1, q_2, ..., q_n)$, Sharma and Taneja (1974) obtain axiomatic characterizations of these two information measures as corollaries of Theorem 3.5.3.

3.6. Information in Physics

Science may be described as the art
of systematic oversimplification.
Karl Popper (1902–1994)

It is possible to divide all information theory applications in physics into two big categories: particular and comprehensive approaches. In a particular approach, information theory is used to derive additional properties of physical objects or to analyze existing physical theories. Applications of Shannon's information theory are examples of particular approaches. They turned out to be important in physics, as well as information processing and communication technology. For instance, Brillouin (1956) applies Shannon's information theory to different problems in physics. He analyses measurement and observation in physics, functioning of oscillators and resonators, problems of space and time, quantum physics and classical physics. In contrast to particular approaches, the authors of comprehensive approaches try to develop the whole physics based on information theory. Frieden's theory (Frieden, 1998) is an example of a comprehensive approach.

Understanding that physicists study physical systems not directly but only through information they get from these systems has created a school of thought about the role of information processing in physical processes and its influence on physical theories. According to one of the outstanding physicists of the 20th century John Archibald Wheeler (1911–2008), it means that every physical quantity derives its ultimate significance from information. He called this idea "It from Bit" where It stands for things, while Bit impersonates information as the most popular information unit (Wheeler, 1990; 1994). For Wheeler and his followers, space-time itself must be understood and described in terms of a more fundamental pregeometry without dimensions and classical causality. These features of the physical world only appear as emergent properties in the ideal modeling the physical reality based on information about complex interactions of very simple basic elements, such as subatomic particles.

This well correlates with one of the main neurobiological assumptions, which implies that things and the world perceived by people are constructed in their brains. This construction is based on the information received from these things and the whole world. Thus, whatever people know, in particular, all scientific knowledge, depends only on information they have.

The most fundamental approach to physics based on information was developed by von Weizsäcker, his coauthors, and his followers in the, so-called, *ur*-theory (von Weizsäcker, 1958; von Weizsäcker, Scheibe and Süssmann, 1958; Castell, et al, 1975–1986; Lyre, 1998). The main idea is that what physicists learn about nature comes from observation and experiments through measurement. According to von Weizsäcker, measurement is information extraction from physical objects and processes. Thus, physical objects and processes are entirely characterized by the information that can be acquired from them. In such a way, physics reduces to predicting measurement outcomes.

Information from measurement comes in the form of data and represents values of possible measurement outcome alternatives. It is possible to code these data in the form of binary digits consisting of 0's and 1's. Each binary digit is, in some sense, a value of a binary alternative called *ur* by von Weizsäcker. More exactly, *ur*-alternatives or, simply, *urs* are empirically decidable binary alternatives. Binary alternatives, in turn, are traditionally considered as bits of information that gives knowledge about results of experiments. *Urs* represent potential information, and measurement is a transition from potential to actual information (Lyre, 2003).

Data obtained from measurements give information that is used to produce (derive) knowledge about physical objects and processes. In such a way, it is possible to imply that that physical objects and processes are reduced to or even made of information. For instance, Born (1953) wrote that an observation or measurement did not refer to a natural phenomenon as such, but to its aspect from, or its projection on, a system of reference, which as a matter of fact, is the whole apparatus used. This projection is accomplished by information transmission. Born advocates that there are objects that exist independently of people but people can deal only with information about these objects.

Based on these ideas, von Weizsäcker and his coauthors used multiple quantization to build the classical Weyl equation, Dirac equation and Klein-Gordon equation. In the *ur*-theory, Maxwell equations emerge as algebraic identities.

There is a popular approach in science that advocates (and even tries) reducing all natural sciences (chemistry, biology, etc.) to physics. Von Weizsäcker goes even further when he suggests reducing physics to information science.

Information-theoretic perspective allows researchers to achieve new understanding of physical phenomena even without such radical changes. For instance, Harmuth (1992) applies Shannon's information theory to space-time physics. From the very beginning, Harmuth assumes that the entropy of a system can be reasonably used as a measure of the amount of information a physical system contains, but it is difficult to make sense of such an idea unless the amount of information in the physical system is finite. If the positions and orientations of molecules can be specified to any degree of precision, then there is no limit to the number of bits needed to describe the state of a finite volume of a gas. The main assumption of Harmuth's approach is that each measurement gives only finite information, i.e., finite number of bits, about physical objects and events. Principles of quantum mechanics restrict information flow so that it is possible to transmit only finite information in finite time. This brings Harmuth to the conclusion that the conventional model of the physical space-time manifold as a continuum is inadequate and it is necessary to use a discrete model. He builds such a model and develops physics based on it, studying physical time and motion in this setting, building difference counterparts of the classical Schrödinger equation, Dirac equation, and Klein-Gordon equation.

Like von Weizsäcker, Frieden (1998) developed a new approach to physics as a whole scientific discipline. It is based on Fisher information, explaining that physics as a knowledge system about the universe is built from knowledge acquired through information reception and processing. According to this conception, the observer is included into the phenomenon of measurement as his properties as an information receiver (cf. Section 3.1) influence what data are collected and how they are collected and interpreted. Frieden (1998) demonstrates how, using principles based on Fisher information, to derive a great deal of contemporary physical theory, including Newtonian mechanics, statistical mechanics, thermodynamics, Maxwell's equations, Lorentz transformation, general relativity, EPR experiment, Schrodinger equation, Klein-Gordon equation, Dirac equation, and explanation of the fundamental physical constants.

The basic contention of Frieden is that physicists think that they study material systems. Actually, they study information that they are able to get from these systems. In this situation some part of information present in the source (the observed and measured physical system), that is, in the physical system, is inevitably lost when observing the source and measuring its properties (parameters). Moreover, the random errors can contaminate the measurement (observation) data. Thus, as a rule, the Fisher information $\mathbf{I}$ in the data obtained in the measurement is different from the Fisher information $\mathbf{J}$ in the source.

This approach is supported by some physicists who think that people will never know complete truth about what the real universe is in general and at the most fundamental level, in particular. All theories in physics are invented by people as attempts to possibly explain unexplained aspects of what has been observed, or unexplained aspects of previous theories. There has never been a completely true theory or model of the universe. Thus, the purpose of physics is not to fully describe what is but to explain what is observed.

For instance, the *weak holographic principle* (cf. (Smolin, 1999)) asserts that there are no things in nature, only processes, and that these processes are merely the exchange of data across two-dimensional screens. According to this theory, the three-dimensional world *is* the flow of information. This idea is also reflected in the new direction in physics called process physics (Cahill, 2002; 2002a; 2002b).

Frieden grounds his approach by three axioms:

Axiom 1 (The law of conservation of information change). When an act of observation perturb a source and the data gathered from that source and the perturbed information is $\delta\mathbf{I}$ for the source and $\delta\mathbf{J}$ for the data, then $\delta\mathbf{I} = \delta\mathbf{J}$.

Axiom 2 (The law of integral representation). There exist sequences of functions

$i_n(x)$ and $j_n(x)$ such that $\mathbf{I} = \int \sum_{n \in \omega} i_n(x) \, dx$ and $\mathbf{J} = \int \sum_{n \in \omega} j_n(x) \, dx$.

Axiom 3 (The microscopic zero condition). There is a number k such that $0 < k \le 1$ and $i_n(x) - kj_n(x) = 0$.

The microscopic zero condition implies the following *macroscopic zero condition*:

There is a number k such that $0 < k \le 1$ and $\mathbf{I}(x) - k\,\mathbf{J}(x) = 0$.

The cornerstone of Frieden's approach is the Extreme Physical Information (EPI) Principle. Thus, it is possible to call this approach by the name *EPI theory*. The EPI Principle is introduced with the purpose to express scientific laws in the conventional form of differential equations and probability distribution functions. Examples of such laws are the Schrödinger wave equation and the Maxwell-Boltzmann distribution law.

The Extreme Physical Information Principle builds on a natural idea that observation and measurement are never completely accurate, while any information channel is imperfect. Thus, the Extreme Physical Information Principle states that the difference $\mathbf{J} - \mathbf{I}$ the Fisher information in the source and the Fisher information in the measurement data has to reach an extremum. For most situations, this extremum is a minimum, meaning that there is a tendency for any observation to describe its source in the best possible way.

The *loss*, or *defect*, of information $\mathbf{K} = \mathbf{I} - \mathbf{J}$ is physical information of the system that gives this information.

The Extreme Physical Information Principle is a principle describing how information about physical and other systems is acquired and corresponding knowledge is built. As such, it neither provides nor derives a detailed ontological model of reality as it is. Instead, this principle describes the rules of how people get information from physical systems and how they derive knowledge about reality from this information. For instance, certain primitive object properties, such as position, velocity, mass, charge, etc., are assumed to exist and measurement provides information about their properties. This information is used to derive knowledge about these systems. In this way, EPI theory differs from theories that require a detailed ontology, such as the classical quantum mechanics, quantum electrodynamics, gauge theories, Cooper pair model, Higgs mass theory, or string theory. For instance, string theory assumes a detailed model for all particles, stating that they are composed of strings, which can be open or closed, and have various shapes. EPI theory makes no such stipulations about particles, or indeed about reality, other than that it exists and is accessible by observation. The key aim of EPI theory is to learn the correct probability law and the dynamics of that assumed and observed reality.

Fisher information is a powerful new method for deriving laws governing many aspects of nature and human society. Frieden and coauthors have also used the EPI Principle to derive some established principles and new laws of biology, the biophysics of cancer growth, chemistry, and economics.

There are also information-theoretical approaches to physics and consequently, to the whole world that are not based on probability theory and statistical considerations. An interesting approach called information mechanics is suggested by Kantor (1977; 1982). In his theory, he introduces an important concept of the universe U_O of an observer O. Thus, instead of the absolute universe of the conventional physics, relative universes are considered. Namely, the *universe U_O of an observer O* is all about which O can have information. The main principle of information mechanics is that O knows nothing more than information received from U_O. As a result, it is necessary for the relative universe physics, to include only assumptions in the form of principles or postulates about information. Information is defined by Kantor as the logarithm with the base 2, i.e., log_2, of the number of possibilities and is measured in bits.

Information mechanics has the following postulates (Kantor, 1977):
1. Information is conserved.
2. Information is communicable.
3. Information is finitely accessible.

One more approach that takes information as the basic essence is process physics (Cahill, 2002; 2002a; 2002b). The fundamental assumption of this theory is that reality has to be modeled as self-organizing semantic or relational information based on self-referentially limited neural networks (Müller, et al, 1995). This modeling was motivated by the discovery that such stochastic neural networks are foundational to known quantum field theories. In process physics time is a distinct non-geometric process, while space and quantum phenomena are emergent and unified. Quantum systems and processes are described by fractal topological defects embedded in and forming a growing three-dimensional fractal process-space called the quantum foam.

Siegfried (2000) also studies the role of information in physics, making emphasis on quantum information theory. Gershenfeld (1992) shows how the conventional information theory can be efficiently applied to dynamics.

Thus, it is possible to conclude that in all cases considered in statistical information theories (Shannon's theory, Fisher's theory, etc.) information changes knowledge of the receiver. Consequently, all statistical information theories are subtheories of the general theory of information.

Chapter 4

Semantic Information Theory

You can have great ideas. But if you cannot communicate them,
they have no place in science. It's like a rainbow in the darkness...
Carlos Gershenson

Gershenson's statement emphasizes that there is a difference between communication of a text and communication of ideas contained in this text. Communication of ideas involves understanding and understanding is based on semantics. However, Shannon writes (1948):

"Frequently the messages have meaning .... These semantic aspects of communication are irrelevant to the engineering problem. The significant aspect is that the actual message is one selected from a set of possible messages".

The statistical theory of information concentrates its efforts on technical problems of text transmission, while semantic information theory is concerned with communication of ideas. In this context, text is understood as a sequence of symbols, and engineers are interested how to better transmit these symbols from the source to the receiver. Here better means without distortion, with the highest speed, etc.

Weaver suggested (Shannon and Weaver, 1949), the problem of communication has three issues:

1. The technical problem: How accurately can communication symbols be transmitted?
2. The semantic problem: How precisely do the transmitted symbols retain the meaning that the sender intended?
3. The effectiveness problem: How precisely do the desired effect and actual effect produced by the received signal correspond?

As many others, Weaver believed that Shannon provided a solution to the first issue of the communication problem.

However, as Mingers writes (1996), this approach is like measuring the volume of a container without knowing what it contains. In a similar mode, von Weizsäcker writes (2006), "what colloquially in human communication is understood as information is not the syntactically definable multiplicity of forms of a message but what a competent listener can understand in the message." Text in any language is not an arbitrary sequence of symbols, but a sequence that is built according to definite rules and has meaning.

Shannon's deliberate exclusion of semantic aspects from his theory led several authors, such as Bar-Hillel and Carnap (1952; 1958), MacKay (1969), Nauta (1970), Dretske (1981), and Barwise and Seligman (1997), to elaborate alternative information theories. These theories were based on interesting and promising premises, but were not generally accepted and were not able to compete with Shannon's information theory because they did not solve practical problems in a constructive way.

In contrast to statistical theories of information, semantic theories of information study meaning of information. In some cases meaning of information is understood as the assumption (cf., for example, (Bateson and Ruesch, 1951)), that every piece of information has the characteristic that it makes a positive assertion and at the same time makes a denial of the opposite of that assertion. However, meaning is a more complicated phenomenon and to understand it in the context of information theory, we start this chapter (Section 4.1) with a study of three communicational aspects, or dimensions, of information: syntax, semantics and pragmatics. Then we present semantic information theories, that is, theories that make emphasis on the semantics of information. In Section 4.2, we give an exposition of the first semantic information theory developed by Bar-Hillel and Carnap, as well as its later developments and improvements. In Section 4.3, we reflect on knowledge oriented information theories.

4.1. Three Dimensions of Information

> *Well building hath three conditions.*
> *Commodity, firmness and delight.*
> Henry Wotton (1568–1639)

Analyzing the concept of information, it is natural to come to the conclusion that information has three major aspects or issues: *syntax*, *semantics*, and *pragmatics*. To better understand properties information structures, let us look what syntax, semantics and pragmatics are in general and how they are applied to information theory.

These three areas, syntax, semantics and pragmatics, were, at first, developed and studied in linguistics. Historically, linguists has found and investigated various structures in natural languages. These structures were classified and linguistical knowledge was organized and divided into separate fields. As a result, the standard representation of a general language structure was formed as a three-dimensional space. That is, any natural or artificial language is described and studied in three dimensions or projections: syntax, semantics, and pragmatics. Somewhat later the theory of signs, semiotics, was also organized along these lines forming three branches: semantics, syntax, and pragmatics (cf. (Pierce, 1961; Morris, 1938; 1946)).

The word *syntax* originated from the ancient Greek words *syn* (συν), which means "together", and *táxis* (τάξις), which means "arrangement". Now syntax means the system of the rules that govern the structure of sentences usually determined grammaticality. The term *syntax* can also be used to refer to these rules themselves, as in "the syntax of a language".

In semiotics, *syntax* represents relations of signs to each other in formal structures. For instance, the rule that the predicate (often in the form of a verb) goes after the subject (often in the form of a noun) in a sentence belongs to the syntax of English.

Syntactics is the theoretical discipline that studies syntax of languages, i.e., the formal relations of signs to one another. It also studies and expresses not only direct relationships between signs but also in what way these relations are generated and induced by semantics and pragmatics. In the same way, pragmatics studies relations that appear through semantics and syntactics. This is obviously displayed in communication when the expressive means of language are used for achievement of the beforehand purposes. In some cases such purposes are latent (that we can see in poetry). In other cases, they become foremost (as we can see in advertising). Thus, even the elementary syntactic structures must have the rational convincing force. For instance, combination of two predicates or propositions, as well as combination of a name and a predicate, induces a new predicate or proposition.

In semiotics, *semantics* represents relations of signs to the objects to which the signs are applicable, i.e., to their *denotata*. The word *semantics* originated from the Greek word *sēmantikos*, which means *giving signs* or *significant*, and is related to the Greek word *sēma*, which refers to aspects of meaning, as expressed in language or other systems of signs. Semantics as a theoretical field is a study of this meaning understood as relations between signs and the things they refer to, their *denotata*.

There are many meanings of the term *meaning*. Even in theory, there are different approaches to interpretation of this term and there is no generally accepted consensus. For instance, Ogden and Richards (1953) found 23 meanings of this term.

Semanticists generally recognize two sorts of meaning that an expression (such as the sentence "Information is power") may have: extentional meaning and intentional meaning.

Extentional meaning is the relation that the expression has to things and situations in the real world, as well as in possible worlds.

Intentional meaning is the relation the signs have to other signs, such as the sorts of mental signs that are conceived of as concepts.

In semiotics, extentional meaning is presented by the relation between a sign and its object, which is traditionally called *denotat* or *denotation*. Intentional meaning presented by the relation between a given sign and the signs that serve in practical interpretations of the given sign is traditionally called sign *connotation*. However, many theorists prefer to restrict the application of semantics to the denotative aspect, using other terms or completely ignoring the connotative aspect.

As a result, *semiotic semantics* related to symbols, e.g., letters, words and texts, consists of two components. One of them, *relational semantics*, expresses intentional meaning and represents relations between symbols. The other one, *denotational semantics*, expresses extentional meaning and represents relations between symbols and objects these symbols represent.

In addition, there are other kinds of semantics.

Logical semantics is related to propositions and involves truth values of these propositions.

Linguistic semantics is related to words and texts and is expressed by relations between them.

Semantics for computer applications falls into three categories (Nielson and Nielson, 1995):

- *Operational semantics* is the field where the meaning of a construct is specified by the computation it induces when it is executed on a machine. In particular, it is of interest *how* the effect of a computation is produced.

- *Denotational semantics* is the field where meanings are modeled by mathematical objects that represent the effect of executing the constructs. Thus, *only* the effect is of interest, not how it is obtained.

- *Axiomatic semantics* is the field where specific properties of the effect of executing the constructs represent meaning and are expressed as *assertions*. Thus, there are always aspects of the executions that are ignored.

Semantics deals with relations between signs and objects that signs designate. Syntax studies the relations between signs. The pragmatics (also called *dectics*) investigates the attitudes of people who use language to signs. In real functioning and evolutionary changing, a language is developing simultaneously in these three interconnected dimensions. Some linguists assume that this development is also continuous. Really, semantics studies and expresses not direct attitudes of signs to objects but what these attitudes relate and display through syntax and pragmatics. Similar things exist in quantum physics where the measurement of parameters of particles is changing these parameters.

Pragmatics is the study of the ability of natural language users (e.g., speakers or writers) to communicate more than that, which is explicitly stated. To distinguish semantic and pragmatic meanings of a message (or sentence), communication researchers use the term the *informative intent*, also called the *sentence meaning*, and the term the *communicative intent*, also called the *sender meaning* or *speaker meaning* when it is an oral communication (Sperber and Wilson, 1986).

In semiotics, *pragmatics* represents relations of signs to their impacts on those who use them, e.g., relations of signs to interpreters. This impact exists when a sign makes sense to the interpreter. *Sense* of a message (information) is defined by Vygotski (1956) as a system of psychological facts emerging in the brain that received a message (information).

The ability to understand another sender intended meaning is called *pragmatic competence*. An utterance describing pragmatic function is described as metapragmatic. As pragmatics deals with the ways people reach their goals in communication, pragmatics explains that it is possible to reach the same goal with different syntax and semantics. Suppose, a person wants to ask someone else stop smoking. This can be achieved by using several utterances. The person could simply say, "Stop smoking, please!", which has direct and clear semantic meaning. Alternatively, the person could say, "Oh, this room needs more air conditioners", or "We need more fresh air here", which infers a similar meaning but is indirect and therefore requires pragmatic inference to derive the intended meaning.

The structure of a sign includes various relations that can be also divided into three groups:
- relations to other signs, forming the structure of the sign,
- relations to objects, which determine the references of signs to things of the external world,
- relations to subjects or interpreters, where the communicative effect of the sign is manifested.

Each of these three kinds of relations represents one of the three dimensions of the sign, namely, the syntactic dimension, the semantic dimension, and the pragmatic dimension. These dimensions together make semiosis possible and thus constitute the sign as such (Morris, 1938; 1946; 1964).

Each of these three dimensions expresses a certain part not only of a language but also of speech and communication activity. Thus, language has syntax, semantics and pragmatics because information has syntax, semantics and pragmatics, while language is a system for representing, processing, storing, and exchanging information (communication).

All these considerations show that the three-dimensional structure of languages and signs results from the fact that information has exactly the same dimensions.

Syntax of information refers to aspects of information representation.

Semantics of information refers to aspects of information meaning.

Pragmatics of information refers to aspects of information action (influence).

It is possible to consider semantics, syntax, and pragmatics on different levels of information representation. There is a subsymbolic level of information, which is represented by signals. The next level consists of sets of separate abstract symbols, such as letters or binary digits. Higher levels are related to natural and artificial languages. For instance, the level of natural languages is higher than the level of formal logic.

The three-dimensional structure of information has its roots in the structure of the world as a whole. Namely, each of the dimensions is related to the corresponding component of the existential triad discussed in Chapter 2. Syntax reflects relations in the world of structures. Semantics represents the physical world, relating expressions to things and situations in the physical world. Pragmatics exhibits intentional and effective structures from the mental world.

A similar situation exists for languages. The main functions of a natural language are communication and discourse, aiming at improvement of the efficiency of people interaction. These two functions are based on and realized through other functions of the language, which include informing, modeling, rejection, influence, formation and expression of ideas. Because of such a peculiarity of languages, any natural language represents the whole world. As Wittgenstein wrote (1922), the internal structure of reality shows itself in language. On the one hand, language reflects the real world in which people live. On the other hand, it is formed under the influence of nature and social forces in the mental space of society. As a consequence, the structure of language reflects basic features of society and nature. For instance, the lexicon of language contains only such words that are used in society. Natural languages are linear as texts in natural languages are formed as linear sequences of words. The cause of this peculiarity is certain characteristics of the nervous system and mentality of a person. A similarity between language and the world is explicated in a structural isomorphism between a natural language and the real world.

Indeed, the global structure of the world (the "existential triad") is reflected in the triadic structure of language (*syntax, semantics,* and *pragmatics*). This reflection is a structural similarity in the sense of Section 2.1. The world of structures in this similarity corresponds to syntax, which covers purely linguistic structures. The physical world is embodied in semantics, which connects language structures and the real objects. The mental world (as the complex essence) gives birth to the pragmatics of language as pragmatics refers to goals and intentions of people.

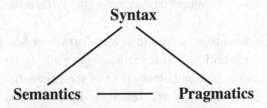

Figure 4.1. The triadic structure of language

Existence of the existential cycle described in (Burgin and Schumann, 2006) can be also observed in natural languages, or more precisely, in the virtual world generated by a language (Burgin and Milov, 1998). In the opinion of experts in the field of linguistics and philosophy of language, the following regularity can be explicated in the history of poetic and philosophy of language: the language imperceptibly directs theoretical ideas and poetic impulses on one of the axes of it three-dimensional space (Stepanov, 1985). The first direction, which is taken, goes along semantics, then it turns to syntactics, and at last, to pragmatics. After achieving the third component of the linguistic "existential cycle", the language again and again repeats it.

This peculiarity of natural languages has a historical explanation. Languages emerged and have been used as a tool for communication. Communication is information exchange. Thus, specific aspects of information influence many features of language. A manifestation of this influence is the three-dimensional structure of language: syntax, semantics and pragmatics. Another example is given by the functions of language. Roman Jakobson (1971) identified six functions of language, only one of which is the traditional system of reference:

- The *referential function* of language conveys information about some real phenomenon.
- The *expressive function* of language describes (or provides information about) feelings of the speaker or writer.
- The *conative function* of language attempts to elicit some behavior from the addressee.
- The *phatic function* of language builds a relationship between both parties in a conversation.
- The *metalingual function* of language self-references.
- The *poetic function* of language focuses on the text independent of reference.

All these functions involve information transmission and reception and thus, are typical to information flow. Namely, we have:

- The *referential function* of information flow conveys cognitive information about some real phenomenon.
- The *expressive function* of information flow conveys affective information and describes (or provides cognitive information about) feelings of the information sender.
- The *conative function* of information flow attempts to elicit some behavior from the addressee by conveying effective information.
- The *phatic function* of information flow builds a relationship between both parties, the sender and receiver, in a communication by conveying cognitive and affective information between both parties.

- The *metalingual function* of information flow self-references, i.e., establishes connections of the information portion to itself.
- The *poetic function* of information flow focuses on information independent of reference.

These parallel traits of natural languages and information flow show that information, in general, realizes the same functions as language, while information studies have reflected all three dimensions of information. It is generally assumed that the classical statistical theory of information deals only with syntax of information representation and transmission (cf. Chapter 3). Many researchers wrote on semantic information (cf., for example, (Bar-Hillel and Carnap, 1952; 1958; Shreider, 1967; Dretske, 1981; Hintikka, 1968; 1970; 1971; 1973; 1973a; 1984; Dodig-Crnkovic, 2005; Allo, 2007)). Several researchers concentrate their attention on pragmatic information (cf., for example, (Yovits, 1975; Kornwachs, 1996)). Janich (1998) developed a theory of information that is exclusively related to purpose-oriented human actions. In this theory, information is defined as a predicate that qualifies standard request dialogues where linguistic utterances are speaker-, listener-, and form-invariant. This makes it possible to reproduce these situations on the basis of anthropomorphic artificial devices.

Yasin (1970), Von Weizsäcker (1985/2006), Antonov, (1988), and Capuro and Hjorland (2003) are concerned with all three kinds: syntactic information, semantic information, and pragmatic information. However, Von Weizsäcker suggested that semantic information is only measurable as pragmatic information.

The most developed information theory that incorporates all three levels is developed by Nauta (1970), who considers syntax, semantics and pragmatics of information in the context of the *meta-semiotic* information theory. In the terminology of Nauta, the highest level of information theory called the meta-semiotic information theory can be divided into the three disciplines of syntax, semantics and pragmatics.

In contrast to many other authors, Nauta does not want to oppose a new information theory to existing theories, but develops a conceptual framework for existing and future information theories. He treats information as a connection between the social sciences, humanities and natural sciences, applying a cybernetic interpretation to the structural and conceptual tools of semiotics (the theory of signs), such as syntax, semantics, pragmatics, and semiosis. In this context, semiosis is viewed as a sign process when a sign perceived by an individual prompts this individual to react and behave in the form of a specific behavioral pattern.

Semiosis takes place in an i-system R, which consists of a *receptor/receiver*, *emitter*, and *filters*. The receptor receives an information carrier and passes it to a filter. When the filter recognizes a sign in this carrier, it changes the interior state of the i-system R and also stimulates the next filter to prepare a goal-directed *behavioral pattern*, which goes as output to the environment via the emitter as an action.

To make possible these operations, a sign has to possess the following features:

1. An internal and/or external structures (syntactic component), which allow the filter to recognize it as a sign, find its meaning and form the output action – reaction to the sign.
2. A meaning for the i-system (semantic component), which allows the filter to change the internal state of R.
3. A stimulus that activates the i-system (pragmatic component), which allows the filter to form a behavioral pattern and to produce and trigger the corresponding action.

According to Nauta, information theories that do not take these features into account are incomplete.

Nauta divides all information theories into three basic classes:

1. The *zero-semiotic information theories* form the lowest level, in which signs and symbols are represented by signals and considered as formal entities without inner structure. Shannon's information and communication theory, which only takes into account communication engineering or signaling aspects, is an example of zero-semiotic information theories.

2. The *semiotic information theories* form the next level, which embraces the analysis of signals, signs and symbol semiosis. Semiotic information theories also provide a framework for the description of genetic and intrinsic information.

3. The top level, *meta-semiotic information theories*, can be divided into the three sublevels of syntax, semantics and pragmatics.

In his study, Nauta (1970) focuses mainly on the third type, the so-called meta-semiotic information theory, and defines the necessary framework for the formulation of a theory for all the three fields. From this perspective, the syntactic information theory is the most advanced of the three disciplines. The situation in the semantic information theory is seen as less promising because it provides hardly anything beyond a phenomenological description. Pragmatic information causes Nauta the most difficulties since he does not consider economic theories of information (cf. Chapter 6), as well as the algorithmic information theory (cf. Chapter 5). These theories provide different scientific measures for pragmatic information.

An interesting question is whether information in nature and technical systems also has these three dimensions — syntax, semantics and pragmatics. On the first glance, the answer is no as these three dimension are related to an essentially human construct such as language. However, a deeper analysis of the situation allows us to find syntax, semantics and pragmatics of information in nature.

Indeed, syntax of information refers to aspects of information representation, e.g., by symbols or electrical signals, and there are definitely these problems for information in nature and technology.

Semantics of information in general refers to aspects of information meaning. Meaning is usually related to understanding and we may ask what understanding we can find in inanimate nature. However, people have used to say that computers correctly or incorrectly understand their programs. It means that a computer can correctly or incorrectly interprete and perform what is written in the program. This process usually goes in two stages: at first, the program is compiled in the machine code and then this code is executed. In a similar way, transmission of genetic information in a cell goes in two stages: at first, genetic information in DNA is transcribed (encoded) in various RNA molecules and then in the process of translation proteins are built from information encoded in RNA (cf., for example, (Loewenstein, 1999)). Thus, we can see semantics of information in nature and technology. It is related to some kind of pragmatic or operational "understanding".

Pragmatics of information refers to aspects of information action (influence) and as action is the primary essence of information, there is also pragmatics of information in nature and technology. For instance, now many researchers and research institutions study the problem of what genetic information causes definite types of diseases. Another example is the basic question in computer science, which is concerned with the role of information in computation.

Directions in information studies considered in this book represent all three dimensions of information. In this chapter, we consider semantic approaches. The *statistical approach* considered in Chapter 3 reflects the syntactic dimension as it is interested in and based on probability distributions of events or symbols. These distributions represent relations between portions of information where each portion is related to occurrence of some event (symbol). The *algorithmic approach* considered in Chapter 5 reflects the pragmatic dimension as it is interested in building some object (usually, a string of symbols) and is based on algorithmic complexity. Algorithmic complexity estimates the minimum information needed to build an object with given means (algorithms). Chapter 6 exhibits explicitly pragmatic information theories.

Some researchers introduce and study more than three dimensions of information. For instance, Von Weizsäcker (1985/2006) also considers the dimension of time. Mingers (1996), following Morris (1938) and Stamper (1973; 1985; 1987), adopts a semiotic context. In this context, he discusses four levels of symbolic information studies: *empirics*, which is concerned with signal properties and transmission; *syntactics*, which is a study of formal (more exactly, structural) properties of systems of signs; *semantics*, which is aimed at the meaning of signs; and *pragmatics*, which is interested in the use of signs. However, it is possible to decompose empirics into three dimensions: syntactics, semantics, and pragmatics of signals. For instance, probability distribution of signals refers to the syntax of a signal. Semantics describes what symbol this signal represents. The channel coding theorem (cf. Chapter 3), which describes the capacity of a channel, deals with pragmatic problems of signal transmission. This shows that parallel to the syntax, semantics, and pragmatics of information, it is also possible to study syntax, semantics, and pragmatics of information representation. It can be represented by symbols, bringing us to semiotics. It can be represented by signals, involving semiotics of signals. It can be also represented by a language, or linguistic representation, which is a special case of symbolic representations.

Thus, we have six levels/dimensions:

- Syntax of information representation.
- Semantics of information representation.
- Pragmatics of information representation.

and

- Syntax of information.
- Semantics of information.
- Pragmatics of information.

It is possible to treat each of these dimensions/levels in a static form, as well as in a dynamic form, adding, thus, the dimension of time.

It is the consonance of context that makes the world or reality coherent. Hence in addition to the triad of a sign system considered in Section 2.1, a complete understanding of information requires five elements where an *agent* (receiver) is informed by a *sign* coming from another agent (sender) about some *object/domain* in a certain *context* (cf. Figure 4.2).

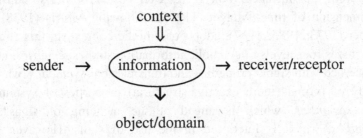

Figure 4.2. The structure of information flow

Semiotics also emphasizes the *pragmatics* of sign-systems, in addition to the more known dimensions of *syntax* and *semantics*. Therefore, a complete (semiotic) understanding of information includes these three dimensions of sign systems:

- Syntax is the characteristics of signs and symbols devoid of meaning, encompassing different aspects of the relation among signs such as their rules of operation, production, storage, and manipulation.
- Semantics is the content or meaning of the sign of an object for an agent, encompassing different aspects of the relation between signs and objects for an agent, in other words, the study of meaning.
- Pragmatics is the operational context of signs and repercussions of sign systems in an environment, reflecting how signs influence the agent and environment.

Here a sign can be a message, text, image picture, letter, digit, etc. An agent can be a woman reading a book, a cell receiving a biochemical message, or a robot processing some visual input, but it is typically understood as a person. A context can include agents knowledge, situation, another agent, person, weather, etc. Moreover, it is not true, that any person or agent, faced with a sign in a certain context, can recognize what the sign is about or to discern all meanings of this sign. It takes intelligence to do so: machine intelligence for simple signs, normal intelligence for customary signs, and unusual intelligence when the signs are extraordinary.

Informatics in the form of information technology essentially deals with the syntax of information, that is, with issues of data manipulation, storage, retrieval, and computation independently of meaning. Other subfields of informatics deal with semantics and pragmatics, for instance, computer science, human-computer interaction, bioinformatics social informatics and science informatics as well.

4.2. Logical Approach of Bar-Hillel and Carnap, Hintikka, and others: Logic in Information

> *Information can tell us everything. It has all the answers.*
> *But they are answers to questions we have not asked,*
> *and which doubtless don't even arise.*
> Baudrillard, Jean

In the semantic information theory, transformations of such infological system as thesaurus, or system of knowledge, are treated as information.

The founders of the semantic approach, Yehoshua Bar-Hillel (1915–1975) and Rudolf Carnap (1891–1970) build semantic information theory as a logical system (Bar-Hillel and Carnap, 1952; 1958). To make comprehension easier, Bar-Hillel and Carnap write (1952) that their theory lies explicitly and wholly within semantics, being a certain ramification of the Carnap's theory of inductive probability (Carnap, 1950).

They consider semantics of sets of *logical propositions*, or predicates that assign properties to entities. Propositions are nonlinguistic entities expressed by sentences. A *basic proposition/statement* assigns one property to one entity. An *ordinary proposition/statement* is built of basic statements by means of logical operations. Universes of entities are considered. A *state description* of such a universe consists of all true statements/propositions about all entities from this universe.

This is related to the possible-world semantics. It is often called Kripke semantics as it became popular after Kripke (1963) used it for building semantics for modal logics. In the possible-world semantics, a (logical) universe W is considered, states of which are fixed consistent assignments of truth values to primitive propositions from a propositional language L. It is possible to treat such states as possible worlds.

Bar-Hillel and Carnap distinguish between information and amount of information within a linguistic framework, treating statements as carriers of information. If p is a proposition, then information In p in this proposition p is defined as a set. This makes information In a set-valued function on propositions. This function must satisfy the following condition:

$$\text{In } p \text{ includes In } q \text{ if and only if } p \text{ L-implies } q \qquad (4.2.1)$$

Here "p L-implies q" means p *logically implies* q. However, as there are many different logics, we assume that L is a logic and "p L-implies q" means p *implies* q *in the logic* L.

Bar-Hillel and Carnap (1958) obtain many properties of their model of information. Here are some examples.

Theorem 4.2.1. In p = In q if and only if p is L-equivalent to q.

Indeed, p is L-equivalent to q if and only if p L-implies q and q L-implies p. Thus, by Condition (4.2.1), In p includes In q and In q includes In p. By the Extentionality Axiom (cf., for example, (Fraenkel and Bar-Hillel, 1958)), sets In p and In q are equal to one another.

Theorem 4.2.2. In p is the minimum set if and only if p is L-true, i.e., a tautology in the logic L.

Indeed, p is L-implied by any proposition q (cf., for example, (Kleene, 1967)). Thus, by Condition (4.2.1), In q includes In p for any proposition q. At the same time, when p is not L-true, there is a proposition q that does not L-imply p. Thus, In p is is not the minimum set because it is not included in In q.

Theorem 4.2.3. In p is the maximum set if and only if p is L-false, i.e., false in the logic L.

Indeed, as any L-false proposition L-implies any proposition, p L-implies any proposition q (cf., for example, (Kleene, 1967)). Thus, by Condition (4.2.1), In p includes In q for any proposition q. At the same time, when p is not L-false, there is a proposition q such that p does not L-imply q. Thus, In p is is not the maximum set because it does not include In q.

Developing their theory, Bar-Hillel and Carnap assert that they ignore many important properties of information. In particular, they are not concerned whether information in a proposition is grounded or not, scientifically valuable or not, and so forth.

Theorem 4.2.4. In p includes In $p \vee q$ and is included in In $p \wedge q$.

Taking some language W, Bar-Hillel and Carnap give two plausible interpretations $\mathrm{Inf}_1 p$ and $\mathrm{Inf}_2 p$ of the set In p:

$\mathrm{Inf}_1 p$ is the set of all sentences in W that are L-implied by p and are not L-true.

$\mathrm{Inf}_2 p$ is the set of all sentences in W that are L-implied by p.

This kind of information has peculiar properties. As a consequence of the given content definition, analytical statements, i.e., statements true for all states, contain no information. For instance, the amount of information carried by the sentence "$17 \times 19 = 323$" is zero because Bar-Hillel and Carnap assume that this equality is always true. However, existence of non-Diophantine arithmetics (Burgin, 1997d; 2007b) show that this is not an analytical sentence and even in the theory of Bar-Hillel and Carnap, it contains non-zero information.

At the same time, self-contradictory statements contain maximum information because they exclude all states of the universe. More particular propositions contain more information than corresponding general propositions.

Two more sets, content and range, are corresponded to statements from L. Each statement p excludes universe states with descriptions that are not consistent with this statement, i.e., in which the statement p is false. The set of all excluded states forms the *content Cont p* of (information conveyed by) the statement p.

This definition has roots in statistical information theory where, in its main version, information is defined as excluded uncertainty, while uncertainty is estimated by the number of possible outcomes of an experiment.

Here are some properties of content.

Theorem 4.2.5. *Cont p = Cont q* if and only if p is L-equivalent to q.

Theorem 4.2.6. *Cont p* is empty if and only if p is L-true, i.e., a tautology in the logic L.

Theorem 4.2.7. *Cont p* contains all states of the universe if and only if p is L-false, i.e., false in the logic L.

Theorem 4.2.8. *Cont p* includes *Cont q* if and only if p L-implies q.

Thus, it is possible to consider *Cont p* as one more interpretation of In p.

Theorem 4.2.9. *Cont p* and *Cont q* do not intersect if and only if p and q are disjunctive.

Theorem 4.2.10. *Cont p∧q = Cont p* $\cup$ *Cont q*.

Theorem 4.2.11. *Cont p∨q = Cont p* $\cap$ *Cont q*.

The opposite of content is the *range* R(p) of a proposition p defined as the set of all universes where this proposition p is true. Properties of range are dual to properties of content. For instance, we have the following results.

Theorem 4.2.12. R(p) includes R(q) if and only if q L-implies p.

Theorem 4.2.13. R(p) = R(q) if and only if p is L-equivalent to q.

Theorem 4.2.14. R(p) is empty if and only if p is L-false.

Theorem 4.2.15. $R(p)$ contains all states of the universe if and only if p is L-true.

Theorem 4.2.16. $R(p \wedge q) = R(p) \cap R(q)$.

Theorem 4.2.17. $R(p \vee q) = R(p) \cup R(q)$.

As it is discussed in Chapter 2, after defining information, it is necessary to find ways to measure it. At first, Bar-Hillel and Carnap introduce a presystematic concept of amount of information $\text{in}(p)$ in a proposition (or sentence) p. According to (Bar-Hillel and Carnap, 1958), $\text{in}(p)$ is a numerical function that satisfies the following conditions.

$$\text{in}(p) \geq \text{in}(q) \text{ if } Cont \ p \text{ includes } Cont \ q \qquad (4.2.2)$$

$$\text{in}(p) = 0 \text{ if } Cont \ p \text{ is empty} \qquad (4.2.3)$$

$$\text{in}(p) > 0 \text{ if } Cont \ p \text{ is not empty} \qquad (4.2.4)$$

$$\text{in}(p) > \text{in}(q) \text{ if } Cont \ p \text{ properly includes } Cont \ q \qquad (4.2.5)$$

These conditions allow Hillel and Carnap to obtain some simple properties of the amount of information $\text{in}(p)$.

Theorem 4.2.18. If $Cont \ p = Cont \ q$, then $\text{in}(p) = \text{in}(q)$.

Theorem 4.2.19. $\text{in}(p) = 0$ if p is L-true.

Theorem 4.2.20. $\text{in}(p) > 0$ if p is not L-true.

Theorem 4.2.21. $\text{in}(p \vee q) \leq \text{in}(p) \leq \text{in}(p \wedge q)$.

The process of formalization continues. Bar-Hillel and Carnap introduce (in an axiomatic way) a measure-function, or m-function, $m(P)$ defined for sets P of propositions that describe states of the considered universe and find its properties.

A *proper* m-function $m(P)$ must satisfy (Bar-Hillel and Carnap, 1952) the following nine conditions (axioms):

A1. The value $m(P) > 0$ for any state description P.

A2. The sum of values $m(P)$ for all state descriptions P is equal to 1.

A3. For any L-false proposition p, $m(p) = 0$.

A4. For any non-L-false proposition p, $m(p)$ is equal to the sum of values $m(P)$ for all state descriptions P that belong to the range $R(p)$ of p.

A5. Replacing individual constants by correlated individual constants does not change the value of the m-function.

A6. Replacing basic propositions (primitive predicates) by correlated basic propositions does not change the value of the m-function.

A7. Replacing basic propositions (primitive predicates) by their negations does not change the value of the m-function.

A8. If propositions q and p does not include common basic propositions, then $m(q, p) = m(q) \cdot m(p)$.

A9. The value of $m(p)$ is not influenced by individuals not mentioned in p.

An m-function $m(P)$ is called *regular* if it satisfies axioms A1–A4 and is called *symmetrical* if, in addition, it satisfies axioms A5.

Here are some properties of m-functions.

Theorem 4.2.22. $0 \le m(p) \le 1$ for any p.

Theorem 4.2.23. $m(p) = 0$ if p is L-false.

Theorem 4.2.24. $m(p) = 1$ if p is L-true.

Theorem 4.2.25. If q L-implies p, then $m(p) \le m(q)$.

Theorem 4.2.26. $m(p, q) \le m(p) \le m(p \vee q)$.

Theorem 4.2.27. $m(p \vee q) = m(p) + m(q) - m(p, q)$.

One of possible interpretations of m-function is inductive or logical probability in the sense of (Carnap, 1950). Carnap defines the concept of *logical* probability as a function of the number of logical state descriptions in the range of a proposition. The simplest case is when all state descriptions are equiprobable. The main principle that connects probability of a statement to information in this statement is:

The more probable a statement is, the less information it conveys

This principle implies that, a measure of information can be defined as a reciprocal (inverse) function of the probability. A number of such functions exist. Choosing two of them Bar-Hillel and Carnap specify two measures of information.

The first one is the content function $\mathrm{cont}(p)$ and the measure of information defined as

$$\mathrm{cont}(p) = m(\neg p) \tag{4.2.6}$$

Theorems 4.2.22–4.2.27 imply the following result.

Theorem 4.2.28. $\mathrm{cont}(p) = 1 - m(p)$.

Theorem 4.2.29. $0 \le \mathrm{cont}(p) \le 1$ for any p.

Theorem 4.2.30. $\mathrm{cont}(p) = 1$ if p is L-false.

Theorem 4.2.31. $\mathrm{cont}(p) = 0$ if p is L-true.

Another measure of information introduced by Bar-Hillel and Carnap (1952) is $\inf(p)$ defined as

$$\inf(p) = \log(1/(1 - \text{cont}(p))) = -\log(1 - \text{cont}(p)) \qquad (4.2.7)$$

Properties of both measures are studied.

Theorem 4.2.28 and formula (4.2.7) imply the following result.

Theorem 4.2.32. $\inf(p) = -\log m(p)$.

Theorems 4.2.24 and 4.2.32 imply the following result.

Theorem 4.2.33. a) $\inf(p) = 0$ if p is L-true;

b) $\inf(p)$ is infinite if p is L-false.

Then several other functions are constructed. For instance, two probabilities measures, inductive m-function $m_I(p)$ and deductive m-function $m_D(p)$, are specified and corresponding information measures, $\text{cont}_I(p)$, $\text{cont}_I(p)$, $\inf_I p$, and $\inf_D p$, are built and studied.

It is possible to see that such functions as $\inf_I p$, $\inf_D p$, *Cont p* and $R(p)$ are tractable only for very small universes where states are described by few propositions. Thus, in practice, this theory can be effective exclusively in the case when we essentially restrict considered universes or represent real situations by simplified models.

Although the approach of Bar-Hillel and Carnap is contrasted with Shannon's information theory in the same way as semantics of a language is contrasted with its syntax, it is rooted in the approach of Hartley and Shannon where information is assigned to messages about events and the chosen information measure depends on the probability of events the message informs about. In this context, a message is a statement about a situation, property of an object or that a definite event happened.

This feature is explicitly formulated by Bateson and Ruesch (1951), who write that every piece of information has the characteristic that it makes a positive assertion and at the same time, makes a denial of the opposite of that assertion.

In such a way, the semantic information theory specifies how to represent and measure changes in knowledge about the state of the universe due to reception of information contained in propositions. Taking the system of knowledge as the inforlogical system, we see that the semantic information theory of Bar-Hillel and Carnap is a specification of a general theory of information.

At the same time, the semantic information theory of Bar-Hillel and Carnap has controversial properties. For instance, Hintikka (1970; 1971; 1973; 1973a) and other researchers (cf., for example, (Bremer, 2003; Sequoiah-Grayson, 2008)) challenge some conclusions of this theory. The main problem is with the statements that claim that logically true sentences do not contain information (Theorems 4.2.24 and 4.2.33), while false sentences contain maximal information (Theorems 4.2.23 and 4.2.33). Indeed, in the setting of the Bar-Hillel and Carnap theory, if all consequences of known premises are known, then any logically true sentence does not give new knowledge and consequently, does not contain information. At the same time, the probability of any inconsistent sentence is equal to 0 and consequently, it contains maximal information. These claims are counterintuitive. Intuitively, many would say, for example, that somebody who utters a contradiction has said nothing at all, i.e., has conveyed zero information. Some researchers tried to formalize this situation when contradiction contains no information. However, this property is also not always true. For instance, there is such a concept as *koan*. Koan is a paradox (i.e., a paradoxical statement or question) used in Zen Buddhism to gain intuitive knowledge through an awakening of deeper levels of spirit (cf., for example, (Cleary, 1993; Fontana, 2001)). Consequently, koan conveys a lot of information to an advanced disciple in Zen Buddhism and conveys much less (if any) information to a beginner, while it always contains information for those who know practice of Zen Buddhism.

Besides, as history of philosophy shows, from Zeno of Elea (ca. 490 BCE–ca. 430 BCE) to Augustine (354–430) to Immanuel Kant (1724–1804) to Wittgenstein (1889–1951), keen minds have expounded philosophical problems in the form of paradoxes and antinomies (Stegmuller, 1969). Antinomies are kinds of contradictions, but in spite of this, they may have a very profound information content although not for everybody but presumable for creative thinkers.

Even more, as Skinner (1969) and Gabbey (1995) write, classic philosophers are often inconsistent. As Gabbey (1995) suggests, it is due to the problem-creating property of philosophical doctrines themselves, particularly of those otherwise deemed to be in some sense "successful" or "progressive". Problems and even inconsistencies is not a deficiency in an original philosophy. It is a symptom of that originality, or a side effect, and therefore explains in part the philosophy's historical impact. Original philosophies do not solve problems or meet philosophical needs without *ipso facto* creating fresh problems and difficulties of their own. At the same time, philosophical systems without problems attract little interest among contemporaries, nor do they engage the attention of posterity.

In addition, history of science shows that many scientific discoveries came as paradoxes, contradicting dominating scientific dogmas. One example of a paradoxical situation inconsistent with the classical physics is given by physical objects, such as electrons, photons, and neutrons, which can be particles in some situations and waves in other situations. Another example of such situations is the discovery of non-Euclidean geometry, which contradicted to the knowledge of many generations of mathematicians and other educated people (Kline, 1967).

Another problem in the logical approach to information theory is that according to the Bar-Hillel and Carnap information theory, if one knows axioms of some theory T, e.g., axioms of the natural number arithmetic called Peano arithmetic, then any true property of natural numbers gives no information to this person. In other words, deductive inferences in logic and mathematics give no information. Hintikka called this situation "the scandal of deduction."

This is really very strange because in reality, when new properties of mathematical structures, in particular, numbers, are discovered, it often gives a lot of information to mathematicians. It is not a simple task to find properties even of such simple objects as natural numbers 1, 2, 3, … In the process of finding these properties and proving their validity, mathematicians created a special mathematical discipline called number theory. Proofs of some results in number theory became sensations. Examples were proofs of the last Fermat theorem and of undecidability of the Diophantine equation solvability (the tenth Hilbert problem). They caused a lot of attention not only in the mathematical community but also from the general public. In other cases, there were proofs that did not attract public attention but brought even more information than the majority of sensational results. An example is given by the proof of existence of non-Euclidean geometries (Burton, 1997).

One more evidence that proofs do give information is existence of many open, i.e., unsolved, problems in mathematics in general and in number theory, in particular. Mathematicians still do not know how to solve these problems. Thus, when such a problem is solved it can give a lot of information to the mathematical community. For instance, after Cantor proved that the set of all real numbers is uncountable, mathematicians were intrigued whether there are uncountable sets of real numbers cardinality of which is less than the cardinality of the set of all real numbers. Only the results of Gödel (1938) and Cohen (1966) demonstrated that the answer to this question depends on the set of axioms used for building the set of all real numbers.

Thus, we see that, in general, many proofs give a lot of information, while the Bar-Hillel and Carnap information theory states the opposite.

According to Hintikka, all this demonstrates that logicians have apparently failed to relate their subject to the naturally assumed properties of information. To remedy this situation, he suggests using the theory of distributive normal forms in a full polyadic first-order language, namely, in the polyadic predicate calculus. Hintikka assumes that there is understanding of the concept *information* in which logical inference can give information, increasing knowledge. To show how it is possible, he interprets logical inference as recognizing logical truth and getting additional information by increasing depth of constituents and checking their consistency in the polyadic predicate calculus.

Taking a first-order expression C, it is possible to build its distributive normal form

$$C = C_1^d \ \& \ C_2^d \ \& \ C_3^d \ \& \ ... \ \& \ C_k^d$$

In this form, C_i^d is a constituent of depth d. *Depth* depends on the number of quantifier changes and the number of quantifiers. For instance, the expression $\forall x \exists y \forall w \exists v \ P(x, y, w, v)$ where all variables x, y, w, and v are free in $P(x, y, w, v)$ has the depth equal to four, while the expression $\forall x \forall y \exists w \exists v \ Q(x, y, w, v)$ where all variables x, y, w, and v are free in $Q(x, y, w, v)$ has the depth equal to two. Each constituent is the conjunction of basic predicates that characterize possible worlds. All these predicates are true in the possible worlds allowed by the constituent.

In general, the information of a first-order expression C is determined by the possibilities concerning the possible worlds that C excludes or by contrast, concerning the worlds C allows. This is the meaning of the distributive normal form of C. In a logical form, the different possible worlds, or mutually exclusive possibilities, are represented by constituents in the distributive normal form of C.

Based on the theory of constituents developed in (Hintikka, 1964), Hintikka introduces surface and depth information. Explication of depth information and surface information involves details of formulating constituents for the predicate calculus. A restriction on the number of quantifier layers in the definition of depth, is the condition required for the construction of constituents for the polyadic predicate calculus, and hence the condition that allows the calculation of surface information. The notion of a constituent is (for the polyadic predicate calculus) relative to the depth of the formula.

Depth information is defined similarly to the constructions used by Bar-Hillel and Carnap (1952). To define information, a probability measure p is utilized. If C is an expression of the first-order polyadic predicate calculus, then it is possible to measure information in this expression C in two ways (Hintikka, 1984). The definition depends what properties of information measures are preserved.

Taking two probabilistically independent expressions (propositions or predicates) A and B, it is possible to demand additivity of the information measure. As a formal expression of this idea, we have the following equality

$$\mathrm{Inf}(A \,\&\, B) = \mathrm{Inf}(A) + \mathrm{Inf}(B)$$

This leads us to the following information measure definition

$$\mathrm{inf}(C) = -\log p(C)$$

It is also possible to demand additivity of the information measure with respect to implication. As a formal expression of this idea for arbitrary expressions (propositions or predicates) A and B, we have the following equality

$$\mathrm{Inf}(A \,\&\, B) = \mathrm{Inf}(A) + \mathrm{Inf}(A \supset B)$$

This leads us to another information measure definition

$$\text{cont}(C) = 1 - p(C)$$

The measure of probability assigned to an expressions (proposition or predicate) C gets distributed between all the consistent constituents C. The type of this distribution determines what kind of information is obtained.

Using the probability distribution for the constituents of C, the probability of C is defined as

$$p(C) = \sum_{i=1}^{k} p(C_i^d) \tag{4.2.8}$$

when $C = C_1^d \,\&\, C_2^d \,\&\, C_3^d \,\&\, \dots \,\&\, C_k^d$ and all C_i^d are constituents of the depth d.

Taking that the probability of all false, or inconsistent, constituents is equal to zero, we obtain the depth probability and two forms of the *depth information* of the sentence C as

$$\text{cont}(C) = 1 - p(C)$$

and

$$\inf(C) = -\log p(C)$$

Surface information is defined relative to trivially inconsistent constituents in the sense of (Hintikka, 1971; 1973). For instance, a constituent C_i^d is *trivially inconsistent* if the conjunction of basic predicates that constitute this constituent C_i^d contains two predicates one of which is the negation of the other one.

Hintikka defines surface information of depth d, utilizing surface probability $p_d(C)$ of depth d. The measure of surface probability of depth d assigned to an expressions (proposition or predicate) C gets distributed between all constituents of C that have depth d. The surface probability p_d of C is defined by the same formula (4.2.8) where the surface probability p_d of the constituents is used instead of the probability p. However, in contrast to depth probability p, zero probability is assigned only to every constituent of the polyadic predicate calculus that is trivially inconsistent. All consistent constituents and non-trivially inconsistent constituents have a positive probability measure p_d.

In such a way, it is possible to obtain two forms of the *surface information* of the sentence C on the level d as

$$\text{cont}_d (C) = 1 - p_d (C)$$

and

$$\inf_d (C) = - \log p_d (C)$$

As a result, logical inference or deduction of logical expressions can give information if inference/deduction is considered as the truth recognition by means of expression decomposition into a distributive normal form and elimination of trivially inconsistent constituents.

A more general but less formalized approach to eliminating controversial properties of the Bar-Hillel and Carnap information theory is suggested by Voishvillo (1984; 1988). He argues that when it is claimed that true sentences do not contain information, while false sentences contain maximal information, some implicit assumptions are made. Namely, it is supposed that the receiver of information has a given logic L in his system of knowledge (thesaurus). Usually L is the classical logic of propositions or the classical logic of predicates (Kleene, 1967; Bergman, et al, 1980; Manin, 1991). Another assumption is that the receiver has all this logic, which is infinite. However, the real situation contradicts not only to the conjecture that knowledge systems of an individual can contain actual infinity, but also to the assumption that actual infinity exists. Thus, Voishvillo suggests, it is more realistic to consider information based on the actual knowledge of the receiver.

Another approach that gives tools for measuring information obtained in inference/deduction is algorithmic information theory considered in the next chapter. The approach of Voishvillo is similar to the algorithmic approach in information theory because algorithmic size, or algorithmic complexity, of an object depends on prior information.

One more problem with the majority of approaches in semantic information theory is the logical monism. It means that the authors assume that there is only one true logic and semantic information has to be determined based on this logic. However, as Allo (2007) explains, according to logical pluralism, there are more than one true logics and different logics are better suited for different applications. Thus, when a formal account of semantic information is elaborated, the absolute validity of a logic cannot be taken for granted.

Reality is diverse and it is necessary to consider different logics and logical varieties (Burgin, 1997f) to correctly describe it. In particular, it is possible that two possible worlds W_1 and W_2 from the model of Bar-Hillel and Carnap are described by different logics L_1 and L_2. For instance, L_1 is a classical logic, while L_2 is an intuitionistic logic (cf., for example, (Kleene, 2002)). Then it is known that not all true sentences from L_1 are true in L_2 and/or not all false sentences from L_1 are false in L_2. As a result, such true sentences from L_1 will carry non-zero information with respect to L_2, i.e., in the world W_2, while such false sentences from L_1 will carry maximal information with respect to L_2, i.e., in the world W_2. This provides a solution to the paradoxes that claim that logically true sentences do not contain information (Theorems 4.2.24 and 4.2.33 from (Bar-Hillel and Carnap, 1958)), while false sentences contain maximal information (Theorems 4.2.24 and 4.2.33 from (Bar-Hillel and Carnap, 1958)).

This also supports the Ontological Principle O1. Indeed, logical pluralism implies that semantic information content of a message depends on the logic used to determine this information content. Logic that is used depends on the system that receives information.

Here is an anecdote that illustrates this situation.

Sherlock Holmes and Dr. Watson go for a camping trip, set up their tent, and fall asleep. Some hours later, Holmes wakes his friend, saying "Watson, look up at the sky and tell me what you see. Watson replies, "I see millions of stars." "What does that tell you?" continues Holmes. Watson ponders for a minute and then replies, "Astronomically speaking, it tells me that there are millions of galaxies and potentially billions of planets. Astrologically speaking, it tells me that Saturn is in Leo. Astrophysically speaking, it tells me that our universe has been developing for several billions of years. Meteorologically, it seems we'll have a beautiful day tomorrow. Theologically, it's evident the Lord is all-powerful and we are small and insignificant. Time-wise, it appears to be approximately a quarter past three. And what does it tell you, Holmes?" Holmes is silent for a moment, and then speaks, "Someone has stolen our tent."

Moreover, it is possible that even one and the same system may use more than one logic (Burgin, 1997f; 2004c; Dijkstra, et al, 2007). Thus, a message may have different meanings even for one system depending on what logic is utilized to explicate the meaning. In this case, it is more efficient to use logical varieties for meaning explication instead of a logical calculus. Different components of a logical variety represent different logics and the choice of a relevant logic depends on a situation in which it is necessary to use logic.

Logical approach of Bar-Hillel and Carnap is related to the situational approach. In the theory of Israel and Perry (1990), a proposition forms not information itself but only *information content*. Sentences and texts are *information carriers* in a general case. In the world knitted together by *constraints*, which are contingent matters of facts where one type of situations involves another, facts and situations also carry information. Facts and situations that indicate something, e.g., another situation, are called *indicating*. *Connecting facts* connect indicating situations with the specific object information is about.

Israel and Perry base their theory on ten principles:

(A) Facts carry information.

(B) The informational content of a fact is a true proposition.

(C) The information a fact carries is relative to its constraint.

(D) The information a fact carries is not an intrinsic property of it.

(E) The informational content of a fact can concern remote things and situations.

(F) The informational content can be specific; the propositions that are informational contents can be about objects that are not parts of the indicating fact.

(G) Indicating facts contain information such that it is relative only to connecting facts; the information is incremental, given those facts.

(H) Many different facts, involving variations in objects, properties, relations and spatiotemporal locations, can indicate one and the same information content — relative to the same or different constraints.

(I) Information can be stored and transmitted in a variety of forms.

(J) Having information is good; creatures whose behavior is guided or controlled by information (by their information carrying states) are more likely to succeed than those which are not so guided.

Israel and Perry admit themselves that there is a certain tension between Principle (J) and other principles. However, the reason they give for this and later eliminate this tension is only a minor one. Actually, history of science and mathematics show that there are cases when having information is not good. For instance, Girolamo Saccheri (1667–1733) had information that Euclidean geometry is the only one possible geometry. As a result, when in his research, he encountered non-Euclidean geometries, he did not believe that he found something new and important and missed one of the most important discoveries in mathematics for all times and all nations, namely, the discovery of non-Euclidean geometries (Burton, 1997).

Another problem with the approach of Israel and Perry is to consistently define what a fact is. Many people, especially, judges, lawyers and historians, know how difficult is to find what really happened, i.e., to determine the fact. Some historians even claim that there are no fact in history, but there are only interpretations.

In their approach, Israel and Perry do not define information but introduce the unit of information represented by a parametric state of affairs and called *infon*. In addition, they formalize what it means that a fact, which is a state of affairs, carry information a proposition P is true relative to a constraint C. This allows them to explain when it is natural to speak about information flow and to add one more principle.

(K) There are laws of information flow.

This shows how the situational approach to information theory was transformed into the theory of information flow of Barwise and Seligman (cf. Section 7.1).

4.3. Knowledge-base Approach of Mackay, Shreider, Brooks, Mizzaro, and others: Knowledge from Information

> *Information is the seed for an idea,*
> *and only grows when it's watered.*
> Heinz V. Bergen

Many researchers relate information to knowledge and connected information to changes in knowledge. This approach reflects many features of information observed by people in everyday life, as well as in science. For instance, the effectiveness of information management can be measured by the extent of knowledge creation and innovation in organization (Kirk, 1999). In addition, it allows one to solve methodological problems related to knowledge and information. For instance, Eliot's question, "Where is the knowledge we have lost in information?" (Eliot 1934; cf. also Section 2.5) has a relevant interpretation from the knowledge oriented approach to information. Namely, Eliot emphasizes the problem of extracting knowledge from the received information and how much more knowledge people can get from information they have. Naturally, knowledge that was possible to obtain but that was not obtained can be considered as lost.

Let us consider some knowledge-based theories of information.

One of the researchers who worked in this paradigm was Donald MacKay (1922–1987). He developed his approach assuming that the information process is the cornerstone for increase in knowledge. In his theory, the information element is embedded by the information process into knowledge as a coherent representation of reality.

Mackay writes (1969):

"Suppose we begin by asking ourselves what we mean by information. Roughly speaking, we say that we have gained information when we know something now that we didn't know before; when 'what we know' has changed."

Analysis of information is primarily based on intentional communication through language when information changes in the cognitive system of the receiver. At the same time, it is indicated that this approach can also apply to non-intentional signs and symbols. Communication is considered in a situation when a sender has a meaning to transmit, there is a message (statement, question, command) intended to transmit the meaning, and a receiver who is in a particular "state of readiness." This state of readiness may be interpreted as a system, called a conditional probability matrix (C.P.M.), of conditional probabilities for different possible patterns of behavior in different circumstances. The intention of the sender is not to produce some actual behavior, but, through conveying information to the receiver, to alter the settings or state of the conditional probability matrix of the receiver.

There are three tentatively different meanings and correspondingly three portions of information involved in such a communication: the *intended meaning* of the sender and *intended information*; the *received* or *understood meaning* of the receiver and the *received* or *understood information*; and the *conventional* or *literal meaning* of the message and *conventional information*. There is a complex interplay of these meanings and corresponding information. For instance, the conventional meaning of a spoken message might be completely negated by a tone of voice or expression in an ironic or sarcastic comment. In addition to the *intended information* of the sender, there is the *embedded information* of the message. Sometimes what is expressed in the message is not exactly what the sender intends to transmit. The difference can essential when the sender does not have good skills in the language she/he uses for communication. It is possible that *received information* is different from *understood information*. Very often people understood less information than receive reading a book or scientific paper. Besides, there is also *accepted information* discussed in Section 2.4. Conventional information is essentially relative and can vary from one community to another.

The received meaning cannot be identified with either the behavior brought about, i.e., the change to the conditional probability matrix, or the final state of the conditional probability matrix. According to MacKay (1956), meaning is the selective function of the message on an ensemble of possible states of the conditional probability matrix. In this context, the three types of meaning are represented by the *intended selective function*, the *actual selective function*, and the selective function on a conventional symbolic representational system, correspondingly. Thus, meaning is treated as a selection mechanism that may give information to a system if it changes the system's state.

Two messages may be different but have the same selective function (meaning) for a particular receiver. The same message repeated still has the same conventional meaning even though it brings about no change of state the second time. However, repeating a message can imply changes in the intended meaning and/or in the understood meaning. For instance, when the receiver receives the same message the second time, she/he can imply that something was wrong with her/his response, e.g., it did not reached the sender. A message can be meaningless if it has no selective function for someone, for example, if it is in an unknown language or is an inappropriate choice or combination of words. Note that this concept of meaning is relational as the selective function is always relative to a particular domain. Thus, two messages that are different may have the same selective function in one domain but not in another.

In this context, the *selective information content* is the size or extent of the change brought about by a particular selective operation. This obviously depends on the prior state of the conditional probability matrix. According to MacKay, a repeated message can be meaningful but have no information content for the receiver since no change of her/his state will take place because the conditional probability matrix of the receiver will already be in the selected state.

A similar, although less clear, definition is offered by Pratt (1977). He uses Boulding's (1956) concept of the individual's *Image* or *world-view* instead of MacKay's conditional probability matrix and indicates that when someone has understood a message she or he has become *informed*, or rather, "in-formed." That is, their Image has become re-formed or altered. Thus, "in-formation" is the transformation of the Image on the receipt of a message and it occurs at a particular point in time. This, however, does not resolve the problem with the nature of *meaning*. Pratt quotes Boulding as starting that "the meaning of a message is the change it causes in the Image," but this coincides with Pratt's definition of information. Thus, both Pratt and Boulding use a transformational definition of information. Note that in a similar way it is possible to consider a group or social world-view (Image) instead of an individual world-view (Image).

Luhmann (1990) has also further developed MacKay's theory by connecting it to a sophisticated, phenomenologically based, theory of meaning-constituting systems. The goal of Luhmann was to move the discussion of meaning away from the perspective of the conscious intentions of individual subjects (as did MacKay), assuming that meaning is primary and should be defined without reference to the subject's intentions since the subject is already a meaning-constituted entity. Luhmann takes MacKay's idea that meaning is not the content, but a function for *selection*. Meaning functions on two levels — the psychic (individual) level, where meaning frames and organizes our experiences and knowledge, and the social (group or society) level, where it makes possible and structures intersubjective experience and communication. In such a way, meaning connects these two levels and makes possible their differentiation.

MacKay introduced (1969) a two-dimensional quantization of information that takes into account the following two aspects:

1. The inner structure of the information element as a "logical a priori" aspect (structural information).

2. The so-called "weight of evidence" of the individual structural elements as an "empirical a posteriori" aspect (metrical information).

The unit of structural information is called *logon*. It is assumed that any portion of information can be divided into logons. Accordingly logon content, as a convenient term for the structural information-content, of a portion of information I is defined as the number of logons in I.

The unit of metrical information is called *metron*. This unit, *metron*, is defined (MacKay, 1969) "as that which supplies one element for a pattern. Each element may be considered to represent one unit of evidence. Thus, the amount of metrical information in a pattern measures the weight of evidence to which it is equivalent."

MacKay (1969) tries to make this vague definition a little bit more exact, explaining that "the amount of metrical information in a single logon, or its metron-content, can be thought of as the number of elementary events which have been subsumed under one head or 'condensed' to form it."

As a result, the descriptive information is represented as an information vector in an information space. At the same time, the idea that information is some change in the receiver has also been proposed by MacKay (1956, 1961, 1969) and used by Bednarz (1988) through explicit incorporation of meaning into information theory.

A similar concept of information is utilized in the approach that was developed by Shreider (1963; 1965; 1967) and also called the semantic theory of information. The notion of a thesaurus is basic for his theory.

As any thesaurus is a kind of infological systems, this approach is also included in the general theory of information.

The word *thesaurus* is derived from 16th-century New Latin, in turn from Latin *thesaurus*, which has roots in the ancient Greek word thesauros ($\theta\eta\sigma\alpha\upsilon\rho\delta\varsigma$), meaning "storehouse" or "treasury" That is why the medieval rank of *thesaurer* was a synonym for treasurer.

Shreider builds a mathematical model of a thesaurus as a collection (set) **T** of texts and semantic relations between these texts. A description d(**T**) of a thesaurus **T** and rules of its transformation also consists of texts in some language.

In natural languages, texts are sequences of words. It is possible to build different formal models of texts. Here we consider some of them. A mathematical model of a text on the syntactic level is built on three levels.

At first, the level of words is constructed. In a general case, some alphabet A is taken and a *word* (which is a *first-level text*) is defined as a mapping the nodes of a graph G (cf. Appendix A) into A. For instance, if A is the English alphabet, then the mapping f of the graph (4.3.1) nodes gives us the word *text* when $f(1) = t, f(2) = e, f(3) = x$, and $f(4) = t$. The graphic form of this word is given in Diagram (4.3.2).

$$1 \longrightarrow 2 \longrightarrow 3 \longrightarrow 4 \qquad (4.3.1)$$

$$t \longrightarrow e \longrightarrow x \longrightarrow t \qquad (4.3.2)$$

In general, it is assumed that any mapping of the nodes of any graph into an alphabet is a word of some language. However, in conventional formal languages and programming languages, letters in words are linearly ordered (cf., for example, (Burgin, 2005; Hopcroft, et al, 2001)). It means that only connected graphs without cycles and branching are used for building words. In European languages, there are even more restrictions on correct building mappings from graphs to alphabets in order to represent words. However, taking other languages, such as mathematical language, chemical language or language of images, we see that it is possible to use more general graphs and their mappings for building words. For instance, the mathematical formula

$$3\, x_1^{\,2}\, x_2^{\,5} y_2^{\,4}$$

has the following graph representation

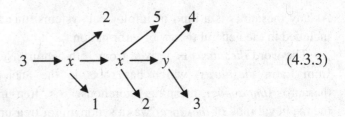

$$(4.3.3)$$

The language of the category theory, in which very sophisticated diagrams are used (cf., for example, (Herrlich and Strecker, 1973)), demands very complex graphs for representation of categorical expressions.

On the second level of this syntactic model, sentences (which are *second-level texts*) are constructed from words and punctuation symbols. We take a vocabulary V of words and their forms and a set P of punctuation symbols and define a *sentence* as a mapping the nodes of a graph G into the union of V and P.

For instance, if V is the English vocabulary, then the mapping f of the graph (4.3.4) nodes gives us the sentence *"What time is it?"* when $f(1) = What$, $f(2) = time$, $f(3) = is$, $f(4) = it$, and $f(5) = ?$. The graphic form of this sentence is given in Diagram (4.3.5).

$$1 \longrightarrow 2 \longrightarrow 3 \longrightarrow 4 \longrightarrow 5 \qquad (4.3.4)$$

$$What \longrightarrow time \longrightarrow is \longrightarrow it \longrightarrow ? \qquad (4.3.5)$$

On the third level of this syntactic model, texts (which are *third-level texts* of the model) are constructed from sentences. We take a set S of sentences and define a *text* as a mapping the nodes of a graph G into the S.

This hierarchy of texts can grow, including many levels more, because texts have a much more sophisticated hierarchy in natural and artificial languages. For instance, when we have a very long text, for example, a book, then sentences are used to form paragraphs. Paragraphs are used to form sections. Sections are used to form chapters. At the same time, a section may consist of subsections. A subsection may be divided into even smaller components. This division continues to paragraphs, which consist of sentences. In a similar way, such a text as a novel or treatise may consist of several books. Each book may include several parts, while each part consists of chapters and so on.

However, in the model of Shreider (1965; 1967), only the first level is considered, i.e., texts are constructed from the elements of an alphabet. At the same time, it is also possible to build texts from words, taking vocabulary as an alphabet and ignoring the third and higher levels of texts.

After Shreider discusses the concept of information, explains its relation to a thesaurus and constructs a theoretical model, he goes to the next step, finding how to measure information. If m is a message represented by a text T_m, then Shreider defines the quantity of information $I(m, \mathbf{T})$ in m for a thesaurus $\mathbf{T}$ is equal to the extent of the transformation of the thesaurus $\mathbf{T}$ under the influence of the message m (text T_m). This implies that any message m is associated with an operator A_m acting in the thesaurus $\mathbf{T}$.

One of the consequences of this approach is that those messages that do not change the thesaurus have zero information for this thesaurus. For instance, incomprehensible text does not contain information. This well correlates with the concept of pseudoinformation in the general theory of information (cf. Section 2.4).

It is possible that two texts m and w carry the same quantity of information $I(m, \mathbf{T}) = I(w, \mathbf{T})$ although they contain different information.

Definition 4.3.1. Two texts m and w are *synonymous* if

$$A_m = A_w$$

Note that what is synonymous with respect to one thesaurus $\mathbf{T}$ can be not synonymous with respect to another thesaurus $\mathbf{Q}$.

We can derive some properties of the relation "*to be synonymous*". As the equality is an equivalence relation, we have the following result.

Proposition 4.3.1. The relation "*to be synonymous*" is an equivalence relation.

Definition 4.3.2. When one message w comes after another m, it is called the *sequential composition* $m \circ w$ of messages.

The sequential composition of messages (texts) in a natural way defines the *sequential composition* of corresponding operators. The latter is the ordinary sequential composition of mappings of the thesaurus $\mathbf{T}$.

Proposition 4.3.2. If texts m and w are synonymous and sequential compositions $m \circ v$ and $w \circ v$ are defined for a text v, then texts $m \circ v$ and $w \circ v$ also are synonymous.

The semantic approach of Shreider is related to the statistical approach of Shannon (cf. Chapter 3) in the following way. In the statistical approach, the thesaurus **T** consists of a list of events and their probabilities. Accepted messages change probabilities of events in the list and thus, initiate transformations of the thesaurus **T**. According to the Shannon's information theory, some events may acquire zero probability. It virtually excludes these events. At the same time, the main assumptions of the Shannon's information theory forbid adding new events to **T**. This shows one more limitation of the statistical approach because in real life a message can bring us information about a new unexpected and even unknown event. As a result, we will need to expand the list of events and change the thesaurus **T**.

This interpretation of the thesaurus allows one to measure the extent of the transformation of the thesaurus **T** caused by a message, estimating changes in probabilities of the events in the list. For instance, if a probability of some event a was known to be equal to p and after receiving a message m, this probability changes to q, then using the logarithmic measure, we have

$$I(m, \mathbf{T}) = \log_2 (q/p)$$

In particular, if the message informs that the event a happened, then $q = 1$, and we have the Shannon-Hartley information entropy

$$I = \log_2 (1/p) = - \log_2 p$$

It is necessary to remark that Shreider's notion of information is more general than the one used in (Bar-Hillel and Carnap, 1958) as he assumes that information is a transformation of a thesaurus and it is possible to consider any system of propositions as a thesaurus.

An interesting approach to meaning is suggested by Menant (2002). He bases his study of the meaning of information on the assumption that information is a basic phenomenon in our world. Different kinds of information (and of infological systems) induce existence of different types of meaning in the sense of Menant. It is possible to consider meaning for a given infological system. This results in emergence of an individual meaning when a person or more generally, a system receives some information. For instance, a mathematical theorem has different meaning for a student, professional mathematician, and engineer.

Although meaning, as we have seen, is not an inherent attribute of information, many observations show that meaning is closely connected to information that is received, searched for, disseminated, and processed by people. To explain this peculiarity, it is possible to suggest the Filtering Hypothesis based on main assumptions of the general theory of information. First, meaning in a general understanding is connected only to cognitive information. Second, to answer the question why a message/text that has no meaning for an individual *A* has no information for this individual, we have to assume that, according to the general theory of information, a message has information for *A* if and only if it changes the knowledge system of *A*.

The human mind does not accept everything that comes from senses. It has filters. One of such filters tests communication for meaning and rejects meaningless messages. Thus, meaningless messages cannot change the knowledge system of the individual *A* and consequently, have no information for this individual. Thus, we come to the following conclusion.

Individual Filtering Hypothesis. Meaning plays the role of a test criterion for information acceptance by an individual.

Consequently, a message without meaning gives no information for this individual.

The same is true for social information where meaning also plays the role of a test criterion for an information filter. As a result, we have a corresponding hypothesis on the social level.

Social Filtering Hypothesis. Meaning plays the role of a test criterion for information acceptance by a social group.

However, either it is irrelevant to expand this connection between information and meaning beyond personal and social information or it is necessary to extend the notion of meaning so as to grasp other kinds of information.

In his approach to information, Brookes (1980) assumes that information acts on knowledge structures, suggesting, what he calls "the fundamental equation of information science":

$$K(S) + \Delta I = K(S + \Delta S) \qquad (4.3.6)$$

This equation means that action of a portion of information ΔI on a knowledge structure $K(S)$ changes this structure and it becomes the structure $K(S + \Delta S)$. The effect of this change is shown as ΔS.

Brookes connects his approach to the Popper triad of the world (cf. Section 2.1), claiming that information in documents is the reified and objective instance of Popper's World 3 (i.e., the World of knowledge and information). Equation (4.3.6) explicates relations between knowledge and (cognitive) information, demonstrating that what is usually called information is only cognitive information.

The equation (4.3.6) has been amended by Ingversen (1992) to include potential information pI. In addition, Ingversen changed equation (4.3.6) to the system of transitions

$$pI \rightarrow K(S) + \Delta I \rightarrow K(S + \Delta S) \rightarrow pI' \qquad (4.3.7)$$

Gackowski (2004) derives a similar equation but additionally includes data into it. In his approach, knowledge K can be defined as the logical sum of data D known by decision makers, the additional information I they may acquire, and the rules R of reasoning known to them. He synthesizes this understanding in the following formula:

$$K = D + I + R$$

An interesting theory of epistemic information is constructed by Mizzaro (1996; 1998; 2001). In this theory, he further develops approaches of MacKay and Brookes. The basic assumption of Mizzaro's theory is:

Information is a change in a knowledge system

The initial assumption is that the world is populated by *agents* that act in the *environment* called the *External World*. A general definition specifies an agent as a system (a being) that (who) acts on behalf of another system (being). Mizzaro takes the definition of an agent from the book (Russell and Norvig, 1995). Namely, an *agent* is anything (or anybody) that can be viewed as perceiving its environment through *sensors* and acting upon this environment through *effectors*. A human agent has eyes, ears, and other organs for sensors, and hands, legs, mouth, and other body parts for effectors. A robotic agent uses cameras and infrared range finders as sensors and various body parts as effectors. A software agent has communication channels both for sensors and effectors.

This gives us the following structure of an agent.

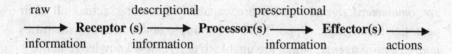

Figure 4.3. The triadic structure of an agent

Ruben (1992) asserts that for living beings, all interactions with the environment take place through the processing of matter-energy and/or processing of information.

The general structure of the world in the form of the Existential Triad gives us three classes of agents:

- *Physical agents.*
- *Mental agents.*
- *Structural* or *information agents.*

People, animals, and robots are examples of physical agents. Software agents and *Ego* in the sense of psychoanalysis (cf., (Freud, 1949)) are examples of mental agents. The head of a Turing machine (cf., for example, (Burgin, 2005)) is an example of a structural agent.

Physical agents belong to three classes:

- *Biological agents.*
- *Artificial agents.*
- *Hybrid agents.*

People, animals, and birds are examples of biological agents. Robots are examples of artificial agents. Hybrid agents consist of biological and artificial parts (cf., for example, (Venda, 1990)).

Mizzaro (2001) classifies agents according to three parameters. With respect to perception, he identifies *perceiving agents* with various perception levels. *Complete perceiving agents*, which have complete perception of the world, constitute the highest level of perceiving agents. Opposite to perceiving agents, Mizzaro identifies *no perception agents*, which are completely isolated from their environment. This does not correlate with the definition of Russell and Norvig (1995) but it is consistent with a general definition of an agent.

With respect to reasoning, there are *reasoning agents* with various reasoning capabilities. Reasoning agents derive new knowledge items from their knowledge state. The highest level of reasoning agents are *omniscient agents*, which are capable of making actual all their potential knowledge. Opposite to reasoning agents, Mizzaro identifies *nonreasoning agents*, which are unable to derive new knowledge items.

With respect to memory, there are *permanent memory agents*, which are capable of never loosing any portions of their knowledge. *No memory agents* are opposite to permanent memory agents because they cannot keep their knowledge state. *Volatile memory agents* are between these two categories. People are volatile memory agents.

Wooldridge (2000) considers four types of agents:
- Situated agents, which are embedded in their environment
- Goal directed agents, which have goals that they try to achieve
- Reactive agents, which react to changes in their environment
- Social agents, which can communicate with other agents (including humans)

In his artificial intelligence course taught in 2000, the author of this book gave the following classification of agents:

1. *Reflex* (or *tropistic*, or *behavioristic*) agents, which realize the simple schema action $\rightarrow$ reaction.
2. *Model based* agents, which have a model of their environment.
3. *Inference based* agents, which use inference in their activity.
4. *Predictive* (or *prognostic*) agents, which use prediction in their activity.
5. *Evaluation based agent*, which use evaluation in their activity.

Some of these classes are also considered in (Russell and Norvig, 1995).

Note that prediction and/or evaluation do not necessarily involve inference.

In addition, according to the dynamic criterion there are *static*, *mobile*, *effector mobile*, and *receptor* (*sensor*) *mobile* agents. Mobility can be realized on different levels and in different degrees.

According to the interaction criterion there are *deliberative* (*proactive*), *reactive* and *inactive* agents.

According to the autonomy criterion there are *autonomous*, *dependent* and *controlled* agents. Control is considered as the highest level of dependence.

According to the learning criterion there are *learning*, *remembering* and *conservative* agents. Conservative agents do not learn at all. Remembering agents realize the lowest level of learning – remembering or memorizing.

According to the cooperation criterion there are competitive, individualistic and collaborative agents. Competitive agents do not collaborate but compete. Individualistic agents do not interact with other agents.

A *cognitive agent* has a system of knowledge K. Such an agent perceives information from the world and it changes the initial knowledge state, i.e., the state of the system K.

In his approach, Mizzaro does not consider knowledge in general, but prefers to use the term a *knowledge state* (KS) of a (cognitive) agent. It is postulated that knowledge of an agent consists of atomic components called *knowledge items* (KI) and the number of these items in a portion of knowledge gives an adequate measure of this knowledge. This allows Mizzaro to consider and utilize natural set-theoretical operations, such as union, difference, complement, and intersection, as well as set-theoretical relations, such as inclusion, emptiness, and membership, on the set of all knowledge states. Agents that may have only finite knowledge states will be called *finite agents*.

It is possible to take elementary knowledge mathematically modeled in Section 2.5 as a knowledge item. It provides for a reasonable formalization of the concept of a knowledge state introduced by Mizzaro. Another option is to consider propositions and/or predicates from a logical language as knowledge items. One more option is to consider situations as knowledge items (cf., for example, (Dretske, 1981)).

In this context, a modification in the knowledge state or knowledge transformation is described by the difference between the initial knowledge state K_I and the final knowledge state K_F. This difference consists of two parts: what is added to K_I is denoted by K_A and what is removed from K_I is denoted by K_R. The removal happens, for example, when the agent forgets something. Both systems K_A and K_R also consist of knowledge items. Thus, we have the formal definitions:

$$K_A = K_F \backslash K_I$$

and

$$K_R = K_I \backslash K_F$$

Based on these concepts, Mizzaro proposes formalized definitions of data, knowledge and information. As he writes, a *datum* is a difference in the world. It is something physical, which can be observed. There are different ways to define the amount $|d|$ of a datum d but they are not specified in this theory.

Knowledge, according to Mizzaro, exists only inside agents and is what the knowledge items are made of. Thus, either a book is an agent or it cannot contain knowledge.

However, even without specifying what knowledge is, it is possible to give formal definitions of some important concepts related to knowledge.

Definition 4.3.3. The *quantity of knowledge* $|K|$ in the knowledge state K is the number of knowledge items in K.

Definition 4.3.4. Given an agent A in an initial knowledge state K_I, receiving a datum d, and ending in a final knowledge state K_F, the *variation of the quantity of knowledge* $\Delta|K|$ caused by receiving the datum d is defined as

$$\Delta|K| = |K_F| - |K_I| = |K_A| - |K_R|$$

As always, $|K|$ denotes the number of elements in the set K.

It is interesting to remark that classical logical tools, such as formal deduction logical calculi or formal grammars, perform only addition of knowledge items, such as propositions, predicates or texts. More modern tools of knowledge generation, such as non-monotonic logics, in which the set of deduced statements does not necessarily grow, and can shrink, when new statements are added to the set of premises (cf. (Ginsberg, 1987; Rankin, 1988; Gabbay, et al, 1994)), grammars with prohibition (Burgin, 2005c), and corrections grammars (Carlucci, Case and Jain, 2007), perform both addition and deletion of knowledge items.

Information is defined in a much more rigorous way than in other knowledge-based theories of information.

Definition 4.3.5. Given an agent A in an initial knowledge state K_I, receiving a datum d, and ending in a final knowledge state K_F, the *information* (or more exactly, *portion of information*) *brought by d for A* is defined as the ordered pair

$$I(d, K_I, K_F) = \Delta K = (K_A, K_R)$$

We also call the ordered pair (K_A, K_R) by the name *information explication* of information brought by d for A as it is a specific form of information representation.

Taking some group G of agents, we denote the set of all information explications of agents from group G by **IE**(G). In particular, **IE**(A) is the set of all information explications of a cognitive system/agent A.

Note that information brought by data d to an agent A depends on the initial state K_I of the knowledge system K of the agent A.

In the context of the general theory of information, transformations in knowledge states caused by receiving data (messages) are represented by local information operators. Namely, the transformation in the knowledge state K_I of an agent A caused by receiving the datum (message) d is represented by the local information operator $O_I(d, K_I, K_F)$. Note that according to the mathematical model of information constructed and studied in Section 7.3.1, a local information operator is a projection of some global information operator described in Section 7.3.1. Consequently, the theory of Mizzaro spaces and local information operators in these spaces is a localization of the mathematical (formalized) stratum of the general theory of information.

Definition 4.3.5 implies the following result.

Proposition 4.3.3. If an agent A that starts with an initial knowledge state K_I and after receiving data d, as well as after receiving data c, ends in the same final knowledge state K_F, then d and c carry the same information for the agent A.

Indeed, we have

$$I(d, K_I, K_F) = (K_A, K_R) = (K_F \backslash K_I, K_I \backslash K_F).$$

Definition 4.3.6. Given an agent in an initial knowledge state K_I, receiving a datum d, and ending in a final knowledge state K_F, the quantity $|I(d, K_I, K_F)|$ of information (*of information portion*) $I(d, K_I, K_F)$ that A gets by receiving the datum d is defined as

$$|I(d, K_I, K_F)| = |\Delta K| = |(K_A, K_R)| = |K_A| + |K_R| \qquad (4.3.8)$$

Definition 4.3.6 demonstrates that logical deduction can give information for an agent if deduced sentences do not belong to the initial knowledge state of this agent. In particular, a finite agent can get new information from the majority of logical true sentences in contrast to the situation in the approach of Bar-Hillel and Carnap considered in the previous section.

In addition, formula (4.3.8) also shows that information is measured by structural work (cf. Section 2.2) realized by mental work performed by the agent A in order to add and remove knowledge items. It is similar to measuring energy by physical work it can produce.

This relation between information, or more exactly, cognitive information, and knowledge is well understood in the area of information management. For instance, Kirk (1999) writes, "the effectiveness of information management can be measured by the extent of knowledge creation or innovation in organizations." Innovations are based on knowledge how to do something important and useful. Thus, the effectiveness of information management is measured by the extent of knowledge creation.

According to laws of set theory (cf., (Kuratowski and Mostowski, 1967) and Appendix A), sets K_A and K_R are disjoint as $K_A = K_F \backslash K_I$ and $K_R = K_I \backslash K_F$. Indeed, it looks natural to assume that a cognitive agent does not do inefficient work, such as at first, adding a knowledge item KI and then eliminating the same knowledge item KI. However, in real life, people and other cognitive agent are not perfect and do a lot of inefficient work. Reasoning, as a rule, is not a monotone process (cf. (Ginsberg, 1987; Rankin, 1988; Gabbay, et al, 1994)). Consequently, to be able to study information processing in a general case, we need a more encompassing definition of information explication.

It is interesting that while at first Mizzaro (1996; 1998) writes about information in general, later (cf. (Mizzaro, 2001)) he starts calling it *epistemic information*. This type of information is a subclass of cognitive information considered in the general theory of information (cf. (Burgin, 2001; 2001a; 2002) and Section 2.2). Epistemic information changes only knowledge, while cognitive information changes knowledge, beliefs, ideas and other elements from cognitive infological systems.

As Lenski (2004) remarks, Mizzaro's theory of information is an interesting approach to define information, which essentially tries to capture the result of an information seeking process through its traces in the result of the process, i.e., the effect it has on given structures. It may thus be characterized as a post-process definition of information as opposed to in-process definitions, which seek to capture the very process of getting informed.

Mizzaro's approach also forms a suitable base for the development of a rigorous mathematical theory of cognitive/epistemic information. To do this, we put Mizzaro's theory into the context of the general theory of information and describe the first steps in developing such a theory. In particular, we see that properties of a cognitive agent used in Mizarro's approach are valid for any cognitive system, i.e., for a system that has a subsystem of knowledge. For instance, science, as a system, has these properties. That is why in what follows, we use the term *cognitive system* interchangeably with the so popular now but more restricted term (*cognitive*) *agent*.

Let us consider a universal set W of knowledge items (units). As it was mentioned before, it is possible to take the set W_C of elementary knowledge units mathematically modeled in Section 2.5 as a universal set W. Another possibility for W is the set W_L of propositions and/or predicates from some logical language L (cf. Section 4.2). One more possibility for W is the set W_S of situations possible in a world U.

The set W is called universal because we assume that the following axiom is true.

MA1 (the Internal Representation Axiom). For any cognitive system (agent) A, knowledge states KS_A of A are subsets of the set W.

It is possible to interpret W as the base of knowledge that agents are able to have about their environment.

Another aspect of universality of the set **W** may be in the possibility to describe all possible (existing) worlds taking knowledge only from **W**. For instance, when **W** is the set W_L of propositions and/or predicates from some logical language *L*, then it is possible to build all descriptions of all possible worlds by combining elements from W_L. This possibility is reflected in the following axiom.

MA2 (the External Representation Axiom). For any environment (situation) *D*, there is a subset W_D of the set **W** that contains all accessible knowledge about *D*.

Taking these two axioms as the foundation, we develop a theory of cognitive systems/agents called the *theory of Mizzaro spaces*.

Definition 4.3.7. Subsets of **W** are called *Mizzaro spaces*.

In some cases, only specific subsets of **W** are used in the theory. For instance, if elements of **W** are propositions and we build our model in the context of classical logical calculi, then only consistent subsets of propositions are acceptable.

In the context of the general theory of information, a modification in the knowledge state (of an agent) or knowledge transformation is called a *local information operator*. When the initial knowledge state of an agent *A* is K_I and the final knowledge state after transformation caused by reception of a message *d* is K_F, then we denote the corresponding local information operator by $O(d, K_I, K_F)$. Note that any local information operator is the restriction on one point of the state space of some information operator studied in Section 7.3.1.

Definition 4.3.8. Given an agent *A* is called *finite* if all knowledge states of *A* contain finite number of knowledge items.

It looks like it might be sufficient to consider only finite agents. However, if knowledge is represented by logical statements and it is assumed (as in the theory of Bar-Hillel and Carnap, cf. Section 4.2) that any knowledge system contains all logical consequences of all its elements, then an agent with such knowledge system is infinite.

In a similar fashion, in information algebras, portions of information are represented by closed subsets of sentences from a logical language *L* (Kohlas and Stärk, 2007). In conventional logics, closed with respect to such operators as deduction sets are infinite because any sentence *p* implies $p \vee q$ for any sentence *q* from *L*, which is, as a rule, infinite (cf., for example, (Shoenfield, 1967)). Thus, in the context of classical logic and information algebras any portion of information has infinitely many representations. Consequently, such a portion generates a system with the infinite number of knowledge items.

We also need a more general information explication. Given an agent in an initial knowledge state K_I, receiving a datum d, and ending in a final knowledge state K_F, we denote by CK_A the set of all knowledge items added to the agent's knowledge states before it comes to the final knowledge state K_F and by CK_R the set of all knowledge items removed from the agent's knowledge states before it comes to the final knowledge state K_F.

Note that it is possible that items from the set K_I are added in the procedure of data (message) processing, i.e., $K_I \cap CK_A \neq \varnothing$ in some cases. In a similar way, it is possible that $K_F \cap CK_R \neq \varnothing$ because some knowledge items may be, at first, removed and then once more added in the procedure of data (message) processing. Thus, to reflect a general situation, we assume that sets CK_A and CK_R may have common elements.

Definition 4.3.9. Given a cognitive system/agent A in an initial knowledge state K_I, receiving a datum d, and ending in a final knowledge state K_F, the *extended information explication* of information brought by d for A is defined as the ordered pair

$$IE(d, K_I, K_F) = \Delta K = (CK_A, CK_R)$$

If the transition from K_I to K_F is performed by a local information operator $O(d, K_I, K_F)$, then $IE(d, K_I, K_F)$ is the extended information explication
of this operator.

The extended information explication $IE(d, K_I, K_F) = (CK_A, CK_R)$ reflects structural work performed by A after receiving the datum d. The information explication $I(d, K_I, K_F) = (K_A, K_R)$ reflects minimal structural work, i.e., a work of adding and removing knowledge items in which nothing is added and then deleted, performed by receiving the datum d, performed by A after receiving the datum d.

It is possible to consider any transformation as a function. For instance, a deterministic finite automaton (cf., for example, (Burgin, 2005)) changes the state when it receives a symbol (data) and this change is performed according to the transition function of this automaton. Corresponding local information operators are called *functional information operators*.

At the same time, it is possible to treat a transformation as a process. For instance, a Turing machine or inductive Turing machine (Burgin, 2005) starts a process of computation when it receives its data, which are usually given in the form of a word (for a one-tape (inductive) Turing machine) or several words (for multitape (inductive) Turing machine). Corresponding local information operators are called *processual information operators*.

It is possible to ask whether it is really necessary to study information conveyed by a single message using processes or it is sufficient to have only a functional representation. If we analyze real situations, we find several reasons in favor of the process representation.

First, functions are correspondences (relations). In the case of information reception, a functional information operator is a relation between the initial state K_I and the final state K_F. Such a relation does not show what is going on in the process of transformation of K_I into K_F and how this transformation is performed. At the same time, an information explication reflects inner changes induced by the transformation process, while an extended information explication does the same only with a higher precision.

Second, even a simple message, such as *yes* or *no*, can trigger a long sequence of reasoning and inference. This reasoning may refer, for example, to consequences of the event that happened if the message informs that this even really happened. For instance, Andrew asks Bernice to marry him in an e-mail. Then he receives an e-mail from with one word *yes*. After this, Andrew starts thinking when he will see Bernice, how to organize their wedding ceremony, and what their future life will be like.

. Third, a message can be long and complicated. For instance, a book is a message to its reader. Even when messages are (represented by) logical propositions, it is possible that proposition is a disjunction or conjunction of 100 other propositions, e.g.,

$$p = p_1 \wedge p_2 \wedge p_3 \wedge \ldots \wedge p_{99} \wedge p_{100} \qquad (4.3.9)$$

Fourth, information reception is a process, in which processing of a part of a message may depend on processing of another part of the same message. For instance it is possible to consider a situation when the proposition p_6 in the formula (4.3.) states "this message contains incorrect statements p_{20}, p_{23}, and p_{27}, which have to be disregarded". The system (agent) A that receives the message p starts processing with the part p_1 and then successively goes from p_i to p_{i+1} processing p. After processing p_6, A disregards statements p_{20}, p_{23}, and p_{27}. However, in the process of reception, A comes to the statement p_{88}, which asserts "statement p_6 is invalid and has to be disregarded". This makes A to return back and process statements p_{20}, p_{23}, and p_{27}. This shows that information reception is a complicated (in some cases non-monotone) process.

If information reception is pure, i.e., other processes do not impact the knowledge state of the receiver, then the final knowledge state does not contain knowledge items that were not in the initial knowledge state or were not added to it in the process of reception. Likewise, pure information reception implies that all knowledge items that were in the initial knowledge state either are removed or are present in the final knowledge state. A formal description of these conditions for a pure information reception has the following structure:

$$K_F \subseteq CK_A \cup K_I$$

and

$$K_I \subseteq CK_R \cup K_F$$

At the same time, assuming that we cannot remove from a set what does not belong to this set, we come to the following property of extended information explications:

$$CK_R \subseteq CK_A \cup K_I$$

In addition, it is natural to suppose that if a knowledge item was added and was not removed, then it belongs to the final knowledge state. Formally, it implies the following inclusion:

$$CK_A \subseteq CK_R \cup K_F$$

Taking some group G of cognitive systems/agents, we denote the set of all extended information explications of cognitive systems/agents from the group G by $\mathbf{EIE}(G)$. In particular, $\mathbf{EIE}(A)$ is the set of all extended information explications of a cognitive system/agent A.

The set $\mathbf{IE}(G)$ is a subset of the set $\mathbf{EIE}(G)$. In addition, Definitions 4.3.5 and 4.3.9 give the following result.

Proposition 4.3.4. For any group G of cognitive systems/agents, there is a natural projection $p: \mathbf{EIE}(G) \to \mathbf{IE}(G)$.

Extended information explications and information explications are two types of information representation.

Definition 4.3.10. A local information operator $O(d, K_I, K_F)$ is called *reduced* if

$$CK_A \cap CK_R = \varnothing$$

and

$$CK_A \cap K_I = \varnothing$$

Informally it means that a reduced local information operator does not perform such superfluous actions as at first adding a knowledge item and then removing the same item, or at first removing a knowledge item from a current knowledge state and then adding the same item once more, or adding a knowledge item that already belongs to the initial knowledge state of the cognitive system/agent or adding the same knowledge item several times.

Theorem 4.3.1. A local information operator $O(d, K_I, K_F)$ is reduced if and only if

$$CK_A = K_A$$

and

$$CK_R = K_R$$

Proof. Sufficiency. Let us assume that $CK_A = K_A$ and $CK_R = K_R$ for a local information operator $O(d, K_I, K_F)$. Then by the laws of set theory (Kuratowski and Mostowski, 1967), we have

$$CK_A \cap CK_R = K_A \cap K_R = (K_F \backslash K_I) \cap (K_F \backslash K_I) = \varnothing$$

as $K_A = K_F \backslash K_I$ and $K_R = K_I \backslash K_F$.

In addition,

$$CK_A \cap K_I = K_A \cap K_I = (K_F \backslash K_I) \cap K_I = \varnothing$$

Thus, the local information operator $O(d, K_I, K_F)$ is reduced.

<u>Necessity</u>. If the local information operator $O(d, K_I, K_F)$ is reduced, then $CK_A \cap CK_R = \varnothing$ and $CK_A \cap K_I = \varnothing$. Consequently,

$$K_F = (K_I \backslash CK_R) \cup CK_A$$

and

$$CK_R \subseteq K_I.$$

Thus,

$$K_R = K_I \backslash K_F = K_I \backslash [(K_I \backslash CK_R) \cup CK_A] =$$
$$[K_I \backslash (K_I \backslash CK_R)] \cap [K_I \backslash CK_A] = CK_R \cap K_I = CK_R$$

as $CK_R \subseteq K_I$ and $CK_A \cap K_I = \varnothing$.

In addition,

$$K_A = K_F \backslash K_I = [(K_I \backslash CK_R) \cup CK_A] \backslash K_I =$$
$$[(K_I \backslash CK_R) \backslash K_I] \cup [CK_A \backslash K_I] = CK_A$$

as $(K_I \backslash CK_R) \backslash K_I = \varnothing$ and the equality $CK_A \cap K_I = \varnothing$ implies $CK_A \backslash K_I = CK_A$.

Theorem is proved.

Corollary 4.3.1. There is a one-to-one correspondence between reduced local information operators and information explications.

The functional mode of operation of a device (system or agent) presupposes that everything, that is, the whole action, is done (performed) in one step of operation. However, adding and removing the same knowledge item demand, at least, two steps. Besides, adding to a set one of its elements corresponds to the identity operator because this operation does not change the set. This gives us the following result.

Proposition 4.3.5. Any functional local information operator is reduced.

Corollary 4.3.1 and Proposition 4.3.5 give us the following result.

Corollary 4.3.2. There is a one-to-one correspondence between functional local information operators and information explications.

Extended informations explications allow us to define logical operations over Mizzaro spaces. To do this, we consider any finite group G of agents as one agent, taking the union of the common knowledge states of the participants of this group as the common knowledge state of the agent G.

Let us consider two extended information explications $IE(d, K_{1I}, K_{1F}) =$ (CK_{1A}, CK_{1R}) for an agent A and $IE(c, K_{2I}, K_{2F}) = (CK_{2A}, CK_{2R})$ for an agent H.

Definition 4.3.11. a) The *disjunction* $IE(d, K_{1I}, K_{1F}) \lor IE(c, K_{2I}, K_{2F})$ of the extended information explications $IE(d, K_{1I}, K_{2F})$ and $JE(c, K_{2I}, K_{2F})$ is equal to

$$(CK_{1A} \cup CK_{2A}, CK_{1R} \cup CK_{2R})$$

b) The *conjunction* $IE(d, K_{1I}, K_{1F}) \land IE(c, K_{2I}, K_{2F})$ of the extended information explications $IE(d, K_{1I}, K_{1F})$ and $IE(c, K_{2I}, K_{2F})$ is equal to

$$(CK_{1A} \cap CK_{2A}, CK_{1R} \cap CK_{2R})$$

c) The *negation* $\rceil IE(d, K_{1I}, K_{1F})$ of the extended information explication $IE(d, K_{1I}, K_{1F})$ is equal to

$$(CK_{1R}, CK_{1A})$$

Informally, these operations are interpreted in the following way. Disjunction is realized in the case when there are two agents A and H, one of them, say A, receives data d and the other, say H, receives data c. Then their mutual information explication describes independent processing of information in these data and is equal to the disjunction $IE(d, K_{1I}, K_{1F}) \lor IE(c, K_{2I}, K_{2F})$. The union of the knowledge states of all agents from the group is called the *combined knowledge state* of this group. When a group consists of one agent, any combined knowledge state of this group is a knowledge state of this agent. A group of agents that perform their information processing independently is called a *separated group*.

Conjunction is also performed by two separate agents A and H when one of them, say A, receives data d and the other, H, receives data c. However, in this case, these agents process information in data in a coordinated way and make only those transformations that are implied by both sets of data. This kind of information processing is described by the conjunction $IE(d, K_{1I}, K_{1F}) \land IE(c, K_{2I}, K_{2F})$. A group of agents that perform their information processing in accord with one another is called a *coordinated group*.

Negation of information means performing, when it is possible, operations opposite to those that the initial information implies.

Proposition 4.3.6. If $IE_1(d, K_{1I}, K_{1F})$ and $IE_2(c, K_{2I}, K_{2F})$ are the extended information explications of reduced local information operators $O_1(d, K_{1I}, K_{1F})$ and $O_2(c, K_{2I}, K_{2F})$, then $IE_1(d, K_{1I}, K_{1F}) \wedge IE_2(c, K_{2I}, K_{2F})$ is the extended information explication of a reduced local information operator.

Proof. Let us assume that $IE_1(d, K_{1I}, K_{1F})$ and $IE_2(c, K_{2I}, K_{2F})$ are the extended information explications of reduced local information operators $O_1(d, K_{1I}, K_{1F})$ and $O_2(c, K_{2I}, K_{2F})$. Taking $CK_{3A} = CK_{1A} \cap CK_{2A}$ and $CK_{3R} = CK_{1R} \cap CK_{2R}$, we show that $(CK_{3A}, CK_{3R}) = IE(d, K_{3I}, K_{3F})$ is an information explication of a reduced operator $O_3(d, K_{3I}, K_{3F})$.

Let us take $K_{3I} = (K_{1I} \cap K_{2I})$ and $K_{3F} = (K_{3I} \cup K_{3A})\backslash K_{3R}$. By Definition 4.3.10, we have $CK_{1A} = K_{1A}$, $CK_{2A} = K_{2A}$, $CK_{1A} \cap CK_{1R} = \varnothing$ and $CK_{2A} \cap CK_{2R} = \varnothing$ as both operators $O_1(d, K_{1I}, K_{1F})$ and $O_2(c, K_{2I}, K_{2F})$ are reduced. Thus,

$$CK_{3A} \cap CK_{3R} = (K_{1A} \cap K_{2A}) \cap (K_{1R} \cap K_{2R}) =$$
$$(CK_{1A} \cap CK_{2A}) \cap (CK_{1R} \cap CK_{2R}) =$$
$$(CK_{1A} \cap CK_{1R}) \cap (CK_{2A} \cap CK_{2R}) =$$
$$(K_{1A} \cap K_{1R}) \cap (K_{2A} \cap K_{2R}) = \varnothing$$

Consequently, the first condition from Definition 4.3.10 is valid for the extended information explication (CK_{3A}, CK_{3R}).

By our construction, we have

$$CK_{3A} \cap K_{3I} = (K_{1A} \cap K_{2A}) \cap (K_{1I} \cap K_{2I}) =$$
$$(K_{1A} \cap K_{1I}) \cap (K_{2A} \cap K_{2I}) = \varnothing$$

Consequently, the second condition from Definition 4.3.10 is valid for the extended information explication (CK_{3A}, CK_{3R}) and thus, $O_3(d, K_{3I}, K_{3F})$ is a reduced operator.

Proposition is proved.

Proposition 4.3.7. If $IE(d, K_I, K_F) = (K_A, K_R)$ is the extended information explications of a reduced local information operator $O(K_I, K_F)$, then $\rceil IE(d, K_I, K_F)$ is the extended information explications of a reduced local information operator $O(K_F, K_I)$.

The proof follows from the formulas

$$K_F = (K_A \cup K_I)\backslash K_R$$

and

$$K_I = (K_R \cup K_F) \backslash K_A$$

It is possible to define a partial order for extended information explications over Mizzaro spaces.

Definition 4.3.12. An extended information explication $IE(d, K_{1I}, K_{1F}) = (CK_{1A}, CK_{1F})$ is *less than* or *equal to* an extended information explication $IE(c, K_{2I}, K_{2F}) = (CK_{2A}, CK_{2F})$ if

$$CK_{1A} \subseteq CK_{2A} \quad \text{and} \quad CK_{1R} \subseteq CK_{2R}$$

This relation is denoted by $IE(d, K_{1I}, K_{1F}) \le IE(c, K_{2I}, K_{2F})$.

Informally, this relation means that the first extended information explication corresponds to a local information operator that does less changes than the local information operator to which the second extended information explication corresponds.

Properties of the inclusion of sets allow us to prove the following result.

Proposition 4.3.8. The relation $\le$ is a partial order in the set of all extended information explications.

Indeed, the inclusion relation $\subseteq$ of sets is a partial order (Kuratowski and Mostowski, 1967). Thus, the relation $\le$ is a partial order in the set of all extended information explications because it is induced by the inclusion relation.

Let us consider extended information explications $IE(d, K_{1I}, K_{1F})$ and $IE(c, K_{2I}, K_{2F})$ of local information operators $O(d, K_{1I}, K_{1F})$ and $O(c, K_{2I}, K_{2F})$.

Proposition 4.3.9. If the local information operator $O(d, K_{1I}, K_{1F})$ is reduced, $K_{1I} K_{2I}$ and $IE_1(d, K_{1I}, K_{2I}) \le IE_1(c, K_{1J}, K_{2J})$, then the local information operator $O(c, K_{2I}, K_{2F})$ is also reduced.

Proof is similar to the proof of Proposition 4.3.6.

Proposition 4.3.10.

a) $IE(d, K_{1I}, K_{1F}) \vee JE(c, K_{2I}, K_{2F}) = JE(c, K_{2I}, K_{2F}) \vee IE(d, K_{1I}, K_{1F})$.

b) $IE(d, K_{1I}, K_{1F}) \wedge JE(c, K_{2I}, K_{2F}) = JE(c, K_{2I}, K_{2F}) \wedge IE(d, K_{1I}, K_{1F})$.

c) $IE(d, K_{1I}, K_{1F}) \wedge JE(c, K_{2I}, K_{2F}) \le IE(d, K_{1I}, K_{1F}) \le JE(c, K_{1I}, K_{1F}) \vee IE(d, K_{2I}, K_{2F})$.

d) If $JE(c, K_{1I}, K_{1F}) \le IE(d, K_{2I}, K_{2F})$ and $K_{1I} \subseteq K_{2I}$, then $\rceil IE(c, K_{1I}, K_{1F}) \le \rceil IE(d, K_{2I}, K_{2F})$.

Propositions 4.3.7 and 4.3.9 allow us to prove the following result.

Theorem 4.3.2. With respect to disjunction and conjunction, the set **I(W)** of all potential extended information explications, i.e., of all pairs (X, Y) where X, Y are subsets of **W**, forms a lattice with the largest element **(W, W)** and the smallest element $(\varnothing, \varnothing)$.

As the disjunction and conjunction of finite potential extended information explications are finite, we have the following result.

Corollary 4.3.3. The set **FI(W)** of all potential finite extended information explications forms a lattice with respect to disjunction and conjunction and this lattice is a sublattice of the lattice **I(W)**.

Different people can have the same mentality. This is formalized in the following concept.

Definition 4.3.13. Two groups of agents G_1 and G_2 are called *mentally equivalent* if they have the same set of combined knowledge states and any message causes the same changes in their combined knowledge states if it is accepted when they both are in the same knowledge state.

In particular, two agents A and B are *mentally equivalent* if they have the same set of knowledge states and any message causes the same changes in their knowledge states if it is accepted when they both are in the same knowledge state.

The concept of mental equivalence allows us to consider useful properties of agent systems.

MA3s (the S-Completeness Axiom). For any separated group of agents, there is a mentally equivalent agent.

MA3c (the C-Completeness Axiom). For any coordinated group of agents, there is a mentally equivalent agent.

Theorem 4.3.2 gives us following results.

Corollary 4.3.4. If Conditionss MA1, MA3s, and MA3c are true, then the set **EIE(W)** of all extended information explications forms a lattice with respect to disjunction and conjunction, and the set **FEIE(W)** of all finite extended information explications is its sublattice.

Let us consider two messages d and c and two local information operators $O_1(d, K_{1I}, K_{1F})$ and $O_2(c, K_{2I}, K_{2F})$ corresponding to the same agent A who (that) receives these messages. If messages come one after the other, c after d, so that $K_{1F} = K_{2I}$, then it is possible to define the *sequential composition* $O_1(d, K_{1I}, K_{1F}) \circ O_2(c, K_{2I}, K_{2F})$ of these operators.

Definition 4.3.14. $O_1(d, K_{1I}, K_{1F}) \circ O_2(c, K_{2I}, K_{2F}) = O_1(d \circ c, K_{1I}, K_{2F})$.

Composition of operators in a natural way induces the *sequential composition of extended information explications*. Namely, we have

$$IE_1(d, K_{1I}, K_{1F}) \circ IE_2(c, K_{2I}, K_{2F}) = (K_{1A} \cup K_{2A}, K_{1R} \cup K_{2R})$$

where $IE_1(d, K_{1I}, K_{1F}) = (K_{1A}, K_{1R})$ and $IE_2(c, K_{2I}, K_{2F}) = (K_{2A}, K_{2R})$.

As the sequential composition of local information operators is an associative partial operation, we have the following result.

Proposition 4.3.11. The sequential composition of extended information explications is a monotone associative partial operation in the set $\mathbf{I(W)}$ of all potential extended information explications.

Note that the sequential composition of extended information explications coincides with the disjunction of the same extended information explications when the sequential composition is defined. However, there is an essential difference between the sequential composition and disjunction because the former is a partial operation in a general case, while the latter is always a total operation.

Let us consider a group G of cognitive systems/agents.

Theorem 4.3.3. The set $\mathbf{CEIE}(G)$ of extended information explications and knowledge states of cognitive systems/agents from the group G is a category with knowledge states as objects, extended information explications as morphisms and their sequential composition as the composition of morphisms.

<u>Proof</u>. Taking two knowledge states K_1 and K_2, we define the set of morphisms $\mathrm{Mor}_{\mathbf{CEIE}(G)}(K_1, K_2)$ from K_1 into K_2 in the following way:

$$\mathrm{Mor}_{\mathbf{CEIE}(G)}(K_1, K_2) = \{(K_A, K_R); K_A, K_R \subseteq W, K_R \subseteq K_1 \cup K_A,$$
$$K_2 \subseteq (K_1 \cup K_A) \backslash K_R\}$$

Then the sequential composition of extended information explications is the composition of morphisms in $\mathbf{CEIE}(G)$ because it is associative by properties of the union and Theorem 4.3.2. The extended information explication $(\varnothing, \varnothing)$ belongs to any set $\mathrm{Mor}_{\mathbf{CEIE}(G)}(K, K)$ and is the identity element with respect to the sequential composition of extended information explications. Indeed, we have

$$IE(d, K_I, K_F) \circ (\varnothing, \varnothing) = (K_A \cup \varnothing, K_R \cup \varnothing) = (K_A, K_R)$$

for any K_I, $K_F \subseteq W$. Consequently (cf. Appendix D), **CEIE**(G) is a category.

Theorem is proved.

Definition 4.3.15. Given an agent in an initial knowledge state K_I, receiving a datum d, and ending in a final knowledge state K_F, the *full quantity* $\| I(d, K_I, K_F) \|$ *of information* $I(d, K_I, K_F)$ caused by receiving the datum d is defined as

$$\big| IE(d, K_I, K_F) \big| = \big| \Delta K \big| =$$
$$\big\| I(d, K_I, K_F) \big\| = \big\| \Delta K \big\| =$$
$$\big| (CK_A, CK_R) \big| = \big| CK_A \big| + \big| CK_R \big|$$

The full quantity of information in the datum d for an agent A is measured by the amount of structural/mental work, i.e., structural work realized by mental work, performed by receiving the datum d. The extended information explication $IE(d, K_I, K_F)$ shows the structural/mental work performed in the transition from the initial knowledge state K_I to the final knowledge state K_F. The full quantity of information in the datum d for an agent A reflects the amount of the efficient structural/mental work, i.e., a work of adding and removing knowledge items in which nothing is added and then deleted, performed by receiving the datum d.

When the corresponding local information operator is reduced, then $IE(d, K_I, K_F) = I(d, K_I, K_F)$ is an information explication. It shows the minimal structural/mental work necessary to achieve the final knowledge state

K_F from the initial knowledge state K_I and corresponds to the reduced local information operator that performs this work. The quantity of information in the datum d for an agent A estimates this minimal structural/mental work.

The full quantity of information is additive with respect to conjunction and monotone with respect to disjunction and negation of extended information explications.

Proposition 4.3.12. For any extended information explications $IE(d, K_{1I}, K_{1F})$ and $IE(c, K_{2I}, K_{2F})$, we have:

a) $\left| IE(d, K_{1I}, K_{1F}) \vee IE(c, K_{2I}, K_{2J}) \right| = \left| IE(c, K_{2I}, K_{2J}) \right| +$
   $\left| IE(d, K_{1I}, K_{1F}) \right|$

b) $\left| IE(d, K_{1I}, K_{1F}) \wedge IE(c, K_{2I}, K_{2F}) \right| \leq \left| IE(c, K_{2I}, K_{2F}) \right|$

and

   $\left| IE(d, K_{1I}, K_{1F}) \wedge IE(c, K_{2I}, K_{2F}) \right| \leq \left| IE(d, K_{1I}, K_{1F}) \right|$

c) $\left| \neg IE(d, K_{1I}, K_{1F}) \right| = \left| IE(d, K_{1I}, K_{2I}) \right|$

To conclude, it is necessary to remark that the theory of Mizzaro spaces allows further development in several directions.

Thus, analyzing the main concepts and constructions of semantic information theories, we see that these theories are special cases of the general theory of information. Systems of propositions/predicates are infological systems in semantic information theories based on logic and developed by Bar-Hillel, Carnap, Hintikka, Voishvillo, Israel, Perry, and others. A system of knowledge, or thesaurus, is the basic infological system in semantic information theories oriented at knowledge transformations and developed by MacKay, Shreider, Brookes, Gackowski, Mizzaro, and others. Information is considered or/and measured as transformations incurred by these infological systems.

Chapter 5

Algorithmic Information Theory

It's not what our message does to the listener —
but what the listener does with our message
[that determines our success in communication].
Hugh Mackay, The Good Listener

As we have seen in Chapter 3, statistical information theory is based on probability theory, which gives tools to measure information. Indeed, all statistical measures of information, such as Shannon's quantity of information (entropy) $H[m] = -\sum_{i=1}^{n} p_i \log p_i$ of the message m or Fisher information $I(\theta) = \mathbf{E}\left[[(\partial/\partial\theta) \ln f(X; \theta)]^2\right]$, use probabilistic constructions and concepts, namely, in these formulas, p_i is the probability of the event (outcome of an experiment, case or situation) i, f is the probability density function of a random variable X and θ is an unobservable parameter upon which the probability distribution of X depends. In spite of all its success and popularity, statistical information theory was not able to solve many problems of information processing. The same is true for semantic information theories.

Analysis of such unsolved problems brought Ray Solomonoff, Andrey Kolmogorov (1903–1987) and Gregory Chaitin to ideas that invert this relation. They suggested that information theory must precede probability theory, and not to use probability theory as the base because information is a more fundamental concept than probability. In addition, they limited the scope of information processes, taking into account only constructive processes controlled by algorithms. This correlates with the main principles of the general theory of information. These principles imply (cf. Section 2.2) that information is related to transformation and

361

constructive transformations (computations, in particular) are directed by algorithms.

Such ideas gave birth to algorithmic information theory. It is based on the concept of Kolgmogorov or algorithmic complexity of objects. Kolgmogorov complexity has a deep and sophisticated theory, which provides means to measure the intrinsic information related to objects via their algorithmic description length. It has been applied to a wide range of areas including theory of computation, combinatorics, inductive reasoning and machine learning. This theory explicates an important property of information that connects information to means used for accessing and utilizing information. Information is considered not as some intrinsic property of different objects but is related to algorithms that use, extract or produce this information. In this context, a system (person) with more powerful algorithms for information extraction and management can get more information from the same carrier and use this information in a better way than a system that has weaker algorithms and more limited abilities. This correlates with the conventional understanding of information. For instance, system (person) that (who) has a code C can read codified in C texts, while those who do not have this code cannot read such texts. As a result, efficiency or complexity and not uncertainty (cf. Chapter 3) or diversity (cf. Chapter 4) of algorithms becomes a measure of information. Efficiency is a clue problem and a pivotal characteristic of any activity. Inefficient systems are ousted by more efficient systems. Problems of efficiency are vital to any society, advanced technology or individual. Consequently, measures of efficiency and complexity provide means for measuring information.

The aim of this chapter is to provide a broad perspective on algorithmic information theory with its main peculiarities and constructions. In comparison with the majority of sources on algorithmic information theory, we present this theory on several levels. At first, the conventional level of recursive algorithms is considered in Sections 5.1 and 5.2. Many believe that the recursive information size (often called Kolmogorov complexity or descriptive complexity) of an object is a form of absolute information of the individual object. However, discovery of super-recursive algorithms and emergence the theory of super-recursive algorithms demonstrated a relative nature of the recursive information size. That is why the recursive algorithmic approach to information is upgraded to the next, super-recursive level of the algorithmic universe in Section 5.3. This is a necessary step as super-recursive algorithms are essentially more efficient in processing information than recursive algorithms and give more adequate models for the majority of information systems. Finally the highest, axiomatic level is achieved in a form of an axiomatic algorithmic information theory considered in Section 5.5. Section 5.4 contains a relativistic version of algorithmic information theory.

Theorem 5.3.6 and Corollaries 5.3.9–5.3.15 show that more powerful algorithms, or a system with more powerful algorithms, need less information to do the same thing, i.e., to understand a text, to write a novel, to make a discovery or to organize successful functioning of a company. These results mathematically prove that the quantity of information that a carrier has for a system depends on that system. It gives additional supportive evidence for Principle O1. Informally, these results mean that a system with more powerful algorithms can extract more information from the same information carrier (e.g., a book, unknown plant, new star, etc.) or information presentation (e.g., a text, message, signal, etc.) than a system with weaker algorithms. This also shows how powerful algorithms reduce information size of objects.

5.1. Information, Algorithms, and Complexity

> *... the simplest explanation is always the best.*
> Umberto Eco, Foucalt's Pendulum

> *Simplicity is very subjective.*
> Lubos Lumo Motl

Objects considered in algorithmic information theory are strings of symbols because the most habitual representation of information uses symbols. It is possible to interpret such strings as words or texts in some language. It means that information is presented and processed in the symbolic form and all systems are represented by their symbolic (semiotic) models. Exact models have mathematical structure. The main question is how much information we need to reconstruct (compute) a given string (word). Thus, the traditional approach in algorithmic information theory treats only symbolic information. This question relates information to complexity because measure of necessary information appears here as a measure of complexity of the string reconstruction.

Reconstruction/computation of a string of symbols is an action that is realized as a process. Its complexity depends on means that are used for reconstruction. To make this idea precise a concept of an algorithm is used. Namely, strings are reconstructed (built) by algorithms. Algorithms are working in the domain of strings and this domain usually consists of all finite strings in some alphabet. In this context, an algorithm (it is also possible to say, automaton or computer) takes one string of symbols z and eventually produces another string x. The input string is a carrier of information about the output string, i.e., string that we are going to reconstruct/compute. It is possible to consider the input string z as the program that has been given to the algorithm/machine for computing x. This program provides information about x for an algorithm (computing device). In such a way, researchers come to information size (complexity) of a string of symbols, which is the theory's fundamental concept. Note that very often, information content of a string is called Kolmogorov complexity. Namely, the information content $C(x)$ of a string x is the minimum quantity of information needed to reconstruct this string. In the conventional approach, such quantity of input information is measured by the size of information carrier and as carriers are strings of symbols the volume of a string z is the length $l(z)$ of this string. Thus, the length of the shortest program for calculating the output string x gives the measure of information needed to reconstruct/compute this string.

This was the first developed form of such complexity measures. It is known under the name *Kolmogorov complexity* or *constructive complexity* or *algorithmic complexity* of an object such as a piece of text and is a measure of the computational resources needed to specify the object. Consequently, the most popular name for the algorithmic information theory has been the theory of Kolmogorov complexity.

At first, the theory of Kolmogorov complexity was developed for such class of algorithms as the class of all Turing machines. Traditionally this theory is divided into three parts. The first part studies complexity *per se*. The second part concentrates on randomness. The third part makes emphasis on information.

The history tells us, this approach was defined and studied independently by three mathematicians: Solomonoff who studied problems of inductive learning (1960; 1964), Kolmogorov who studied problems of foundations for probability and information (1965), and Chaitin who studied complexity of computations (1966). The construction of complexity based on shortest programs (descriptions) but lacking the Invariance Theorem can be also attributed to A.A. Markov (1964). It is necessary to mention that ideas of Solomonoff on algorithmic complexity were described by him in 1960 and then presented by Minsky in 1962, while ideas of Kolmogorov on algorithmic approach to complexity and randomness were suggested in 1963 (cf. (Li and Vitanyi, 1997)). In addition to Kolmogorov's long-life interest in probability theory with "randomness" as its central concept, he worked in such areas as the theory of algorithms (cf., for example, (Kolmogorov, 1953) where a new model of algorithms is suggested) and information theory (cf., for example, (Kolmogorov, 1957; Gelfand, Kolmogorov, and Yaglom, 1956)). All this brought Kolmogorov to the concept of algorithmic or Kolmogorov complexity. As Chaitin published his results later than two other mathematicians, some authors do not consider Chaitin a co-author of Kolmogorov complexity although he worked more and contributed more to this area than Solomonoff and Kolmogorov.

The original *Kolmogorov complexity* $C(x)$ of a word x is taken to be equal to the size of the shortest program (in number of symbols) for a universal Turing machine U that without additional data, computes the string x and terminates. This measure is called *absolute* Kolmogorov complexity because Kolmogorov complexity has also a relative form $C(x|y)$.

Besides, *Kolmogorov complexity* usually denoted by $K(x)$ or $C(x)$ is known under many other names. Although the majority of researchers use the name Kolmogorov complexity as the name for this measure, other authors prefer a different name. For instance, Gell-Mann (1994) trying to avoid the conflicting word "complexity" calls it simply *algorithmic information content*. In the literature, we encounter such names as *algorithmic information, program-size complexity, shortest program length, algorithmic randomness, stochastic complexity,* and *algorithmic entropy.* Even one and the same author can call this concept by different names. For instance, Chaitin uses such names as *information content* denoted by I(x), *information-theoretic complexity, complexity,* and *randomness.* Lewis (2001) calls this measure *algorithmic* (AC) or *KCS* (Kolmogorov-Chaitin-Solomonoff) *complexity.* The term *constructive complexity* is used in (Burgin, 2005) as a possible name for this concept. In the fundamental treatise *An Introduction to Kolmogorov Complexity and its Applications* (Li and Vitanyi, 1997), both names, *Kolmogorov complexity* and *algorithmic complexity,* are used. One more natural name for this measure is *descriptive complexity* (cf., for example, (Gács, 1993)).

Different names reflect different approaches to the concept. When we want to know how difficult it might be computing or constructing some object x with recursive algorithms, *Kolmogorov* or *algorithmic complexity* is an appropriate name. When the question is how much information we need to build or compute x with algorithms, *information size* of x is a better name. When we consider probabilistic aspects of x, e.g., randomness, the best name might be *algorithmic entropy.*

The name *information size* also well correlates with the common understanding that information is measured in bits. When we have a word or text z in the binary alphabet $\{0, 1\}$, the length of this word z is equal to the number of bits in z, the length is the traditional measure of information. Consequently, C(x) measures (in this case) the minimal information (in bits) that we or machines need to compute/construct x by algorithms in some class.

One of the principal goals of the Kolmogorov complexity introduction was to ground probability theory and information theory, creating a new approach based on algorithms. This goal was achieved. Algorithmic information theory and its applications to probability theory have become very popular. However, new theories did not substitute either the classical probability theory, which was grounded before by the same Kolmogorov (1950) on the base of measure theory, or Shannon's information theory.

Kolmogorov, or algorithmic, complexity has become an important and popular tool in computer science, programming, probability theory, statistics, and information theory. It has found applications in medicine, biology, neurophisiology, physics, economics, hardware and software engineering. In physics, problems of quantum gravity are analyzed based on algorithmic complexity of the given object. In particular, the algorithmic complexity of the Schwarzschild black hole is estimated (Dzhunushaliev, 1998: Dzhunushaliev and Singleton, 2001). Mora and Briegel (2004) apply algorithmic complexity to quantum mechanics. Yurtsever (2000) considers relations between quantum mechanics and algorithmic randomness based on algorithmic complexity. Benci, et al, (2002) apply algorithmic complexity to chaotic dynamics. The inclusion of algorithmic complexity and randomness in the definition of physical entropy allows Zurek (1989; 1991) to get a formulation of thermodynamics. Kreinovich and Kunin (2004) use Kolmogorov complexity to study problems in mechanics. Brudno (1983) used Kolmogorov complexity of strings to define orbit complexity for symbolic dynamical systems.

Tegmark (1996) discusses what can be the algorithmic complexity of the whole universe. The main problem with this discussion is that the author identifies physical universe with physical models of this universe. To get valid results on this issue, it is necessary to define algorithmic complexity for physical systems because conventional algorithmic complexity is defined only for such symbolic objects as words and texts (Li, and Vitanyi, 1997). Then it is also necessary to show that there is a good correlation between algorithmic complexity of the universe and algorithmic complexity of its model used by Tegmark.

In biology, algorithmic complexity is utilized for estimation of protein identification (Dewey, 1996; 1997). In neurophysiology, algorithmic complexity helps to measure characteristics of brain functions (Shaw, et al, 1999). In economics, a new approach to understanding of the complex behavior of financial markets using algorithmic complexity is developed by Mansilla (2001). Algorithmic complexity has been useful in the development of software metrics and solving evaluation problems of software engineering (Lewis, 2001; Burgin, and Debnath, 2003; Debnath, and Burgin, 2003). Bush and Todd

(2005) use Kolmogorov complexity for building a vulnerability analysis framework that is capable of identifying new types of vulnerabilities before attackers innovate their attacks. Bush and Kulkarni (2001) apply Kolmogorov complexity to network management. Crosby and Wallach (2003) use algorithmic complexity to study low-bandwidth denial of service attacks that exploit algorithmic deficiencies in many common applications' data structures. Veldhizen (2005) analyses software reuse from the perspective of Kolmogorov complexity.

In addition, as Li and Vitanyi write (1997), applications of Kolmogorov complexity have encompassed a plethora of areas, including randomness of individual finite objects or infinite sequences, tests for randomness, incompleteness results in mathematics, computer science and software engineering, information theory of individual objects, universal probability, general inductive reasoning, inductive inference, prediction, mistake bounds, computational learning theory, inference in statistics, the incompressibility method, combinatorics, graph theory, Kolmogorov 0-1 Laws, probability theory, theory of parallel and distributed computation, time and space complexity of computations, average case analysis of algorithms such as HEAPSORT, language recognition, string matching, routing in computer networks, circuit theory, formal language and automata theory, parallel computation, Turing machine complexity, complexity of tapes, stacks, queues, average complexity, lower bound proof techniques, structural complexity theory, oracles, logical depth, universal optimal search, physics and computation, dissipationless reversible computing, information distance and picture similarity, thermodynamics of computing, statistical thermodynamics and Boltzmann entropy.

Thus, we see that algorithmic information in the form of Kolmogorov complexity is a frequent term in present days' scientific literature, in various fields and with diverse meanings, appearing in some contexts as a precise concept of algorithmic complexity, while being a vague idea of complexity in general in other texts.

However, as it often happens with profound mathematical and scientific theories, algorithmic information theory encountered many problems both in its understanding based on different interpretations and in its application. Almost from the beginning, researchers understood that an attempt to define in this setting an appropriate concept of randomness was unsuccessful. It turned out that the original definition of Kolmogorov complexity was not relevant for that goal. To get a correct definition of a random infinite sequence, it was necessary to restrict the class of utilized algorithms. That is why algorithmic information (Kolmogorov complexity) was defined and studied for various classes of subrecursive algorithms. For instance, researchers discussed different reasons for restricting power of the device used for computation when estimating the minimal complexity. Thus, the concept of algorithmic information (Kolmogorov complexity) has been developing top down: from larger classes to smaller classes of algorithms that were more relevant to computational problems. As a result many versions of Kolmogorov complexity have been introduced.

When Kolmogorov complexity is defined for the class of Turing machines that compute symbols of a word x, we obtain *uniform complexity* $KR(x)$ studied by Loveland (1969).

When Kolmogorov complexity is defined for the class of prefix functions or prefix-free Turing machines, we obtain *prefix complexity* or *prefix-free complexity* $K(x)$ studied by Gács (1974), Levin (1974), and Chaitin (1975).

When Kolmogorov complexity is defined for the class of monotonous Turing machines, we obtain *monotone complexity* $Km(x)$ studied by Levin (1973).

When Kolmogorov complexity is defined for the class of Turing machines that have some extra initial information, we obtain *conditional Kolmogorov complexity* $CD(x)$ studied by Sipser (1983).

Let $t(n)$ and $s(n)$ be some functions of natural number variables.

When Kolmogorov complexity is defined for the class of recursive automata that perform computations with time bounded by some function of a natural variable $t(n)$, we obtain *time-bounded Kolmogorov complexity* $C^t(x)$ studied by Kolmogorov (1965) and Barzdin (1968).

When Kolmogorov complexity is defined for the class of recursive automata that perform computations with space (i.e., the number of used tape cells) bounded by a function of a natural variable $s(n)$, we obtain *space-bounded Kolmogorov complexity* $C^s(x)$ studied by Hartmanis (1983).

When Kolmogorov complexity is defined for the class of multitape Turing machines that perform computations with time bounded by some function $t(n)$ and space bounded by some function $s(n)$, we obtain *resource-bounded Kolmogorov complexity* $C^{t,s}(x)$ studied by Daley (1973).

At the same time, researchers extended algorithmic complexity not only to more restricted classes of subrecursive algorithms but also to more powerful classes of subrecursive algorithms. Burgin (1995; 2004a; 2005) introduced and studied algorithmic complexity for inductive Turing machines. It is called *inductive Kolmogorov complexity*. Schmidhuber (2002) introduced and studied algorithmic complexity for general and monotonic Turing machines. Svozil (1996), Vitanyi (1999; 2001), Berthiaume, et al (2000), and Gács (2001) introduced and studied algorithmic complexity for quantum computations. This changed the opinion that the Kolmogorov complexity of an object is a form of absolute information of the individual object.

Even earlier, Chaitin (1976; 1977) introduced algorithmic complexity for infinite computations in the case when the output of a machine or algorithm was an infinite set.

Thus, in the process of algorithmic information theory development, different authors introduced different versions of Kolmogorov complexity. This caused a necessity to build a unified theory of algorithmic information. Mathematics suggests an approach for unification, namely, it is necessary to find axioms that characterize all theories in a specific area and to develop the theory in an axiomatic context. This approach worked well in a variety of fields:

- in geometry (Euclid's axioms and later Hilbert's axioms (Hilbert, 1899));
- in algebra (actually the whole algebra exists now as an axiomatic discipline);
- in set theory, according to Fraenkel and Bar-Hillel (1958), the most important directions in the axiomatic set theory are: Zermelo-Fraenkel (**ZF**), von Neumann (**VN**), Bernays-Godel (**BG**), Quine (**NF, ML**) and Hao Wang (**Σ**) axiomatic theories;
- in topology (Hausdorff's axioms (Hausdorff, 1927));

- in probability theory (Kolmogorov's axioms (Kolmogorov, 1933)).

Axiomatic approach to Kolmogorov complexity was introduced by Burgin (1982; 1990), further developed in the book (Burgin 2005) and applied to software metrics (Burgin and Debnath, 2003; 2003a; Debnath and Burgin, 2003).

In what follows, we denote by $C_P(x)$ algorithmic complexity of an object (word) x in a class **P** of algorithms (automata).

Properties of algorithmic complexity change when it is considered in different classes of algorithms. For instance, taking the original Kolmogorov complexity $C(x)$, or $C_T(x)$ where **T** is the class of all Turing machines, for the most words, we have

$$C_T(x) = l(x) + O(1)$$

At the same time, in the class **P** of all prefix-free Turing machines, Kolmogorov complexity $C_P(x)$ becomes larger than in the class **T** of all Turing machines, and for the most words, we have

$$C_P(x) = l(x) + C_P(l(x)) + O(1) = l(x) + O(\log l(x))$$

As Katseff and Sipser (1981) write, "a number of different measures of program size complexity have been introduced. Unfortunately, these measures do not have unique properties. The reasons for this are, in general, not well understood and there is disagreement as to which is most appropriate to use in different situations."

This shows that to deal with this variety of algorithmic complexity measures, we need a much more general approach. Besides, it is necessary to include complexity of constructive objects represented or constructed by more powerful super-recursive algorithms. It is has been done on the basis of axiomatic dual complexity measures and utilization of abstract classes of algorithms. These constructions follow and develop the dual complexity measure approach (Burgin, 1982; 1990; 1991; 1992b).

An essential problem with algorithmic complexity is related to its information theoretical interpretation. It is generally assumed that the algorithmic complexity of a binary string x measures the amount of information in the string x (cf., for example, (Chaitin, 1977; Li and Vitanyi, 1997; Calude, 2002)). Thus, according to the algorithmic information theory, random sequences have maximum complexity as by definition, a random sequence can have no generating algorithm shorter than simply listing the sequence. It means that information content of random sequences is maximal.

Physicists were the first who attracted attention to this peculiarity. Richard Feynman (1918–1988) wrote (1999):

"How can a random string contain *any* information, let alone the maximum amount? Surely we must be using the wrong definition of 'information'?..."

As Gell-Mann (1995) points out, this contradicts the popular understanding that random sequences should contain no information. He writes that this property of algorithmic information content (Kolmogorov complexity) reveals the unsuitability of the quantity as a measure of information, since the works of Shakespeare have less algorithmic information content than random gibberish of the same length that would typically be typed by the proverbial roomful of monkeys.

Raatikainen (1998) thoroughly analyzed this discrepancy and came to the grounded conclusion that the algorithmic complexity and the information content of a linguistic expression have no real connections. In addition, he found that according to common sense and semantic information theory based on logic, if a sentence P logically implies a sentence Q, it is naturally assumed that P contains more information than Q. At the same time, neither the statistical approach nor the algorithmic approach respects this property of information.

To eliminate these contradictions and discrepancies that are prevalent in algorithmic information theory and to solve the problem of correct understanding the meaning of the function $C_K(x)$, which denotes Kolmogorov complexity in the class **K** of algorithms use the general theory of information. Traditionally (cf., for example, (Li and Vitanyi, 1997; Burgin, 2005)), $C_K(x)$ and all its special cases, such as $C(x)$, $C_T(x)$, $C_{IT}(x)$, $K(x)$, $Km(x)$, and $C^t(x)$, are considered as the *measure of information in x* (cf., for example, (Chaitin, 1977; Li and Vitanyi, 1997; Calude, 2002)). The analysis based on the general theory of communication and interaction, as well as on the general theory of information (cf. Section 2.2) shows that it is more adequate to consider $C_K(x)$ and all its special cases as the *measure of information about x* with the special goal to build or reconstruct x. It means that in reality, x is not the carrier of information measured by $C_K(x)$, but the object of this information. Thus, it becomes not surprising that people, or a machine, need more information about a random sequence of letters to reconstruct it than about a masterpiece, such as a poem by Dante or a play by Shakespeare.

This understanding is also supported by the following mathematical reasoning. According to the suggestion of Kolmogorov (1965), information in a word y about a word x is equal to

$$I_T(y: x) = C_T(x) - C_T(x|y)$$

Then taking $y = x$, we have

$$I_T(x: x) = C_T(x) - C_T(x|x) \asymp C_T(x)$$

because $C_T(x|x) = O(1)$. It means that $C_T(x)$ is information in x about itself.

Thus, it is better to call $C_T(x)$, or more generally $C_K(x)$ where **K** is some class of algorithms, not the information content of x, but the *information* (we may also call it, *algorithmic*) *size* of x in the class **K** as it is the measure of information needed to build or reconstruct x by means of algorithms from **K**. It is necessary to understand that information size essentially depends on the system of algorithms with respect to which this size is defined and evaluated. For instance (cf. Section 5.3), for infinitely many words x, their information size in the class of recursive algorithms is essentially larger than their information size in the class of inductive algorithms. The situation with the information size is similar to the situation with the physical weight of physical bodies: the same body has less weight on the Moon than on the Earth.

Naturally, $C_T(x|y)$ is the information size of x when y is given. Thus, information in a word y about a word x, which is equal to

$$I_T(y: x) = C_T(x) - C_T(x|y),$$

shows how much y can decrease the information size of x.

The suggested understanding and the new relevant name of the function $C_T(x)$ eliminates the contradiction with random sequences that was emphasized by Feynman (1999) and Gell-Mann (1995). Really, in a new context, random sequences contain no information but have maximal information size. This resembles a situation when the biggest man can be the most stupid in a group of people.

It is also necessary to remark that $C_K(x)$ is complete as a particular measure of information only with respect to the linguistic structure of the word x. It tells us about the organization of symbols in x, but displays nothing, for example, about meaning of x or about its value.

In what follows, preserving the conventional notation, we use the term *information size* instead of more popular terms *Kolmogorov complexity* or *algorithmic complexity*, making emphasis on the informational issues of this theory.

5.2. Algorithmic Information Theory based on Recursive Algorithms: Recursive Approach

> *But when a rule is extremely complex,*
> *that which conforms to it passes for random.*
> Leibniz

In the study of algorithmic measures of information, researchers make the following assumptions. Algorithms work with words in some alphabet X. It is possible to codify all symbols from X by finite strings consisting of two symbols 1 and 0. This allows us to consider only algorithms that work with words in the alphabet $\{1, 0\}$. In addition, it is practical, in some cases, to interpret such binary words as representations of nonnegative integer numbers (natural numbers) and assume that algorithms work with natural numbers, i.e., compute partial functions from N into N. As before $l(p)$ is the length of the word p.

Definition 5.2.1. The *information size* $C_A(x)$ of an object (word) x with respect to an algorithm A is defined as

$$C_A(x) = \min \{l(p); \ A(p) = x\}$$

in the case when there is a word p such that $A(p) = x$;

otherwise $C_A(x)$ is not defined.

The algorithm A works with information contained in its input. So, if a word x is an input for A, then it is possible to measure information in x

as the length $l(x)$ of the word x. As we know, $l(x)$ is the number of symbols in x. When x is a binary word, i.e., it consists of 0s and 1s, $l(x)$ is the number of bits in x. Thus, $l(x)$ is the popular measure of information.

In this context, the information size $C_A(x)$ of an object (word) x with respect to an algorithm A is the least quantity of information (measured in bits for binary words) A needs to compute x.

It is necessary to understand that, contrary to some claims, $C_A(x)$ is not a measure of all information that A can extract from x but the quantity of information that A needs to compute x. Thus, it is possible to treat $C_A(x)$ as information in x about its own structure. It is proved that there no recursive algorithms that can extract information $C_A(x)$ from x in a general case because the function $C_A(x)$ can be not a recursively computable function. At the same time, there is an inductive algorithm can extract from x this minimal information because the function $C_A(x)$ is an inductively computable function (Burgin, 2005). In addition, it is also possible to extract other information from the word x. For example, if x is a code of a Turing machine T, a recursive algorithm can find the structure of T and reconstruct T from its code.

In this section, we consider only recursive algorithms. By definition (Burgin, 2005), any class **R** of recursive algorithms has the same computational power as the class **T** of all Turing machines. This equivalence makes it possible to consider information size only with respect to Turing machines. In particular, we can assume that A is a Turing machine.

The value $C_A(x)$ essentially depends on the algorithm A. For instance, if A is an algorithm that computes the identity function, i.e., $A(x) = x$, then $C_A(x) = l(x)$. If A is an algorithm that computes a function that is not defined for any x, then $C_A(x)$ is also undefined for all x.

However, it is possible to find an invariant information measure that characterizes strings of symbols with respect to the whole class **T**.

Theorem 5.2.1 (the Recursive Optimality Theorem (Solomonoff, 1964; Kolmogorov, 1965; Chaitin, 1969)). There is a Turing machine U such that for any Turing machine T, there is a constant c_T such that for all words x, we have

$$C_U(x) \le C_T(x) + c_T \tag{5.2.1}$$

Proof. Let us take a universal Turing machine U. By its definition (cf. Appendix B), given a code $c(T)$ of any Turing machine T and a word p as input to U, the machine U computes the same result as T given the word p or U gives no result when T gives no result on input p. For all words p, the pair $< c(T), p >$ can be coded by one word $d(c(T), p)$ so (cf. (Burgin, 2005)) that the length of the code will be equal to the length of the word p plus some constant c_T. Thus, if $T(p) = x$, then $U(p) = x$. Consequently, if $C_T(x) = l(p)$, then $C_U(x) \leq l(p) + c_T$. This implies the necessary inequality as the constant c_T depends only on the Turing machine T.

Theorem is proved.

Theorem 5.2.1 shows that the function $C_U(x)$ is *additively optimal*, that is, optimal up to an additive constant, in the class of all functions $C_T(x)$. Formally, it is denoted by $C_U(x) \preccurlyeq C_T(x)$.

Corollary 5.2.1 (the Recursive Invariance Theorem (Solomonoff, 1964; Kolmogorov, 1965; Chaitin, 1969)). For any two universal Turing machines U and V, the corresponding information sizes $C_U(x)$ and $C_V(x)$ are additively equivalent.

Formally, it is denoted by $C_U(x) \asymp C_V(x)$.

The Invariance Theorem makes it possible to define information size not only for separate algorithms (e.g., Turing machines), but also for any class **R** of recursive algorithms. Such class **R** has a universal algorithm U (cf. (Burgin, 2005)). For instance, in the class of all Turing machines, a universal Turing machine is a universal algorithm (cf. Appendix B).

Definition 5.2.2. The *recursive information size* $C_R(x)$ of an object (word) x is defined as

$$C_R(x) = \min \{l(p); \ U(p) = x\}$$

when there is a word p such that $U(p) = x$; otherwise $C_R(x)$ is not defined.

Here $l(p)$ is the length of the word p and U is a universal algorithm in the class **R**.

In a similar way, it is possible to define information size $C_P(x)$ for any subclass **P** of the class **R** when **P** has a universal algorithm.

Definition 5.2.3. The *information size* $C_P(x)$ of an object (word) x with respect to the class **P** is defined as

$$C_P(x) = \min \{l(p); \ U(p) = x\}$$

when there is a word p such that $U(p) = x$ and U is a universal algorithm in the class **P**; otherwise $C_A(x)$ is not defined.

As before, X^* denotes the set of all words in an alphabet X.

Theorem 5.2.2 (the Totality Theorem). If there is an algorithm in **P** the range of which is X^*, then $C_P(x)$ is a total function on X^*.

Indeed, in this case, the equation $U(p) = x$ has a solution for any word x because U is a universal algorithm in **P**.

Corollary 5.2.2 (Kolmogorov, 1965). $C_T(x)$ is a total function on N^+ (on the set of all words in some alphabet).

Let us suppose that the class **P** is infinite and contains only such algorithms that give as the result only one word or one number. In addition, we assume, without loss of generality, that all algorithms from **P** are working with natural numbers that are represented by words in the alphabet $\{1, 0\}$.

Lemma 5.2.1. For any number n, there is a number z such that for all elements x that are larger than z, the values $C_P(x)$ are larger than n.

Proof. The number of those elements x for which $C_P(x)$ is less than or equal to a given number n is less than 2^{n+1} because there are at most 2^{n+1} programs having the length less than or equal to n and the universal algorithm U computes only one word given one program for computation. So, there is the largest element x such that $C_P(x) \le n$. Then we can take $z = x$. Consequently, for all elements y that are larger than some element z, the values $C_P(y)$ are larger than n.

Definition 5.2.4. a) A partial function $f: X^* \to N^+$ tends to infinity (we denote it by $f(x) \to \infty$, or $f(x) \to \infty$ when $l(x) \to \infty$) if for any number m from N^+ there is a number k such that $f(x) > m$ when $l(x) > k$.

b) A partial function $f: X^* \to X^*$ tends to infinity (we denote it by $f(x) \to \infty$) if the partial function $l(f(x))$ tends to infinity.

Lemma 5.2.1 implies the following result.

Theorem 5.2.3 (the Recursive Unboundedness Theorem). $C_P(x) \to \infty$ when $l(x) \to \infty$.

Proof. Since the number of elements x for which $C_P(x)$ is less than or equal to a given number n is finite by Lemma 5.2.1, so as n tends to infinity, the function $C_P(x)$ does the same.

Corollary 5.2.3 (Kolmogorov). $C_T(x) \to \infty$ when $l(x) \to \infty$.

However, it is possible to find natural boundaries for the function $C_T(x)$.

Theorem 5.2.4 (the Recursive Boundary Theorem). a) For any word x, we have

$$C_T(x) \preccurlyeq l(x)$$

b) For any natural number r, the set X of words with m elements has, at least, $m(1 - 2^{-r+1})$ elements with $C_T(x) > \log m - r$.

Proof. a) Let us consider a Turing machine T that realizes the identity function $e(x) = x$. Then by the definition, $C_T(x) = l(x)$ and by the Invariance Theorem, $C_U(x) \le C_T(x) + c_T = C_U(x) \le l(x) + c_T$.

b) Let $k = \lceil \log m \rceil - r$. The number $2^{k+1} - 1$ of different binary strings of length less than or equal to k is an upper bound on the number of different shortest programs that can result in the value k or less for the information size $C_T(x)$. There are exactly $2^{k+1} - 1$ such words. Then at least, $m - 2^{k+1} + 1$ elements from X can have recursive information size $C_T(x)$ larger than or equal to k. From this, we have

$$m - 2^{k+1} + 1 > m - 2^{k+1} = m(1 - 2^{k+1}/m) =$$
$$m(1 - 2^{\lceil \log m \rceil - r + 1}/m) = m(1 - (2^{\lceil \log m \rceil}/m)\, 2^{-r+1}) \ge m(1 - 2^{-r+1})$$
$$\text{as } (2^{\lceil \log m \rceil}/m) \ge 1.$$

Part b) of the theorem is also proved.

Remark 5.2.1. This result implies Theorem 5.2.3.

Remark 5.2.2. As it was explained, studying properties of the function $C_R(x)$, we can consider only the information size $C_T(x)$ for the class **T** of all Turing machines

Remark 5.2.3. If a natural number n is represented by a binary word x_n, then the length $l(x_n)$ of the word x_n is equal to $\lceil \log n \rceil$. That is why, $l(x)$ is traditionally called the logarithmic term. Many relations in algorithmic information theory are true only up to such a logarithmic term. For instance (cf. (Zvonkin and Levin, 1970)), we have

$$\left| I_T(y:x) - I_T(x:y) \right| \preccurlyeq 12 l(C_T(xy))$$

Let us consider computational properties of recursive information size $C_R(x)$. To show that recursive information size $C_R(x)$ cannot be computed by Turing machines, we prove, at first, a stronger result.

Theorem 5.2.5 (the Recursive Lower Boundary Theorem). If $f(x)$ is a partial recursive function whose values are smaller than the corresponding values of $C_R(x)$ whenever defined, then $f(x)$ is a bounded function.

Proof. Let us take a recursive enumeration $h(z)$ of the domain of the function $f(x)$, that is, an enumeration by a Turing machine. As it is proved in the theory of recursive functions (cf., for example, (Rogers, 1987)), such enumeration exists. Then we define the function $g(m)$ to equal to the first number $h(z)$ for which $f(h(z)) = m$. Then by our assumption, $m < C_R(h(z)) = C_R(g(m))$. As $g(m)$ is computable by some Turing machine, by the Invariance Theorem, we have

$$m < C_R(g(m)) < l(m) + c_T$$

This inequality can be true only for a finite number of numbers m because $l(m) < m$. Thus, the values of the function $f(x)$ are bounded by some constant from above.

Theorem is proved.

Corollary 5.2.4 (the Recursive Non-computability Theorem (Kolmogorov, 1965)**).** Recursive information size $C_R(x)$ is not computable by Turing machines or by any kind of recursive algorithms.

Proof. If we assume that $C_R(x)$ is computable by some Turing machine, then the function $f(x) = C_R(x) - 1$ is a partial recursive function, i.e., it is also computable by some Turing machine, and all its values are smaller than the corresponding values of $C_R(x)$. Thus, by Theorem 5.2.5, $f(x)$ is a bounded function. At the same time, by Theorem 5.2.3, $f(x)$ tends to infinity when x grows. This contradiction to the Lower Boundary Theorem (Theorem 5.2.5) completes the proof of the Recursive Non-computability Theorem.

Thus, information size is not recursively computable for classes of recursive algorithms and Turing complete programming languages. However, contrary to popular belief, there exist individual Turing machines, recursive algorithms and programming languages for which information size (Kolmogorov complexity) is directly and easily computable.

Corollary 5.2.5 (Zvonkin and Levin, 1970). If h is an increasing computable function that is defined in a decidable set V and tends to infinity when $l(x) \to \infty$, then for infinitely many elements x from V, we have $h(x) > C(x)$.

Let **P** be a recursively enumerable class of recursive or subrecursive algorithms that contains a universal algorithm U and given a number of an algorithm from **P**, it is possible to build a description (coding) of this algorithm.

Theorem 5.2.6 (the Computability Theorem (Burgin, 1982)). Recursive information size $C_P(x)$ is an inductively computable function, namely, it is computable by some inductive Turing machine of the first order.

Proof. We assume that all algorithms in **P** work with words in the alphabet $\{0, 1\}$ and all such words are enumerated, forming the sequence $x_1, x_2, \ldots, x_n, \ldots$. This is not a restriction as any alphabet can be coded in the alphabet $\{0, 1\}$ (see, for example, (Rogers, 1987)). Such a codification is used in contemporary computers and calculators.

Let us take the universal algorithm U, a recursive numbering $A_1, A_2, \ldots, A_n, \ldots$ of all algorithms in **P**, and build an inductive Turing machine M of the first order such that it contains a copy of the machine U, a counter C that can add 1 to its previous result, a Turing machine G such that given a number i, G builds codes $c(A)$ for all algorithms $A_1, A_2, \ldots, A_i$. Such a Turing machine G exists because **P** is a recursively enumerable class and given a number of an algorithm from **P**, it is possible to build a description (coding) of this algorithm. A technique for building such a machine M is described in (Burgin, 2005). To prove the theorem, we show how the machine M computes $C_P(x)$, given x as its input.

Computation of M goes in cycles according to the number that the counter C produces.

In the first cycle, M simulates U with the input $(c(A_1), x_1)$, which in its turn simulates one step of the algorithm A_1 with the input x_1. If U, and consequently, A_1, gives the result z, then z is compared to x. When $z = x$, the machine M writes the number $l(x_1)$ in its output register (tape). When $z \neq x$, the machine M writes nothing in its output register (tape). When one step of the algorithm A_1 with the input x_1 does not give a result, the machine M writes nothing in its output register (tape). After this the counter C adds 1 to its previous result and M goes to the second cycle.

In the cycle number k, the machine M simulates U with all inputs $(\mathbf{c}(A_i), x_j)$, which in its turn simulates k steps of the algorithm A_i with the input x_j for all $i, j = 1, 2, \ldots, k$. If U, and consequently, A_i, gives the result z, then z is compared to x and $l(x_j)$ is compared to the number m written in the output register of M. When $z = x$ and $l(x_j) < m$, the machine M writes the number $l(x_j)$ in its output register. In all other cases, the machine M writes nothing in its output register. This process is repeated for all $i, j = 1, 2, \ldots, k$. After this the counter C adds 1 to its previous result and M goes to the next cycle.

As $\mathbf{P}$ is a class of recursive or subrecursive algorithms, the machine M will write $C_{\mathbf{P}}(x)$ in its output register and after this the content of the output register will not be changing. By the definition of inductive Turing machines (cf. Appendix B), M computes $C_{\mathbf{P}}(x)$.

Theorem is proved, as x is an arbitrary number.

This result also follows from the fact that limiting recursive functions are realized by inductive Turing machines of the first order (cf. (Burgin, 2005)) and the theorem of Kolmogorov that states that $C_{\mathbf{T}}(x)$ is a limiting recursive function (cf. (Zvonkin and Levin, 1970)).

Information size $C_{\mathbf{R}}(x)$ has a property of continuity, or more exactly, of fuzzy continuity studied in scalable topology (Burgin, 2004; 2006), digital topology (Rosenfeld, 1979; 1986), and utilized in the theory of programming (Hamlet, 2002) and in calculus (Burgin, 2008).

Let q and r be some non-negative real numbers.

Definition 5.2.5. a) A function $f: \mathbf{R} \to \mathbf{R}$ is called (q, r)-continuous at a point $a \in \mathbf{R}$ if for any sequence $l = \{a_i \in \mathbf{R}; i = 1, 2, 3, \ldots\}$, for which a is an q-limit, the point $f(a)$ is an r-limit of the sequence $\{f(a_i) \in \mathbf{R}; i \in \omega\}$.

b) A function $f: \mathbf{R} \to \mathbf{R}$ is called (q, r)-continuous in $\mathbf{R}$ if it is (q, r)-continuous at any point $a \in \mathbf{R}$.

For functions defined only for natural numbers, this definition is transformed to the following one.

Definition 5.2.6. A function $f: \mathbf{N} \to \mathbf{N}$ is called (d, c)-continuous in $\mathbf{N}$ where d and c are natural numbers if the inequality $|x - y| \le d$ implies the inequality $|f(x) - f(y)| \le c$.

Example 5.2.1. The function $f(x) = 7x$ is $(1, 7)$-continuous but it is not $(1, 2)$-continuous.

Remark 5.2.3. In digital topology (Rosenfeld, 1979; 1986), only the case of (1, 1)-continuity is considered. However, for recursive information size, this is not enough.

Theorem 5.2.7 (the Recursive Continuity Theorem (Katseff and Sipser, 1981)). There is a natural number c such that recursive information size $C_T(x)$ is (1, c)-continuous.

<u>Proof</u>. To prove the theorem, it is necessary to show that there is a natural number c such that for any natural number n, we have $|C_T(n + 1) - C_T(n)| \leq c$. As $C_T(n) = C_U(n)$ for some universal Turing machine U, we take this machine U and assume, without loss of generality, that U works with natural numbers represented by binary sequences. Then it is possible to build a Turing machine A such that A simulates U until the result is obtained and U halts. After this, the machine A adds 1 to the result of U. By the definition, we have $C_U(n) = C_A(n + 1)$.

It is also possible to build a Turing machine B such that B simulates U until the result is obtained and U halts. After this, the machine B subtracts 1 from the result of U. By the definition, we have $C_U(n) = C_B(n - 1)$.

By Theorem 5.2.1, $C_U(n) \leq C_A(n) + c_A$ for some natural number c_A and $C_U(n) \leq C_B(n) + c_B$ for some natural number c_B. Let us take $c = \max \{c_A, c_B\}$. Then we have the following relations:

$$C_U(n) = C_A(n + 1) \tag{5.2.2}$$

$$C_U(n + 1) \leq C_A(n + 1) + c \tag{5.2.3}$$

$$C_U(n + 1) = C_B(n) \tag{5.2.4}$$

and

$$C_U(n) \leq C_B(n) + c \tag{5.2.5}$$

In the case when $C_U(n + 1) > C_U(n)$, we take the equality (5.2.2) and subtract it from the inequality (5.2.3), obtaining $C_U(n + 1) - C_U(n) \leq c$. In the case when $C_U(n + 1) < C_U(n)$, we take the equality (5.2.4) and subtract it from the inequality (5.2.5), obtaining $C_U(n) - C_U(n + 1) \leq c$. Consequently, $|C_U(n + 1) - C_U(n)| \leq c$ or $|C_T(n + 1) - C_T(n)| \leq c$.

Theorem is proved.

Note that if we take another universal Turing machine V, it may change the number c (making it larger or smaller), but the result of Theorem 5.2.7 still remains true.

Traditionally (cf., for example, (Li and Vitaniy, 1997)), researchers also consider the exact increasing lower boundary $mC_R(x) = \min \{C_R(y); y \geq x\}$ of the information size $C_R(x)$. Function $mC_R(x)$ is the minorant of $C_R(x)$ and has the following properties.

Theorem 5.2.8 (the Recursive Minorant Theorem) (Kolmogorov).
a) $mC_R(x)$ is a total increasing function;

b) $mC_R(x)$ is not recursively computable;

c) $mC_R(x)$ is inductively computable;

d) $mC_R(x) \rightarrow \infty$ when $l(x) \rightarrow \infty$.

Proof. (a) Since $C_R(x)$ is a total function, $mC_R(x)$ is also a total function. By definition, $mC_{IT}(x)$ is increasing.

(b) If $mC_R(x)$ is a recursively computable function, then by the Lower Boundary Theorem (Theorem 5.2.5), $mC_R(x)$ has to be a constant function. This contradiction completes the proof of the part (b).

Part (c) follows from Lemma 5.2.1.

Theorem is proved.

There are connections and similarities between Shannon entropy and recursive information size. These relations demonstrate that recursive information size is a kind of information measures related to statistical measures of information.

Let us consider a word $x = y_1 y_2 \ldots y_m$ with $l(y_1) = l(y_2) = l(y_m) = r$. In addition, the frequency of occurring in x is given for all words of the length r. Namely, the frequency of the ith word is q_i (here $i = 1, 2, 3, \ldots, 2^r$). Taking these frequencies, it is possible to find Shannon entropy of the word x. Namely,

$$H[x] = -\sum_{i=1}^{2^r} q_i \log q_i$$

Theorem 5.2.9 (Kolmogorov). $C_T(x) \preccurlyeq n(H[x] + g(n))$ where $g(n) = c_r \cdot ((\ln n)/n)$ and c_r is a constant.

Properties of logarithm imply that when n tends to infinity, $g(n)$ tends to zero. It means that when n grows, we can ignore, to some extent, the second term in the right part of the inequality.

It is possible to find a proof of this result in (Zvonkin and Levin, 1970) or in (Li and Vitaniy, 1997). It is also possible to show that a closer link between Shannon entropy and recursive information size cannot be established.

Another similarity is related to inequalities that both measures of information satisfy.

Theorem 5.2.10 (the Linear Connections Theorem). Any linear inequality that is true for recursive information size $C_R(x)$ is also true for Shannon entropy up to an additive logarithmic term and vice versa.

It is possible to find a proof of this result in (Hammer, et al, 2000).

However, there are properties of information size and Shannon entropy that do not coincide. For instance, Muchnik and Vereshchagin (2006) proved that this is the case for $\forall\exists$-assertions, exhibiting an example where the formal analogy between Shannon entropy and information size (Kolmogorov complexity) fails.

Benci, et al, (2001) introduced a measure of information content (or more correctly, of information size) utilizing compression algorithms.

Definition 5.2.7. A data compression algorithm Z is called *lossless* if there is an algorithms D such that for any string of symbols (word) w, we have $D(Z(w)) = w$.

Informally, a lossless data compression algorithm preserves enough information to reconstruct the original string.

Proposition 5.2.1. Any lossless data compression algorithm determines a one-to-one mapping on the set of all strings in its working alphabet.

Let us consider a string of symbols (word) w and a lossless data compression algorithm Z.

Definition 5.2.8. The *information content* (*information size*) of the word w with respect to the algorithm Z is equal to the length $l(Z(w))$ of the word $Z(w)$.

Usually information content of a word is defined with respect to optimal lossless data compression algorithms. It is possible to find examples of optimal lossless data compression algorithms in (Benci, et al, 2001).

As in the case with algorithmic information via Kolmogorov complexity, it is more relevant to call the length $l(Z(w))$ of the word $Z(w)$

by the name the *compression information size* of the word *w* with respect to the algorithm *Z*.

The advantage of the compression information size in comparison with the conventional recursive information size is that the compression information size is a recursively computable function, while the conventional recursive information size is not recursively computable (cf. Corollary 5.2.4).

5.3. Algorithmic Information Theory based on Inductive Algorithms: Inductive Approach

> *Life is like riding a bicycle.*
> *To keep your balance you must keep moving*
> Albert Einstein (1879–1955)

Taking a class **SR** of super-recursive algorithms (e.g., inductive Turing machines, limit partial recursive functions, grid automata and so on), we can define super-recursive information size $C_{SR}(x)$ (Burgin, 2006).

Definition 5.3.1. The *information size* $C_A(x)$ of an object (word) *x* with respect to a super-recursive algorithm *A*, e.g., an inductive Turing machine, is defined as

$$C_A(x) = \min \{l(p); \ A(p) = x\}$$

when there is a word *p* such that $A(p) = x$; otherwise $C_A(x)$ is not defined.

For some classes of super-recursive algorithms and in particular, for the class **IT** of all inductive Turing machines of the first order, it is possible to find an invariant measure that characterizes strings of symbols with respect to the whole class **IT**, i.e., invariant in the class of inductive information size $C_A(x)$ where *A* is an arbitrary inductive Turing machine of the first order. In what follows, inductive Turing machine always means inductive Turing machine of the first order, inductively computable means computable by an inductive Turing machine of the first order, and inductively decidable means decidable by an inductive Turing machine of the first order.

Theorem 5.3.1 (the Inductive Optimality Theorem (Burgin, 1982)**).** There is an inductive Turing machine *U* such that for any inductive Turing machine M there is a constant c_T such that for all words x, we have

$$C_U(x) \leq C_M(x) + c_M$$

or

$$C_U(x) \preccurlyeq C_M(x)$$

Proof. Let us take a universal inductive Turing machine U. By its definition (cf. Appendix B), given a code $c(M)$ of any inductive Turing machine M and a word p as input to U, the machine U computes the same result as M given the word p or U gives no result when M gives no result on input p. For all words p, the pair $< c(M), p >$ can be coded by one word $d(c(M), p)$ so (cf. (Burgin, 2005)) that the length of the code will be equal to the length of the word p plus some constant c_M. Thus, if $M(p) = x$, then $U(p) = x$. Consequently, if $C_M(x) = l(p)$, then $C_U(x) \leq l(p) + c_M$. This implies the necessary inequality as the constant c_M depends only on the inductive Turing machine M.

Theorem is proved.

The function $C_U(x)$ is additively optimal in the class of all inductive information sizes $C_M(x)$.

Corollary 5.3.1 (the Inductive Invariance Theorem (Burgin, 1982)). For any two universal inductive Turing machines U and V, the corresponding information sizes $C_U(x)$ and $C_V(x)$ are additively equivalent, i.e., $C_U(x) \asymp C_V(x)$.

Such an Invariance Theorem makes it possible to define information size not only for separate super-recursive algorithms (e.g., inductive Turing machines), but also for the class **IT** of all inductive Turing machines of the first order and study inductive information size.

Definition 5.3.2. The *inductive information size* $C_{IT}(x)$ of an object (word) x is defined as

$$C_{IT}(x) = \min \{l(p); \ U(p) = x\}$$

when there is a word p such that $U(p) = x$ and U is a universal inductive Turing machine of the first order; otherwise $C_A(x)$ is not defined.

In a similar way, we define super-recursive information size $C_{SR}(x)$.

Here we assume, without loss of generality, that all considered inductive Turing machines are working with natural numbers that are represented by words in the alphabet $\{1, 0\}$.

Theorem 5.3.2 (the Super-recursive Totality Theorem (Burgin, 1990)). $C_{SR}(x)$ is a total function on X^*.

The proof is similar to the proof of the Theorem 5.2.2 and is based on the property that super-recursive algorithms can simulate whatever Turing machines can do (compute). As Turing machines can compute/realize the identity function, the same is true for any class of super-recursive algorithms.

Corollary 5.3.2 (Burgin, 1990). $C_{IT}(x)$ is a total function on N (on the set X^* of all words in some alphabet).

Although the inductive information size $C_{IT}(x)$ of x often is essentially smaller than the recursive information size $C_R(x)$ of x, the function $C_{IT}(x)$ also tends to infinity when the length of the word x grows without bounds.

Theorem 5.3.3 (the Inductive Unboundedness Theorem (Burgin, 1990)). $C_{IT}(x) \to \infty$ when $l(x) \to \infty$.

The proof is similar to the proof of the Theorem 5.2.3.

However, $C_{IT}(x)$ grows slower than any total increasing inductively computable function.

Theorem 5.3.4 (the First Inductive Lower Boundary Theorem). If f is a total strictly increasing inductively computable function, then for infinitely many elements x, we have $f(x) > C_{IT}(x)$.

Proof. Let us assume that there is some element z such that for all elements y that are larger than z, we have $f(x) \le C_{IT}(x)$. Because $f(x)$ is an inductively computable function, there is an inductive Turing machine T that computes $f(x)$. It is done in the following way. Given a number x, the machine T makes the first step, producing $f_1(x)$ on its output tape. Making the second step, the machine T producing $f_2(x)$ on its output tape. After n steps, T has $f_n(x)$ on its output tape. Since the function is inductively computable, this process stabilizes on some value $f_n(x) = f(x)$, which is the result of computation with the input x. Taking the function $h(m) = \min \{x; f(x) \ge m\}$, we construct an inductive Turing machine M that computes the function $h(x)$.

The inductive Turing machine M contains a copy of the machine T. Utilizing this copy, M finds one after another the values $f_1(1), f_1(2), ..., f_1(m + 1)$ and compares these values to m. Then M writes into the output tape the least x for which the value $f_1(x)$ is larger than or equal to m. Then

M finds one after another the values $f_2(1)$, $f_2(2)$, ..., $f_2(m + 1)$ and compares these values to m. Then M writes into the output tape the least x for which the value $f_2(x)$ is larger than or equal to m. This process continues until the output value of M stabilizes. It happens for any number m due to the following reasons. First, $f(x)$ is a total function, so all values $f_i(1)$, $f_i(2)$, ..., $f_i(m + 1)$ after some step $i = t$ become equal to $f(1), f(2), ... , f(m + 1)$. Second, $f(x)$ is a strictly increasing function, that is, $f_i(m + 1) > m$. In such a way, the machine M computes $h(m)$. Since m is an arbitrary number, the machine M computes the function $h(x)$.

Since for all elements y that are larger than z, we have $f(y) \leq C_{IT}(y)$, there is an element m such that $IC(h(m)) \geq f(h(m))$ and $f(h(m)) \geq m$ as $f(x)$ is a strictly increasing function and $h(m) = \min \{x; f(x) \geq m\}$. By definition, $C_T(h(m)) = \min \{l(x); T(x) = h(m)\}$. As $T(m) = h(m)$, we have $C_T(h(m)) \leq l(m)$. Thus, $l(m) \geq C_T(h(m)) \succcurlyeq C_{IT}(h(m)) \geq m$. However, it is impossible that $l(m) \succcurlyeq m$. This contradiction concludes the proof of the theorem.

Remark 5.3.1. Although Theorem 5.3.4 can be deduced from Theorem 5.3.5, we give an independent proof because it demonstrates another technique, which displays essential features of inductive Turing machines.

Corollary 5.3.3. If a total inductively computable function $f(x)$ is less than $C_{IT}(x)$ almost everywhere, then $f(x)$ is a bounded function.

We can prove a stronger statement than Theorem 5.3.4. To do this, we assume for simplicity that inductive Turing machines are working with words in some finite alphabet and that all these words are well ordered, that is, any set of words contains the least element. It is possible to find such orderings, for example, in (Li and Vitaniy, 1997).

Theorem 5.3.5 (the Second Inductive Lower Boundary Theorem). If h is an increasing inductively computable function that is defined in an inductively decidable set V and tends to infinity when $l(x) \to \infty$, then for infinitely many elements x from V, we have $h(x) > C_{IT}(x)$.

Proof. Let us assume that there is some element z such that for all elements x that are larger than z, we have $h(x) \leq C_{IT}(x)$. Because $h(x)$ an inductively computable function, there is an inductive Turing machine T that computes $h(x)$. Taking the function $g(m) = \min \{x; h(x) \geq m$ and

$x \in V$}, we construct an inductive Turing machine M that computes the function $g(x)$.

As V is an inductively decidable set, there is an inductive Turing machine H that given an input x, produces 1 when $x \in V$, and produces 0 when $x \notin V$. It means that H computes the characteristic function $c_V(x)$ of the set V.

The inductive Turing machine M contains a copy of the machine H and a copy of the machine T. Utilizing this copy of T, the machine M computes the value $h_1(1)$ and compares it to m. Utilizing this copy of H, the machine M computes the value $c_{V1}(1)$. If $h_1(1)$ is larger than m and $c_{V1}(1) = 1$, then M writes 1 into the output tape. Otherwise, M writes nothing into the output tape. After this, M finds the values $h_2(1)$ and $h_2(2)$ and compares these values to m. Concurrently, M finds the values $c_{V2}(1)$ and $c_{V2}(2)$. Then M writes into the output tape the least x for which the value $h_1(x)$ is larger than or equal to m and at the same time, $c_{V2}(x) = 1$. This process continues. Making cycle i of the computation, M computes the values $h_i(1)$, $h_i(2)$, ..., $h_i(i)$ and compares these values to m. We remind here that $h_i(j)$ is the result of i steps of computation of T with the input j. Concurrently, M computes the values $c_{Vi}(1)$, $c_{Vi}(2)$, ..., $c_{V\,i}(i)$. Then M writes into the output tape the least x for which the value $h_i(x)$ is larger than or equal to m and at the same time, $c_{Vi}(x) = 1$. Such cycle is repeated until the output value of M stabilizes.

Each value $c_{Vi}(x)$ stabilizes at some step t because $c_V(x)$ is a total inductively computable function. In a similar way, each value $h_i(x)$ stabilizes at some step q because $h(x)$ is an inductively computable function defined for all $x \in V$. Thus, after this step $p = \max \{q, t\}$, the value $h_i(x)$ becomes equal to the value $h(x)$. In addition, there is such a step t when a number n is found for which $h(n) \geq m$. After this step, only such numbers x can go to the output tape of M that belong to V and are less than or equal to n.

This happens for any given number m due to the following reasons. First, $h(x)$ is defined for all elements from V total function, so those values $h_i(1)$, $h_i(2)$, ..., $h_i(m + 1)$ for which the argument of h_i belongs to V after some step $i = r$ become equal to $h(1)$, $h(2)$, ..., $h(m)$. Second, $h(x)$ is an increasing function that tends to infinity.

This shows that the whole process stabilizes and by the definition of inductive computability, the machine M computes $g(m)$. Since m is an arbitrary number, the machine M computes the function $g(x)$.

To conclude the proof, we repeat the reasoning from the proof of Theorem 5.3.4. Since for all elements y that are larger than z, we have $f(x) \leq C_{IT}(x)$, there is an element m such that $C_{IT}(g(m)) \geq h(g(m))$ and $h(g(m)) \geq m$ as $h(x)$ is an increasing function and $g(m) = \min \{x; h(x) \geq m\}$. By the definition, $C_T(g(m)) = \min \{l(x); T(x) = g(m)\}$. As $T(m) = g(m)$, we have $C_T(g(m)) \leq l(m)$. Thus, $l(m) \geq C_T(h(m)) \succcurlyeq C_{IT}(h(m)) \geq m$. However, it is impossible that $l(m) \succcurlyeq m$. This contradiction concludes the proof of the theorem.

Corollary 5.3.4. If an inductively computable function $f(x)$ is defined in an inductively decidable set V and has values smaller than the corresponding values of $C_{IT}(x)$ everywhere on V, then $f(x)$ is a bounded function.

Corollary 5.3.5. If h is a total increasing inductively computable function that tends to infinity when $l(x) \to \infty$, then for infinitely many elements x, we have $h(x) > C_{IT}(x)$.

Corollary 5.3.6. If h is an increasing inductively computable function that is defined in an recursive set V and tends to infinity when $l(x) \to \infty$, then for infinitely many elements x from V, we have $h(x) > C_{IT}(x)$.

Since the composition of two increasing functions is an increasing function and the composition of a recursive function and an inductively computable function is an inductively computable function, we have the following result.

Corollary 5.3.7. If $h(x)$ and $g(x)$ are increasing functions, $h(x)$ is inductively computable and defined in an inductively decidable set V, $g(x)$ is a recursive function, and they both tend to infinity when $l(x) \to \infty$, then for infinitely many elements x from V, we have $g(h(x)) > C_{IT}(x)$.

Corollary 5.3.8 (the Inductive Non-computability Theorem (Burgin, 1995)**).** The function $C_{IT}(x)$ is not inductively computable. Moreover, no inductively computable function $f(x)$ defined for an infinite inductively decidable set of numbers can coincide with $C_{IT}(x)$ in the whole of its domain of definition.

Indeed, if we assume that $C_{IT}(x)$ is computable by some inductive Turing machine, then the function $f(x) = C_{IT}(x) - 1$ is an inductively computable function, i.e., it is also computable by some inductive Turing machine, and all its values are smaller than the corresponding values of $C_{IT}(x)$. This contradicts the Second Lower Boundary Theorem and completes the proof of the Non-computability Theorem for inductive Turing machines of the first order.

By Theorem 5.2.6, the function $C_R(x)$ is inductively computable. Consequently, Theorem 5.3.5 implies the following result.

Theorem 5.3.6 (the Efficiency Theorem (Burgin, 2005)**).** For any increasing recursive function $h(x)$ that tends to infinity when $l(x) \to \infty$ and any inductively decidable set V, there are infinitely many elements x from V for which $h(C_R(x)) > C_{IT}(x)$.

Corollary 5.3.9. In any inductively decidable set V, there are infinitely many elements x for which $C_R(x) > C_{IT}(x)$.

Corollary 5.3.10. In any recursive set V, there are infinitely many elements x for which $C_R(x) > C_{IT}(x)$.

Corollary 5.3.11. In any inductively decidable (recursive) set V, there are infinitely many elements x for which $\ln_2(C_R(x)) > C_{IT}(x)$.

If $\ln_2(C_R(x)) > C_{IT}(x)$, then $C(x) > 2^{IC(x)}$. At the same time, for any natural number k, the inequality $2^n > k \cdot n$ is true almost everywhere. This and Corollary 5.3.8 imply the following result.

Corollary 5.3.12. For any natural number k and in any inductively decidable (recursive) set V, there are infinitely many elements x for which $C_R(x) > k \cdot C_{IT}(x)$.

Corollary 5.3.13. There are infinitely many elements x for which $C_R(x) > C_{IT}(x)$.

Corollary 5.3.14. For any natural number a, there are infinitely many elements x for which $\ln_a(C_R(x)) > C_{IT}(x)$.

Corollary 5.3.15. There are infinitely many elements x for which $\ln_2(C_R(x)) > C_{IT}(x)$.

All these results show that, with respect to a natural extension of the Kolmogorov complexity or recursive information size, inductive Turing machines may be much more efficient than any kind of recursive algorithms. Informally, it means that in comparison with recursive algorithms, super-recursive programs for solving the same problem might be shorter, have lower branching (i.e., less instructions of the form IF *A* THEN *B* ELSE *C*), make less reversions and unrestricted transitions (i.e., less instructions of the form GO TO *X*) for infinitely many problems solvable by recursive algorithms.

Another proof of higher efficiency of inductive Turing machines in comparison with conventional Turing machines and other recursive algorithms is given by their ability to solve such problems that cannot be solved by conventional Turing machines. For instance, Lewis (2001) demonstrates limits of software estimation, using boundaries implied by the theory of recursive information size (Kolmogorov complexity). Inductive Turing machines are able to make many estimations that are inaccessible for conventional Turing machines. It is possible only because inductive Turing machines have lower recursive information size for those problems than conventional Turing machines.

Theorem 5.3.7 (the Inductive Computability Theorem). The function $C_{IT}(x)$ is computable by an inductive Turing machine of the second order.

Proof. We consider only inductive Turing machines of the first order that work with words in the alphabet $\{0, 1\}$ and all such words are enumerated, forming the sequence $x_1, x_2, ..., x_n, ...$ This is not a restriction as any alphabet can be coded in the alphabet $\{0, 1\}$.

Let us take the universal inductive Turing machine of the first order *U* (cf. Appendix B), a recursive numbering $M_1, M_2, ..., M_n, ...$ of all inductive Turing machine of the first order that work with words in the alphabet $\{0, 1\}$, and build an inductive Turing machine *M* of the second order such that it contains a copy of the machine *U* , a counter *C* that can add 1 to its previous result, a Turing machine *G* such that given a number *i*, *G* builds codes $c(M_r)$ for all algorithms $M_1, M_2, ..., M_i$. A technique for building such a machine *M* is described in (Burgin, 2005). To prove the theorem, we show how the machine *M* computes $C_{IT}(x)$, given *x* as its input.

Computation of *M* goes in cycles according to the number that the counter *C* produces.

In the first cycle, *M* simulates *U* with the input $(c(M_1), x_1)$, which in turn, simulates the machine M_1 with the input x_1 . Because *M* has the second order, it can get the result of *U* whenever this result is produced

(Burgin, 2005). If U, and consequently, M_1, gives the result z, then z is compared to x. When $z = x$, the machine M writes the number $l(x_1)$ in its output register (tape). In all other cases, the machine M writes nothing in its output register. After this the counter C adds 1 to its previous result and M goes to the second cycle.

In the cycle number k, the machine M simulates U with all inputs $(\mathbf{c}(A_i), x_j)$, which in its turn simulates the machine M_i with the input x_j for all $i, j = 1, 2, ..., k$. If U, and consequently, A_i, gives the result z, then z is compared to x and $l(x_j)$ is compared to the number m written in the output register of M. When $z = x$ and $l(x_j) < m$, the machine M writes the number $l(x_j)$ in its output register. In all other cases, the machine M writes nothing in its output register. This process is repeated for all $i, j = 1, 2, ..., k$. After this the counter C adds 1 to its previous result and M goes to the next cycle.

In such a way, the machine M will write $C_{IT}(x)$ in its output register in some cycle and after this the content of the output tape stops changing. By the definition of inductive Turing machines (cf. Appendix B), M computes $C_{IT}(x)$.

Theorem is proved as x is an arbitrary number.

As in the case of recursive information size (Section 5.2), inductive information size is also a fuzzy continuous function.

Theorem 5.3.8 (the Inductive Continuity Theorem). There is a natural number c such that inductive information size $C_{IT}(x)$ is $(1, c)$-continuous.

The proof is similar to the proof of the Theorem 5.2.7 and is based on the Inductive Optimality Theorem (Theorem 5.3.1).

Note that if in the proof of Theorem 5.3.8, we take two different universal Turing machines, we may have two different numbers c, but the result of Theorem 5.3.8 still remains true.

As in the case of the recursive information size, it is interesting to consider the exact increasing lower boundary $mC_{IT}(x) = \min \{C_{IT}(y); y \geq x\}$ of the inductive information size $C_{IT}(x)$. Function $mC_{IT}(x)$ is the minorant of $C_{IT}(x)$ and has the following properties.

Theorem 5.3.9 (the Inductive Minorant Theorem). (a) $mC_{IT}(x)$ is a total increasing function;

(b) $mC_{IT}(x)$ is not inductively computable;

(c) $mC_{IT}(x) \to \infty$ when $l(x) \to \infty$.

<u>Proof</u>. (a) Since $C_{IT}(x)$ is a total function, $mC_{IT}(x)$ is also a total function. By definition, $mC_{IT}(x)$ is increasing.

(b) If $mC_{IT}(x)$ is an inductively computable function, then by Theorem 5.3.5, for infinitely many elements x, we have $mC_{IT}(x) > IC(x)$. However, by the definition of $mC_{IT}(x)$, we have $mC_{IT}(x) \leq C_{IT}(x)$ everywhere. This contradiction completes the proof of the part (b).

Part (c) follows from Lemma 5.2.1.

Theorem is proved.

5.4. Conditional Information Size as a Relative Information Measure: Relativistic Approach

> *It is rumored that someone once asked Dr. Bellman*
> *how to tell the exercises apart from the research problems,*
> *and he replied: "If you can solve it, it is an exercise;*
> *otherwise it's a research problem."*
> Donald E. Knuth, *The Art of Computer Programming*

To define *conditional information size* (conditional Kolmogorov complexity), we consider algorithms that work with two input words/strings.

Definition 5.4.1. The *relative* to a given word y *information size* $C_A(x|y)$ of an object (word) x with respect to an algorithm A is defined as

$$C_A(x|y) = \min \{l(p); A(p, y) = x \}$$

when there is a word p such that $A(p, y) = x$; otherwise $C_A(x|y)$ is not defined.

Function $C_A(x|y)$ is called conditional information size.

Definition 5.4.2. The *relative* to a given word y *recursive information size* $C_T(x|y)$ of an object (word) x is defined as

$$C_T(x|y) = \min \{l(p); U(p, y) = x\}$$

when there is a word p such that $U(p, y) = x$ and U is a universal Turing machine; otherwise $C_T(x|y)$ is not defined.

It is necessary to remark that the function $C_T(x|y)$ is often called conditional algorithmic or Kolmogorov complexity and is denoted by $C(x|y)$ or $K(x|y)$. The function $C_T(x|y)$ intuitively represents the quantity of information necessary for computation/construction of x when y is already given. Thus, it is called the *information size* of x with respect to y.

Absolute information size is a particular case of relative information size, namely:

$$C_T(x) = C_T(x|\Lambda) = \min \{l(p); \ U(p, \Lambda) = x\}$$

It is possible to understand the conditional information size $C_T(x|y)$ as the algorithmic (information) distance from y to x. This distance reflects complexity of converting y to x by means of algorithms from **T**.

In a similar way, we define conditional inductive information size, or information size in some general class of algorithms.

Definition 5.4.3. The *relative* to a given word y *inductive information size* $C_{IT}(x|y)$ of an object (word) x is defined as

$$C_{IT}(x|y) = \min \{l(p); U(p, y) = x\}$$

when there is a word p such that $U(p, y) = x$ and U is a universal inductive Turing machine of the first order; otherwise $C_{IT}(x|y)$ is not defined.

Many properties of the conditional information size $C_T(x|y)$ are similar to properties of the absolute information size $C_T(x)$. That is why, we give them without proofs.

Theorem 5.4.1 (The Relative Optimality Theorem). There is a (inductive) Turing machine U such that for any (inductive) Turing machine T there is a constant c_T such that for all words x, we have

$$C_U(x|y) \le C_T(x|y) + c_T \qquad (5.4.1)$$

The function $C_U(x|y)$ is *additively optimal*, that is, optimal up to an additive constant, in the class of all functions $C_T(x)$.

Corollary 5.4.1 (The Relative Invariance Theorem). For any two universal (inductive) Turing machines U and V, the corresponding information sizes $C_U(x|y)$ and $C_V(x|y)$ are additively equivalent.

Theorem 5.4.2 (The Relative Totality Theorem). If there is an algorithm $A(p, y) = x$ in a class **P** of algorithms the range of which for a given y is X^*, then as function of x, $C_P(x|y)$ is a total function on X^*.

Many other properties of conditional information size are similar to properties of absolute information size (absolute Kolmogorov complexity). In addition, there are natural relations between conditional and absolute information sizes.

Theorem 5.4.3 (The Upper Boundary Theorem). There is a constant c such that for all words x and y, we have

$$C_T(x|y) \leq C_T(x) + c \qquad (5.4.2)$$

and

$$C_{IT}(x|y) \leq C_{IT}(x) + c \qquad (5.4.3)$$

Proof. Taking a universal Turing machine U, we can build a Turing machine T such that for all words x and y, T computes output x on input (z, y) if and only if U computes output x on input (z, ε) where ε is the empty word. Then $C_T(x|y) = C_T(x)$. By Theorem 5.4.1, there is a constant c_T such that for all words x, we have $C_T(x|y) \leq C_T(x|y) + c_T = C_T(x) + c_T$.

In a similar way, taking a universal inductive Turing machine U, we can build an inductive Turing machine T such that for all words x and y, T computes output x on input (z, y) if and only if U computes output x on input (z, ε) where ε is the empty word. Then $C_T(x|y) = C_{IT}(x)$. By Theorem 5.4.1, there is a constant c_{IT} such that for all words x, we have $C_{IT}(x|y) \leq C_T(x|y) + c_{IT} = C_{IT}(x) + c_{IT}$.

Theorem is proved as we can take $c = \max \{c_T, c_{IT}\}$.

Conditional information size allows one to define an algorithmic measure of information content. According to ideas of Kolmogorov (1968), information (or better to say, the *algorithmic measure of information* or *information content*) $I_Q(y: x)$ in a word y about a word x with respect to a class of algorithms **Q** is equal to the difference of the absolute information size of x and its information size relative to y, formally, we have

$$I_Q(y: x) = C_Q(x) - C_Q(x|y)$$

This function shows to what extent information in y changes the information size of x. This perfectly correlates with the general theory of information where information cause change and measure of information is a measure of such a change.

Theorem 5.4.3 implies the following result.

Theorem 5.4.4 (The Nonnegativity Theorem). $I_Q(y: x)$ is a nonnegative function up to some constant, i.e.,

$$I_Q(y: x) \geqslant 0$$

In the statistical theory of information, one of the fundamental properties is symmetry. For algorithmic information, we have a similar property where symmetry is true up to a logarithmic term.

Theorem 5.4.5 (The Symmetry Theorem) (Kolmogorov, Levin). $|I_T(y: x) - I_T(x: y)| \leqslant O(\min\{\log C_T(x), \log C_T(y))$.

It is possible to find a proof of this result in (Li and Vitaniy, 1997).

However, all relations in the algorithmic theory of information are usually considered up to a constant term. That is why many experts thought that symmetry up to a logarithmic term is insufficient (cf., for example, (Gammarman and Vovk, 1999)). A solution to this problem was given by Gács (1974), Levin (1974), and Chaitin (1975) who introduced prefix complexity $K(x)$ or, in our terms, $(\mathbf{P}, l(p))$-information size $C_P(xy)$ where $\mathbf{P}$ is the class of all prefix-free Turing machines. In this case, we have

$$I_P(y: x) = C_P(x) - C_P(x|y, C_P(x))$$

and

$$I_P(y: x) \asymp I_P(x: y)$$

5.5. Dual Complexity and Information Measures: Axiomatic Approach in Algorithmic Information Theory

> *The truth knocks on the door and you say,*
> *"Go away, I am looking for the truth,"*
> *and so it goes away. Puzzling.*
> Robert Pirsig, *Zen and the Art of Motorcycle Maintenance*

We see that information size, or algorithmic complexity, can be defined for different classes of algorithms, resulting in different measures. However, all these measures are constructed by a similar technique. As a result, it is possible to axiomatize this approach. The result of this axiomatization is called dual complexity measures (Burgin, 2005). As before, we are going to call these measures by the name *dual information size* as they reflect information necessary to compute (construct) a given object. These measures give much more opportunities to estimate information size of words and infinite strings than conventional types of information size (Kolmogorov complexity).

Dual information size is a property of objects that are constructed and processed by algorithms. On the other hand, it is possible to interpret dual information size as a property of classes of algorithms. Here we consider only static dual information size for algorithms. It measures information necessary to construct (build) an object by means of some system of algorithm **A**. That is why, we call the dual to a static complexity measure (static direct information size) α complexity measure of an object x by the name $(\mathbf{A}, \alpha)$-*information size* of x.

Dual information size is constructed from direct information size by the minimization operation. For instance, a natural direct information size for algorithms/programs is the length of their description (symbolic representation). A natural direct information size for data is also the length of their description. It is possible to measure length in such natural units of information as bits and bytes. When we take the dual to this measure in the class of recursive algorithms, we obtain Kolmogorov complexity or recursive information size considered in previous sections.

Thus, to get an axiomatic description of dual information size, we need to give an axiomatic description of direct complexity measures. In this, we follow the approach suggested by Blum (1967; 1967a) and further developed by Burgin (1982; 2005).

All kinds of direct information sizes are divided into three classes: static information size, functional information size, and processual information size. *Static information size* depends only on an algorithm/program that is measured. Direct static information size usually reflects information in the algorithm/program representation. Examples of a direct static information size are the length or numbers of algorithms/programs in a Gödel enumeration. As it is discussed in Section 5.1, the length of a text (of an algorithm) gives information in bits. If we say that a memory has the capacity 10 gigabytes, it means that it is possible to store 8×10^9 bits of information in this memory. *Functional information size* depends both on an algorithm/program that is measured and on the input. Examples of a direct functional information size are such popular measures as time of a computation or space used in a computation. *Processual information size* depends on an algorithm/program, its realization, and on the input. Examples of a direct processual information size are to process branching or the number of data transactions between memories of different type used in this process.

In contrast to static information size, functional information size and processual information size are dynamic characteristics. Here we consider only static information size as even this class gives us much more types of dual information size than the classical Kolmogorov complexity or recursive information size.

All types of direct static information sizes of algorithms are functions of the form $c: A \to N$ where $A = \{A_i; i \in I\}$ is a class of algorithms (programs or automata/machines) and N is the set of all natural numbers. Such measures are direct as they estimate algorithms, programs or machines. We introduce several axioms to distinguish definite classes of information size and to characterize their properties.

Very often, one algorithm A can be a part/component of another algorithm B. This relation between algorithms is denoted by $A \subseteq B$.

Composition Axiom. If $A \subseteq B$, then $c(A) \leq c(B)$.

Informally, the Computation Axiom means that the direct static information size of a system component cannot be larger than the direct static information size of the whole system.

Let $B = \{B_j; j \in J\}$ be a class of algorithms.

Computational Axiom. The function $c(A)$ is total and computable in B.

Recomputational Axiom. For any number n, it is possible to compute all indices i such that $c(A_i) = n$.

Reconstructibility Axiom. For any number n, it is possible to build all algorithms A from A for which $c(A) = n$.

Weak Reconstructibility Axiom. For any number n, it is possible to compute indices of all algorithms A_i from A for which $c(A_i) = n$.

Cofiniteness Axiom. The set $c^{-1}(n)$ is finite for all numbers n from N.

Lemma 5.5.1. If a function $c(A)$ satisfies the Cofiniteness Axiom and the set **A** is infinite, then this function tends to infinity.

Remark 5.5.1. A function $c(A)$ can satisfy the Cofiniteness Axiom, but it can be non-computable even in such powerful class as the class **R** of recursive algorithms, in particular, the class of all Turing machines. However, $Sc(i)$ can be inductively computable (Burgin, 2005).

Definition 5.5.1. a) A function $Sc: \mathbf{A} \to N$ is called a *direct static information size* of algorithms from **A** if it satisfies the Composition Axiom.

b) A direct static information size is called *reconstructible* (computable, *recomputable*, *weakly reconstructible*, *cofinite*) if it satisfies the Reconstructibility (Computational, Recomputational, Weak Reconstructibility or Cofiniteness, respectively) Axiom.

Remark 5.5.2. This approach to information size of algorithms/ programs reflects the condition that direct static information size depends on structural features of algorithms/programs.

Remark 5.5.3. When all algorithms from **A** have indices (are enumerated by natural numbers), it is possible to consider a function $Sc: I \to N$ (a function $Sc: N \to N$) instead of the function $Sc: \mathbf{A} \to N$. We also call this function a *direct static information size* of algorithms from **A** when it satisfies the Composition Axiom.

Example 5.5.1. Let **A** consists of algorithms generated by a Turing machine W with two input tapes. One tape is used for data, while the content of the second tape is considered as a program for computation. Each program for the machine W is an algorithms from **A**. Then the length $l(p)$ of this program p is a computable, reconstructible, cofinite direct static information size of Turing machines from **A**.

Example 5.5.2. Let **A** consists of all programs, which are written in some programming language (e.g., Java, C^{++} or FORTRAN). Then the length $l(p)$ of a program p as the number of letters (or as the number of words) in p is a direct static information size, which is computable, reconstructible, and cofinite.

Both these measures satisfy even a stronger form of the Composition Axiom.

Additive Composition Axiom. If $A, C \subseteq B$, then $c(B) \geq c(A) + c(B)$.

Strong Composition Axiom. If A, $C \subseteq B$, then $\mathbf{c}(B) > \max \{\mathbf{c}(A),$ $\mathbf{c}(C)\}$.

Remark 5.5.5. Additive Composition Axiom is not necessarily satisfied when algorithms/programs allow nesting. The following example demonstrates how nesting violates this axiom.

Example 5.5.3. Let us consider the following program F:

WHILE P DO A

A: IF Q THEN GO TO C ELSE D

C: IF R THEN GO TO E ELSE H

Thus, we have the program F and its parts A and C. Taking such static complexity as the length l of a program that is measured in the number of symbols in the program, we have $l(F) = 47$, $l(A) = 36$, and $l(C) = 17$. Consequently, $l(F) = 47 < l(A) + l(C) = 53$. This violates Additive Composition Axiom.

Software metrics give different examples of direct static information size.

Example 5.5.4. When the length of a line in a program is bounded (and this is true for all programming languages as compilers demand this restriction), then the software metrics LOC (cf. (Halstead, 1977)) is a finite computable reconstructible direct static information size.

Example 5.5.5. Describing a program formally as a sequence of operators and operands, we see that the length of program $N(P)$ (Halstead, 1977) is also a direct static information size, namely, the length $l(P)$ of a program. For a programming language in which the numbers n_1 of the unique operators and n_2 of the unique operands are finite, $N(P)$ is a cofinite reconstructible direct static information size. However, some languages (at least, potentially) operate with infinite alphabets of operands, for example, with all real numbers or all words in some alphabet. There are also theoretical models in which there are infinitely many unique operators. In such cases, $N(P)$ is not a finite direct static information size. If the sets of operands and operators are computable, then this measure is also computable.

Example 5.5.6. When it is defined, the volume of the program $V(P)$ (Halstead, 1977) is always a cofinite reconstructible direct static information size.

Example 5.5.7. Representing a program formally as a structure of operators and operands, we can demonstrate that the cyclomatic number $V(P)$ (McCabe, 1976) is a computable reconstructible direct static information size. However, this measure does not satisfy Cofiniteness Axiom as, for instance, it is possible to build infinitely many programs with only one cycle.

It is necessary to remark that some nice properties of the traditional definition of axiomatic static complexity measures representing direct static information size are lost in the suggested approach. For instance, axiomatic static complexities are not closed with respect to enumerations. It means that when we enumerate algorithms and take the induced function on these numbers, the new function does not necessarily satisfy the Composition Axiom.

Example 5.5.8. Let **A** consists of all Turing machines. Each Turing machine T has a description/coding $c(T)$. Then the length $l(c(T))$ of this description $c(T)$ is a direct static information size of Turing machines.

Definition 5.5.2. A class $\mathbf{A} = \{A_i;\ i \in I\}$ of algorithms is called *constructible* if there is a set **B** of *basic algorithms* and all algorithms from **A** are built by combining algorithms from **B**. The set **B** is called a *base* of **A**.

For instance, all programs in procedural programming languages, such as ALGOL, FORTRAN, COBOL, C^{++}, and Java, are built from a system of operators or instructions.

It is usually assumed that the base **B** is finite and when a finite number of elements from **B** is given, it is possible to construct only a finite number of algorithms from **A**.

Proposition 5.5.1. For a constructible set **A** of algorithms with a finite base, Composition Axiom implies Cofiniteness Axiom.

We prove this statement by induction.

There are recursive relations between arbitrary total recursively computable (or simply, recursive) functions. These relations are true for many types of direct static information sizes. At first, we prove boundedness from below.

Let us assume that $\mathbf{A} = \mathbf{R}$.

Lemma 5.5.2. If $g(x)$ and $h(x)$ are total recursively computable functions, $g^{-1}(x)$ and $h^{-1}(x)$ are recursively computable multifunctions (relations) and for any n, both sets $g^{-1}(n)$ and $h^{-1}(n)$ are finite, then there is a increasing recursively computable function $f(x)$ such that it tends to infinity and $f(g(x)) \leq h(x)$ for almost all x.

Proof. We informally describe an algorithm of computation of such a function $f(x)$. Then by the Church-Turing Thesis (cf., for example, (Rogers, 1987)), $f(x)$ is a recursively computable function.

An algorithm for computation of the function $f(x)$:

1. Compute $h(\varepsilon)$ where ε is an empty word. It is possible to do this because $h(x)$ is a recursively computable function.

2. If $h(\varepsilon) = r$, find all elements $x_1, x_2, \ldots , x_n$ for which $h(x_i) \leq r$ and choose from these values the least number $p = h(x_j)$ for some x_j. It is possible to do this because $h^{-1}(x)$ is a recursively computable multifunction and for any n, the set $h^{-1}(n)$ is finite.

3. Find the largest element x_j for which $p = h(x_j)$. It is possible to do this because $h(x)$ is a recursively computable function.

4. Find the largest value (say t) of all values $g(x)$ with $x \leq x_j$. It is possible to do this because $g(x)$ is a recursively computable function.

5. Define $f(k) = p$ for all $k \leq t$.

6. Find the least number $q > p$ such that $q = h(x_j)$ for some x_j. Then go to the step 3 with q instead of p, continuing this process with u instead of t in the step 4 and the condition $t < k \leq u$ instead of the condition $k \leq t$ and with q instead of p in the step 5.

In such a way, we build the necessary increasing function $f(x)$. By Lemma 5.5.1, both functions $g(x)$ and $h(x)$ tend to infinity. So, by its construction, the function $f(x)$ also tends to infinity.

Lemma is proved.

Corollary 5.5.1. If $g(x)$ and $h(x)$ are total recursively computable functions, $g^{-1}(x)$ and $h^{-1}(x)$ are recursively computable multifunctions (relations) and for any n, both sets $g^{-1}(n)$ and $h^{-1}(n)$ are finite, then there is an increasing recursively computable function $f(x)$ such that $f(g(x)) \leq h(x)$ and $f(h(x)) \leq g(x)$ for almost all x.

Lemma 5.5.2 and Corollary 5.5.1 imply the following result.

Theorem 5.5.1 (the First Boudedness Theorem for direct measures). For any direct static information sizes $c(A)$ and $b(A)$ that satisfy the Computational, Recomputational, and Cofinite axioms, there is a recursively computable total increasing function $f(x)$ such that $f(\mathbf{c}(A)) \leq \mathbf{b}(A)$ and $f(\mathbf{b}(A)) \leq \mathbf{c}(A)$ for almost all A from $\mathbf{R}$.

Now we prove boundedness from above.

Lemma 5.5.3. If $g(x)$ and $h(x)$ are total computable functions, $g^{-1}(x)$ and $h^{-1}(x)$ are recursively computable multifunctions (relations) and for any n, both sets $g^{-1}(n)$ and $h^{-1}(n)$ are finite, then there is an increasing recursively computable function $f(x)$ such that $f(g(x)) > h(x)$ for all x.

Proof. Let us take the function $f(n) = \max \{h(x); \exists\, z$ such that $g(z) = n$ and for all y, the equality $g(y) = g(z)$ implies $y < z$, and $x \leq z \} + 1$. It is an increasing function. In addition, using the initial conditions from the lemma, we can build an algorithm of computation for this function $f(x)$.

At first, using computability and finiteness of $g^{-1}(x)$, we find all x_j such that $g(x_j) = n$. Then we take the largest of them z. Then we can compute the function $f(n) = \max \{h(x); x \leq z \} + 1$.

By the Church-Turing Thesis, $f(x)$ is a computable function and by its construction $f(g(x)) > h(x)$ for all x.

Lemma is proved.

Corollary 5.5.2. If $g(x)$ and $h(x)$ are total recursively computable functions, $g^{-1}(x)$ and $h^{-1}(x)$ are recursively computable multifunctions (relations) and for any n, both sets $g^{-1}(n)$ and $h^{-1}(n)$ are finite, then there is a recursively computable strictly increasing function $f(x)$ such that $f(g(x)) > h(x)$ and $f(h(x)) > g(x)$ for all x.

Lemma 5.5.3 and Corollary 5.5.2 imply the following result.

Theorem 5.5.2 (the Second Boudedness Theorem for direct measures). For any direct static information sizes $\mathbf{c}(A)$ and $\mathbf{b}(A)$ that are defined for algorithms from the class $\mathbf{A}$ and satisfy the Computational, Recomputational, and Cofinite axioms, there is a recursively computable total strictly increasing function $f(x)$ such that $f(\mathbf{c}(A)) > \mathbf{b}(A)$ and $f(\mathbf{b}(A)) > \mathbf{c}(A)$ for all A from $\mathbf{A}$.

Theorem 5.5.2 implies corresponding results from (Blum, 1967; Hartmanis and Hopcroft, 1971; Burgin, 1982).

Now we can axiomatically define dual information size.

Let $\mathbf{A} = \{A_i; i \in I\}$ be a class of algorithms, A be an algorithm that works with elements from I as inputs and $\mathbf{Sc}: I \to N$ be a direct static information size of algorithms from a class $\mathbf{A}$. Elements of I are usually treated as programs for the algorithm A. In addition, developing the theory of such a direct static information size as Kolmogorov complexity, researchers assume for simplicity that I consists of natural numbers in a form of binary sequences. These numbers may be indices enumerating algorithms from **A** or codes of these algorithms (cf., for example, (Burgin, 2005)). In what follows, we consider only computable recomputable cofinite direct static information size.

Definition 5.5.3. The *dual* to **Sc** *complexity measure* or $(A, \mathbf{Sc})$- *information size* $\mathbf{Sc}_A^{\circ}$ of an object (word) x with respect to the algorithm A is the function from the codomain (the set of all outputs) Y of A that is defined as

$$\mathbf{Sc}_A^{\circ}(x) = \min \{\mathbf{Sc}(p); p \in I \text{ and } A(p) = x\}.$$

Naturally when there is no such p that $A(p) = x$, the value of $\mathbf{Sc}_A^{\circ}$ at x is undefined.

When $\mathbf{Sc}(x)$ measures information of the word or text x, the *information size* $\mathbf{Sc}_A^{\circ}(x)$ estimates minimal information necessary to compute or build x by the algorithm A. For instance, when words in the alphabet $\{0, 1\}$ are considered and $\mathbf{Sc}(x)$ is the length of x, then $\mathbf{Sc}(x)$ gives the number of bits in x.

If $\mathbf{L}_A^{\circ}(x)$ is the dual to the length $l(p)$ of program/algorithm description p information size with respect to a algorithm A, then

$$\mathbf{L}_A^{\circ}(x) = \min \{l(p); p \in I \text{ and } A(p) = x\}.$$

Let M and T be some algorithms.

Proposition 5.5.2. If $M(x) > T(x)$ for (almost) all x, then $\mathbf{L}_T^{\circ}(x) > \mathbf{L}_M^{\circ}(x)$ for (almost) all x for which both $\mathbf{L}_M^{\circ}(x)$ and $\mathbf{L}_T^{\circ}(x)$ are defined.

The most interesting case is when A is a universal algorithm V for the class **A**. Let $\mathbf{c}: \mathbf{A} \to X^*$ be some coding of algorithms from **A**.

An algorithm V is called *universal* for the class **A** if for any $A \in \mathbf{A}$ and any x given the pair $(\mathbf{c}(A), x)$ as its input, the result of V is equal to the result of A applied to x (cf. Appendix B).

Examples of universal algorithms are a universal Turing machine and a universal inductive Turing machine (cf., for example, (Burgin, 2005)). Universal algorithms allow us to build optimal forms of the dual information size.

Let **H** and **G** be two sets of functions.

Definition 5.5.4. A function $f(n)$ is called (asymptotically) **H**-optimal in **H** if there is such $h \in \mathbf{H}$ that $f(n) \leq h(g(n))$ for any $g \in \mathbf{G}$ and (almost) all $n \in N$.

If there is such $h \in \mathbf{H}$ that $f(n) \leq h(g(n))$ for almost all $n \in N$, we denote this relation by $f(x) \preccurlyeq_{\mathbf{H}} g(x)$. In the case, when **H** consists of such functions that add some constant to the argument, for example, $f(n) = g(n) + c$, we write simply $f(x) \preccurlyeq g(x)$ or $g(x) \succcurlyeq f(x)$. This relation is basic for the theory of Kolmogorov complexity (Li and Vitaniy, 1997), as well as for the theory of inductive algorithmic complexity (Burgin, 2005).

Lemma 5.5.5. Relations $g(x) \succcurlyeq f(x)$ and $f(x) \preccurlyeq g(x)$ mean that there is a constant number c such that $f(x) \leq g(x) + c$ for all x.

Let $\mathbf{H} = \mathbf{H}(h) = \{h_k(n) = h(h(n) + k),\ k \in N\}$ and **A** be a class of algorithms with a universal algorithm U.

Theorem 5.5.3 (the Axiomatic Optimality Theorem (Burgin, 1982))**.** For any axiomatic static complexity measure $\mathbf{Sc}(p)$ on **A** and for some recursively computable function $h(n)$, there is an $\mathbf{H}(h)$-optimal dual measure $\mathbf{Sc}^{\mathrm{o}}(x)$.

Proof. Let A be some algorithm from the class **A**. At first, we consider the dual information size $L^{\mathrm{o}}_A(x)$ and the dual information size $L^{\mathrm{o}}(x)$ dual to the length $l(p)$ of program/algorithm description p with respect to A and to general class **A** of algorithms that work with words or natural numbers, respectively. These measures are defined by the formulas: $L^{\mathrm{o}}_A(x) = \min\ \{l(p);\ p \in I$ and $A(p) = x\}$ and $L^{\mathrm{o}}(x) = L^{\mathrm{o}}_U(x) = \min\ \{l(p);\ p \in I$ and $U(p) = x\}$ where U is a universal algorithm in **A**.

If $L^{\mathrm{o}}_A(x) = l(p)$ for some x and $A(p) = x$, then $U((\mathbf{c}(A), p)) = x$. By the definition, $l((\mathbf{c}(A), p)) = l(p) + k_A$ where k_A is some number. Then by the definition of $L^{\mathrm{o}}(x)$, we have $L^{\mathrm{o}}(x) \leq l((\mathbf{c}(A), p)) = l(p) + k_A = L^{\mathrm{o}}_A(x) + k_A$ for any x.

We define $\mathbf{Sc}^{\mathrm{o}}(x) = \mathbf{Sc}^{\mathrm{o}}_U(x)$. Then by Theorem 5.5.2, there is a recursively computable total strictly increasing function $h_1(x)$ such that $h_1(L^{\mathrm{o}}_U(x)) > \mathbf{Sc}^{\mathrm{o}}_U(x)$. By Proposition 5.5.2, we have $\mathbf{Sc}^{\mathrm{o}}(x) \leq h_1(L^{\mathrm{o}}(x)) \leq h_1(L^{\mathrm{o}}_A(x) + k_A)$ as h_1 is an increasing function. At the same time, by Theorem 5.5.2, there is a recursively computable total strictly increasing function $h_2(x)$ such that $L^{\mathrm{o}}_A(x) \leq h_2(\mathbf{Sc}^{\mathrm{o}}_A(x))$. Taking $h(x) = \max\ \{h_1(x), h_2(x)\}$, we have

$$\mathbf{Sc}^o(x) \le h(\mathrm{L}^o_A(x) + k_A)$$

and

$$\mathrm{L}^o_A(x) \le h(\mathbf{Sc}^o_A(x))$$

Thus, $\mathbf{Sc}^o(x) \le h(h(\mathbf{Sc}^o_A(x)) + k_A)$. This inequality means that $\mathbf{Sc}^o(x)$ is an $\mathbf{H}(h)$-optimal dual information size.

Theorem is proved.

The result of Theorem 5.5.3 spares a researcher and a student:

1) to prove optimality for different versions of dual information size, e.g., for recursive information size and for inductive information size;

2) to prove optimality for other specific types of dual information size.

Optimality for recursive information size and its versions is additive (cf. Section 5.2). However, there are other kinds of optimality, which depend on the measure $\mathbf{Sc}^o(x)$. It is possible to find an example of such measures in (Manin, 1991) where instead of the length $l(x)$ of the word x representing some number n, this number n is taken as a direct static measure $\mathbf{Sc}(x)$ of x, i.e., $\mathbf{Sc}(n) = n$. As a result, for the corresponding dual dual information size $\mathbf{Sc}^o(x)$, we have a different type of optimality. Namely, $\mathbf{Sc}^o(x) \le k_A \cdot \mathbf{Sc}^o_A(x)$ for any Turing machine A. Calude (2005) also introduces invariance (optimality) principle for information sizes with a multiplicative constant.

Definition 5.5.5. a) $f(n) \preccurlyeq_{\mathbf{H}(h)} g(n)$ $(f(n) \preccurlyeq^a_{\mathbf{H}(h)} g(n))$ if there is a function $h \in \mathbf{H}$ such that $f(n) \le h(g(n))$ for all $n \in \mathbf{N}$ (for almost all $n \in \mathbf{N}$).

b) Functions $f(n)$ and $g(n)$ are called (*asymptotically*) $\mathbf{H}(h)$-*equivalent* if $f(n) \preccurlyeq_{\mathbf{H}(h)} g(n)$ and $g(n) \preccurlyeq_{\mathbf{H}(h)} f(n)$ $(f(n) \preccurlyeq^a_{\mathbf{H}(h)} g(n)$ and $(g(n) \preccurlyeq^a_{\mathbf{H}(h)} f(n)))$. It is denoted by $f(n) \asymp_{\mathbf{H}(h)} g(n)$ (by $f(n) \asymp^a_{\mathbf{H}(h)} g(n)$, correspondingly).

For instance, Theorems 5.5.1 and 5.5.2 show that any axiomatic static complexities $\mathbf{c}(A)$ and $\mathbf{b}(A)$ that satisfy the Computational, Recomputational, and Cofinite axioms are $\mathbf{TR}$-equivalent, i.e., $\mathbf{c}(A) \asymp_{\mathbf{TR}} \mathbf{b}(A)$, where $\mathbf{TR}$ is the class of all recursive functions. However, we have a stronger result for optimal functions.

Theorem 5.5.4 (the Axiomatic Invariance Theorem (Burgin, 1982)). Any two (asymptotically) $\mathbf{H}(h)$-optimal functions are (asymptotically) $\mathbf{H}(h)$-equivalent.

This means that optimal dual measures of information size are in some sense invariant.

Theorems 5.5.3 and 5.5.4 imply existence and uniqueness of optimal/invariant measures for many dual measures of information size: Kolmogorov complexity, uniform complexity, prefix complexity, monotone complexity, process complexity, conditional Kolmogorov complexity, time-bounded Kolmogorov complexity, space-bounded Kolmogorov complexity, conditional resource-bounded Kolmogorov complexity, time-bounded prefix complexity, resource-bounded Kolmogorov complexity, etc. We do not need to prove these theorems for each case separately because it is sufficient only to check conditions from theorems 5.5.3 and 5.5.4 and then to apply these theorems.

The dual information size that corresponds to a universal algorithm V from $\mathbf{A}$ gives an invariant characteristic of the whole class $\mathbf{A}$.

Definition 5.5.6. The *dual* to **Sc** *complexity measure* or *information size* $\mathbf{Sc}^{\circ}_{\mathbf{A}}$ of an object (word) x with respect to the class $\mathbf{A}$ is defined as

$$\mathbf{Sc}^{\circ}_{\mathbf{A}}(x) = \min \{\mathbf{Sc}(p); p \in I \text{ and } V(p) = x\}.$$

Naturally when there is no such p that $A(p) = x$, the value of $\mathbf{Sc}_{\mathbf{A}}^{\circ}$ at x is undefined.

Because algorithm V is universal for the class $\mathbf{A}$, this condition is equivalent to the condition that there is no such algorithm A from $\mathbf{A}$ and such p that $A(p) = x$.

When $\mathbf{Sc}(x)$ measures information in the word or text x, the information size $\mathbf{Sc}^{\circ}_{\mathbf{A}}(x)$ estimates minimal information necessary to compute or build x by algorithms from the class $\mathbf{A}$.

By the definition, $\mathbf{Sc}^{\circ}_{\mathbf{A}}(x) = \mathbf{Sc}_{V}^{\circ}(x)$ for a universal algorithm V for the class $\mathbf{A}$. However, as a rule, $\mathbf{A}$ has several universal algorithms. In such a case, the function of $\mathbf{Sc}^{\circ}_{\mathbf{A}}(x)$ is defined not in unique way. Nevertheless, as the Invariance Theorem shows, the definition of $\mathbf{Sc}^{\circ}_{\mathbf{A}}(x)$ is invariant with respect to certain transformations.

Proposition 5.5.3. For any algorithm A and any axiomatic static complexities $\mathbf{Sc}(x)$ and $\mathbf{Sb}(x)$ the condition $f(\mathbf{Sc}(x)) \leq \mathbf{Sb}(x)$ for almost all

x ($f(\mathbf{Sb}(x)) > \mathbf{Sc}(x)$ for all x) implies the condition $f(\mathbf{Sc}^o_A(x)) \le \mathbf{Sb}^o_A(x)$ for almost all x ($f(\mathbf{Sb}^o_A(x)) > \mathbf{Sc}^o_A(x)$ for all x).

Indeed, if $\mathbf{Sb}^o_A(x) = \mathbf{Sb}(q)$, then $f(\mathbf{Sc}(q)) \le \mathbf{Sb}(q)$ and $A(q) = x$. By the definition, if $\mathbf{Sc}^o_A(x) = \mathbf{Sc}(r)$, then $A(q) = x$ and $\mathbf{Sc}(r) \le \mathbf{Sc}(p)$ for all $p \in I$ such that $V(p) = x$. In particular, we have $\mathbf{Sc}(r) \le \mathbf{Sc}(q)$. Consequently, $f(\mathbf{Sc}^o_A(x)) = f(\mathbf{Sc}(r)) \le f(\mathbf{Sc}(q)) \le \mathbf{Sb}(q) = \mathbf{Sb}^o_A(x)$ because $f(x)$ is an increasing function.

Inequality $f(\mathbf{Sb}^o_A(x)) > \mathbf{Sc}^o_A(x)$ is proved in a similar way.

Proposition is proved.

Remark 5.5.6. If we can choose different algorithms from **A** to build the element x, the dual measure of the information size with respect to the class **A** is defined in a different way.

Definition 5.5.7. The *dual* to **Sc** *complexity measure* or (**A**, **Sc**)-*information size* $\mathbf{Sc}^o_A$ of an object/word x with respect to the class **A** with selection is defined as

$$\mathbf{Sc}^o_{SA}(x) = \min \{\mathbf{Sc}(p); p \in I, A \in \mathbf{A}, \text{ and } A(p) = x\}.$$

Naturally when there is no such algorithm A from **A** and such p that $A(p) = x$, the value of $\mathbf{Sc}_A^o$ at x is undefined.

When **A** is a immense class of algorithms, such as the class of all Turing machines or the class of all inductive Turing machines, the information size $\mathbf{Sc}^o_{SA}(x)$ is constant and minimal for any word x, for example, $\mathbf{L}^o_{SA}(x) = 0$ for all x. However, in practice, we need to study algorithms and programs that satisfy some additional conditions, e.g., algorithms that solve a given problem. This restriction on the class **A** can make the function $\mathbf{Sc}^o_{SA}(x)$ non-trivial. For instance, let us take all algorithms that count occurrence of some letter (say a) in a given text as the class **A**. Then $\mathbf{Sc}^o_{SA}(n) = n$ because to contain n letters a, the text must have, at least, n letters.

Lemma 5.5.4. $\mathbf{Sc}^o_{SA}(x) \le \mathbf{Sc}^o_A(x) \le \mathbf{Sc}(x)$.

In general, both functions $\mathbf{Sc}^o_{SA}(x)$ and $\mathbf{Sc}^o_A(x)$ are defined for all elements x from the domain $\cup_{A \in \mathbf{A}} C(A)$ where $C(A)$ is the range of the algorithm A. In particular, when all algorithms from **A** have a common domain X, then, as a rule, both functions $\mathbf{Sc}^o_{SA}(x)$ and $\mathbf{Sc}^o_A(x)$ are defined for all elements x from X. For example, when **A** is the set of all partial recursive functions, both functions $\mathbf{Sc}^{\circ}{}_{SA}(x)$ and $\mathbf{Sc}^{\circ}{}_{A}(x)$ are defined for all natural numbers. This is a consequence of the following more general result.

Let the class **A** contains an identity algorithm E that computes the function $e(x) = x$.

Theorem 5.5.5 (the Axiomatic Totality Theorem). $\mathbf{Sc}^{\circ}{}_{A}(x)$ is a total function on N^+ (on the set of all words in some alphabet).

Dual information sizes are usually interpreted as complexity of problem solution with the help of algorithms from **A**. More exactly, the problem under the consideration is construction or computation of a word x by means of algorithms from **A**.

It is necessary to remark that besides different kinds of the generalized Kolmogorov complexity (Burgin, 1990), there are other dual information sizes. As an example of another kind of a dual measure of the information size, we can take Boolean circuit complexity, which is also a nonuniform complexity measure (Balcazar, et al, 1988). There are two direct and two dual measures for such automata as Boolean circuits:

-   The *cost* or *size* $c(A)$ of a Boolean circuit A is the number of gates it has. It is a direct static information size of Boolean circuits.
-   The Boolean *cost* $c(f)$ of a Boolean function f is the size of the smallest circuit computing f:

    $$c(f) = \min \{c(A); A \text{ defines the function equal to } f\}$$

It is a dual information size.

-   The *depth* of a Boolean circuit is the length of the longest path in the graph of this circuit. It is a direct static reconstructible information size of Boolean circuits.
-   The Boolean depth $d(f)$ of f is the depth of the minimal depth circuit computing f:

    $$d(f) = \min \{d(A); A \text{ defines the function equal to } f\}$$

It is a dual information size.

Thus, we see that not all dual information sizes coincide with the Kolmogorov complexity or some its kind. There are other examples of such measures.

There are also other approaches leading to dual measures of the information size. For instance, Gell-Mann (1994) introduced the concept of crude complexity of a system. It is possible to find many other examples of direct and dual complexity measures in (Burgin, 1990; Burgin and Chelovsky, 1984; 1984a; Cardoso, et al, 2000; Burgin and Debnath, 2003). Communication complexity $cc(f)$ (cf. (Hromkovic, 1997)) is a dual information size on the set **P** of all protocols.

Let $Sc_A^{\circ}(x)$ and $Sc_B^{\circ}(x)$ be dual to **Sc** measures of the information size with respect to classes **A** and **B**, respectively. If $A \subseteq B$, then any algorithm universal for **B** is also universal for **A**. We assume that such an algorithm is used to build dual measures of the information size with respect to **A** and **B**. This implies the following results.

Theorem 5.5.6 (the Axiomatic Monotonicity Theorem). If $A \subseteq B$ and $Sc_A^{\circ}(x)$ is defined for x, then $Sc_B^{\circ}(x)$ is also defined for x and $Sc_B^{\circ}(x) \leq Sc_A^{\circ}(x)$.

Corollary 5.5.7. If $A \subseteq B$ and $Sc_A^{\circ}(x)$ is defined for all x, then $Sc_B^{\circ}(x)$ is also defined for all x and $Sc_B^{\circ}(x) \leq Sc_A^{\circ}(x)$ for all x.

Dual information sizes with respect to the class **A**, i.e., information sizes determined by a universal algorithm, have invariance properties, defining minimal resources that are necessary in **A** to build/compute objects from Y where the set Y contains such objects that can be computed by algorithms from **A**.

Analysis of the main concepts and constructions of algorithmic information theory show that this theory is a special case of the general theory of information. In this case, infological systems consist of symbolic/linguistic objects and are changed (transformed) by algorithms from a given class, e.g., by Turing machines. Algorithmic information size shows how much information specific classes of algorithms need to compute/construct a symbolic/linguistic object. Algorithmic information measures the potency of information contained in some linguistic object/text, e.g., a program or algorithm, to build/reconstruct another (linguistic) object using algorithms from a given class, e.g., recursive algorithms, such as Turing machines, or super-recursive algorithms, such as inductive Turing machines.

Chapter 6

Pragmatic Information Theory

Not everything that can be counted counts,
and not everything that counts can be counted.
Albert Einstein

As we can see (cf. Chapter 4), pragmatics of information refers to aspects of information action (influence) and utilization. For instance, Kornwachs (1996) defines pragmatic information as an impinging entity, one that is able to change the structure and the behavior of systems. Weaver (Shannon and Weaver, 1949) also considered the pragmatic, or he called it effectiveness, problem of communication as finding to what extent desired and actual effect produced by the received message correspond.

Pragmatic aspects of information are embedded into the situational context of the system that receives, produces, sends or finds information, e.g., a person or organization. When information becomes a resource or commodity, it is natural to find the value or in a more restricted sense, cost of this resource/commodity. Individual aspects, such as need for information, goals and motivation, play an important part in pragmatic information evaluation. Value and cost represent pragmatic aspects of information. Under certain circumstances the monetary value that the system, e.g., an individual or organization, is ready to pay for a piece of information may provide a measure for pragmatic information.

In Section 6.1, we consider economics of information, which studied the role of information in the economic activity and developed measures for estimation of economic value and cost of information and information sources.

412

In Section 6.2, we consider such important characteristics as value, cost, and quality of information. Information quality, as researchers from MIT emphasize, is a survival issue for both public and private sectors: companies, organizations, and governments with the best information have a clear competitive edge.

In Section 6.3, we present elements of the qualitative information theory developed by Mazur.

6.1. Economic Approach of Marschak: Cost of Information

Better egg today than a hen tomorrow.

Proverb

Jacob Marschak (1898–1977), one of the most important economists in the middle of the 20th century, was the first who studied economical aspects of information, becoming one of the founders of the economic of information (cf. (Arrow, 1980/81)). His approach was based on a specific "economic" point of view of decision-making and the role of information in this process. He treated information economics as an extension of decision theory. In a number of papers, he, his collaborators and followers developed a pragmatic economically oriented theory of information (Arrow, Marschak and Harris, 1951; Marschak, 1954; 1959; 1964; 1971; 1972; 1974; 1976; 1980; 1980a; MacQueen and Marschak, 1975).

Marschak starts his construction of economical measures of information with an explanation that Shannon's entropy $H(X)$ is an important measure of information, which clearly does not depend on the particular uses to which the information will be put. However, the user is more interested in how much a portion of information is worth for her/him and how much she/he is willing to pay for this portion of information. In a general case, the answers will vary from user to user for the same portion of information. To estimate information as a commodity, economics is not interested in a portion of information that a single message contains but rather in information coming from a given source or channel.

A major concern of information economics, for example, is the effect which information has upon a person's expectations and the resulting decisions she or he may contemplate or enact. Most economists are even prone to regard only those data as genuine information that are capable of changing a person's expectation about a certain events and situations. Since these expectations are best quantified by means of probabilities assigned to future events (states of the world, including nature, actions of competitors, etc.), the relationship between information and changing probabilities becomes crucial.

A natural approach of an economist is to associate the value, or worth, of information with the average amount earned with the help of this information. Namely, if a person, or company, A has access to a certain kind of information that earns the amount V_0 on the average, while another kind of information will allow this person (company) to earn the amount V_1 on the average, then the demand price for the new information is taken equal to $V_1 - V_0$.

At the same time, the supply price of information is the lowest price its supplier is willing to charge. In this context, the value of information is related to its demand price, while the cost of information is related to its supply price.

Marschak assumes that the supply price cannot, in the long run, be below the cost of information to the supplier or she/he will lose the economical motive to supply it. However, it is necessary to take into account that the supplier can have other, not economical, reasons to supply this information. For instance, the supply price of information can be very low or even nil for political reasons. Many countries usually supply some kinds of information to their allies for no cost but only because there is an agreement for information exchange. For instance, during the World War II, USA and England exchange information on nuclear power without any payment although the cost of obtaining this information by research was very high. As we know England was not able to invest necessary amount of money to develop this project independently.

In definite situations, the supply price of information can be very low or nil even for economical reasons. For instance, many Internet companies, such as Google, Amazon or Yahoo, provide information for free although the cost of this information is not nil for these companies. The reason for this is to attract people who would come to their site. Google and Yahoo are interested in advertising – the more people use their sites, the higher price for advertisements they may charge. Amazon, on the other hand, provides free information about products it sells. Actually any company gives free information, promoting and advertising their products or/and services.

To find the demand price for the new information, Marschak utilizes a *payoff function* $f(x, a)$ where the variable x reflects changes in the situation (state of the world) where the user/customer is operating to make a profit and the variable a reflects actions of the user. Rationality conditions imply that the goal of the user is to maximize the expected value of the payoff. Information is used to make a rational decision. Thus, the user acquires the role of a decision-maker. As von Neumann and Margenstern demonstrated (1944), such payoff functions, also called criterion functions, exist when simple conditions of decision-maker rational behavior are satisfied.

Usually payoffs have monetary values. However, there are other kinds of payoffs important for information evaluation. For instance, when a slow and fast military communication systems are compared (cf. (Page, 1957)), payoff can be measured in military damage, lost battle or in killed people.

Having a payoff function, it is important to understand the role of information in decision-making. In the considered model, information gives knowledge of how the set X of all possible situations (states of the corresponding environment) is *partitioned*, i.e., how X is divided into a group of non-intersecting subsets. It is assumed that each element of partition, i.e., a subset of X, yields, or dictates, an appropriate action. Situations that belong to the same element of a given partition are equivalent with respect to actions as they dictate the same action. Each partitioning is called an *information structure*. This structure represents the situation (state of the world) where the user is operating. Thus, it is assumed that the variable x in the payoff function $f(x, a)$ takes values in the set of all partitions of X.

It is interesting that this information structure is similar to information structures studied in other directions of information theory.

For instance, a partition is a kind of classifications studied in the theory of information flow by Barwise and Seligman (1997), such that elements of these classifications do not have specific names (types) (cf. Section 7.1). In turn, a classification is a kind of named sets used to mathematically model communication and other information processes (cf. Chapters 2 and 3).

In general, information ranking depends on the chosen payoff function. However, it is possible to order partitions on the base of inclusion. Namely, one partition P is a *subpartition* of the other partition Q if each of the subsets of the former partition is contained in some subset of the latter. In this case, we say that the partition P is *finer* than the partition Q. Marschak asserts that the finer partition is the higher rank it has independently of the payoff function. It is also possible to regard each partition, or information structure, P as a function (operator) P that corresponds the message

$$y = P(x) \qquad .$$

to each state of the world represented by an element x from X.

All these considerations make it possible to build a mathematical model for information evaluation. It is assumed that the following objects are given:

- the set X of all possible situations (states of the world);
- a probability distribution $p(x)$ on X (when X is finite, $p(x)$ is the probability of the element x from X);
- the set Ac of possible actions or decisions about actions;
- a payoff function $f(x, a)$;
- a rule of action $g(y)$, i.e., a function that associates a message y with some action from Ac.

Then the expected payoff for a partition P is determined by the following formula:

$$U = U(g, P; f, p) = \mathrm{E}\, f(x, g(P(x))) = \sum_{x \in X} p(x) \cdot f(x, g(P(x)))$$

When the rule of action g is optimal, i.e., $g = g^*$, then it is maximized over the set of all action rules. Namely,

$$U(g^*, P; f, p) = \max \{ U(g, P; f, p);\ g \text{ is a rule of action} \} = V(P; f, p)$$

This allows one to determine the *value of information* about P, or the value of the information structure P. Namely, the demand price $dp(I)$ is equal to

$$dp(I) = V(P_1; f, p) - V(P_0; f, p)$$

where P_0 is the information structure before definite information I comes from the information source and P_1 is the information structure after this information I comes from the information source.

It is necessary to remark that such an approach inherently assumes that information supplied gives true knowledge. Otherwise, applications of the payoff function in particular and this model in general will be misleading. To deal with the situation when acquired knowledge may be false, Marschak considers faulty (false) information, which is interpreted as noise and information loss in the channel.

Note that due to noise in the communication channel, true information may become false. Such information is studied in many directions of information theory, e.g., in statistical information theory (cf. Chapter 3). In spite of such evidence for false information existence, some researchers continue to insist that misinformation (false information) is a kind of information.

It is demonstrated that in some cases, the cost of information can be proportional to the Shannon's entropy (Marschak, 1959) and that information value can coincide with the channel capacity (Kelly, 1956). It is conjectured that the latter coincidence is due to the logarithmic nature of the payoff function chosen by Kelly.

A large amount of research has evolved from the works of Marschak and his collaborators (cf. (Hirshliefer and Rilley, 1979; McCall, 1982; Phlips, 1988; Laffont, 1989)). It ranges from an examination of information streams (e.g., predecision information versus post-decision information) to different types of auctions, price dispersions, predatory pricing, signals and "signaling theory", credit rationing, antitrust implications, different kinds of economic equilibria, contingent markets and constraining contract clauses, competition among agents, even to cheating and misinformation.

Another approach to the cost of information is considered in (Burgin, 1996a) and is based not on expectations but on actual changes caused by information.

It is also possible to build a mathematical model for the economical value of information based not on a payoff function but on an *efficient incentive function* $f(y)$ (Good, 1952; McCarthy, 1956). This function shows a fee a forecaster receives from a client when the event y happens. Natural properties of such functions are:

- the expected fee is the largest when the forecaster's estimates are perfect;
- the expected fee is nil if the forecaster does not know more than his client.

In addition, Cockshott and Michaelson (2005) suggest a computational approach to the value of information. In their theory, the information value of an information portion I is measured by the number of cycles of a universal Turing machine (cf. Appendix B) needed to extract or produce this portion of information. According to the authors, such a measure reflects the thermodynamic cost of information extraction and/or production. At the same time, Cockshott and Michaelson use *information utility* as a measure of the uses to which information can be put.

It is possible to formalize information value in the sense of Cockshott and Michaelson using dual measures to time complexity for classes of algorithms. Such dual measures and more general dual measures to computational complexity are introduced and studied in (Burgin, 1982).

Gackowski (2004) introduces several economical attributes of information, which reflect its economical value. He suggests that in a business environment, the single most important and cumulative measure of information quality is the *expected cost effectiveness of information/data*. To define this measure, Gackowski describes several economic characteristics of information/data.

Utility value $V(I)$ *of information/data* I is determined by the following formula:

$$V(I) = V_R(D + I) - V_R(D)$$

Here $V_R(D + I)$ and $V_R(D)$ are the cost (e.g., monetary values) of actions or business operations with and correspondingly, without the additional information I.

Net business utility value $V_N(I)$ *of information/data I* is determined by the following formula:

$$V_N(I) = V(I) - C(I)$$

Here $C(I)$ is the procurement (acquisition) cost (e.g., monetary value) of the additional information I.

Cost effectiveness $C_E(I)$ *of information/data I* is determined by the following formula:

$$C_E(I) = V(I)/C(I) \tag{6.1.1}$$

Any ratio given by the formula (6.1.1) that is less than 1 means an inappropriate (source of) information. In a profit-oriented environment, even when the ratio is equal to 1, the corresponding information is considered useless.

It is also possible to define cost effectiveness of information/data when the value of the additional results attributed to the information I cannot be expressed in monetary units.

Qualitative cost effectiveness $C_E(I)$ *of information/data I* is determined by the following formula:

$$C_E(I) = Q(I)/C(I) \tag{6.1.2}$$

Here $Q(I)$ is the number of results attributed to the additional information I and $C(I)$ is the procurement (acquisition) cost (e.g., monetary value) of the additional information I.

However, people are mostly interested not in the *a posteriori* value but in what is possible to gain in future. That is why expected values of the considered characteristics are more adequate for economic applications.

Expected cost effectiveness $EC_E(I)$ *of information/data I* is determined by the following formula:

$$EC_E(I) = EV(I)/EC(I) \tag{6.1.3}$$

Here the *expected utility value* $EV(I)$ *of information/data I* is determined by the formula:

$$EV(I) = V(I) \cdot [1 - \text{risk factor or failure rate}]$$

and the *expected procurement cost* $EC(I)$ *of information/data I* is determined by the formula:

$$EC(I) = C(I) \cdot [1 + \text{relative average cost overruns}]$$

Economics of information is a pragmatic direction in the general theory of information because it develops approaches to and models information measurement based on expected or actual changes caused by information.

Economics of information studies information as a resource from the economical perspective. At the same time, information perspective is used for studies in economics. For instance, Samuelson and Swinkels (2006) study evolution of utility functions and the role of information in this process.

6.2. Mission-Oriented Approach: Value, Cost, and Quality of Information

Our greatest glory is not in never falling,
but in rising every time we fall.
Confucius

Some researchers assume that information, to be information, has to have value, has to be used for decision-making, and has to be designed to lead to action (cf., for example, (Machlup and Mansfield, 1980)). This means that value of information is considered as a necessary property of information. That is why the concept of information value was formalized and studied in information theory by many researchers. As we have seen in the previous section, the most popular approach to the concept of information value is based on the assumption that information is used for decision-making and information value reflects the improvement of decision-making gained from receiving information. In other words, the value of information is determined by its importance to the decision-maker or to the outcome of the decision being made.

The approach of Harkevitch (1960) is based on utility theory. Positing that from the pragmatic perspective, the value of information is its usefulness in achieving some goal, he defines the pragmatic measure of information as the gain in the probability distributions of the receiver's actions, both before and after receipt of a message in a pre-defined ensemble. Different goals can assign different values to the same portion of information. Thus, value of information is defined for a mission oriented system R. If I is some portion of information, then the *value* of this information is equal to the caused by I change of the probability $p(R, g)$ of achievement of a particular goal g by the system R. Thus, if $p_0(R, g)$ is the probability of achievement of g by R before R receives information I and $p_1(R, g)$ is the probability of achievement of g by R after R receives information I, then the value $J(I)$ of I for R with respect to the goal g is equal to

$$J(I) = \log_2 (p_1(R, g)/p_0(R, g)) = \log_2 p_1(R, g) - \log_2 p_0(R, g) \quad (6.2.1)$$

This approach is a particular case of the general theory of information. Indeed, if we consider objective probability, then the corresponding infological system is the state space of the world in which system R functions. If we consider subjective probability, then the corresponding infological system is the belief space in which probabilities for different events involving system R are represented. In both cases, information appears as a change in the corresponding infological system.

At the same time, the Hartley quantity of information (cf. Chapter 3) can be deduced from the formula (6.2.1). Indeed, if before R has received information I, there were N possible results with equal probability to achieve any of them, and after R receives information I, there is only one result g, then $p_0(R, g) = (1/N)$ and $p_1(R, g) = 1$. As a result, formula (6.2.1) gives us

$$\log_2 p_1(R, g) - \log_2 p_0(R, g) = \log_2 1 - \log_2 (1/N) = \log_2 N \quad (6.2.2)$$

Bongard (1963; 1967) develops his approach based on an extension of the Shannon's measure of information. As in other pragmatic approaches to information measurement, he considers a system R that uses information for solving problems by a sequence of experiments with an object. In this process, the system R can get information from other sources. Received information can change the number of necessary experiments and thus, to alter complexity of the problem that is solved. This change is used to find the value of information.

To formalize this idea, it is assumed that a problem a_i has the unique answer b_i with the probability p_i. At the same time, the system R that makes experiments tries the answer b_i with the probability q_i. Then the average number of experiments is equal to $1/q_i$ and uncertainty of the problem is defined as

$$\log(1/q_i) = -\log q_i$$

Thus probability of this situation is equal to p_i and uncertainty for a collection of problems $A = \{a_i; i = 1, 2, 3, ..., n\}$ is equal to

$$H(A) = -\sum_{i=1}^{n} p_i \log_2 q_i$$

This allows Bongard (1963; 1967) to consider information received by the system R, or more exactly, the value of this information equal to

$$I = \sum_{i=1}^{n} p_{1i} \log_2 q_{1i} - \sum_{i=1}^{n} p_{0i} \log_2 q_{0i} \qquad (6.2.3)$$

Here p_{1i} and q_{1i} are probabilities after R receives information I and p_{0i} and q_{0i} are probabilities before R receives information I ($i = 1, 2, 3, ..., n$).

Another approach to the value of information also stems from utility theory where a decision-maker aims for maximization of expected utility based on known information. The formalization uses a *probability space* $(\Omega, \mathbf{F}, \mu)$ where Ω is interpreted as the set of states of the world, subsets of Ω from $\mathbf{F}$ are called *events*, $\mathbf{F}$ is a σ-algebra of events, i.e., $\mathbf{F}$ has Ω as a member, is closed under complementation (with respect to Ω) and union, and μ is the decision-maker's probability measure. Information is modeled by partitions of the set Ω of states of the world (cf. (Marschak, 1959; Aumann, 1974; Hirshliefer and Rilley, 1979)). Each partition is a finite set of pairwise disjoint elements from $\mathbf{F}$ the union of which is equal to Ω.

It is possible to interpret a partition of the set Ω as organization of the states in the world. Note that a partition is a classification in the sense of Barwise and Seligman (1997), only they do not have specific names (types) for the classes from the partition (cf. Chapter 7). Formalizing this relation, we obtain the following proposition.

Proposition 6.2.1. Any classification determines a definite partition and any partition determines different but equivalent classifications.

Partitions are estimated by a function $f: \mathbf{P} \to \mathbf{R}$ from then set $\mathbf{P}$ of all partitions of the set Ω into the set of real numbers $\mathbf{R}$. This function indicates the value of a partition for the decision-maker. As partitions model information, or give the information structure, f is called an *information function* when it satisfies the maximality condition related to strategies of the decision-maker and his expected utility. Information functions represent the value of information. Different properties of information functions have been studied.

When the decision-maker has information that the situation she is dealing with is represented by a partition P and then receives information that the situation is actually represented by a partition Q, which is a refinement of P, then it possible to treat the difference $f(Q) - f(P)$ as the value of additional information to the decision-maker. Observe that the value of a given portion of information may be negative in some cases.

Value of information has different interpretations. For instance, utility is often identified with money obtained by the decision-maker. In this case, the value of additional information to the decision-maker is the maximal price she/he will be willing to pay for this additional information. Note that information value is not necessarily cost although contemporary society tends to estimate all other values in money.

There was a lot of research based on more simplified models, as well as empirical research were users of information were asked to estimate the value of information they use. The actual value of a piece of information essentially depends on the user's cognitive and psychological needs and preferences. That is why the interest in such user behavior has also propelled more integrative research efforts. From this perspective, it is important to understand situations, where the user develops information needs, and analyze the usage of the same information once it has been obtained and interpreted by the user. The goal of such an integrative research is to develop more holistic theories about human information seeking and usage.

The most popular approach is based on the assumption that the value of information has to reflect an outcome of choice in uncertain situations (cf. (Hirshliefer and Rilley, 1979; McCall, 1982). The outcome is estimated as the expected value of the income that resulted from making a decision. In this situation, the value depends on the following factors or determinants (Hilton, 1981):

- (*Action flexibility*) The structure of the decision-maker's actions.
- (*Initial uncertainty*) The extent of uncertainty of the decision-maker.
- (*Payoff function*) What is at stake as an outcome of the decision, i.e., what are tentative losses when a wrong decision is made.
- (*Quality of information*) Such attributes of information as timeliness, accuracy, and clearness.
- (*Price of information*) The price of information under consideration.
- (*Price of substitutes*) The price of the next-best substitute of this information.

In this context, for example, a portion of information has no value, or more exactly, the value is nil, when there are no costs associated with making the wrong decision or there are no actions that can be taken in light of this information.

There are several mathematical definitions of the value in terms of the outcome function, utility function, prior probability and posterior probability. They allow one to finds properties of information value determinants.

Here are some examples.

Theorem 6.2.1 (Hilton, 1981). There is no general monotone relationship between action flexibility and information value.

It is necessary to understand that the lack of a general monotone relationship between action flexibility and information value does not preclude such a relationship from holding in a special class of decision problems.

Similar to information value, risk aversion is an important attribute of utility function, characterizing tentative losses incurred by a wrong decision. However, the following result was established.

Theorem 6.2.2 (Hilton, 1981). There is no general monotone relationship between the degree of absolute or relative risk aversion and information value.

Uncertainty can be characterized in several ways. For instance, such characteristics as the number of system states with non-zero probability or the variance of the prior are intuitive measures of uncertainty. For these and several other measures of uncertainty, negative results were also proved.

Theorem 6.2.3 (Gould, 1974). There is no general monotone relationship between the degree of uncertainty in the prior and information value.

The approach to information value based on the decision-making utility has been applied to a variety of fields and situations. Many studies treated the value of weather information for agriculture production and management. Considered issues included applications to stock prices of wheat; markets of orange juice; haymaking; irrigation frequency; bud damage and loss; production of peas, grain, grapes, and soybeans; wool; and fruit (cf. (Bradford and Kelejian, 1977; Roll, 1984; Johnson and Holt, 1986)). Teisberg and Weiher (2000) studied the value of information coming from geomagnetic storm forecasts, making an estimate of the net economic benefits of a satellite warning system. Bernknopf, et al, (1997) studied the social value of geologic map information. Babcock (1990) studied the value of weather information in market equilibrium, while Macauley (1997) considered general problems of weather information evaluation.

Another approach to the value of information is not to theoretically define this value but to study how users of information, e.g., decision-makers, evaluate information. It is a behavioral direction in information value studies. Such studies have been performed in accounting. An important function of the accountant is that of choosing among alternative information sources utilized by decision-makers in selecting actions (cf. (Schepanski and Uecker, 1983)). Different normative mathematical models, such as the Bayesian, conservative, and moderately conservative models, have been developed for information evaluation (Demski, 1980). It is usually assumed in normative models that the preferences for information are consistent with the axioms of utility theory. To be useful, a model has to correctly represent practical situations. Thus, researchers started to investigate the adequacy of normative models of information evaluation as positive models of accountants' behavior in making information-source choices (cf. (Schepanski and Uecker, 1983)).

In the empirical research on the value of information, people are usually asked how much they are willing to pay for information they need or/and use. For instance, Marshall (1993) examined the positive impact of information on corporate decision making. In assessing the monetary value of information supplied by special libraries, Griffiths and King (1993) drew upon 27 studies performed with 16 companies and 7 government agencies, as well as 4 national surveys of professionals, with a total of over 10,000 responses. Koenig (1992) investigated the correlation between expenses of information services and corporate productivity.

However, the cost of information is only one form of the value of information. For instance, Griffiths and King (1993) also assessed the value of information in terms of time savings and work quality.

Practitioners who are interested in risk estimation suggest a *customer lifetime value* analysis (cf., for example, (Hill, 2004)). This approach requires modelling of the future costs and benefits associated with a customer's contract or other interactions with the company/organization. The company is interested in decreasing the risk of such interactions and estimate information value based on the criterion of risk aversion. Additionally, company's pay-offs include the costs of undertaking the information quality improvement activity as well.

An alternative approach is based on customer segmentation into groups of individuals equivalent with respect to some criterion related to customer's interactions with the company/organization. One of the reasons for such a grouping is that organisations do it for the customer base when it is too costly to treat each customer individually or it is possible to separate the most important groups of customers. For instance, it is, presumably, very useful to identify a very small segment (say, 0.3% of the customers) and treat them differently since the pay-off from identifying those three in a thousand customers would have to "pay" for the other 997 subjected to the partitioning process. Then the entropy of the class distribution is used as a proxy measure for information value. There is some empirical evidence to support this approach (cf., for example, (Vetschera, 2000)). In fact, entropy and measures derived from it are often used as a grouping criterion in formal decision analyses (Shih 1999).

In addition, there was an empirical research on the value of information in qualitative terms. For instance, in their study of 12 major corporations, Owens and Wilson (1997) found that senior executives view the creation of an information culture in an organization as a critical step toward ensuring continued success.

Another approach to the quality of information for a company, which appears online or in print, is based on five criteria: the scope of coverage, authority, objectivity, accuracy and timeliness. This approach identifies incidents of questionable, false or fraudulent information as reported in the news or trade literature, provides examples of Web sites that illustrate good or bad information, and suggests strategies that help one to detect bad information.

As information has become the precious resource, primary driving force and effective mechanism of contemporary society, many critical issues in data and information quality have been plaguing information systems for many years. As a result, deficiencies with information quality impose significant costs on individual companies, organizations, and the whole economy, resulting in the estimated costs to the US economy at $600 billion per year (cf., (Hill, 2004)).

In spite of these, researchers have relatively recently begun to address information quality as a discipline in its own right, and a body of information quality literature has just begun to appear. Numerous researchers started developing means to evaluate information quality, as well as quality of information sources (such as mass media), carriers (such as web sites), depositories (such as databases and knowledge bases), and retrieval systems (such as search engines) (cf., for example, (Dedeke, 2000; Eppler, 2001; Hu and Feng, 2006)). There are several books on information quality (Al-Hakim, 2006; Fisher, et al, 2006; Huang, et al, 1999; Lee, et al, 2006; Olson, 2003; Piattini, et al, 2002; Pipino, et al, 2002; Redman, 2001; Wormell, 1990; Wang, et al, 2001).

Information quality is often considered as a measure of the reliability and effectiveness of data that carry this information, especially, in the context of decision making. According to (Strong, et al, 1997; 1997a) information quality issues impact every level of the organizations and companies:

- On the *strategic level*, absence of high quality information results in less effective strategic business and organizational decisions
- On the *tactical level*, absence of high quality information results in compromised decision making, inability to reengineer, and mistrust between internal divisions and organizations
- On the *operational level*, absence of high quality information results in customer dissatisfaction, increased costs, lowered employee job satisfaction and morale and loss of revenue.

Researchers have identified organisational need in a high quality information supply. For instance, Ballou and Tayi (1989) prescribed a method for periodic allocation of resources to information quality improvement proposals (e.g., for maintenance of the data assets in the organization). There are two approaches to resource allocation. A *budgetary approach* presupposes that a fixed budget for information quality research and management has to be shared among a set of other proposals. At the same time, an *investment approach* is based on evaluation of proposals taking into account the expected value returned. It also assumes that managers are seeking and win the largest budget they can justify to their organization.

Hundreds of tools have been produced for evaluating quality in practice since 1996 (cf. (English, 1999; Hu and Feng, 2006)). For instance, Eppler (2001) gives a review of twenty information quality frameworks that have appeared in the literature from 1989 to 1999 in sixteen various application contexts. DeLone and McLean (2003) describe twenty-three information quality measures. Knight and Burn (2005) analyze ten approaches to information quality. Here we consider only some of the existing directions to information quality.

Developed approaches to information quality represent theoretical views, as well as practitioners' understanding. Various parameters have been included in information quality, such as relevancy, accessibility, usefulness, timeliness, clarity, comprehensibility, completeness, consistency, reliability, importance, and truthfulness (Hu and Feng, 2006). Parallel to studies on information quality, researchers assess data quality. These studies are intrinsically related to information quality studies as data are primary sources of information and the most frequent representation of information processed by people. However, data are important not by themselves but exclusively by the virtue of information they carry. As Strong, et al (1997) write, organizational [and other, M.B.] databases reside in the larger context of information systems. From this perspective, data are a kind of information representation, and thus, the quality of information depends on the quality of its representation. For instance, it is possible to consider any organization as an information processing engine. Information on many issues in organizations is represented by data in various databases: personnel databases, production databases, billing and collection databases, sales management databases, customer databases, supply chain databases, accounting databases, financial databases, and so on. Even more data exist on the Internet. An assortment of parameters have been included in data quality, such as accuracy, format, readability, timeliness, accessibility, ergonomic, precision, and size of data arrays (Hu and Feng, 2006).

The World Wide Web has become an important information, knowledge and communication resource. As a result, data and information quality for the World Wide Web has been studied by various authors without making difference between data, information and the web site itself (Strong, et al, 1997; Zhu and Gauch, 2001; Kahn, et al, 2002). As the general theory of information explains, this absence of differentiation is caused by the situation when the real goal is to achieve high quality in information representation.

Both data and web pages are information carriers and representations, while the quality of information for a user depends on the quality of its representation. For instance, information accessibility to some extent depends on whether information is presented in the language known to the user or not. Information quality is mostly understood as satisfaction of the user needs. Thus, the quality of information depends on the different perceptions and needs of the information users and is therefore relative. High quality information would therefore meet the specific requirements of its intended use. However, in spite of all distinctions between users, their goals and requirements, information quality researchers have tried to find common and comprehensive quality characteristics.

It is necessary to remark that parameters of web page quality are useful (may be, with some modifications) for evaluation of information quality in any information space in the sense of information geometry (cf. Section 7.3.3).

Informally, *quality* is defined as an essential and distinguishing attribute of something or someone, as a degree of excellence or worth, and as a characteristic property that molds the apparent individual nature of something or someone. Information quality studies are oriented at the first of these definitions, i.e., information quality is perceived as an essential and distinguishing attribute of information. There is also a pragmatic approach to information quality. In it, information quality is treated as fitness for a purpose. This makes information quality very subjective and relative to specific goals and intentions. As Adams (2003) writes, in pragmatic terms, information quality means that information items geared towards one set of consumers may be perceived as poor quality when located by a different set of consumers.

It is necessary to remark that the term *information quality* is often substituted by its acronym IQ. However, a more traditional interpretation of IQ is the intelligence quotient. Thus, we do not use IQ as a substitute for the term *information quality*.

Quality is a very complex property in general. Thus, to evaluate information quality, it necessary to have an efficient framework that provides a systematic and concise set of criteria to which information can be evaluated. Such a framework must be able to provide the basis for identification of information quality problems, for information quality measurement and for proactive management of information representation and information carriers (Eppler and Wittig, 2000).

Analyzing information processes in society, we can see that information quality is influenced by three basic factors:
~ the perception, abilities and needs of the user
~ information
~ the process of accessing, utilizing and transforming information

At the same time, processes of accessing, utilizing and transforming information involve three entities:

~ the user

~ information

~ information processing system, e.g., information retrieval system

Information quality, as a complex property, comprises many properties, components, and parameters. To organize this sophisticated system of characteristics, researchers build a hierarchical representation of information quality, introducing *information quality dimensions* on the first level and information quality characteristics on the second level of this hierarchy. There are two types of information quality dimensions: *direct information quality dimensions* and *structural information quality dimensions*, also called *information quality areas*, *categories* or *levels* of information quality.

There are several types of structural information quality dimensions: the semiotic dimensions, action dimensions, and interaction dimensions.

The *semiotic dimensions* reflects relations in the following structure:

$$\text{Subject} \xrightarrow{\text{process}} \text{Object} \qquad (6.2.3)$$

The *action dimensions* correspond to the elements of the following diagram:

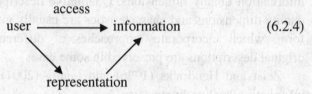

$$(6.2.4)$$

The *interaction dimensions* are related to the components of the following schema:

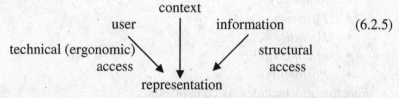

$$(6.2.5)$$

Hill (2004) makes a distinction between *statistical information*, which changes probability distribution reflecting user knowledge of the world, and *pragmatic information*, which changes user decisions and course of action. From the general information theory perspective, statistical information is pure cognitive information, while pragmatic information is effective information (cf. Section 2.6).

Statistical information corresponds to the semantic level of Shanks and Darke (1998) who contemplate three abstract levels (dimensions) that build upon each other:

- *Syntactics* is concerned with form of information representation.
- *Semantics* is concerned with meaning of information.
- *Pragmatics* is concerned with use of information.

The semantic level characteristics, such as completeness and accuracy, assess the mapping between the real world and its representation. It is possible to treat this mapping as a *communication process* between the real-world and the representation, with the information system as a *channel* and the real-world as the *information source* (cf. Figures 3.1 and 3.9). Attributes of the real world systems (e.g., customers) are *messages*.

Let us consider some of information quality studies, describing, at first, direct information quality dimensions and then structural information quality dimensions. Qualitative descriptions of information quality dimensions and characteristics are usually given in a synthesized form, which incorporates approaches of different authors although original descriptions are preserved in some cases.

Zeist and Hendricks (1996) and Leung (2001) consider six direct information quality dimensions:

- *Functionality*
- *Reliability*
- *Efficiency*
- *Usability*
- *Maintainability*
- *Portability*

Functionality includes six components:

  o *Suitability* is the quality of having the properties that are right for a specific purpose.

- o Accuracy shows to what extent information is correct and reliable, thus including two subcomponents:
  - Correctness shows to what extent information represents the real situation.
  - Reliability reflects to what extent it is possible to believe in this information,
- o Interoperability reflects the ability to exchange and use information.
- o Traceability reflects the ability to trace the history, application or location of an item or activity by means of recorded identification.
- o Compliance shows to what extent information is presented in accordance with established guidelines.
- o Security shows to what extent the access to information is appropriately restricted to maintain its security.

Reliability has five components:

- o Maturity
- o Recoverability
- o Availability shows the extent to which information is physically available.
- o Degradability shows whether the information carrier can be broken down by external impact.
- o Fault tolerance

Efficiency of the webpage content reflects:

- o Temporal characteristics
- o Resource behavior

Usability includes nine components:

- o Understandability reflects easiness/complexity of content understanding and comprehension.
- o Learnability reflects easiness/complexity of learning how to use information.
- o Operability reflects easiness/complexity of operation with information representation.
- o Luxury
- o Clarity

o   Helpfulness
o   Explicitness
o   Customizability indicates the ability for information (information representation) to be changed by the information user or provider.
o   User-friendliness reflects characteristics of information representation (e.g., on the webpage) that are helpful for solving the user problem.

Maintainability includes six components:

o   Analyzability
o   Changeability
o   Stability
o   Testability reflects easiness/complexity of testing.
o   Manageability reflects easiness/complexity of managing information representation.
o   Reusability of information (e.g., of the webpage content)

Portability includes four components:

o   Adaptability
o   Conformance shows to what extent information is presented in accordance with established guidelines.
o   Replaceability reflects easiness/complexity of component replacement.
o   Installability reflects easiness/complexity of installation.

Alexander and Tate (1999) consider six direct information quality dimensions:

• *Authority*
• *Accuracy*
• *Objectivity*
• *Currency*
• *Orientation*
• *Navigation*

Authority includes two components:

o   Extent to which information is validated
o   Visibility of the webpage author

Accuracy also includes two components:

- o Reliability reflects to what extent it is possible to believe in this information.
- o Absence of errors

Objectivity reflects to what extent information content is unbiased, unprejudiced and impartial.

Currency reflects to what extent information (e.g., the web content) is up-to-date.

Orientation warrants that there is a clear target audience.

Navigation reflects to what extent information/data is/are easily found and linked to.

Kahn, et al, (2002) consider two direct information quality dimensions each of which has two subdimensions:

- *Product quality*
  - o Soundness
  - o Usefulness
- *Service quality*
  - o Dependability
  - o Usability

Soundness includes five components:

- ▪ Absence of errors
- ▪ Conciseness reflects the extent to which information is compactly represented without being overwhelming, i.e., brief in representation, yet sufficiently complete and to the point.
- ▪ Representability
- ▪ Completeness reflects the extent to which information is not missing and is of sufficient breadth and depth for the task at hand.
- ▪ Consistency of representation shows the extent to which information is presented in the same format and compatible with related, e.g., previous, data.

Usefulness includes seven components:

- ▪ Amount (of information/data) appropriateness ensures that there is not too much or little information/data on the website or when this information/data is/are unavailable.

- Adequacy/relevancy reflects to what extent information is applicable and helpful for the task in hand.
- Understandability shows the extent to which information is clear, easily comprehended and its representation has no ambiguities.
- Interpretability is suitability of information representation for explication of necessary knowledge.
- Objectivity reflects to what extent information content is unbiased, unprejudiced and impartial.
- Accuracy shows to what extent information is correct and reliable.
- Comprehensiveness

Dependability includes three components:

- Timeliness reflects to what extent information content is up-to-date for the task at hand.
- Security shows to what extent the access information is restricted appropriately to maintain its security.
- Traceability of the web page reflects the ability to trace the history, application or location of information presented on the web page.

Usability includes six components:

- Believability shows to what extent information content (on the webpage) is true and credible.
- Accessibility shows to what extent information (on the website) is readily available and downloadable.
- Maintainability indicates the effort needed to make specified modifications.
- Reputation shows how presented information is regarded with respect to its source and/or content.

- Value augmentation shows to what extent information is beneficial (adds value) and provides advantages from its use.
- Ease/complexity of manipulation (e.g., speed)

Klein (2002) considers the same six direct information quality dimensions in the context of the web:

- *Accuracy* ensures that the source and author of the information on the webpage is obtainable.
- *Amount* (*of information/data*) *appropriateness* ensures that there is not too much or little information/data on the website or when this information/data is/are unavailable.
- *Completeness* shows when necessary information is missing, lack of depth and website incomplete when compared with other sites.
- *Relevance* shows when the website purpose is too broad or biased.
- *Timeliness* reflects to what extent information on a webpage is current (e.g., the date when webpage was published must be known).

Several approaches to information quality use structural information quality dimensions.

Strong, et al, (1997) consider four structural data/information quality dimensions:

- *Intrinsic quality*
- *Accessibility*
- *Contextual quality*
- *Representational quality*

Intrinsic quality includes four components:

- o Accuracy shows to what extent information is correct and reliable.
- o Objectivity reflects to what extent information content is unbiased, unprejudiced and impartial.
- o Believability shows to what extent information content (on the webpage) is true and credible.
- o Reputation shows how presented information is regarded with respect to its content.

Accessibility includes seven components (cf. Figure 2 in (Strong, et al, 1997)):

- o   Access complexity
- o   Resource sufficiency
- o   Access security
    - ▪   Privacy
    - ▪   Confidentiality
- o   Interpretability is suitability of information representation for explication of necessary knowledge.
- o   Understandability shows the extent to which access tools are clear, easily comprehended and have no ambiguities.
- o   Conciseness reflects the extent to which information is accessible for analysis.
- o   Consistency shows the extent to which information is presented in the same format and compatible with related information.

Contextual quality includes five components:

- o   Relevancy reflects to what extent information is applicable and helpful for the task in hand.
- o   Value augmentation shows to what extent information is beneficial (adds value) and provides advantages from its use.
- o   Timeliness reflects to what extent information content is up-to-date for the task at hand.
- o   Completeness reflects the extent to which information is not missing and is of sufficient breadth and depth for the task at hand.
- o   Amount (of information/data) appropriateness ensures that there is not too much or little information/data on the website or when this information/data is/are unavailable.

Representational quality includes four components:

- o   Interpretability is suitability of information representation for explication of necessary knowledge.
- o   Understandability shows the extent to which information is clear, easily comprehended and its representation has no ambiguities.

- o Conciseness of representation reflects the extent to which information is compactly represented without being overwhelming, i.e., brief in representation, yet sufficiently complete and to the point.
- o Consistency of representation shows the extent to which information is presented in the same format and compatible with related, e.g., previous, data.

Katerattanakul and Siau (1999) use the same four information quality dimensions as Strong, et al, (1997) but give a different interpretation, assigning other information quality characteristics to these dimensions. As a result, they have four dimensions:

- *Intrinsic quality*
- *Accessibility*
- *Contextual quality*
- *Representational quality*

Intrinsic quality includes five components:

- o Accuracy of the content shows to what extent information content is correct and reliable.
- o Absence of errors in the content
- o Accuracy of the hyperlinks on the webpage shows to what extent these hyperlinks are correct and reliable.
- o Workability of the hyperlinks on the webpage
- o Relevance of the hyperlinks on the webpage reflects to what extent the hyperlinks are applicable and helpful for the task in hand.

Accessibility shows what navigational tools used to access and move around on the website are provided.

Contextual quality warrants provision of the author's information.

Representational quality includes five components:

- o Organization
- o Visual settings
- o Typographical features
- o Consistency shows the extent to which information is presented in the same format and compatible with related information.
- o Vividness/attractiveness

In contrast to other researchers, Alter (2002) treats information quality as a component (dimension) of information usefulness and considers four direct dimensions of *information usefulness*:

- Quality
- Temporal dimension
- Accessibility
- Presentation
- Security

The *quality* dimension has three components:
  - Accuracy shows the extent to which the information represents what it is supposed to represent, e.g., bias and random error lead to inaccuracy.
  - Precision reflects the fineness of detail.
  - Completeness shows the extent to which the available information is appropriate for the task.

The *temporal* dimension has two components:
  - Age is the amount of time that had passed since the information was produced.
  - Timeliness reflects the extent to which the age of the information is appropriate for the task.

The *accessibility* dimension has two components:
  - Availability reflects the extent to which the age of the information is available in the information system.
  - Admissibility reflects to whether laws, regulations, or culture require or prohibit the use of the information.

The *presentation* dimension has two components:
  - Level of summarization is a comparison between the number of individual items on which the data are based and the number of individual items in the data presented.
  - Format is the way the information is organized and expressed.

The *security* dimension has two components:
  - Encryption is converting data into a coded form that unauthorized people cannot decode.
  - Access restriction reflects who can access what information under what circumstances.

Shanks and Corbitt (1999) use a semiotic-based framework of information quality, which has three levels of hierarchy and four semiotic dimensions:

- The *syntactic* dimension reflects consistency of information in the sense that it should be well-defined and have formal syntax.
- The *semantic* dimension has two components and four subcomponents:
  - Completeness reflects the extent to which information is not missing and is of sufficient breadth and depth for the task at hand.
  - Accuracy includes four components:
    - Comprehensiveness
    - Non-ambiguity
    - Meaningfulness
    - Correctness shows to what extent information represents the real situation.
- The *pragmatic* dimension includes two components and four subcomponents:
  - Usefulness shows the extent to which information is applicable and helpful for the task of the user.
  - Usability shows the extent to which information is clear and easily used and includes four components.
    - Timeliness reflects to what extent information content is up-to-date for the task at hand.
    - Conciseness of representation reflects the extent to which information is compactly represented without being overwhelming, i.e., brief in representation, yet sufficiently complete and to the point.
    - Accessibility
    - Reputation shows how presented information is regarded with respect to its content.

- The *social* dimension/level includes two components:
  o Shared understandability shows the extent to which information is clear to and easily comprehended by different social groups.
  o Awareness of bias

Naumann and Rolker (2000) consider three structural information quality dimensions:
- Subject dimension (criteria)
- Object dimension (criteria)
- Process dimension (criteria)

Subject criteria have six components:
  o Believability shows to what extent information content (on the webpage) is true and credible.
  o Conciseness of representation reflects the extent to which information is compactly represented without being overwhelming, i.e., brief in representation, yet sufficiently complete and to the point.
  o Understandability of content shows the extent to which information content is clear, easily comprehended and its representation has no ambiguities.
  o Interpretability is suitability of information representation for explication of necessary knowledge.
  o Relevancy of information reflects to what extent information is applicable and helpful for the task in hand.
  o Value augmentation shows to what extent information is beneficial (adds value) and provides advantages from its use.

Objective criteria have five components:
  o Completeness reflects the extent to which information is not missing and is of sufficient breadth and depth for the task at hand.

- o Security shows to what extent the access information is restricted appropriately to maintain its security.
- o Objectivity reflects to what extent information content is unbiased, unprejudiced and impartial.
- o Timeliness reflects to what extent information content is up-to-date for the task at hand.
- o Content author verifiability

Process criteria have five components:

- o Accuracy shows to what extent information is correct and reliable.
- o Hyperlink activity level
- o Availability
- o Consistency shows the extent to which information is presented in the same format and compatible with related information.
- o The retrieval response time of a webpage

Dedeke (2000) considers five structural information quality dimensions:

- • *Ergonomic quality* reflects the ease of navigation on the webpage.
- • *Accessibility* has six components:
  - o Data accessibility
  - o Data sharing
  - o Data convertibility
  - o System availability
  - o Technical accessibility
  - o Technical security shows to what extent technology maintains information security.
- • *Transactional quality* has six components:
  - o System responsiveness
  - o Controllability
  - o Error tolerance
  - o Efficiency shows the extent to which information are able to quickly meet the information needs of the user.

- o Adaptability of the content
- o System feedback
- *Contextual quality* has four components:
  - o Relevancy reflects to what extent information is applicable and helpful for the task in hand.
  - o Completeness reflects the extent to which information is not missing and is of sufficient breadth and depth for the task at hand.
  - o Appropriateness shows the extent to which the content is appropriate according to what the users are requiring.
  - o Timeliness reflects to what extent information content is up-to-date for the task at hand.
- *Representational quality* has six components:
  - o Consistency shows the extent to which information is presented in the same format and compatible with related information.
  - o Conciseness of representation reflects the extent to which information is compactly represented without being overwhelming, i.e., brief in representation, yet sufficiently complete and to the point.
  - o Structure
  - o Interpretability is suitability of information representation for explication of necessary knowledge.
  - o Readability
  - o Contrast of information on the web page

Eppler and Muenzenmayer (2002) consider two structural information quality dimensions each of which has two subdimensions:

- *Content quality* is concerned with the quality of the presented (on the web) information and has two components:
  - o Relevancy reflects to what extent information is applicable and helpful for the task in hand.
  - o Soundness
- *Media quality* is concerned with the quality of the medium used to deliver the content and also has two components:
  - o Optimized processing
  - o Infrastructure reliability

Relevancy has four components:

- Comprehensiveness
- Accuracy shows to what extent information is correct and reliable.
- Clearness reflects complexity/easiness of understanding.
- Applicability

Soundness has four components:

- Correctness shows to what extent information represents the real situation.
- Consistency shows the extent to which information is presented in the same format and compatible with related information.
- Currency reflects to what extent information (e.g., the web content) is up-to-date.
- Conciseness of representation reflects the extent to which information is compactly represented without being overwhelming, i.e., brief in representation, yet sufficiently complete and to the point.

Optimized processing has four components:

- Timeliness reflects to what extent information content is up-to-date for the task at hand.
- Convenience
- Traceability reflects the ability to trace the history, application or location of an item or activity by means of recorded identification.
- Interactivity

Infrastructure reliability has four components:

- Accessibility
- Security shows to what extent the access information is restricted appropriately to maintain its security.

- ▪ Maintainability indicates the effort needed to make specified modifications.
- ▪ Retrieval speed

O'Brien (1997) considers three structural data/information quality dimensions:

- *Time*
- *Content*
- *Form*

Time includes four components:

- o Currency reflects to what extent provided information is up-to-date.
- o Timeliness reflects to what extent information content is up-to-date for the task at hand.
- o Frequency shows whether information is provided as often as it is needed.
- o Time period shows whether information is provided about past, present, or/and future period of time.

Content includes six components:

- o Relevance reflects to what extent information is related to the needs of a specific recipient for a definite purpose.
- o Accuracy displays to what extent there are no errors in information and its representation.
- o Completeness reflects to what extent needed information is provided.
- o Conciseness ensures that only needed information is provided.
- o Scope can be broad or narrow, representing internal or/and external focus.
- o Performance exposes activities accomplished, progress made, or resources accumulated.

Form includes five components:

- o Order exhibits to what extent information is organized, e.g., arranged in a predetermined sequence.

o Detail indicates whether information is general (summarized) or provided in detail.
o Clarity reflects complexity/easiness of understanding.
o Presentation reflects whether information has the form of a narrative, text, numerical system, graphical image, picture, etc.
o Media component reflects whether information carrier is a printed paper document, computer file, video display, Web site or other media.

Yasin (1970) considers such parameters of information quality as utility, value, range, cost, accessibility, significance, usability, timeliness, and reliability. They are defined in the following way.

Significance of an information portion I is defined with respect to a chosen problem P and reflects to what extent information I is necessary for solving P.

Utility of an information portion I is defined with respect to a chosen problem P and reflects how I is used for solving P.

Usability of an information portion I is the frequency its utilization for solving problems from some class **P**. Usability depends both on the class **P** of problems and on how often these problems are solved.

The *range* of an information portion I is defined with respect to a chosen problem P and reflects how I is used for solving P.

Other characteristics are defined in the standard way.

As Knight and Burn (2005) write, despite the sizable body of literature on information quality, relatively few researchers have tackled the difficult task of quantifying conceptual definitions of information quality and suggesting methods for its estimation. Nevertheless, information quality has many measures and includes different measured characteristics as it is fairly certain that no one "magic number" can serve as a measurement for all characteristics of information that might be considered important. Some of these characteristics are related to directly measurable numerical values, while numerical values related to others are calculated from measurements. To achieve a comprehensive information quality representation by characteristics and to build good measures for these characteristics, it is practical to use the general theory of evaluation and measurement (Burgin and Kavunenko, 1994). According to this theory, the process of measurement/evaluation has three main stages: *preparation*, *realization*, and *analysis*.

The first stage in evaluation preparation demands to determine a specific criterion for evaluation. Such a criterion describes the goal of evaluation. Criteria of information quality include such properties as reliability, adequacy, exactness, completeness, convenience, user friendliness, etc. However, such properties are directly immeasurable and to estimate them, it is necessary to use corresponding indicators or indices. With respect to information quality such indicators are called *general information quality measures* or *metrics*. However, an indicator can be too general for direct estimation. This causes necessity to introduce more specific properties of the evaluated object. To get these properties, quantifiable tractable questions are formulated. Such properties play the role of *indices* for this criterion. Thus, the second stage of evaluation consists of index selection that reflects criteria. Sometimes an index can coincide with the corresponding criterion, or a criterion can be one of its indices. However, in many cases, it is impossible to obtain exact values for the chosen indices. For instance, we cannot do measurement with absolute precision. What is possible to do is only to get some estimates of indices. Consequently, the third stage includes obtaining estimates or *indicators* for selected indices. In the case of information, these indicators have form of *constructive* (*procedural*) *information quality measures*.

For instance, in the work of Zhu and Gauch (2000), we can see that such an information quality characteristic of a Web page as *currency* is a criterion, the index corresponding to this criterion is *how recently the Web page was updated*, and the corresponding indicator is *the time stamp of the last modification of the website*. In the same manner, *popularity* criterion of a Web page is turned into the index that shows *how many other Web pages have cited this particular Web page*. However, this index is not sufficiently constructive. Thus, it is specified by the following indicator: *the number of links to the considered Web page*.

In the work of Eppler and Muenzenmayer (2002), we can see that such an information quality characteristic as *convenience* is a criterion, the index corresponding to this criterion is *difficult navigation paths*, and the corresponding indicator is *the number of lost/interrupted navigation trails*.

Thus, to achieve correct and sufficiently precise evaluation, preparation demands the following operations:

1. Choosing evaluation criteria.
2. Corresponding characteristics (indices) to each of the chosen criteria.
3. Representing characteristics by indicators (estimates).

This shows that a complete process of evaluation preparation has the following structure:

Criterion → Index → Indicator

Creation of information quality measures must include the following three stages:

1. Setting goals specific to needs in terms of purpose, perspective, and environment.
2. Refinement of goals into quantifiable tractable questions.
3. Construction of an information quality measure and data to be collected (as well as the means for their collection) to answer the questions.

Information quality measures are useful only when there are corresponding procedures/algorithms of measurements. Thus, we need additional stages for the measure development.

1. Designing procedures/algorithms for data collection in the measurement.

2. Designing procedures/algorithms for computing measurement values.

3. Designing procedures/algorithms for analyzing measurement results.

Realizing this process, Eppler and Muenzenmayer (2002) correspond the following Web-indicators (measures) to the information quality criteria they consider (cf. Table 6.1).

Table 6.1. A correspondence between information quality criteria and their Web-indicators

Information quality criteria	Web-indicators
Accuracy	User ratings
clearness/clarity	User ratings
Applicability	The number of orphaned, i.e., not visited or linked, pages or user ratings
correctness	User ratings
consistency	The number of pages with style guide deviations
currency	Last mutation
conciseness	The number of deep (highly hierarchic) pages
timeliness	The number of heavy (over-sized) pages/files with long loading time
convenience	Difficult navigation paths, i.e., the number of lost/interrupted navigation trails
traceability	The number of pages without the author or source
interactivity	The number of forms The number of personalized pages
accessibility	The number of broken links The number of broken anchors
security	The number of weak log-ins
maintainability	The number of pages with missing metainformation
speed	Server and network response time
comprehensiveness	User ratings

Zhu and Gauch (2000) analyzed information quality criteria and other metrics used by different Internet systems that provide rating services, such as Internet Scout, Lycos, Argus Clearinghouse, WWW Virtual Library, and Internet Public Library. They identified sixteen metrics/measures of information retrieval on the Web used in those systems: subject, breadth, depth, cohesiveness, accuracy, timeliness, source, maintenance, presentation, quality of writing, availability, authority, currency information-to-noise ratio, cohesiveness, and popularity. They selected six of those metrics, which are widely used and amenable to automatic analysis: availability, authority, currency information-to-noise ratio, cohesiveness, and popularity. These metrics are defined in a constructive way as indices, allowing efficient measuring procedures:

- *Currency metric* tells how recently the Web page was updated.
- *Availability metric* is characterized by the number of broken links on the Web page.
- *Information-to-noise ratio* is defined as the proportion of useful information contained in a Web page of a given size.
- *Authority metric* reflects the reputation of the organization that produced the Web page.
- *Popularity metric* reflect how many other Web pages have cited this particular Web page.
- *Cohesiveness* reflects the degree to which the content of the Web site is concentrated on one topic.

To make the chosen metrics/measures more constructive, Zhu and Gauch (2000) go from metrics as indices to metrics as indicators. In this new context, they have the following correspondence between criteria and indicators:

- *Currency metric* is determined by the time stamp of the last modification of the Web site.
- *Availability metric* is defined by the formula BL/NL where BL is the number of broken links and NL is the total number of links on the Web page.

- *Information-to-noise* ratio is defined by the formula NT/WS where NT is the total length of tokens after pre-processing and WS is the size of Web page.
- *Authority metric* reflects opinions of users estimated by assigning a score to the Web site.
- *Popularity metric* is determined by the number of links to the Web page.
- *Cohesiveness* reflects the extent to which the major topics of the Web site are related.

We see that there is no standard for information quality characteristics, indices and indicators. Different authors can give dissimilar interpretations and definitions to the same dimensions or/and characteristics. The same characteristic may be included into different dimensions. For instance, Zeist and Hendricks (1996) include accuracy into the functionality dimension, Alexander and Tate (1999) identified accuracy as a separate dimension, while Strong, et al, (1997) include accuracy into the intrinsic quality dimension.

At the same time, some characteristics are more popular than others. For instance, analyzing ten approaches to quality of information presented on the World Wide Web, Knight and Burn (2005) show that accuracy is the most popular characteristic (it is used in eight different works), while usefulness surprisingly is one of the least popular characteristics (it is used only in three different works).

Information quality is measured and quality information is used for different purposes: to choose reliable source of information, to search and prepare information for data mining and knowledge acquisition, to improve (change, organize, restructure, provide better access to information, add necessary information, etc.), to maintain, and to update information representation (such as web sites), information utilization, and information itself. For instance, search effectiveness was essentially improved when the currency, availability, information-to-noise ratio, and web page cohesiveness were incorporated as search criteria in the centralized information search (Zhu and Gauch, 2000). Implementation of availability, information-to-noise ratio, popularity, and cohesiveness as search criteria also had positive impact on site selection (Zhu and Gauch, 2000).

Several approaches to information quality estimation use Shannon's theory to build measures for information quality. For instance, Kononenko and Bratko (1991) use Shannon's entropy and information, considering such information quality measure as the *relative information score* defined by the following formula:

$$R(Z;Y) = I(Z;Y)/H(Z)$$

Similar measures are developed in the context of decision making. Hill (2004) grounds this approach on the following model. Suppose that we have, a priori, a particular amount of uncertainty about possible events in or states of the real world W that have a probability distribution X. Then, receiving a message or making a decision Y, allows us to revise our probability distribution and thus, to reduce our uncertainty.

The considered model is often used in economic studies of information. It considers customers partitioning processes made by decision-makers. This model is inherent for many organizational processes, e.g., for direct marketing, campaign management, fraud detection, intrusion monitoring, loan approval, and credit management.

In this context, an ensemble of customers is considered and W is the set of all states customers can be. Each customer has a state $w \in W$ and a random variable W is used to represent uncertainty about the customer state. At the same time, X is the set of states that the decision system can represent and each customer is assigned a representation state. A random variable X is used to represent uncertainty about the customer representation state. The set X is obtained from communication and it is assumed that X is a direct and categorical message of W (Lawrence, 1999). The numbers of states in W and X are both equal to m and communication ensures a one-to-one correspondence between W and X. However, a state may be garbled for some customer during communication, causing semantic errors.

After representations in X are obtained, a classification Y of customers with the classification state space Y and n classes is built. Later in the process of realisation, states of the classification Y are related to the correct classification Z having the classification state space Z with n classes. It is assumed that the correct classification Z can be obtained with perfect and complete information about the world.

This gives us the following diagram:

The real world W

Communication $\downarrow$

A representation X

Decision $\downarrow$

A classification Y

Action $\downarrow$

A realization Z

If we denote the communication process that delivers information about W by C, the decision-making process by C, and the realization process by A, then we come to the following diagram:

$$W \xrightarrow{\quad C \quad} X \xrightarrow{\quad D \quad} Y \xrightarrow{\quad A \quad} Z$$

The mutual information $I(W; X)$ is a measure of semantic information as it tells us how much our prior uncertainty about the real world state is reduced by observing our representation X.

To build a measure of *pragmatic IQ effectiveness* that takes into account flaws in the decision-process, as well as latent uncertainty, an alternative realisation state space Y^* is defined so that it always equals the classification that the decision-process would have made with perfect information about the real-world W. In essence, $Y^* = D(W)$, where $D(x)$ is a decision function. In this way, the communication process can be successful, getting $Y^* = Y$, even when the decision sub-process fails, i.e., $Y^* \neq Z$. To each state space $V \in \{W, X, Y, Y^*, Z\}$ a random variable $V \in \{W, X, Y, Y^*, Z\}$ is corresponded.

Using statistical theory of information, Hill (2004) introduces several information quality measures related to the grouping/partitioning process in decision making.

The *classifier effectiveness* $E_C(Z; Y, W)$ answers the question of the grouping process efficiency and is defined by the following formula:

$$E_C(Z; Y, W) = I(Z; Y)/I(Z; W)$$

The *pragmatic information quality effectiveness* $E_P(Y^*; Y)$ of the grouping process answers the question to what extent flaws in the decision process obviate the need for improved information quality and is defined by the following formula:

$$E_P(Y^*; Y) = I(Y^*; Y)/H(Y^*)$$

If the value of $E_C(Z; Y, W)$ is small, while the value of $E_P(Y^*; Y)$ is already large, then the possibility to achieve improvements in information quality to have much effect on the overall performance is rather small.

The *actionability* α of an attribute answers the question what attributes are most influential on decision-making and is defined by the following formula:

$$\alpha = I(Y; A)/H(Y)$$

The *semantic information quality effectiveness* $E_S(W; X)$ of the grouping process answers the question to what extent the representation of the customer reflect the real world and is defined by the following formula:

$$E_S(W; X) = I(W; X)/H(W)$$

It is possible to find values of this measure on a per attribute basis by using a partial specification A instead of X, and A^* (defined as the real-world value of the attribute) instead of W.

Besides, the entropy $H(Z)$ of the correct classification Z answers the question how much information each process requires.

As we see, information quality is often considered as a component of information value and information is evaluated from perspectives that take into account information action. This is the basic feature of information as the potency to change user actions, needs and abilities. Consequently, all studies in information quality are comprised by the general theory of information as a wide-ranging framework.

6.3. Transformational Approach of Mazur: Impetus of Information

> *Information is a source of learning.*
> *But unless it is organized, processed,*
> *and available to the right people in a format for decision making,*
> *it is a burden, not a benefit.*
> William Pollard

In contrast to the majority of directions in information theory, which concentrate their efforts on developing various information measures and applying a quantitative approach, Marian Mazur (1970) develops an essentially qualitative approach based on principles of cybernetics. That is why he starts exposition of the theory with introduction of concepts of a control system and controlled system.

Definition 6.3.1. A *controlled system R* is a system such that another system causes intended changes in R.

For instance, a car is a controlled system for its driver. A plane is a controlled system for its pilot.

Definition 6.3.2. A *control system Q* is a system such that causes intended changes in another system.

For instance, the driver is a control system in a car. The pilot is a control system in a plane.

Definition 6.3.3. A *control chain C* is a system through which one system causes intended changes in another system.

For instance, the wheel, brakes and different switches are parts of the control chain in a car.

Definition 6.3.4. A *control circuit D* is a feedback loop that consists of a control system Q, controlled system R and control chains C_1 and C_2 (cf. Figure 6.1).

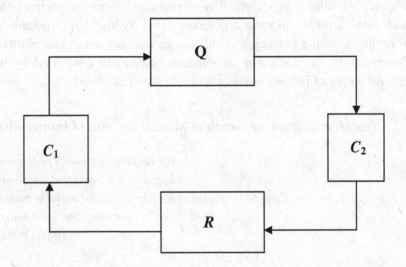

Figure 6.1. A general control circuit

We can see many control circuits around us. For instance, the driver and the car she is driving is a control circuit. There is special science that studies control circuits and their behavior. It is called cybernetics.

In a control chain, one system is acting on another one. The first system is called the (*action*) *source*, while the second system is called the (*action*) *receptor*. It is assumed that the source has output ports to send messages (signals) and the receptor has input ports to receive messages (signals). This gives the general structure of a control chain presented in Figure 6.2.

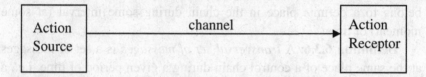

Figure 6.2. The general structure of a control chain

Note that the general structure of a control chain is the same as (isomorphic to) the classical communication triad (cf. Diagram (2.2.1)).

Definition 6.3.5. A *message m* is an identifiable physical state of a part of a control chain.

In this context, an *action* in a control chain consists of a finite number of messages.

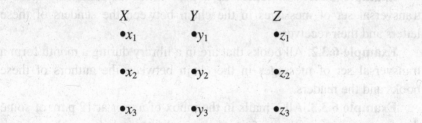

Figure 6.3. A collection of messages x_i, y_i and z_i ($i = 1, 2, 3$) in a control chain with three transversal set of messages X, Y, and Z

To define information, Mazur considers *cross-sections* of a control chain as sets of parts (components) of this chain. Each cross-section is called a *place* of the corresponding control chain. Usually all considered systems consist of a finite number of parts (components). Thus, any cross-section also is finite.

Two types of chains are distinguished: *chains with a static structure*, that is, chains in which its parts (components) and relations between them are not changing, and *dynamic chains* in which their parts (components) and/or relations between them are changing. In the first case, cross-sections do not depend on time — they are always the same. In contrast to this, defining cross-sections in a dynamic chain, it is necessary to take time into account. Consequently, a *cross-section* of a dynamic control chain is a set of parts (components) of this chain that belong to a definite place in the chain during some interval (at some moment) of time.

Definition 6.3.6. A *transversal set of messages* is a set of messages at the same place of a control chain during a given period of time, i.e., a set of states of parts (components) from one cross-section.

When the considered period of time is sufficiently long, this definition applies to two forms of transversal sets: the *upper transversal* and *lower transversal* sets. The upper transversal set consists of all messages that were at a given place of a control chain, at least, at some moment of the given period of time. The lower transversal set consists of all messages that were at a given place of a control chain all time during the given period of time.

Example 6.3.1. All letters at a post office at some day form a transversal set of messages in the chain between the senders of these letters and their receivers.

Example 6.3.2. All books that are in a library during a month form a transversal set of messages in the chain between the authors of these books and the readers.

Example 6.3.3. All e-mails in the inbox of a user at 12 p.m. of some day form a transversal set of messages in the chain between the senders of these letters and this user.

If we take the graphical example from Figure 6.3, then the following sets are transversal sets of messages:

$$\{x_1, x_2\}; \ \{x_2, x_3\}; \ \{x_1, x_3\}; \ \{x_1, x_2, x_3\};$$
$$\{y_1, y_2\}; \ \{y_2, y_3\}; \ \{y_1, y_3\}; \ \{y_1, y_2, y_3\};$$
$$\{z_1, z_2\}; \ \{z_2, z_3\}; \ \{z_1, z_3\}; \ \{z_1, z_2, z_3\}.$$

Definition 6.3.7. A *longitudinal set of messages* is a finite ordered set of messages where each message but the first one emerges from the previous message.

If we take the graphical example from Figure 6.3, then the following sets are longitudinal sets of messages:

$$\{x_1, y_1\}; \{y_1, z_1\}; \{x_1, y_1, z_1\};$$
$$\{x_2, y_2\}; \{y_2, z_2\}; \{x_2, y_2, z_2\};$$
$$\{x_3, y_3\}; \{y_3, z_3\}; \{x_3, y_3, z_3\}.$$

Definition 6.3.8. An *original message* is a message from the transversal set of messages at the output port of the source.

If we take the graphical example from Figure 6.3, then x_1, x_2, and x_3 are original messages.

Definition 6.3.9. An *image message* is a message from the transversal set of messages at the input port of the receptor.

If we take the graphical example from Figure 6.3, then z_1, z_2, and z_3 are image messages.

Definition 6.3.10. a) A *message association* is a pair of messages either from a longitudinal set of messages or from a transversal set of messages.

b) A pair of messages from a transversal set of messages is called an *information association.*

c) A *code association* is a pair of messages from a longitudinal set of messages.

If we take the graphical example from Figure 6.3, then z_1 and z_3 is an information association and x_1 and z_1 is a code association.

Definition 6.3.11. A *transformation* is a process or function that transforms one message from a message association into another message from the same association.

A transformation is denoted by

$$T: x \to y$$

Definition 6.3.12. *Information*, or more exactly, a *portion of information*, is a transformation of one message from an information association into another message from the same association.

Information (a portion of information) is denoted by

$$I: x \to y$$

We call y the *image* and x the *source* of the information I.

Example 6.3.4. Taking the graphical example from Figure 6.3, we have the following information portions:

$$I_1: x_1 \rightarrow x_2, I_2: y_2 \rightarrow y_3, \text{ and } I_3: x_2 \rightarrow x_3$$

This concept gives us several types of information:

- *Trivial information* is a trivial transformation, i.e., a transformation in which the second message is not different from the first one.
- *Non-trivial information* is a non-trivial transformation, i.e., a transformation in which the second message is different from the first one.
- *Identical information* is a trivial transformation, i.e., a transformation in which the second message is the same as (identical to) the first one.

There is a difference between trivial information and identical information because in a trivial information, there are no transformations, while in identical information, transformations take place but their output is the same as (or identical to) their input. For instance, you take the number 3 and do nothing with it. It will be trivial information. However, if you take the number 3, add the number 2, and subtract 2 from the result, it will be identical information.

It is possible to define the sequential composition of information as the sequential composition of corresponding transformations. It is defined for portions of information I and J when the image of I coincides with the source of J.

Definition 6.3.13. For portions of information

$$I_1: x_1 \rightarrow x_2 \text{ and } I_3: x_2 \rightarrow x_3,$$

the resulting portion of information (called the *sequential composition* of I_1 and I_2) is obtained by application of I_1 and then by application of I_2.

Proposition 6.3.1. The sequential composition of trivial (identical) portions of information is a trivial (identical) portion of information.

Definition 6.3.14. A portion of information $I^{-1}: x_1 \rightarrow x_2$ is called inverse to a portion of information $I: x_1 \rightarrow x_2$,

Definition 6.3.15. *Information chain* is a chain of messages from a transversal set of messages.

Definition 6.3.16. *Informing* is a transformation of a portion of information from a chain of original messages into a portion of information from a chain of image messages.

According to Mazur (1970), these definitions imply the following result.

Proposition 6.3.2. The resulting information for the sequential composition of information I and its inverse information I^{-1} is a trivial portion of information.

To illustrate this proposition, Mazur (1970) gives an example of such portions of information and their composition. The portion of information I is the statement "Stockholm lies 1500 km to the North of Budapest" and the inverse portion of information I^{-1} is the statement "Budapest lies 1500 km to the South of Stockholm". However, in contrast to Proposition 6.3.2, the sequential composition of the portion of information I and its inverse portion of information I^{-1} is not a portion of trivial information.

For information and informing as an operation in sets of information portions, Mazur introduces and studies a diversity of types and kinds, such as *simulating informing, disimulating informing, confusing informing, transinforming, comparative transinforming, compensation transinforming, equivalent transinforming, exclusive transinforming, pseudoinforming, pseudoinformation, simulating pseudoinformation, parainformation, paratransinformation, paradisinformation, disimulating pseudoinformation, useful information, superfluous information,* and *parasitic information,* to mention but a few.

All this shows that Mazur's theory becomes a specific direction of the general theory of information if we take the set of information associations as the infological system IF(R). Mazur identifies information with transformation, while in the general theory of information, information is only a potency to perform transformation. The latter understanding is more efficient because it allows one to consider transformation as actualized information. In essence, information is related to transformation as energy is related to work (Burgin, 1998/1999).

Chapter 7

Dynamics of Information

The universe is dynamic. When we are creative,
we are the most alive and in touch with it.
Brad Dourif

According to the general theory of information, information, like energy, is a dynamic essence. Different researchers understood this feature of information. For instance, Nonaka explains (1996), "information is the flow, and knowledge is the stock." As we have seen, information controls and directs the majority (if not all) processes in the world. Information expresses itself in a variety of processes. That is why information dynamics is so important. It is necessary to know and understand how information behaves in definite situation, what is possible to do with information and what information can do. As a result, many researchers paid main attention to information dynamics. This Chapter contains an exposition of dynamic theories of information developed by different authors.

In Section 7.1, theories of information flow constructed by Fred Dretske, Jon Barwise and Jerry Seligman are presented. Dretske built a qualitative theory of information actions and transformations. Barwise and Seligman formalized and further developed Dretske's approach, extensively using tools from set theory, logic and category theory. In Section 7.2, the operator theory of information elaborated by Alexander Chechkin is portrayed. Concepts of data and information are formalized and mathematically modeled in a set theoretical setting. The given description of information and data dynamics is based on structures of ultraoperators and ultramappings.

Section 7.3 presents information algebra and geometry. The mathematical component of the general theory of information is expounded in Subsection 7.3.1. This mathematical theory is based on functional models, utilizing multidimensional linear spaces, spaces of semantic networks, and spaces of named set chains with operators acting in these spaces. These information spaces are used as state spaces and phase spaces of infological systems. Operators represent portions of information, acting on representation spaces of infological systems. These operators are called information operators. Note that it is also possible to develop a mathematical component of the general theory of information based on category theory.

A theory of abstract information algebras is presented in Subsection 7.3.2. In some cases, abstract information algebras can model operations and processes with information carriers, such as data, texts, and documents. Two basic operations, combination and focusing, are mathematically modeled in abstract information algebras, reflecting important situations in information processing and management. First, information often comes from different sources and is usually combined. This process is modeled by information combination. Second, information retrieval aims to extract those parts out of a piece of information that are relevant to specific questions. This process is modeled by information focusing.

Abstract information algebras capture a variety of formalisms in computer science, which seem to be different on the surface, but interconnected in their essence: relational databases, multiple systems of formal logic, distributed computations and numerical problems of linear algebra.

To represent information dynamics researchers have developed not only algebra but also logical calculi. For instance, van Rijsbergen (1986) suggested to treat information retrieval as a form of plausible, or uncertain, inference and formulated a logical uncertainty principle. Later a calculus based on these ideas was developed (cf. (van Rijsbergen and Laimas, 1996)).

Information geometry is discussed in Subsection 7.3.3, demonstrating application of geometrical methods and structures to information studies.

7.1. Information Flow in the Approach of Dretske, Barwise and Seligman: Information Process

> *... they began running when they liked,*
> *and left off when they liked,*
> *so that was not easy to know when the race was over.*
> Lewis Carroll (1832–1898)

Dretske (1981) developed an information theory that combines the main ideas of the syntactic approach with the semantic information content. He perceives information as an objective essence that exists in the world. Meaning is partly generated by a receiver but information has independent existence. Information flows due to the causal connections between some "object *a being F*" and another "object *b being G*". Here *F* and *G* are some classes, types or properties. In his approach, Dretske ignores average amounts of information associated with symbols but considers information content given a framework of natural laws and the circumstances of the situation. His theory is based on the following fundamental principle.

Xerox-Principle. If *A* carries the information that *B* and *B* carries the information that *C*, then *A* carries the information that *C*.

Dretske gave such a name to this natural property of information due to the following situation that illustrates this principle. When Xerox copies a text that contains some information, then as long as the copy is sufficiently accurate, it carries the same information as the original.

The Xerox principle is an expression of the property of information content transitivity, i.e., there is a transition of information content from *B* to *A*. However, a detailed analysis shows that there are situations when the Xerox principle is not true. For instance, an individual *A* receives a letter informing her that her aunt will come to visit her next month. This aunt is a meticulous person and always wants to make communication more reliable. That is why, she makes a copy of her original letter to *A*, using a Xerox machine, and sends it to *A* at the same day. After some time, *A* receives an identical copy of the first letter. However, this copy does not have any information for *A* because *A* already knows everything that it is in this letter, namely, that her aunt will come to visit her next month. Thus, the copy does not contain the same information as the original.

However, if A receives the copy of the original letter before she receives the original letter, then the copy will convey information to A, while the original will not have this information for A.

This correlates well with the opinion of Stonier (1996) that the meaning of two identical messages might be increased somewhat as a result of being repeated.

Information flow is an intrinsic component of any communication process. That is why Dretske introduces conditions on communication to specify basic properties of information flow (information transmission).

The **first communication condition** tells that

(**A**) If a signal carries the information that some entity s is F, then the signal carries as much information about s as would be generated by s being F.

The **second communication condition** tells that

(**B**) If a signal carries the information that some entity s is F, then s is F.

The **third communication condition** specifies the first communication condition.

(**C**) The quantity of information the signal carries about s is (or includes) the quantity of information generated by s being F.

Communication conditions reflect semantic aspects of information as they connect information and meaning. However, to exist, information does not require, according to Dretske, an interpretive process, although it is a necessary condition for acquiring knowledge.

Importance of communication conditions for Dretske's theory of information implies necessity to thoroughly analyze these conditions. The first condition looks very natural but it is easy to show that it is violated in many cases. Indeed, it is possible that a signal carries information that s is F and also information that s is H. Thus, this signal carries more information than only information that s is F. This contradicts the first communication condition.

The third condition is considered as a specification of the first condition. Indeed, it remedies the situation when in addition to the possibility that "information the signal carries about s is the quantity of information generated by s being F", it is also allowed that "information the signal carries about s includes the quantity of information generated by s being F".

The second condition actually is an assertion that information conveys only true knowledge. In turn, this means that only true information in the conventional sense is allowed. Dretske assumes there is no false information but there is meaning without truth. He writes (1981), "information is what is capable of yielding knowledge, and since knowledge requires truth, information requires it also". However, it is persuasively demonstrated in Sections 2.4 and 2.5 that it is necessary to consider false information and false knowledge to reflect situations that exist in society and nature, as well as to encompass conventional usage of words *information* and *knowledge*.

In this context, the situation "*a is F*" *carries the information* about the situation "*b is G*" if the conditional probability of b being G given a being F is 1. Knowledge is treated by Dretske as a belief that a is F caused by the information that a is F, given some natural laws. The natural laws, as well as laws of logic belong to the framework of information flow. Elements of this framework, e.g., logically true sentences, cannot carry information. Even natural laws, according to Dretske, have an informational measure of zero.

On the one hand, our experience shows that information is not an indivisible absolute concept, because we can acquire varying degrees of information about an object. On the other hand, however, if information is only knowledge that a is F, as in (Dretske, 1981), then such information cannot come in degrees. It is an all or nothing affair. This understanding well correlates with the representation of knowledge on the microlevel described in Section 2.5 but only for elementary units of knowledge. Indeed, an elementary knowledge has the form of a statement that an object a has a property F or that the value of a property P is equal to F for an object a.

This inconsistency (related to information indivisibility) has two sources. First, as it is explained in fuzzy set theory, s can be F only to some degree (cf., for example, (Zimmermann, 2001)). Thus, it is necessary to use fuzzy statements, to get a more relevant information theory. Second, as it is discussed in Chapter 2, information comes in portions.

According to Dretske (1981), information is always relative to a receiver's background knowledge. However, what makes the amount of information a relative property is not what the receiver knows about the channel over which she gets information or about the source of this information, but what the receiver knows about the situation about which she is getting information. In accordance with the popular approach (cf., for example, (Mackay, 1969; Shreider, 1965; Brookes, 1980; Mizzaro, 2001) and Section 4.3), Dretske also assumes that information is "something that is required for knowledge". Thus, Dretske's information concept is different from meaning, but basically related to cognitive systems and knowledge. The relation between knowledge and information is a recursive but not a circular one. In order to learn that a is F a person should know, at least, something about s, without knowing that a is F. On the other hand, the information that a is F "causes K's belief that a is F". "Knowledge is information-produced belief" (Dretske, 1981) Thus, in the case of "genuine cognitive systems," as distinct from "mere processors of information," knowledge is specified with regard to information, meaning, and belief; or, in other words, with regard to interpretation during the learning process. Computers have, at least so far, no capability of using information. It means nothing to them. That is why, according to Dretske (2000), they can only manipulate symbols.

Developing Dretske's ideas on information as a relative property, Barwise and Perry (1983) built a situation theory and situation semantics. This theory is based on the concept of regularities between types of situations, which allow an information flow to take place. Linguistic regularities, as considered by Dretske, are special cases of information flow. According to the situation theory, information is always information about something and thus, dependent on constraints between situations. Situations contain information about other situations, inducing the constraint or *informational relation* between situations. Such a constraint always has a direction from the situation carrying the information to the situation about which is this information. Thus, it is a kind of a named set. The elementary information in situation theory is an element of the semantic closure of that situation that functions as the information carrier. As a result, it is possible to take the situation theory of Barwise and Perry (1983) as a base for a comprehensive information theory.

The theory of information flow developed by Kenneth Jon Barwise (1942–2000) and Jerry Seligman (Barwise, 1997; Barwise and Seligman, 1997) is a formalization and development of the approach introduced by Dretske (1981). In the introduction to their book, Barwise and Seligman acknowledge how it is difficult to talk about information based on the existing situation in information theory. Even related concepts, such as information flow and information content can be misleading and confusing. They assume that the least dangerous is the metaphor of carrying information because it is possible to make sense of one thing carrying different items of information on different occasions and for different people. Thus, they adopt the form "*x* carries/bears/conveys the information that *y*" as their primary means of making claims about the conveyance of information.

The theory of Barwise and Seligman is based on four principles.

First Principle of information Flow. Information flow results from regularities in distributed systems.

Second Principle of Information Flow. Information flow critically involves both types and their particulars.

Third Principle of information Flow. It is by virtue of regularities among connections that information about some component of a distributed system carries information about other components.

Fourth Principle of information Flow. The regularities of a given distributed system are relative to its analysis in terms of information channels.

Any theoretical science, such as theoretical physics or theoretical biology, and mathematics work at the level of types or patterns as this allows one to compress information about the theory domain. For instance, when meteorologists estimate probability of rain, they treat rain as some atmospheric phenomenon and state of weather. Other states of weather, such as clear, windy, shower, etc., also are types. Some rain in New York that happened on May 5, 2008 is a particular event of the type *rain*. In a similar way, when physicists describe behavior of an electron or proton by corresponding equations, electron and proton are types. In astronomy, planet is a type and the Earth is its particular. Events of probability theory are types of events and not particular events.

To reflect this peculiarity of science and mathematics, Barwise and Seligman make *classification* the basic concept of the theory of information flow.

Definition 7.1.1. A *classification A* is a triad (T_A, P_A, κ_A) where T_A is a set of objects also denoted by tok(A) and called *tokens* of A, P_A is a set of (usually, symbolic) objects also denoted by typ(A) and called *types* of A, and κ_A is a binary relation between T_A and P_A that tells one which tokens are classified as being of which types.

A classification A is represented by the following diagram:

In a classification $A = (T_A, P_A, \kappa_A) = (\text{typ}(A), \kappa_A, \text{tok}(A))$, objects (tokens) from A are related (belong) to types (classes) from Γ.

There are four ways of thinking of classifications although Barwise and Seligman consider only two of them. First, one may think of token as objects from the physical world and types as more abstract, e.g., from the mental world. It is assumed that tokens are given and need classification, while types are things used to classify tokens. For instance, in the first version of the information flow theory developed by Barwise (1993), only such tokens as situations were considered and classified by their types.

Second, another situation is when tokens are more abstract than types, which can belong to the physical world. In this case, abstract concepts are classified by means of those things they denote. For instance, it is possible to distinguish the concept of a dog (as a token) from the concept of a horse (as another token) by corresponding to these concepts some real dogs and horses, which play the role of types.

Third, one more situation emerges when we classify less abstract concepts by means of more abstract concepts. For instance, *animal* is a common type for the tokens *dog, horse, monkey, elephant*, and *tiger*.

Four, it is also possible to classify some physical objects (tokens) using other physical objects as types. For instance, it is possible to classify faculty at a university by chairs of their departments or by deans of their schools.

Sometimes tokens of a classification are simply mathematical, or more exactly, linguistic, objects, while in other cases, an additional structure is assigned to tokens so as to relate them to other tokens of other classifications.

Classifications are used in all areas and especially, in science. There are many famous classifications, such as classifications of subatomic particles, classifications of chemical elements (e.g., the periodic table), classifications in biology, etc. Many classifications are utilized for decision-making and information quality assessment (cf. Chapter 6).

Example 7.1.1. There are different classifications of subatomic particles. According to their structure, subatomic particles are divided into two groups: *elementary particles* and *composite particles*. Elementary particles are particles with no measurable internal structure.

In turn, elementary particles are classified according to their spin. *Fermions* have half-integer spin, while *bosons* are have integer spin. This classification has a continuation. In the Standard Model, there are 12 types of elementary fermions: six *quarks* and six *leptons*. Besides, in the Standard Model, there are 12 types of elementary bosons — four confirmed (*photons*, *gluons*, *W-bosons*, and *Z-bosons*) and eight unconfirmed.

There are also different types of composite particles. *Hadrons* are defined as strongly interacting composite particles. They are either composite fermions called *baryons* or composite bosons called *mesons*.

Remark 7.1.1. A classification is a special case of a named set. Namely, classifications are set-theoretical named sets (cf. Appendix A). Thus, all classifications are named sets, while there are many types of named sets that are not classifications because all classifications are built from sets, i.e., classifications are set-theoretical named sets. Named sets in which relations are algorithms instead of functions or binary relations are examples of named sets that are not set-theoretical and not classifications. Indeed, it is possible to compute (realize) any computable binary relation or function by many different algorithms. For instance, there are infinitely many Turing machines that build (compute) the same enumerable relation (cf. Appendix B and (Burgin, 2005)). Thus, there are much more algorithmic named sets than computable set-theoretical named sets and computable classifications.

Remark 7.1.2. Barwise and Seligman (1997) denote the classification relation κ_A by the standard logical symbol $\models$, which is usually interpreted as truth-functional entailment (cf., for example, (Bergman, et al, 1980; Kleene, 1967) and Appendix C). That is, if a and b are some propositions or predicates, then $a \models b$ means "if a is true in some interpretation, then b is true in the same interpretation". In the context of classifications, truth-functional entailment means that if a type T is interpreted as a set Q, a token a is interpreted as an element from the same universe (class) that all elements from Q belong, and we have $(t, a) \in \kappa_A$, then a belongs to Q.

Remark 7.1.3. The construction of a classification, as Barwise and Seligman write (1997), has been used in various contexts prior to their book. They attribute the first formal appearance of such structures to Birkhoff (1940) who called them polarities. Another construction that is also a classification is a Chu space (cf. Appendix A). However, classifications were built much earlier. Any system of concepts is a classification. Any language contains many classifications. In mathematics, we encounter a diversity of classifications. For instance, taking fractions as tokens and rational numbers as types, we have a classification. In general such a cognitive operation as abstraction is an operation of building a classification. Actually, a classification coincides with a binary relation in the form of Bourbaki (1960).

Different relation between tokens and types are introduced and studied.

If a is a token of a classification $A = (T_A, P_A, \kappa_A)$, i.e., $a \in \text{tok}(A)$), then the *type set* of a is the set

$$\text{typ}(a) = \{t \in \text{typ}(A); (t, a) \in \kappa_A\}$$

Similarly, if t is a type of a classification $A = (T_A, P_A, \kappa_A)$, i.e., $t \in \text{typ}(A)$), then the *token set* of t is the set

$$\text{tok}(t) = \{a \in \text{tok}(A); (t, a) \in \kappa_A\}$$

Types t_1 and t_2 are *coextensive in A*, written $t_1 \sim_A t_2$, if $\text{tok}(t_1) = \text{tok}(t_2)$.

Tokens a_1 and a_2 are *indistinguishable in A*, written $a_1 \sim_A a_2$, if $\text{typ}(a_1) = \text{typ}(a_2)$.

Barwise and Seligman introduce important classes and types of classifications.

Definition 7.1.2. A classification *A* is called:

1) *separated* provided there are no distinct indistinguishable tokens in *A*;

2) *extensional* provided that all coextensive in *A* types are identical.

Remark 7.1.4. As it is demonstrated by Burgin (1997), ordinary sets and multisets are named sets because elements of each set and of each multiset always have names. Thus, ordinary sets and multisets also are classifications. General classifications are multisets (Hickman, 1980; Blizard, 1991), while separated classifications are ordinary sets as according to set theory all elements in sets are distinguishable (Cantor, 1895; Fraenkel and Bar-Hillel, 1958).

One more important class of classifications form state spaces.

Definition 7.1.3. A *state space* is a classification *A* in which each token is of exactly one type.

Definition 7.1.4. The types of a state space *A* are called *states* of tokens. If a token *a* has the type *t*, then it is assumed that *a is in the state t*.

Definition 7.1.5. A state space is *complete* if each state is the state of some token.

Using classifications, Barwise and Seligman (1997) give a new representation of propositions. Namely, a proposition is a triad $p = (A, a, t)$ where *A* is a classification, *a* is a token and *t* is a type. Then *p* would be true if $(a, t) \in \kappa_A$.

In the process of modeling information systems by classifications, it is necessary to formally represent relations between these systems. The main relation between classifications is *infomorphism*.

Let us consider two classifications $A = (T_A, P_A, \kappa_A)$ and $C = (T_C, P_C, \kappa_C)$.

Definition 7.1.6. An *infomorphism f* is a pair (f^*, f°) where f^*: $P_A \to P_C$ and $f^\circ: T_C \to T_A$ are functions that satisfy the following condition:

$$f^* (c) \ \kappa_A \ \alpha \text{ if and only if } c \ \kappa_C \ f^\circ (\alpha)$$

In other words, the following diagram is commutative.

$$
\begin{array}{ccc}
 & f^* & \\
P_A & \longrightarrow & P_C \\
\kappa_A \uparrow & & \uparrow \kappa_C \\
T_A & \longleftarrow & T_C \\
 & f^\circ &
\end{array}
\qquad (7.1.1)
$$

We use the following denotation of an infomorphism $f: A \rightleftarrows C$. Here *A* is the *domain* and *C* is the *codomain* of the infomorphism f.

Classifications and their infomorphisms form a category in which classifications are objects and infomorphism are morphisms (cf. Appendix D).

Classifications and their infomorphisms are used to build a model of distributed systems.

Definition 7.1.7. An abstract *distributed system* is a pair

$$A = (\text{cla}(A), \text{inf}(A))$$

where cla(A) is a set of classifications and inf(A) is a set of infomorphisms with domains and codomains in cla(A).

Each component of an abstract distributed system is a classification (set-theoretical named set), which represents a real system. Tokens from the classification represent elements and/or components of the real system, while types from the classification represent types of the elements and/or components of the real system.

As Barwise and Seligman suggest, information flow in an (abstract) distributed system is relative to some conception of that system as an information channel.

Definition 7.1.10. A *channel Ch* is an indexed family $\{f_i: A_i \rightleftarrows C; \ i \in I\}$ of infomorphisms (cf. Diagram (7.1.2)) with a common codomain *C* called the *core* of the channel *Ch*. When $|I| = n$, e.g., $I = \{1, 2, 3, \dots, n\}$, the channel *Ch* is called an *n*–ary channel.

$$
\begin{array}{ccc}
 & f_i{}^* & \\
P_{Ai} & \longrightarrow & P_C \\
\uparrow \kappa_{Ai} & & \uparrow \kappa_C \quad C \\
A_i & & \\
T_{Ai} & \longleftarrow & T_C \\
 & f_i{}^\circ &
\end{array}
\qquad (7.1.2)
$$

The channel Ch is interpreted by Barwise and Seligman as a representation of how the whole system C is formed/built of its parts A_i. However, this concept of a channel does not correlate with the conventional concept of a channel in communication theory, e.g., in Shannon's theory of information/communication.

Definition 7.1.11. The tokens from C in a channel Ch are called *connections* of the channel Ch. A connection c *connects* the tokens $f_i{}^\circ(c)$ for all $i \in I$ (cf. Diagram (7.1.2)).

An important relation between distributed systems and channels is covering.

Definition 7.1.12. A channel $Ch = \{f_i: A_i \rightleftarrows C; i \in I\}$ *covers* a distributed system $A = (\mathrm{cla}(A), \mathrm{inf}(A))$ if $\mathrm{cla}(A) = \{A_i; i \in I\}$ and for each infomorphism $f: A_i \rightleftarrows A_j$ from $\mathrm{inf}(A)$, the following diagram is commutative

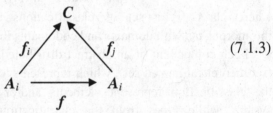

$$
(7.1.3)
$$

Barwise and Seligman (1997) find various properties of the introduced constructions. One of the central results of their book is the following theorem.

Theorem 7.1.1. Every distributed system has a minimal cover, which is unique up to isomorphism.

To prove this result the authors, at first, give an outline of a general category-theoretical proof based on existence of colimits in the category of classifications and their infomorphisms. Then they directly construct such a minimal cover.

It is interesting that the second part of Theorem 7.1.1 is given in the book as a separate statement (Proposition 6.6) with a separate proof.

The main tool of the information flow theory is logic. However, Barwise and Seligman build a generalization of the conventional construction of logic. Usually, a logical theory is a set of sentences and some kind of entailment,

$$\Gamma \vdash \alpha \qquad\qquad (7.1.4)$$

between sets of sentences and individual sentences. This entailment relation $\vdash$ is used for inference of new sentences from old ones in a given logic. The authors of the information flow theory, following Gentzen, use entailment to connect sets and not only sentences, i.e., Γ and α in Diagram (7.1.4) can be arbitrary sets. Such entailment relations are applied to classifications.

Let us consider a classification $A = (T_A, P_A, \kappa_A)$ and two subsets Γ and Δ of the set of types P_A. The pair (Γ, Δ) is called a *sequent* of this classification.

Definition 7.1.13. a) A token a of A *satisfies* (Γ, Δ) provided that if a is of type α for every $\alpha \in \Gamma$, then a is of type β for some $\beta \in \Delta$.

b) Γ *entails* Δ in A if any token $a \in T_A$ satisfies (Γ, Δ).

c) A sequent (Γ, Δ) in which Γ entails Δ in A is called a *constraint* supported by the classification A.

To each classification, Barwise and Seligman correspond a logic called a *local logic*. As some other structures, local logics are defined two times in the book. The first definition is in Section 2.3 of (Barwise and Seligman, 1997).

Definition 7.1.14. A *local logic* is a triad

$$L = (A, C_L, N_L)$$

where A is a classification, C_L is a set of constraints on tokens in A, and $N_L \subseteq \text{tok}(A)$ is a set of tokens in A that satisfy all constraints from C_L and are called normal tokens.

The set of all local logics forms a logical variety in the sense of (Burgin, 1995c; 1997f). Besides, local logics form a category where morphisms are infomorphisms of local logics defined by Barwise and Seligman (1997).

In more detailed way, a local logic, which connects the classifications with logical theories, is also defined on Section 12.1 of (Barwise and Seligman, 1997).

Definition 7.1.15. A *local logic* $L = (\text{toc}(L), \text{typ}(L), \kappa_L, \vdash_L, N_L)$ consists of three components:

A classification $\text{cla}(L) = (\text{toc}(L), \kappa_L, \text{typ}(L))$,

A *regular theory* $\text{th}(L) = (\text{typ}(L), \vdash_L)$, and

A subset N_L of the set of types $\text{typ}(L)$, called the set of *normal tokens* of L, i.e., tokens that satisfy the constraints of $\text{th}(L)$.

A local logic L is the logic of the corresponding classification $\text{cla}(L)$. Taking a set $\mathbf{W}$ of different classifications, we see that local logics that correspond to the classifications from $\mathbf{W}$ form a logical variety in the sense of (Burgin, 1995c; 1997f; 2004). In particular, taking a distributed system A, we obtain the distributed logic that consists of local logics and also forms a logical variety.

As a result, a local logic has two entailment relations: truth-valued entailment and inference entailment. The first one determines logical semantics of this logic, while the second one defines the deduction syntax of this logic.

Systems of local logics and their infomorphisms form abstract information systems, which can model real information systems and their dynamics. From the mathematical perspective, systems of local logics and their infomorphisms form abstract categories with local logics as objects and infomorphisms as morphisms (cf. Appendix D).

In addition, Barwise and Seligman (1997) construct various operations in the set of classifications. Let us consider two classifications $A = (T_A, P_A, \kappa_A)$ and $C = (T_C, P_C, \kappa_C)$.

Definition 7.1.16. The *sum* $D = A + C$ of classifications A and C is equal to (T_D, P_D, κ_D) where $T_D = T_A \times T_C$, $P_D = P_A \cup P_C$ and the relation κ_D is defined by the following rules:

$$(a, c) \, \kappa_D \, t \text{ if and only if } a\kappa_A \, t$$

$$(a, c) \, \kappa_D \, t \text{ if and only if } c\kappa_C \, t$$

where $a \in T_A$, $c \in T_{AC}$, and $t \in P_A \cup P_C$.

An example of the sum of two classifications is considered below.

Example 7.1.2. Let us consider two classifications $A = (T_A, P_A, \kappa_A)$ and $C = (T_C, P_C, \kappa_C)$ in which the sets of tokens T_A and T_C consist of people, $P_A = \{m, p, b\}$, the symbol m means "knows mathematics", the symbol p means "knows physics", the symbol b means "knows biology", $P_C = \{pv, pp, pg\}$, the symbol pv means "can play violin", the symbol pp means "can play piano", and the symbol pg means "can play guitar". Then the pair (Alice, p) belongs to κ_A if the individual Alice knows physics and pair (Bernice, pp) belongs to κ_C if the individual Bernice can play violin. In this case, the set of tokens T_D of the sum $D = A + C = (T_D, P_D, \kappa_D)$ consists of pairs of people, the set of types P_D is equal to $\{m, p, b, pv, pp, pg\}$, and (Alan, p) belongs to κ_D if the individual Alan knows mathematics, while (Bob, pg) belongs to κ_D if the individual Bob can play guitar.

Remark 7.1.5. The sum of classifications is a special case of a componential binary operation in named sets studied in (Burgin, 2006c).

Boolean operations form an important class of operations with classifications.

Definition 7.1.17. a) The *disjunctive power* $\vee A$ of a classification A is equal to the classification $H = (T_H, P_H, \kappa_H)$ where $T_H = T_A$, the set of types P_H is equal to the set $\mathbf{P}(P_A)$ of all subsets of the set P_A and the relation κ_H is defined by the following rule:

$$a \in \kappa_{DH}t \text{ if and only if } a\kappa_A \, t \text{ for some } t \text{ from } t$$

where $a \in T_A$, $t \in \mathbf{P}(P_A)$ and $t \in P_A$.

b) The *conjunctive power* $\wedge A$ of a classification A is equal to the classification $C = (T_C, P_C, \kappa_C)$ where $T_C = T_A$, the set of types P_C is equal to $\mathbf{P}(P_A)$ and the relation κ_D is defined by the following rule:

$$a \in \kappa_C t \text{ if and only if } a\kappa_A \, t \text{ for all } t \text{ from } t$$

where $a \in T_A$, $t \in \mathbf{P}(P_A)$ and $t \in P_A$.

An example of the disjunctive power of a classification is considered below.

Example 7.1.3. Let us consider the classification $A = (T_A, P_A, \kappa_A)$ from Example 7.1.2. In this case, the set of types P_H of the disjunctive power $H = \vee A = (T_H, P_H, \kappa_H)$ is equal to the set $\{\varnothing, \{m\}, \{p\}, \{b\}, \{m, p\}, \{b, m\}, \{p, b\}, \{m, p, b\}\}$. By Definition 7.1.7, if Alan knows mathematics, then he has types $\{m\}, \{m, p\}, \{b, m\}$, and $\{m, p, b\}$, while if Bob knows neither mathematics nor physics nor biology, he has the type $\varnothing$.

An example of the conjunctive power of a classification is considered below.

Example 7.1.4. Let us consider the classification $A = (T_A, P_A, \kappa_A)$ from Example 7.1.2. In this case, the set of types P_H of the disjunctive power $H = \vee A = (T_H, P_H, \kappa_H)$ is equal to the set $\{\varnothing, \{m\}, \{p\}, \{b\}, \{m, p\}, \{b, m\}, \{p, b\}, \{m, p, b\}\}$. By Definition 7.1.7, if Alan knows mathematics and physics but does not know biology, then he has types $\{m\}$, $\{p\}$, and $\{m, p\}$, but does not have types $\{b, m\}$, $\{p, b\}$, and $\{m, p, b\}$, while if Bob knows neither mathematics nor physics nor biology, he has the type $\varnothing$.

Definition 7.1.18. a) The *flip* of a classification A is the classification $A^\perp$ whose tokens are types of A and whose types are tokens of A and in which $\alpha\kappa_{A^\perp} a$ if and only if $a\kappa_A \alpha$ for all $a \in T_A$ and $\alpha \in P_A$.

b) The *flip* of a pair of functions (of an infomorphism) $f = (f^*, f^\circ)$: $A \rightleftarrows C$ is the pair of functions $f^\perp = (f^\circ, f^*)$: $C^\perp \rightleftarrows A^\perp$ where $f^\perp = (f^\circ, f^*)$.

If you have a cat Pussy and a dog Barky, then in the natural biological classification B, Pussy has the type *cat* and Barky has the type *dog*. If we flip the natural biological classification B, then the token *cat* will have the type "the cat Pussy", while the token *dog* will have the type "the dog Barky".

Remark 7.1.6. The flip of a classification A is the operation of involution of a binary relation. It was used and studied both in the set-theoretical and categorical setting (Burgin, 1970).

Proposition 7.1.2. $f: A \rightleftarrows C$ is an infomorphism if and only if $f^\perp$ is an infomorphism.

<u>Proof</u>. Let us assume that $f: A \rightleftarrows C$ is an infomorphism. Then f is represented by the following commutative diagram:

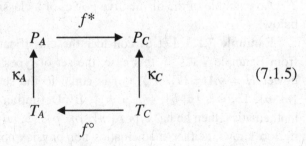

$$(7.1.5)$$

while the pair of functions $f^{\perp} = (f^{\circ}, f^*)$ is represented by the following diagram, which is also commutative.

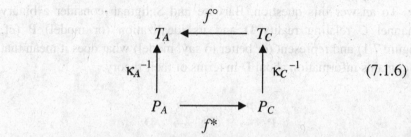

$$(7.1.6)$$

Then the diagram (7.1.5) is commutative if and only if the diagram (7.1.6) is commutative. Commutativity of the corresponding diagram is the condition for a pair of functions to be an infomorphism (cf. Definition 7.1.6). Thus, proposition is proved.

Operations with classifications are extended to operations with local logics.

Barwise and Seligman (1997) discuss various applications of their theory: to modeling information systems and using the mathematical theorems serve as pointers to what constructions are available. In addition, they show how the channel view adds useful insights to existing approaches to distributive system modeling. Besides, a number of explorations are undertaken relating information flow to the theory of information transfer by language use, nonmonotonic reasoning in AI, and quantum experiments.

In addition, Barwise and Seligman apply developed concepts to explore an old problem in the philosophy of science and mathematics that asks what is the source of the great importance of mathematics for science. In an address to the Prussian Academy of Science in Berlin on January 27, 1921, Einstein formulated this problem in the following way:

"How can it be that mathematics, being after all a product of human thought which is independent of experience, is so admirably appropriate to the objects of reality?"

Later this problem was reiterated by Wigner (1960), who also wrote that this enigma "in all ages has agitated inquiring minds".

Barwise and Seligman treat this problem as a question concerning information flow. They formulate the problem in a more general and at the same time, more exact form:

How is it that science, with its use of abstract mathematical models, carries any information at all about the real world?

To answer this question, Barwise and Seligman consider a binary channel C relating reality $\mathbf{D}$ and its idealization (or model) $\mathbf{P}$ (cf. Figure 7.1) and represent (or better to say, model) what does it mean that $\mathbf{P}$ contains information about $\mathbf{D}$ in terms of their theory.

$$\mathbf{P} \xrightarrow{p} C \xrightarrow{d} \mathbf{D}$$

Figure 7.1. Information channel connecting reality to its model (idealization)

It is necessary to remark that although many mathematical constructions used in the theory of information flow were also used in other areas, for example, in pure mathematics or in the theory of concurrent processes, they were not used, as a rule, for information studies. The theory of Barwise and Seligman is an important original direction of information theory as a whole. Other researchers have been using this theory for different purposes and in different areas: to model information integration and ontology mapping (cf., for example, (Kalfoglou and Schorlemmer, 2003)), in psychology and linguistics (Cutting and Ferreira, 1999), for developing models of information interoperability in the grid (Li Weihua and Li Shixian, 2005), for conceptual knowledge organization (Kent, 2000; Malik, 2004) and in the field of semantic databases (Wang and Junkang Feng, 2006). Restall (1996) studies relations between information flow in the sense of Barwise and Seligman and relevant logics.

An original approach to information and information flow (information transfer) was sketched by William Cooper (1978). The goal was to provide a platform upon which a unified theory of language and logic may be developed in an integrative fashion.

The basic primitive notion of the approach is that of a state of *informedness* or *information state*. Taking it as a primitive term, Cooper does not give a precise definition but only explains its meaning on several examples of systems often considered to contain information, such as a library, hologram, DNA molecule (as a storage of hereditary information), the internal memory of a computer, and the human brain. These example shows that an information state, in the sense of Cooper (1978), is a state of system that contains information. In the context of the general theory of information an information state is a state of an infological system. The difference between these two informal definitions is that the former is based on the concept of information, while the second is independent of this concept. As a result, the former definition cannot be used for defining information without the fallacy of the vice circle.

Other primitive notions of the Cooper's approach are *information input*, *test input*, and *information output*. Together with the notion of information state, these three notions are used for defining an information automaton.

A *information automaton* **A** consists of three structures (the linguistic structure, state structure and action structure), that is, **A** = (L, S, δ):

The *linguistic structure* L = (X, Q, Y) where Q is a finite *set of states*, the *input set* X is the union of two sets $X_I \cup \{.\}$ and $X_T \cup \{?\}$, X_I is the *informative input set*, X_T is the *test input set*, the *output set* Y consists of three parts – the *regular output set*, the *set Y_R of yes-no-responses*, and the *null output* Λ of the automaton **A**;

The *state structure* S = (Q, q_0, F) where q_0 is an element from Q that is called the *initial* or *start state* and F is a subset of Q that is called the set of *final* (in some cases, *accepting*) states of the automaton **A**;

The *action structure*, which consists of two functions:

the *transition function* of the automaton **A**

$$\delta: X \times Q \to Q,$$

the *output function* of the automaton **A**

$$\lambda: X \times Q \to Y$$

The action structure satisfies additional conditions:

(a) $\delta(q, <x, ?>) = q$ for any $q \in Q$ and $x \in X_T$;
(b) $\lambda(q, <x, .>) = \Lambda$ for any $q \in Q$ and $x \in X_I$;
(c) $\lambda(q, <x, ?>) \neq \Lambda$ for any $q \in Q$ and $x \in X_T$.

Note that an information automaton is a kind of abstract deterministic automata that takes information items as inputs and produce information items as output (cf., for example, (Burgin, 2005) or (Hopcroft, et al, 2001)).

After introducing the concept of information automata, Copper determines some of them to be language automata and from there proceeds to develop a unified theory of language and logic. As the notion of information is not actually specified and studied in detail, Cooper's approach is not well known in the information science community although this approach has an interesting potential of automata theory application to the study of information flow and information transformations.

7.2. Operator Approach of Chechkin: Information Action

> *You must not give the world what it asks for, but what it needs.*
> Edgar Dijkstra

The approach of Chechkin (1991) is based on two main notions: an object and information about an object. An *object* is one of the most basic notions that can comprise actually anything. A bird, human being, plane, apple, word, book, reader, star — all of them can be objects. For instance, a bird is an object in biology, a star is an object in astronomy and a word is an object in linguistics. Although an object is opposed to a subject, this opposition is relative and what is a subject in one situation will be an object in another situation.

Information about an object is interpreted by Chechkin as a kind of semantic information. In this context, a piece of information about an object A is built from data on this object A. Each datum is a triad (p, D, A) where D is a subset of some universal set X of objects and p is the certainty that A belongs to the domain D. Thus, the datum (p, D, A) tells that the object A belongs to the domain D with the certainty p. A *piece of information* about an object A is such a collection of data that with each datum it contains all simple logical implications of this datum. It is possible to describe the domain D as a subset of the set X. For instance,

D is the set of all whole numbers, which is a subset of the set R of all real numbers. It is also possible to treat the domain D as a property, e.g., D is a whole number. In the second case, the datum (p, D, A) means that the object A has the property D with the certainty p.

Dynamics of information is represented by ultramappings and ultraoperators. That is why we call this approach *operator information theory*.

All these concepts are defined formally as mathematical structures. Below we describe the main constructions, concepts, ideas and some results of this theory.

Let X be some set of objects, which are called *points* in the operator information theory. Properties of points from X are represented (and denoted) by subsets of those points from X that have this property, i.e., each property is a subset of X and each subset of X is a property of points (objects) from X. It is the classical set-theoretical approach to the concept of property (cf., for example, (Shoenfield, 1967)).

Definition 7.2.1. If a is a point from X, then an *elementary datum on* an object a from X is the fact that a has some property H or $a \in H$ or the true proposition "a has some property H".

It is denoted by $H(a)$.

Here the symbol "H" is a predicate symbol of the property H and at the same time, a symbol denoting the subset H of the set X.

The set of all elementary data on a point a from X is denoted by $Data(a)$. The set of all elementary data on all points (objects) from X is denoted by $Data(X)$.

Remark 7.2.1. An elementary datum $H(a)$ on a is naturally represented by the triad (named set) $(a, \in, H)$.

Proposition 7.2.1. If $H \subseteq Q$ and $H(a)$ is an elementary datum on a, then $Q(a)$ is also an elementary datum on a.

Indeed, if $a \in H$, then $a \in Q$ as $H \subseteq Q$.

Corollary 7.2.1. If $Q(a)$ and $H(a)$ are elementary data on a, then $(H \cup Q)(a)$ and $(H \cap Q)(a)$ are also elementary data on a.

This allows one to define logical operations with elementary data on a given point.

Definition 7.2.2. If a is a point from X and $H(a)$ and $Q(a)$ are elementary data on a, then:

The *conjunction* of $H(a)$ and $Q(a)$ is equal to $H(a)$ & $Q(a)$ = $(H \cap Q)(a)$.

The *disjunction* of $H(a)$ and $Q(a)$ is equal to $H(a) \vee Q(a)$ = $(H \cup Q)(a)$.

The *negation* of the elementary datum $H(a)$ is equal to $\rceil H(a)$ = $(X \backslash H)(a)$.

Definition 7.2.2 implies the following result.

Proposition 7.2.2. Logical operations with elementary data on a point satisfy all properties of standard logical operations negation, disjunction and conjunction.

Thus, the set *Data*(a) is a carrier of a universal algebra with these operations.

Definition 7.2.3. If a is a point from X, then two elementary data $H(a)$ and $Q(a)$ on a are *comparable* if either $H \subseteq Q$ or $Q \subseteq H$. In the first case, the elementary datum $H(a)$ is *less general* than the elementary datum $Q(a)$, while the elementary datum $Q(a)$ is *more general* than the elementary datum $H(a)$.

We denote this relation by $H(a) \Leftarrow Q(a)$ and $Q(a) \Rightarrow H(a)$, correspondingly.

Proposition 7.2.3. An elementary datum $Q(a)$ is *more general* than an elementary datum $H(a)$ if and only if the named set $(a, \in, H)$ is a named subset of the named set $(a, \in, Q)$, i.e., $H \subseteq Q$.

As $H \subseteq Q \cup H$ and $Q \cap H \subseteq H$, we have the following result.

Proposition 7.2.4. $Q(a) \vee H(a) \Rightarrow H(a) \Rightarrow Q(a) \wedge H(a)$

In addition, properties of sets imply that if the elementary datum $H(a)$ is more general than the elementary datum $Q(a)$ and the elementary datum $Q(a)$ is more general than the elementary datum $H(a)$, then these data coincide.

Proposition 7.2.5. For any point a from X, the set *Data*(a) of all elementary data on a is a distributive lattice with the maximal and minimal elements, which is isomorphic to the lattice of all subsets of X.

Definition 7.2.4. A family $I(a)$ of elementary data on a point a from X, is called an *elementary piece of information on a* if it satisfies the following conditions:

1. If $H(a) \in I(a)$, then H is not empty as a subset of X.

2. If $H(a) \in I(a)$ and $Q(a)$ is more general than $H(a)$, then $Q(a) \in I(a)$.

3. If $H(a), Q(a) \in I(a)$, then $H(a) \& Q(a) \in I(a)$.

The set of all elementary pieces of information on a point a from X is denoted by $EInf(a)$. The set of all elementary pieces of information on all points from X is denoted by $EInf(X)$.

It is possible to represent many objects used in information technology and engineering as elementary pieces of information.

Example 7.2.1. In relational databases, data are represented by relations. A *database relation* is generally defined as a set of tuples that correspond to a fixed set of attributes (cf., for example, (Elmasri and Navathe, 2000)). It may be treated as a mathematical relation (cf. Appendix A) between objects (or more exactly, names of objects) and values of chosen attributes. A tuple, i.e., an organized set (usually, a sequence) of symbolic elements, usually represents an object by its name and additional information about the object in a form of values of the attributes. For instance, the tuple (Smith, John, 5–10, 155, 35, USA, California, M, 2) represents a person whose name is John Smith, height is 5 ft 10 in, weight is 155 pounds, age is 35 years, nationality is USA, place of living is California, who is married and has two children. This tuple corresponds to the following sequence of attributes: last name, first name, height, weight, age, nationality, marital status, and the number of children.

Considered objects are typically physical entities or conceptual structures. A relation is often presented in of a table organized into rows and columns. Rows correspond to objects and columns correspond to attributes. This representation results in two named sets: (Ob, *Rel*, *T*) and (Ob, *Rel*, Attr). Here Ob is the set of all objects represented in the database, *T* is the set of all tuples, Attr is the set of all attributes used in the database, and *Rel* is the relation between objects and attributes. Usually *Rel* is represented in the matrix or table form. In this case, we have two more two named sets: (Ob, *Rel*, *T*) and (Ob, *Rel*, Attr).

Each tuple that represents an object is an elementary datum in the sense of Chechkin. For instance, the tuple (Smith, John, 5–10, 155, 35, USA, California, M, 2) is an elementary datum on a person whose name is John Smith. This shows that elementary data comprise object representation in relational databases.

Example 7.2.2. Design theory and methodology is an area of engineering research devoted to the formalization and algorithmization of engineering design and methods used by designers. To achieve better organization and utilization of design information, an axiomatic model called the HM model and based on set theory is developed by Salustri and Venter (1992). The basic structure in this model is an object. An *object* is defined as a set of attributes. This allows us to consider each object as an elementary piece of information and each elementary piece of information as an object of the HM model.

Definitions imply the following result.

Proposition 7.2.6. A family $I(a)$ of elementary data on a point a from X, is elementary piece of information on a if and only if family $\mathbf{H} = \{H; H(a) \in I(a)\}$ of subsets of X is a filter (cf., for example, (Kuratowski and Mostowski, 1967) and Appendix A).

Sets $EInf(a)$ of elementary pieces of information on a point a and $Data(a)$ of elementary data on a have additional algebraic structures.

Proposition 7.2.7. For any point a from X, the set $EInf(a)$ of all elementary pieces of information on a is distributive lattice with the largest and smallest elements.

There are also natural relations between sets $EInf(a)$ and $Data(a)$.

Proposition 7.2.8. For any point a from X, the lattice $EInf(a)$ of all elementary pieces of information on a contains a sublattice isomorphic to the lattice $Data(a)$ of all elementary data on a.

Proof. For any elementary datum $H(a)$, we can define $g(H(a)) = \{ Q(a); Q(a)$ is more general than $H(a) \}$. It is an elementary piece of information. Indeed, if $Q(a)$ belongs to the set $g(H(a))$, then $Q(a)$ is not empty because $H(a)$ is not empty, i.e., the first condition from Definition 7.2.4 is true. If $P(a) \in I(a)$ and $Q(a)$ is more general than $P(a)$, then $Q(a)$ is more general than $H(a)$ as $P(a)$ is more general than $H(a)$ and thus, $Q(a) \in g(H(a))$, i.e., the second condition from Definition 7.2.4 is true. If $P(a)$ and $Q(a)$ belong to $g(H(a))$, then $H(a)$ & $Q(a)$ belong to $g(H(a))$ because if a set is a subset of two sets, then it is a subset of their intersection, i.e., the third condition from Definition 7.2.4 is also true.

Proposition is proved.

Definition 7.2.5. A subset $S(a)$ of an elementary piece of information $I(a)$ on a point a from X is called a *support* of $I(a)$ if for any datum $H(a)$ $\in I(a)$, there is a less general than $H(a)$ datum $Q(a)$ from $S(a)$.

Definition 7.2.5 implies the following result.

Proposition 7.2.9. Any support $S(a)$ of $I(a)$ uniquely determines $I(a)$.

Proposition 7.2.10. If $S(a)$ is a support of $I(a)$ and $S(a) \subseteq V(a)$, then $V(a)$ is a support of $I(a)$.

Proposition 7.2.11. If $S(a)$ is a support of $I(a)$ and $|S(a)| = 1$, then $I(a)$ has only one minimal support.

A support of an elementary piece of information $I(a)$ for which $|S(a)| = 1$ is called an *individual support*.

Definition 7.2.6. A subset V of the set $EInf(X)$ of all elementary pieces of information is called a *section* of $EInf(X)$ if for each point x from X, it has exactly one elementary information $I(x)$.

It is possible to consider $EInf(X)$ as a fiber bundle (cf. Appendix G and (Goldblatt, 1979)) with the base X and fibers $EInf(a)$ for all $a \in X$. In this case, sections of the set $EInf(X)$ are sections in the sense of the theory of fiber bundles.

Proposition 7.2.12. There is a duality between classifications $(T_A,$ $P_A,$ $\kappa_A)$ in the sense of Bairwise and Seligman (1997) and sections of elementary pieces of information $I(x)$ in the sense of Chechkin (1991).

Indeed, in a classification $A = (T_A,$ $P_A,$ $\kappa_A)$, an object from T_A is related (belongs) to types (classes) from P_A. At the same time, in an elementary piece of information $I(x)$, the object x from X possesses properties from the set of properties $\mathbf{H} = 2^X = \{H; H \subseteq X)\}$. A set of properties $\mathbf{H}$ defines a type of objects that have these properties. At the same time, "to be of some type" is a property of an object. As a result, we have a correspondence between types and sets of properties. Thus, we obtain a correspondence CI between classifications and elementary pieces of information such that any classification $(T_A,$ $P_A,$ $\kappa_A)$ uniquely defines an elementary piece of information $I(x)$ for each x from the set $X = T_A$, and a correspondence IC any section W of elementary pieces of information $I(x)$ uniquely defines a classification $(X,$ $\mathbf{H},$ $\kappa_A)$. In addition, taking a classification $(T_A,$ $P_A,$ $\kappa_A)$ with $X = T_A$ and $P_A = 2^X$, and at first applying to it the correspondence CI and then the correspondence IC, we obtain the same classification $(T_A,$ $P_A,$ $\kappa_A)$. In a similar way, taking a section W and applying to it at first the correspondence IC and then the correspondence CI, we obtain the same section W. Thus, the correspondence CI determines a one-to-one mapping from the set of all classifications of objects from X with types from the set 2^X onto the set of all elementary pieces of information on objects from X.

It is natural to consider some properties of elements from the set X as concepts. To formalize this, concept lattices are introduced.

Definition 7.2.7. A system L_X of subsets of X is called a *concept lattice* for X if it satisfies the following conditions:

1. $X \in L_X$ and $\varnothing \in L_X$.
2. If $A, B \in L_X$, then $A \backslash B \in L_X$.
3. If $A, B \in L_X$, then $A \cap B \in L_X$.
4. If $A, B \in L_X$, then $A \cup B \in L_X$.

Definition 7.2.8. An element A from a concept lattice L_X is called an *atom* if it does not contain other nonempty elements from L_X.

Definition 7.2.9. Two elements A and B from a concept lattice L_X are called:

    a) *disjunctive* if they do not intersect;

    b) *disjunctive* in L_X if their intersection is either empty or does not belong to L_X.

Proposition 7.2.13. Any two different atoms of L_X are disjunctive in L_X.

Indeed, by Definition 7.2.8, each of these atoms does not contain other nonempty elements from L_X, while each of them contains their intersection. Consequently, this intersection is either empty or does not belong to L_X.

Definition 7.2.10. A subset V of a concept lattice L_X is called a *scale* of L_X if V generates L_X by means of the set-theoretical operations $\cap$, $\cup$, and $\backslash$.

Proposition 7.2.14. Any scale of L_X uniquely defines the concept lattice L_X.

To define the information quantity in an elementary piece of information, Chechkin introduces the concept of a discrete certainty distribution in two forms: local and global.

Let us consider a scale V of a concept lattice L_X.

Definition 7.2.11. A function $p_a: V \to [0, 1]$ is called a (*local*) *discrete certainty distribution* on the scale V for a point a from X if it satisfies the following condition:

$$\sum\nolimits_{v \in V} p_a(v) = 1$$

The concept of a local certainty distribution allows one to define the concept of a global certainty distribution.

Definition 7.2.12. A function $p: V \times X \to [0, 1]$ is called a (*global*) *discrete certainty distribution* on the scale V if for each point a from X the restriction p_a of the function p on a is a discrete local certainty distribution on the scale V for a point a.

Let us consider an elementary datum $Q(a)$. The entropy of the initial certainty distribution is defined by the following formula

$$H[X(a)] = - \sum\nolimits_{v \subset X} p(v, a) \cdot \log_2 p(v, a)$$

The entropy of the secondary certainty distribution is defined by the following formula

$$H[Q(a)] = - \sum\nolimits_{v \subset Q} p_Q(v, a) \cdot \log_2 p_Q(v, a)$$

where

$$p_Q(v, a) = \begin{cases} 0 & \text{when } v \not\subset Q, \\ p(v, a)/(\sum\nolimits_{v \subset Q} p(v, a)) & \text{when } v \subset Q \end{cases}$$

These constructions are similar to the Shannon's entropy.

The concept of entropy allows Chechkin to define the quantity of information $Iq[Q(a)]$ in an elementary datum $Q(a)$ as the measure of the entropy difference between the initial and secondary certainty distributions. Namely, it is defined by the following formula

$$Iq[Q(a)] = H[X(a)] - H[Q(a)]$$

The concept of the quantity of information in an elementary datum allows Chechkin to define the quantity of information in an elementary piece of information.

Let us consider an elementary piece of information $I(a)$ on a point a from X, a concept lattice L_X, and its support $S(a)$ with the property that any Q from $S(a)$ belongs to L_X. Then the quantity of information $Iq[S(a),$ $I(a)]$ that the support $S(a)$ introduces into the elementary piece of information $I(a)$ is defined by the following formula

$$Iq[I(a)] = \sup \{Iq[\wedge_{Q \in S(a)} Q(a)]\}$$

To define a general piece of information, Chechkin introduces the concept of a certainty lattice. Namely, a lattice P with the largest and smallest elements is called a *certainty lattice* if the largest element is interpreted as truth and the smallest element is interpreted as falsehood. In a lattice, some elements are comparable, i.e., an element p is less or equal to an element r if $p \vee r = r$ (cf., for example, (Kurosh, 1963)). We denote this relation by $p \preccurlyeq r$.

When a certainty lattice P is given, then *elementary data* have the form of a triad

$$(p)H(a)$$

or

$$(a, H, p)$$

Here a is an object (point) from X, H is a property of a, and $p \in P$ is the certainty that a has the property H.

It is assumed that logical operations of conjunction $(p)H(a)$ & $(r)Q(a)$, disjunction $(p)H(a) \vee (r)Q(a)$, and negation $\rceil (p)H(a)$ are defined in the set $Data(X)$. In a general case, these operations are partial and have the following properties (Chechkin, 1991):

1. If $p = r$, then $(p)H(a)$ & $(p)Q(a) = (p)(H \cap Q)(a)$ and $(p)H(a) \vee (p)Q(a) = (p)(H \cup Q)(a)$.

2. If certainties p and r are comparable, $p \preccurlyeq r$, and $H = Q$, then $(p)H(a)$ & $(r)H(a) = (r)(H)(a)$ and $(p)H(a) \vee (r)H(a) = (p)(H)(a)$.

3. If $p \neq 0$, then $\rceil (p)H(a) = (0)H(a) = (1)(CH)(a)$ where $CH = X \backslash H$ is the complement of H.

Operations with general data allow Chechkin to define general information.

Definition 7.2.13. A family $I(a)$ of elementary data on a point a from X is called a *general piece of information* or simply, a *piece of information on* (or *about*) a if it satisfies the following conditions:

1. If $(p)H(a) \in I(a)$, then H is not empty as a subset of X.

2. If $(p)H(a) \in I(a)$, $p < r$ and $Q(a)$ is more general than $H(a)$, then $(r)Q(a) \in I(a)$.

3. If $(p)H(a)$, $(r)Q(a) \in I(a)$ and the conjunction $(p)H(a)$ & $(r)Q(a)$ is defined, then $(p)H(a)$ & $(r)Q(a) \in I(a)$.

The set of all pieces of information on a point a from X is denoted by $Inf(a)$. The set of all pieces of information on all points from X is denoted by $Inf(X)$.

Definitions 7.2.4 and 7.2.13 imply the following result.

Proposition 7.2.15. a) There is an injection h of $EInf(a)$ into $Inf(a)$.

b) There is an injection h of $EInf(X)$ into $Inf(X)$.

Indeed, for any elementary piece of information $I(a)$, we can put $h(I(a)) = \{(1)H(a); H(a) \in I(a)\}$. Checking conditions from Definition 7.2.13, we see that $h(I(a))$ is a general piece of information, i.e., the first statement from Proposition 7.2.14 is true.

The second statement from Proposition 7.2.14 is a direct corollary of the first one.

Let us consider two sets with X and Y with the corresponding certainty lattices P_X and P_Y and a mapping $F: X \rightarrow Y$.

Definition 7.2.14. A mapping $F^\wedge: Inf(X) \rightarrow Inf(Y)$ is called an *ultramapping* over F, if a piece of information $I(a)$ is corresponded to the piece of information $I(F(a))$ for any point a from X. The mapping F is called the *support mapping* for $F^\wedge$.

Proposition 7.2.16. For any mapping $F: X \rightarrow Y$ and any certainty lattices P_X and P_Y, there is an ultramapping $F^\wedge: Inf(X) \rightarrow Inf(Y)$ over F for which F is the support mapping.

Let us consider two sets with X and Y with the corresponding certainty lattices P_X and P_Y and concept lattices L_X and L_Y a mapping $F: X \rightarrow Y$.

Definition 7.2.15. An *ultraframe* (also called an *ultraset*) X^* of a set X is the Cartesian product

$$P_X \times L_X \times X = X^*$$

Elements of are triads $(p)H(a)$ where a is an object (point) from X, H is a property from L_X, and $p \in P_X$.

One more basic construction of the operator information theory is the concept of an ultraoperator.

Let us take two ultraframes $X^* = P_X \times L_X \times X$ and $Y^* = P_Y \times L_Y \times Y$.

Definition 7.2.16. A mapping $F^*: X^* \to Y^*$ is an *ultraoperator* over a mapping $F: X \to Y$ if the following diagram is commutative:

$$
\begin{array}{ccc}
& F^* & \\
X^* & \to & Y^* \\
pr_X \downarrow & & \downarrow pr_Y \\
& F & \\
X & \to & Y
\end{array}
$$

Here $pr_X: X^* \to X$ and $pr_Y: Y^* \to Y$ are natural projections of Cartesian products.

The mapping $F: X \to Y$ is called the *support mapping* of the ultraoperator $F^*: X^* \to Y^*$.

Chechkin defines algebra of ultraoperators, introducing such operations as conjunction, disjunction, negation, and (sequential) composition (also called the product) of ultraoperators.

Let us consider two ultraoperators $F^*: X^* \to Y^*$ and $G^*: X^* \to Y^*$ with the ultraframes $X^* = P_X \times L_X \times X$ and $Y^* = P_Y \times L_Y \times Y$ over the same support mapping $F: X \to Y$.

Definition 7.2.17. a) The *conjunction* of ultraoperators F^* and G^* is defined by the following formula:

$$(F^* \,\&\, G^*)((1)H(a)) = F^*((1)H(a)) \,\&\, G^*((1)H(a))$$

a) The *disjunction* of ultraoperators F^* and G^* is defined by the following formula:

$$(F^* \vee G^*)((1)H(a)) = F^*((1)H(a)) \vee G^*((1)H(a))$$

b) The *negation* of the ultraoperator F^* is defined by the following formula:

$$(\daleth F^*)((1)H(a)) = \daleth(F^*((1)H(a)))$$

Let us denote the set of all ultraoperators from X^* into Y^* by **IUOp**(X^*, Y^*).

Proposition 7.2.17. The conjunction (disjunction or negation) of ultraoperators from X^* into Y^* is a total operation in the set **IUOp**(X^*, Y^*) if the conjunction (disjunction or negation, correspondingly) of elementary data is a total operation in the set $Data(Y)$ of all elementary data on all points (objects) from Y.

One of the most important operations with ultraoperators is their sequential composition.

Let us consider two ultraoperators $F^*: X^* \to Y^*$ and $G^*: Y^* \to Z^*$ with the ultraframes $X^* = P_X \times L_X \times X$, $Y^* = P_Y \times L_Y \times Y$, and $Z^* = P_Z \times L_Z \times Z$ over the support mappings $F: X \to Y$ and $G: Y \to Z$.

Definition 7.2.18. The *sequential composition* (also called the *product*) of ultraoperators F^* and G^* is defined as sequential composition of mappings F^* and G^*.

Proposition 7.2.18. The sequential composition of ultraoperators is an ultraoperator.

There is an important class of ultraoperators called standard ultraoperators. These ultraoperators preserve logical operations, namely, conjunction and disjunction, in the set of data on objects (points) from the initial set X.

Definition 7.2.19. An ultraoperator $F^*: X^* \to Y^*$ over a mapping $F: X \to Y$ is called *standard* if for any triads $(p)H(a)$ and $(p)Q(a)$ from , we have

$$F^*((p)H(a) \,\&\, (p)Q(a)) = F^*((p)H(a)) \,\&\, F^*((r)Q(a))$$

and

$$F^*((p)H(a) \vee (p)Q(a)) = F^*((p)H(a)) \vee F^*((r)Q(a))$$

Standard ultraoperators are naturally related to ultramappings.

Denoting the lattice 2^X of all subsets of the set X by $L_{X\max}$, we have the following result.

Proposition 7.2.19. Any standard ultraoperator $F^*: X^* \to Y^*$ with the ultraframe $X^* = P_X \times L_{X\max} \times X$ uniquely determines an ultramapping.

The operator information theory of Chechkin is applied to problems of intelligent system modeling and problems of design and control of antennas, describing how to build an expert system on antennas.

Modern physical theories often use operators and in this respect, the approach of Chechkin (1991) is similar to physics. It is not by chance because physics and information theory more and more exchange ideas and methods in the process of their development. As we have seen in Section 3.6, there are information-theoretical approaches to physics. At the same time, some researcher used methods of physic to build theories of information and information systems. For instance, Lerner (1999; 2003; 2006; 2007) has developed information macrodynamics. He builds information macrodynamics on three levels: the microlevel stochastics, macrolevel dynamics, and hierarchical dynamic network of information flows.

Information macrodynamics represents reality as a field of interacting information systems, which form a collaborative information network. The top-level structure of an information system is similar to the top-level structure of a computer and has the form of the triad

Input — Process — Output

The whole existence and functioning of information systems is based on information processes, which generate the systemic properties of the information network. Information geometry describes information macrostructures in the field of information systems. Information macrodynamics is developed parallel to physics with systemic correspondence between the main concepts. Dynamics of information macrostructures is governed by the minimax Variation Principle (VP) used to select the most probable trajectories of information systems. The Variation Principle implies necessity of receiving *maximum information from minimum available information* and is similar to the principles of minimal action in mechanics, electrodynamics, and optics, as well as to the principle of minimum entropy production in nonequilibrium thermodynamics. The principle of minimum entropy production asserts that the steady state of an irreversible process, i.e., the state in which the thermodynamic variables are independent of the time, is characterized by a minimum value of the rate of entropy production and minimum of viscous dissipation in hydrodynamics. The theorem stating validity of this principle is due to Prigogine (1947; 1961), who proved it by the methods of statistical mechanics for systems with constant coupling coefficients. In the field of hydrodynamics, Helmholtz (1868) proved that flow with negligible inertia is characterized by a minimum of viscous dissipation.

In information macrodynamics, an integral information measure on trajectories of information systems is defined as the path (Feynman) integral (Feynman and Hibbs, 1965). This measure evaluates information contribution of each trajectory into the state parameters of information systems. A macroprocess in the space of information systems is represented by a set of the macrotrajectory segments, which model the multidimensional structure of multiple interactions. Each segment models a piece of macrotrajectory.

Lerner applies information macrodynamics to problems of biology (Lerner, 1997; 2006; 2006c), economics (Lerner, 1993; 2006b), education (Lerner, 1998), and artificial intelligence (Lerner, 1996).

7.3. Information Algebra and Geometry

> *Arithmetic! Algebra! Geometry! Grandiose trinity! Luminous triangle!*
> *Whoever has not known you is without sense!*
> Comte de Lautreamont

Algebra is a mathematical discipline, which studies operations with different abstract entities. At the same time, the term *algebra* is used to denote a mathematical structure with operations in it. That is why, when information is modeled by mathematical structures, it is natural that algebraic structures are introduced and algebraic methods are used more and more. Thus, algebra of information is an algebraic approach to the study of information. Its tools are algebraic structures and methods.

In Section 7.3.1, the mathematical stratum of the general theory of information is illuminated. In it, operators on state spaces or phase spaces of infological systems give a mathematical model of information. Norms and other measures of these operators provide measures for information. In this context, information processes are represented by algebras of information operators.

In Section 7.3.2, information algebras introduced by Kohlas (2003) are described as a mathematical model for situations when information coming from different sources is combined from different elements and when parts of a piece of information that are relevant to specific question are extracted. The word *information* in the term *information algebra* means only that this class of algebras is used in relation to problems of information technology. That is why in some cases, information algebras have interpretations related to information, while in other cases there are no such interpretations.

Considering algebraic methods in information studies, it is necessary to make a distinction between algebra of information and information algebra. Algebra of information is any algebra (algebraic structure) that has an interpretation related to information. At the same time, the term *information algebra* denotes a class of formal algebraic systems introduced by Kohlas (2003) and is similar to terms *Lie algebra*, *associative algebra*, and *Jordan algebra* (cf. (Kurosh, 1963; Cohn, 1965)).

Information geometry presented in Section 7.3.3 is an area where geometrical methods and structures are applied to information studies. There are three main approaches in information geometry: the *theory of information manifolds*, *physically oriented* theories, and *knowledge/database oriented* approach.

7.3.1. *Interpreted information algebra*

> *But there is another reason for the high repute of mathematics:*
> *it is mathematics that offers the exact natural sciences*
> *a certain measure of security which,*
> *without mathematics, they could not attain.*
> Albert Einstein (1879–1955)

In this section, we present the mathematical component of the general theory of information. It is build as an operator theory on state or phase spaces of infological systems. This approach allows us to translate principles of the general theory of information (cf. Sections 2.2 and 2.3) into postulates and find axioms of the general theory of information.

According to the Ontological Principle O2, the essence of information in a broad sense is change (transformation) in a system R. In a similar way, according to the Ontological Principle O2g, the essence of information in the strict sense is change (transformation) in an infological system IF(R) of R. So, building a mathematical model of information, it is necessary to represent the system R and its infological system IF(R) by a mathematical structure and changes in R and in IF(R), by transformations in this structure. This brings us to the conclusion that the basic structure for mathematical representation of information is a system space L, in which the system R, or its infological system IF(R), is represented. Systems that are (potential) receivers of information are represented by points in the space L. Note that elements (points) of the space L can have a sophisticated structure. For instance, L can be a space of functions, of semantic networks or of classifications.

Example 7.3.1.1. State space of a system is an example of a system space L. In control engineering and the theory of automata, a *state space* is a mathematical model that represents a system as a set of input, output and state variables. Dynamics of the system in question is represented by transition rules.

For systems studied in control engineering, variables are related by differential equations that describe evolution of the system. To make this possible and to abstract from the nature of inputs, outputs and states, the variables are expressed as vectors and the differential and algebraic equations are written in matrix form.

For abstract automata, such as finite automata and Turing machines, variables are symbolic systems and evolution of the whole system, i.e., abstract automaton, is described by transition rules or by a transition function.

Example 7.3.1.2. The *state space,* also called the *phase space*, of a physical system is also an example of a system space L. In mathematics and physics, state space is the space in which all possible states of a system are represented, with each possible state of the system corresponding to one unique point in the state space. For mechanical systems, the state space usually consists of all possible values of position and momentum variables. A state is, as a rule, a region of the state space. In thermodynamics, a phase is a region of space where matter lies in a given physical configuration, like for example of a liquid phase, or a solid phase, etc.

Quantum mechanics, in its most general formulation, is a theory of abstract operators (observables) acting on an abstract state space (usually, it is a Hilbert space, cf. Appendix), where the observables represent physically-observable quantities and the state space represents the possible states of the system under study.

In quantum field theory, the phase space is a Hilbert space H, i.e., an infinite dimensional complete vector space with a scalar product, while physical systems are represented by vectors in H. It is assumed that this vector contains all information on the considered physical system (cf., for example, (Bjorken and Drell, 1964; 1965)).

Definition 7.3.1.1. a) A system R is *represented* in a system space L when states or phases of R are corresponded to points of L.

b) The system R is *statically represented* in the space L when states of R are corresponded to points of L. Such a static representation makes L a state space of R.

c) The system R is *processually represented* in the space L when phases of R are corresponded to points of L. Such a processual representation makes L a phase space of R.

Thus, we distinguish two kinds of system representations: state and phase representations. We assume that different states and phases of systems are distinguishable. However, it is possible that different phases are not separable. A state of a system R is a static characterization of this system. A *phase* of a system R is a process that is taking place in this system. It is possible to consider a phase of a system as a dynamic state, i.e., a state in which transition (change) in the system takes place (is going). For instance, static states of a printer are *off* and *on*. *Printing* and *Ready* are dynamic states, or phases, of a printer. Thus, a phase of a system R is a dynamic characterization of this system.

Remark 7.3.1.1. It is possible to treat any state of a system R as a phase of R when nothing changes in R, i.e., the process in this phase is fictitious. This correspondence of states and phases determines an inclusion of the state space $St(R)$ into the phase space $Ph(R)$. For instance, it is possible to consider the class Ob C of all objects from a category C as the state space of the system R and to take the class Mor C of all morphisms from a category C as the phase space of the system R. Then corresponding morphism 1_A to the object A from C, we obtain inclusion of the space Ob *C* into the space Mor *C*. It is possible to do this because any category is determined by its morphisms (Herrlich and Strecker, 1973).

A *state representation* is the triad (named set) $(St(R), r_{st}, L)$. Here $St(R)$ is the set of all possible states of *R*, r_{st} is a binary relation between $St(R)$ and *L*, and *L* is a set with some structure and called the *state representation space*, or simply, the *state space*, of *R*.

A *phase representation* is the triad (named set) $(Ph(R), r_{ph}, L)$. Here $Ph(R)$ is the set of all possible phases of *R*, r_{ph} is a binary relation between $Ph(R)$ and *L*, and *L* is a set with some structure called the *phase representation space*, or simply, the *phase space*, of *R*.

Example 7.3.1.3. An important example of a representation space *L* for a cognitive infological system is the set of all propositions in some logical language *C*. This representation space is used in the information theory of Bar-Hillel and Carnap (cf. Section 4.2), as well as in the multiplicity of works on cognitive and intelligent agents (cf., for example, (Halpern and Moses, 1985)).

It is also possible to consider a propositional calculus (Kleene, 1967; Shoenfield, 1967) or a propositional variety (Burgin, 1995c; 2004c) as a representation space for a cognitive infological system.

Veltman (1996) calls elements from such a representation space by the name *infological state* and studies operations with information states. To define an information state, he considers the powerset $W = 2^A$ of the set **A** of atomic sentences in some propositional language (logic). Then any subset of *W* is an information state. This definition shows that an information state is a knowledge state in the sense of (Mizzaro, 1996; 1998; 2001) when the Mizzaro space is the set of atomic sentences (cf. Section 4.3).

Example 7.3.1.4. Another useful example of a representation space *L* for a cognitive infological system is the set of all predicates in some logical language *D*. It is also possible to consider a predicate calculus (Kleene, 1967; Shoenfield, 1967) or a predicate variety (Burgin, 1995c; 2004c) as a representation space for a cognitive infological system.

Example 7.3.1.5. The category of all classifications from the model of Barwise and Seligman (1997) is an example of a state space *L* (cf. Section 7.1). This space has the algebraic structure of a category with classifications as its objects and infomorphisms as morphisms. Transformations of such categorical spaces are endofunctors.

Example 7.3.1.6. A space of all elementary pieces of information from the model of Chechkin (1991) is an example of a state space L (cf. Section 7.2).

Example 7.3.1.7. A semantic network is a conventional system space for representation of systems of knowledge usually called by the name "thesaurus" (cf. Section 4.3). A semantic network or net is a system of concepts (or more exactly, names of concepts) connected by various relations. Usually, semantic networks use graphic notation for representing knowledge in patterns of interconnected nodes and arcs (Sowa, 1992). That is, a formal representation of a semantic network is a directed graph that consists of vertices which represent concepts and edges which represent semantic relations between the concepts. Examples of semantic relations are: *meronymy* between the concepts A and B means that A is a part of B; *holonymy* between the concepts A and B means that B has A as a part of itself; *hyponymy* (or *troponymy*) between the concepts A and B means that A is kind of B or A is subordinate of B; *hypernymy* between the concepts A and B means that B has A is its superordinate; *synonymy* between the concepts A and B means that B denotes the same object as A; *antonymy* between the concepts A and B means that B denotes something opposite to what A denotes.

Computer implementations of semantic networks support automated systems for reasoning in artificial intelligence and machine translation. However, less formalized versions have long been used in philosophy, psychology, and linguistics. For instance, Anderson and Bower (1973) argue that they can trace the concept of a semantic network all the way back to Aristotle. This concept is extremely popular in cognitive science and artificial intelligence. It has been developed in so many ways, by so many researchers, and for so many purposes that in many instances the strongest connection between recent systems based on networks is only their common ancestry and network structure. As a result, the term "semantic network" as it is used now might therefore best be thought of as the name for a family of structures rather than a single formal construction. Grid automata (Burgin, 2005) provide a mathematical formalism that allows one to unite semantic networks and neural networks in a common unified construction.

Example 7.3.1.8. Chains of named sets are efficiently used for data representation and provide efficient means for working with knowledge (Burgin and Gladun, 1989; Burgin, 1997; Burgin and Zellweger, 2005; Burgin, 2008). Thus, an important case of state spaces for infological systems are systems of named set chains.

Example 7.3.1.9. It is possible to consider infological systems that contain operational and procedural knowledge (Burgin and Kuznetsov, 1994). A natural state space for such infological systems is the space of algorithms, procedures, scenarios, dynamical schemas, etc.

It is possible that the system space L is a mathematical structure, e.g., a linear space, or it can be a set, universe or system of such structures. For instance, when we take the set of all classifications in the sense of Barwise and Seligman (1997) or the set of all semantic networks as L, it has the structure of an algebraic category. In physics, state spaces have the structure of a linear space. At the same time, it is often possible to introduce some structure in the set $St(R)$. This gives us the following concept.

Definition 7.3.1.2. A state representation $(St(R), r_{st}, L)$ (phase representation $(Ph(R), r_{ph}, L)$) is called:

a) *complete* if the domain Dom r_{st} of the relation r_{st} (Dom r_{ph} of the relation r_{ph}) is equal to $St(R)$ (to $(Ph(R))$), otherwise, the representation is called *partial*;

b) *structural* if r_{st} (r_{ph}) preserves the structure of $St(R)$ (of $(Ph(R))$).

Examples of structural representations are:

- Group representations (cf., for example, (Van der Varden, 1971) or (Plotkin, 1994)).
- Ring representations (cf., for example, (Van der Varden, 1971)).
- Category representations (cf., for example, (Cohn, 1965)).
- Representations in varieties of universal algebras (cf., for example, (Cohn, 1965)).
- Representations of groups by automorphisms of algebraic systems (Plotkin, 1966).

Representations of one class of mathematical structures by simpler structures are often used in mathematics: groups are represented by linear transformations of vector spaces, linear transformations of vector spaces are represented by matrices, Boolean algebras are represented by lattices of subsets, etc. Modeling in science, technology and beyond is a representation of some systems (structures) by other systems (structures) when properties of models reflect properties of modeled systems (Burgin and Kuznetsov, 1994).

It is useful to know relations between representations of systems and representations of their subsystems. To find such relations, let us consider the following condition.

Condition A. If a system R is in some state (phase), then any its subsystem Q is also in some state (correspondingly, phase).

Note that in general, it is possible that a system is in a definite state, while some of its subsystems are not in a definite state. For instance, when a table is in some state (e.g., damaged or broken), some electron that is a part of this table, as teaches us quantum physics, is not in a definite state.

Proposition 7.3.1.1. If Condition A is satisfied, then there is a correspondence between all states $St(R)$ (phases $Ph(R)$) of R and states $St(Q)$ (phases $Ph(Q)$) of its subsystem Q.

Proof. If we have a state s from $St(R)$, then by Condition A, when the system R is in the state s, its subsystem Q is in some state q. Thus, we can correspond s to q and do this for all elements from the set $St(R)$.

The proof for a correspondence between the set $Ph(R)$ of phases of R and the set $Ph(Q)$ of phases of Q is similar.

Note that Condition A does not guarantee that this correspondence between states $St(R)$ of R and states $St(Q)$ of Q or the correspondence between phases $Ph(R)$ of R and phases $Ph(Q)$ of Q is a function.

Namely, it is possible that the same state of R corresponds to different states of its subsystem Q . For instance, taking such a system R as a woman and the hair on her head as the subsystem Q of R, we know that the state of R stays the same when the woman looses a single hair on her head. At the same, time essential changes in the state of Q can drastically change the state of R.

Besides, Condition A does not imply existence of the inverse correspondence, i.e., a correspondence between all states $St(Q)$ (phases $Ph(Q)$) of Q and states $St(R)$ (phases $Ph(R)$) of R. Such an inverse correspondence demands additional properties of the subsystem Q.

Let us consider the following condition.

Condition B. If a subsystem Q of R is in some state (phase), then the system R is also in some state (correspondingly, phase).

This condition is not always true, as it is possible that the state of the whole system is changing, while the state of its subsystem stays the same. This has essential consequences for information processes. Namely, information for a subsystem Q of a system R is not always information for the whole system R. The reason is that Q and R may have different (even non-intersecting) infological systems.

Example 7.3.1.10. Let us consider society as the system R and an individual from this society as the system Q. It is possible that knowledge of Q changes, while knowledge of R stays the same. It is also possible that knowledge of R changes, while knowledge of Q stays the same. For instance, the individual Q makes a discovery. He obtains information that changes his knowledge. However, especially, when the discovery does not correlate with the conventional knowledge system, society, as a whole, does not accept this information. As a result, knowledge of the society does not change.

There are also situations when society acquires new information that changes its knowledge, but some individuals either do not get this information or they are very conservative and do not want to accept it. Consequently, knowledge systems of these individuals are not changing.

One more example when information changes in a subsystem do not change the infological system of the whole system is when some atom of the keyboard of a computer changes its state. As a rule, this does not change memory of this computer, which is its infological system.

Condition B allows us to prove the following result.

Proposition 7.3.1.2. If Condition B is satisfied, then there is a correspondence between all states $St(Q)$ (phases $Ph(Q)$) of Q and states $St(R)$ (phases $Ph(R)$) of R.

Indeed, if we have a state r from St(Q), then by Condition B, when the subsystem Q is in the state r, the system R is in some state s. Thus, we can correspond r to s and do this for all elements from the set St(Q).

The proof for a correspondence between the set Ph(Q) of phases of Q and the set Ph(R) of phases of R is similar.

Provided Conditions A and B, Proposition 7.3.1.2 gives us the following result.

Proposition 7.3.1.3. It is possible to extend any state (phase) representation in a space L of a subsystem Q of a system R to a state (phase) representation, called *induced representation*, of the system R.

In particular, taking an infological system IF(R) of R, it is possible to consider any state (phase) representation of IF(R) as a state (phase) representation of the system R. The chosen infological system IF(R) serves as a storage of information that has the type determined by IF(R) and as an indicator of information reception, transformation and action. Thus, the representation space of the infological system IF(R) is naturally called by the name information space. For instance, taking a database **B** as an infological system IF(R), we have such an information space as the system of all data in **B**.

Definition 7.3.1.3. The set U(R) of all points from L that are corresponded to states (phases) of R are called the *state (phase) representation domain* of R in L.

The domain U(R) of L usually is treated as the state space of R where different points correspond to different states of R or as the phase space of R where different points correspond to different phases of R. Very often we have U(R) = L.

Indeed, given a state representation (St(Q), r, L), we can build the state representation (St(R), q, L) by taking the composition $r = r \circ r_{R,Q}$ of the relation r and relation $r_{R,Q}$, which is provided by Proposition 7.3.2. The induced representation is described by the following diagram:

$$\begin{array}{ccc} r_{R,Q} & & r \\ \text{St}(R) \rightarrow & \text{St}(Q) & \rightarrow L \end{array}$$

Transformations of systems have different forms. A transformation can be a process, phase, state transition, multistate transition, i.e., a composition of state transitions, etc. For instance, a state transition of a system R is a transition of R from one state to another.

A *state-transition representation* is the triad $(\text{Trans}(R), r_{sttr}, L)$. Here $\text{Trans}(R)$ is the set of all state transitions of the system R and r_{sttr} is a binary relation between $\text{Trans}(R)$ and the set of pairs $(a, f(a))$ with $a \in L$. Operators on states of R correspond to endomorphisms of L, i.e., mappings of L into itself that preserve a chosen structure in L.

There is a relation between state and phase representations for state transition systems/machines (cf. Appendix B).

Proposition 7.3.1.4. If R is a state transition system, then any state representation $(\text{St}(R), r_{st}, L)$ generates a phase representation $(\text{Ph}(R), r_{ph}, L^2)$ of the same system R.

Proof. Each transition of R from one state to another is a phase. If a state representation $(\text{St}(R), r_{st}, L)$ is given, then each transition is represented by two points of L, namely, by the point that represents the initial state and the point that represents the final state. Thus, we have a representation of all transitions in direct product L^2. This gives us a phase representation $(\text{Ph}(R), r_{ph}, L^2)$.

Proposition is proved.

Remark 7.3.1.2. In the case of state transition systems or machines, the state space $\text{St}(R)$ becomes a subspace of the phase space $\text{Ph}(R)$ and a state representation $(\text{St}(R), r_{st}, L)$ becomes a named subset of the corresponding phase representation $(\text{Ph}(R), r_{ph}, L^2)$.

Proposition 7.3.1.4 and Remark 7.3.1.2 show that in many cases, it is possible to consider either only state representations or only phase representations. Thus, in what follows, we consider only state representations.

Let h be an *effective characteristics/parameter* of the states (phases) of R, i.e., h is a parameter such that if h changes, then the state of R changes.

Definition 7.3.1.4. a) A state representation $(\text{St}(R), r_{st}, L)$ is called *unitary* if $(\text{St}(R), r_{st}, L)$ is a functional named set, i.e., r_{st} is a function (cf. Appendix A).

b) A unitary representation $(\text{St}(R), r_{st}, L)$ is called *faithful* with respect to h if the states of R that have different values of h are related to different points in L.

c) If there is a unitary (faithful) representation $(St(R), r_{st}, L)$, then we say that the system R is *unitarily (faithfully) represented* in the space L with respect to h.

Not all representations are faithful. For instance, if we take a very small material object, we can represent it in the three-dimensional Euclidean space E^3 faithfully with respect to its spatial coordinates. However, this representation will be unfaithful with respect to time, impulse or velocity of this object.

Some subsystems have tighter relations with their supersystems.

Definition 7.3.1.5. A subsystem Q of a system R is called a *component* of R if different states of Q correspond to different states of R.

Example 7.3.1.11. Taking such a system as the Solar system C, we see that both the Earth and the Moon are components of C.

Assuming validity of the Condition A for a component Q of a system R, we have the following result.

Proposition 7.3.1.5. There is a projection of the system $St(R)$ of all states of R onto the system $St(Q)$ of all states of its component Q.

Proposition 7.3.1.5 allows us to prove the following result.

Proposition 7.3.1.6. If a system Q is a component of a system R, Q is faithfully represented in the space L with respect to an effective parameter h, and h is also an effective parameter of R, then the system R is faithfully represented in the space L with respect to h.

In addition to state (static) and phase (processual) representations, there are other types of representations.

Definition 7.3.1.6. a) A system R is *dynamically represented* in the space K if transformations (e.g., transitions of states) of R are corresponded to elements of K.

b) A system R is *transitionally represented* in the space L if transitions of states of R are corresponded to transformations of L.

Let H be a representation space.

Proposition 7.3.1.7. a) If any transformation (transition) of R corresponds to some phase of R, then any phase representation $(Ph(R), r_{ph}, H)$ induces a dynamical representation $(Transf(R), r_{dyst}, H)$. b) If any transition of a state transition system R is represented by a unique phase of R, then there is one-to-one mapping of the set of dynamical representations of R onto some subset of all partial phase representations of R.

Proof. a) Let us consider a phase representation $(\mathrm{Ph}(R),\ r_{\mathrm{ph}},\ H)$ and assume that any transformation of R corresponds to some phase of R. This gives us a mapping $f\colon \mathrm{Transf}(R) \to \mathrm{Ph}(R)$. Then we can define the relation $r_{\mathrm{dyst}} = r_{\mathrm{ph}} \circ f$ (cf. Diagram 7.3.1.1) and get a dynamic representation $(\mathrm{Transf}(R),\ r_{\mathrm{dyst}},\ H)$.

$$
\begin{array}{ccc}
 & r_{\mathrm{ph}} & \\
\mathrm{Ph}(R) & \to & H \\
f \uparrow & \nearrow r_{\mathrm{ph}} & \\
\mathrm{Transf}(R) & &
\end{array}
\qquad (7.3.1.1)
$$

b) Let us consider a state representation $(\mathrm{St}(R),\ r_{\mathrm{st}},\ L)$ of a state transition system R and assume that any transition of R is represented by a unique phase of R. As R is a state transition system, each transition of R from one state to another is a phase. If a state representation $(\mathrm{St}(R),\ r_{\mathrm{st}},\ L)$ is given, then each transition is represented by two points of L, namely, by the point that represents the initial state and the point that represents the final state. Thus, we have a representation $(\mathrm{Trans}(R),\ r_{\mathrm{sttr}},\ L^2)$ of all transitions $\mathrm{Trans}(R)$ in the direct product L^2. By the assumption, there is an injection $h\colon \mathrm{Trans}(R) \to \mathrm{Ph}(R)$. Thus, there is a partial mapping $g\colon \mathrm{Ph}(R) \to \mathrm{Trans}(R)$ defined for elements from the image $\mathrm{Im}\ h$ of h. Then we can define the relation $r_{\mathrm{ph}} = r_{\mathrm{sttr}} \circ g$ (cf. Diagram 7.3.1.2). This gives us a phase representation $(\mathrm{Ph}(R),\ r_{\mathrm{ph}},\ L^2)$.

$$
\begin{array}{ccc}
 & r_{\mathrm{ph}} & \\
\mathrm{Ph}(R) & \dashrightarrow & L^2 \\
g \downarrow \uparrow h & \nearrow r_{\mathrm{sttr}} & \\
\mathrm{Trans}(R) & &
\end{array}
\qquad (7.3.1.2)
$$

Proposition is proved.

Let us consider some important types of state-transition representations.

Definition 7.3.1.7. a) A state-transition representation (Trans(R), r_{sttr}, L) is called *unitary* if (Trans(R), r_{sttr}, L) is a functional named set, i.e., r_{sttr} is a function (cf. Appendix A).

b) A unitary state-transition representation (Trans(R), r_{sttr}, L) is called *faithful* if different state transitions of R are related to different pairs $(a, f(a))$.

c) A unitary state-transition representation (Trans(R), r_{sttr}, L) is called *grounded* if there is a unitary state representation (St(R), r_{st}, L) such that for any state transition T: $st_1R \to st_2R$, we have $r_{sttr}(T) = (r_{st}(st_1R), r_{st}(st_2R))$.

d) If there is a unitary (faithful or grounded) representation (Transf(R), r_{dyst}, f), then we say that the dynamics of the system R is *functionally* (correspondingly, *faithfully* or *groundedly*) *represented* in the space L.

A grounded state-transition representation (Trans(R), r_{sttr}, L) is characterized by the following commutative diagram:

$$
\begin{array}{ccc}
a & \longrightarrow & b \\
r_{st} \uparrow & & \uparrow r_{st} \\
st_1R & \longrightarrow & st_2R
\end{array}
\qquad (7.3.1.3)
$$

Theorem 7.3.1.1. Any faithful state representation of a system R uniquely determines a grounded faithful state-transition representation of the system R.

Proof. Let us assume that a faithful state representation (St(R), r_{st}, L) of the system R in a space L is given and consider a transition tr: $st_1R \to st_2R$. Then taking $a = r_{st}(st_1R)$ and $b = r_{st}(st_2R)$, we can correspond the pair (a, b) to the transition tr. Note as the state representation (St(R), r_{st}, L) is faithful, i.e., r_{st} is a function, elements a and b are uniquely defined. As we consider an arbitrary transition tr, this allows us to build a correspondence r_{sttr}: Trans(R) $\to L^2$. It is determined by the formula $r_{sttr}(tr) = (r_{st}(st_1R), r_{st}(st_2R))$ when tr: $st_1R \to st_2R$. By Definition 7.3.1.7, this representation is grounded and faithful.

Theorem is proved.

Theorem 7.3.1.2. Any grounded state-transition representation of a system R preserves sequential composition of state transitions, i.e., it is a functor from the category $\text{Trans}(R)$ into the category where objects are elements of L and morphisms are transformations of L.

Proof. Let us consider two transitions tr_1 and tr_2 of the system R such that the final state of the transition tr_1 is the initial state of the transition tr_2, i.e., tr_1: $st_1R \rightarrow st_2R$, while tr_2: $st_2R \rightarrow st_3R$. Then it is possible to define the sequential composition $tr = tr_1 \circ tr_2$. of these transitions. Then $r_{sttr}(tr) = (r_{st}(st_1R), r_{st}(st_3R)) = (r_{st}(st_1R), r_{st}(st_2R)) \circ (r_{st}(st_2R), r_{st}(st_3R))$, i.e., the transition tr is mapped by r_{sttr} into the sequential composition of transformations corresponding to tr_1 and to tr_2.

Theorem is proved.

Example 7.3.1.12. Taking such an infological system as the system of knowledge, or thesaurus, we can use the concept space as its state representation space. Many intelligent systems utilize concept spaces for information search. There are also other kinds of knowledge representation, e.g., semantic networks or logical calculi.

As it is demonstrated in Chapter 2, basic types of information, i.e., cognitive information, affective information, and effective information, are related to specific types of infological systems. For instance, cognitive information is related to the center of rational intelligence, which includes the neocortex, affective information is related to the center of emotions, which includes the amigdala, and effective information is related to the center of instincts and will.

It is natural to assume that specific types of infological systems have different types of system spaces better suited for their representation. Thus, we have, for example, cognitive system space L_C, e.g, the set of all semantic networks, affective or emotional system space L_E, and effective system space L_W.

We consider several types of system spaces.

Definition 7.3.1.8. If u is a type of infological systems, then a system space L has the *infological type u* when any infological system of the type u can be represented in L.

Note than one and the same system space L may have several infological types.

Informally, a portion of information I is a potency to cause changes in (infological) systems, i.e. to change the state of this system (cf. Section 2.2). Assuming that all systems involved in such changes are represented by points of some space L, we see that a change in a system is represented by a transition from some point x from L to another point z from L. This gives us a transformation of L, which may be partial. In such a way, I is represented by a (partial) mapping of L into itself. As in many cases, L has a sophisticated structure, we call this mapping an information operator, i.e., I is represented by an information operator $Op(I): L \to L$. When it does not cause misunderstanding, it is possible to denote the information portion I and the corresponding information operator $Op(I)$ by the same letter I.

However, there are situations when reception of information changes the system itself and not the state of this system. It is possible that the new system is represented in another space M. This brings us to the concept of a general information operator as a mathematical construction that represents a portion of information.

To build a mathematical model of information, we consider a class **K** of systems, which are receptors of information, a set **IT** of infological system types, and a set **L** of representation spaces for infological systems of systems from **K**. In addition, for each system R from **K** and each infological system $IF(R)$ that has type t from **IT**, a representation $(St(IF(R)), r_{st}, L)$ with $L \in$ **L** is fixed. Here we use only state representations although it is also possible to develop a similar theory based on phase representations.

Definition 7.3.1.9. a) A *general information operator* I in **L** is a partial mapping

$$I: \bigcup_{L \in L} L \to \bigcup_{L \in L} L.$$

b) A *(particular) information operator* I in **L** is a partial mapping $I: L \to L$.

Examples of information operators are operators studied in the multidimensional structured model of distributed systems and parallel computations (Burgin and Karasik, 1975; 1976; Burgin, 1980; 1982a; 1984).

We denote the set of all general information operators in **L** by **Op L**, the set of all total information operators in **L** by **Opt L**, the set of all information operators in L by **Op** L and the set of all total information operators in L by **Opt** L.

Information operators are related to one of the main concept of quantum physics called *observable*. Observables in physics are information operators. Indeed, in physics, particularly in quantum physics, a system observable is treated as a system state property that can be determined by some sequence of constructive physical operations. These operations are observation (thus, observable) and measurement, often involving analytical transformations and numerical computations. Measurement of an observable in microphysics might involve submitting the system to various electromagnetic fields and eventually reading a value off some gauge. In systems governed by classical mechanics, any experimentally observable value is usually modeled by a real-valued function on the set of all possible system states.

In quantum physics, on the other hand, the relation between system state and the value of an observable is more sophicticated. For instance, in the mathematical formulation of quantum mechanics, states are given by non-zero vectors in a Hilbert space H where two vectors are considered to specify the same state if, and only if, they are scalar multiples of each other. In other words, H is a state space in quantum mechanics and in many other fields of quantum theory. For the case of a system of particles, the space H consists of functions called wave functions. In this context, observables are represented by self-adjoint operators on H.

It is not by chance that observables are represented by self-adjoint operators on a Hilbert space that is the phase space and information is represented by information operators on a the state space of an infological system. Indeed, observables provide a mathematical model for a physical quantity obtained by observation and measurement using physical instruments, while observation and measurement give information about physical systems. Thus, the physical quantity modeled by an observable is information about the physical system. Consequently, as a kind of information, observables are represented by operators.

Operator approach well correlates not only with the basic principles of the general theory of information, but also with ideas of other researchers. For instance, Otten (1975) suggests that information is some change of the state in a system. He proposes a very general framework of communication within which information transfer can take place based on the traditional model of a system that interactively adapts within its environment. Thus, information induces interpretations or transformations of incoming stimuli manifested in internal changes in the state of the system.

Example 7.3.1.13. A proposition p induces an information operator A_p on systems of (logical) worlds used in semantic information theories (cf. Section 4.2). A logical world consists of all true statements/propositions about all entities from this world. If W is a system of logical worlds, then the information operator A_p excludes all worlds where p is not true, i.e., all worlds inconsistent with p.

Example 7.3.1.14. A proposition p induces an information operator A_p on states of a chosen universe U according to semantic information theories (cf. Section 4.2). States of U are consistent assignments of truth values to primitive propositions from a propositional language L. The information operator A_p excludes all states where p is not true, i.e., all states inconsistent with p.

Example 7.3.1.15. A portion of cognitive (epistemic in the sense of (Mizzaro, 2001)) information I induces an information operator A_I on systems of knowledge systems. In the theory of epistemic information, it is assumed that a knowledge system **K** consists of knowledge items. It is possible to take elementary knowledge units built in (Burgin, 2004) (cf. also Section 2.5) or elementary propositions/predicates as such knowledge items. Note that elementary propositions and predicates give a linguistic representation for elementary knowledge units. For instance, the proposition "A cat is a living being" corresponds to the following elementary knowledge unit

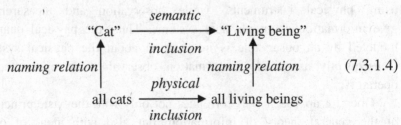

$$(7.3.1.4)$$

and the proposition "*Bobby* is a dog" corresponds to the following elementary knowledge unit

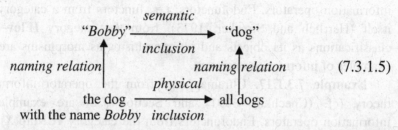

$$(7.3.1.5)$$

Remark 7.3.1.3. An information operator is similar to the energy operator in quantum mechanics where it is possible to derive the three-dimensional time-dependent Schrödinger equation from the Conservation of Energy Law. Analyzing the Schrödinger equation

$$- (\hbar/2m) \, \nabla^2 \, \Psi(r, t) + V(r, t) \, \Psi(r, t) = (i\hbar) \, \partial\Psi(r, t)/\partial t$$

it is possible to see that the first term in the left right side is associated with the kinetic energy and treated as the kinetic energy operator

$$\{K\} = - (\hbar/2m) \, \nabla^2,$$

the second term in the left side is associated with the potential energy and treated as the potential energy operator

$$\{V\} = V$$

and the term in the right side is associated with the total energy and treated as the total energy operator

$$(i\hbar) \, \partial/\partial t$$

Each term is considered as an operator that acts on the wave function $\Psi(r, t)$.

At the same time, it is discussed in Section 3.6 that physicists obtain information from physical objects and can develop physical laws only based on this information. Consequently, physical laws, such as the Schrödinger equation, reflect properties not only of objects themselves but also of information obtained. Thus, in the context of the general theory of information, energy operators are information operators. Moreover, as we have seen, not only energy but also any observable is, in a definite sense, an information operator.

Example 7.3.1.16. Infomorphisms from the theory of information flow (cf. (Barwise and Seligman, 1997) and Section 7.1) are examples of information operators. Endofunctors, i.e., functors from a category into itself (Herrlich and Strecker, 1973), from the category **IFlow** with classifications as its objects and infomorphisms as morphisms are also examples of information operators.

Example 7.3.1.17. Ultramappings from the operator information theory (cf. (Chechkin, 1991) and Section 7.2) are examples of information operators. Endofunctors from the category **IUMap**(X) with the set *Data*(X) of all elementary data on all points from X as the set of its objects and ultramappings as morphisms are also examples of information operators.

Example 7.3.1.18. Ultraoperators from the operator information theory (cf. (Chechkin, 1991) and Section 7.2) are examples of information operators. Endofunctors from the category **IUOp**(X) with general iunformations as its objects and ultraoperators as morphisms are also examples of information operators.

Additional properties separate important classes of information operators.

Let us take a representation space L and a subset H of L.

Definition 7.3.1.10. An information operator I in the space L is called:

a) *closed* in H if $I(H) \subseteq H$, i.e., for any $x \in H$, we have $I(x) \in H$;

b) *covering* H if $I(L) \supseteq H$;

c) *injective* in H if for any $a, b \in H$, $a \neq b$ implies $I(a) \neq I(b)$;

d) *bijective* in H if it is injective in H and covering H;

e) *closed* for an infological system $\mathrm{IF}(R)$ if for any $x \in \mathrm{U}(\mathrm{IF}(R))$, we have $I(x) \in \mathrm{U}(\mathrm{IF}(R))$;

f) *covering* if $I(L) = L$;

g) *injective* if for any $a, b \in L$, $a \neq b$ implies $I(a) \neq I(b)$;

h) *bijective* if it is injective and covering.

Informally, if an information operator I in L is closed in the domain $\mathrm{U}(\mathrm{IF}(R))$, it means that a closed in the domain $\mathrm{U}(\mathrm{IF}(R))$ information operator I does not change the infological system $\mathrm{IF}(R)$, but transforms only its states (or phases).

We denote:

- the set of all closed in H information operators in L by $\mathbf{Op}_H L$;
- the set of all closed in H total information operators in L by $\mathbf{Opt}_H L$;
- the set of all closed for an infological system $IF(R)$ information operators in L by $\mathbf{COp}^R L$;
- the set of all closed for an infological system $IF(R)$ total information operators in L by $\mathbf{COpt}^R L$;
- the set of all closed in H covering information operators in L by $\mathbf{CovOpt}_H L$;
- the set of all closed in H injective information operators in L by $\mathbf{InjOpt}_H L$;
- the set of all closed in H bijective information operators in L by $\mathbf{BOpt}_H L$,

Let us consider a system $\mathbf{L}$ of representation spaces.

Definition 7.3.1.11. A general information operator I in $\mathbf{L}$ is called:

graded if it is closed in L for all $L \in \mathbf{L}$.

system graded if it is closed in $U(IF(R))$ for all $IF(R)$ that have a type u from the set $\mathbf{IT}$ of infological system types and all $R \in \mathbf{K}$;

Definitions imply that any particular information operator is a graded general information operator.

Proposition 7.3.1.8. If an information operator I is closed in subsets H and K of L, then I is closed in $H \cup K$ and $H \cap K$.

Proof. Let us assume that an information operator I is closed in subsets H and K of L, and take some element $a \in H \cup K$. Then either $a \in H$ or $a \in K$. Let us consider the case when $a \in H$. Then $I(a) \in H$ as I is closed in H. Consequently, $I(a) \in H \cup K$.

The case when $a \in K$ is considered in a similar way.

Now let us take an element $a \in H \cap K$. Then $a \in H$ and $a \in K$. By the initial conditions, we have $I(a) \in H$ and $I(a) \in K$. Consequently, $I(a) \in H \cap K$.

Proposition is proved as a is an arbitrary element from $H \cup K$ or from $H \cap K$ in the second case.

Corollary 7.3.1.1. For any information operator A, the system of subsets of the space L in which A is closed is a lattice with the least element $\varnothing$.

Usually spaces from the class **L** have a common structure, e.g., they are linear spaces or categories. This implies that an important property of information operators is that they preserve some basic structure of spaces in **L**, e.g., operators are linear transformations of linear spaces or functors in categories (Herrlich and Strecker, 1973). These considerations bring us to the concept of a coherent information operator.

Let us assume that all spaces from **L** have a definite structure S. For instance, S can be the structure of a linear space or of a graph.

Definition 7.3.1.12. An information operator in **L** is called *coherent* with respect to a structure S in spaces from **L** if it preserves the structure S.

Example 7.3.1.19. Let us assume that all spaces from the class **L** are linear spaces. Then a coherent information operator is a linear information operator when **L** consists of one space or a system of linear information operators otherwise.

Example 7.3.1.20. Let us assume that all spaces from the class **L** are algebraic categories. Then a coherent information operator is a functor or a system of functors (Herrlich and Strecker, 1973).

Example 7.3.1.21. Let us assume that all spaces from the class **L** are diagrams in a given category K. A diagram is a triad of the form (I, Φ, d) where I is a set of vertices, Φ is a set of edges and $d: \Phi \to I \times I$ is a mapping (Grothendieck, 1957). Then a coherent information operator is a morphism or a system of morphisms of diagrams in K.

Coherent mappings are called endomorphisms in algebra, i.e., when mappings preserve algebraic structures (Kurosh, 1963), and continuous mappings in topology, i.e., when mappings preserve topological structures (Kuratowski, 1966).

In what follows, we consider only one system space L and (particular) information operators on it. This is not an essential restriction because it is possible to consider the union of all spaces from **L** as one space.

It is natural to define different operations in the set of all information operators. One of the most important operations is the sequential composition of information operators.

Definition 7.3.1.13. The *sequential composition* (also called *product*) AB of information operators A and B is an information operator C such that the mapping determined by the information operator C is the sequential composition, sometimes called product, of mappings A and B provided Rg $A \cap$ DDom $B \neq \varnothing$.

That is, the sequential composition of A and B is defined only if the range Rg A of A and the definability domain DDom B of B have a non-void intersection. When Rg $A \cap$ DDom $B = \varnothing$, the sequential composition AB of information operators A and B is not defined.

Another option is to define the sequential composition AB in the case when Rg $A \cap$ DDom $B = \varnothing$ as a void information operator, that is, an information operator the definability domain of which is empty.

The sequential composition of information operators corresponds to precise sequential reception of information (of information portions) by one and the same system R. Here precise reception means that the receptor R starts receiving the second portion of information at the moment it stops receiving the first portion of information, or more exactly, the system IF(R) is in the state that is achieved after receiving the first portion of information when R starts receiving the second portion of information. If this condition is not satisfied, then R can get some other information between receptions of the first and the second information portions, and this other information can change the result of reception of the second information portion.

There are also different parallel compositions of information operators. A parallel composition of information operators reflects a situation when two or more systems receive different (but possibly identical) information portions.

In Section 2.6, we have considered three basic systems of the brain: the System of Rational Intelligence, the System of Emotions (or more generally, of Affective States), and the System of Will and Instinct. Each of these systems has a corresponding state space. Information operators that act on these spaces are cognitive, affective and effective information operators, correspondingly. Thus, it is possible to consider, for example, a parallel composition of a cognitive information operator and affective information operator or the sequential information operator and effective information operator.

The sequential composition of information operators has many good properties.

Proposition 7.3.1.9. The sequential composition of coherent information operators is a coherent information operator.

Indeed, if AB is the sequential composition of coherent information operators A and B and V is the structure, e.g., algebraic operations or topology, that is preserved by coherent information operators, then application of A preserves V and when after this, the information operator B is applied, it also preserves V. Consequently, AB also preserves V, i.e., AB is a coherent information operator.

This result corresponds to the theorem in algebra stating that the sequential composition of homomorphisms is a homomorphism of algebraic structures (cf., for example, (Kurosh, 1963)) or to the theorem in topology asserting that the sequential composition of continuous mappings is a continuous mapping of topological spaces (cf., for example, (Kuratowski, 1966)).

Proposition 7.3.1.10. The sequential composition of closed in a domain H information operators is a closed in H information operator.

Indeed, if A and B are information operators for which the sequential composition AB is defined, both A and B are closed in the domain H, and $x \in H$, then $B(x) \in H$, and $A(B(x)) \in H$. Thus, AB is closed in the domain H as x is an arbitrary element from H.

Corollary 7.3.1.2. The sequential composition of closed for an infological system $IF(R)$ information operators is a closed for the infological system $IF(R)$ information operator.

Corollary 7.3.1.3. The sequential composition of closed in H and coherent information operators is a closed in H and coherent information operator.

Properties of the sequential composition of mappings give us the following result.

Proposition 7.3.1.11. The sequential composition of covering (injective or bijective) information operators is a covering (injective or bijective, correspondingly) information operator.

Principles of the general theory of information imply that information depends on the carrier and channel. To formalize this situation, we introduce the concept of a message.

Definition 7.3.1.14. a) A triad $m = (C, c, A)$, where C is the carrier type in some state, c is channel in some state, and A is an information operator, is called a *message*.

b) Reception of a message m by a system R in a state q, which is represented in the state space L of R by $r_{st}(q)$, is the action of A on the element $r_{st}(q)$.

Successive reception of messages is modeled by the sequential composition of information operators from these messages.

Example 7.3.1.22. Let us take a class of texts in some natural language as the carrier type C and regular post service as the channel ch. Then messages are letters send by regular mail.

Example 7.3.1.23. Let us take a class of texts in some natural language as the carrier type C and Internet or an e-mail system as the channel ch. Then messages are e-mails.

Note that the same carriers can go through different channels and even different types of channels.

Example 7.3.1.24. We can also take a class of movies as the carrier type C and TV as the channel ch.

Example 7.3.1.25. We can take a class of movies as the carrier type C and movie theaters as the channel ch.

To each carrier type C and channel c, an algebra $A(C, c)$ of information operators is corresponded. This algebra represents information contained in different messages that go by the channel ch and are contained in carriers having type C.

In this case, it is useful to make a distinction between a system and the system in some state. That is why, the algebra $A(C, c)$ is corresponded to a carrier C and channel c, while in other cases, it is corresponded to a carrier C in some state and channel c in some state.

Proposition 7.3.1.12. Any information operator A from $A(C, c)$ has a potential message for a system R if and only if the domain $U(R)$ is a subset of the definability domain DDom A of A.

Indeed, in this case, the system R has a state q such that A changes R in the state q. Otherwise, A does not act on R.

The algebra A(C, c) is a subalgebra of **Op** L, which has operations of three types: *intrinsic operations*, *induced operations*, and *inherited operations* and the type of A(C, c) is a subtype of the type of the algebra **Op** L.

An example of an *intrinsic operation* is multiplication or sequential composition of information operators defined in the conventional way as their sequential application (cf. Definition 7.3.1.13). It is also possible to define parallel composition, free composition and other compositions of information operators. It is possible to find examples of such operations in Section 4.3 where they are defined for local information operators.

Another type of operations with information operators on the space L is *induced* by operations in L. For instance, when L is a linear space over the field R of all real numbers, it has such operation as addition, subtraction, and multiplication by real numbers (cf. Appendix D). This operation allows one to define addition of operators in the algebra A(C, c). Namely, if A, $B \in$ A(C, c) and $x \in L$, then $(A + B)(x) = A(x) + B(x)$, $(A - B)(x) = A(x) - B(x)$ and $aA(x) = aA(x)$ for any real number a.

Let us assume that H is a subspace of L and information operators A and B are closed in H.

Proposition 7.3.1.13. The sum $A + B$ and difference $A - B$ of total information operators A and B in L, as well as the mapping aA, are closed in H information operators.

Indeed, if x belongs to H, then $A(x)$ and $B(x)$ belong to H as A and B are closed in H information operators. As H is a subspace of L, $(A + B)(x) = A(x) + B(x)$, $(A - B)(x) = A(x) - B(x)$ and $aA(x) = aA(x)$ belong to H for any real number a. As x is an arbitrary element from H, information operators $A + B$, $A - B$ and aA are closed in H.

Let us assume that L is a universal algebra of a type Ω.

Theorem 7.3.1.3. The set **Opt** L of all total information operators in L is a universal algebra of the type Ω.

Proof. Let us take an operation ω from the set Ω and assume that ω has arity n, i.e., $\omega: L^n \to L$ is a mapping from L^n into the representation space L (cf. Appendix D). We show that the operation ω induces the corresponding operation $\omega_I: ($Opt $L)^n \to$ Opt L in the set of total information operators Opt L.

Taking n information operators $A_1, A_2, ..., A_n$ in L, we define the corresponding operation ω_l by the following formula:

$\omega_l(A_1, A_2, ..., A_n)(x) = \omega(A_1(x), A_2(x), ..., A_n(x))$ for any x from L

Thus, we can see that $\omega_l(A_1, A_2, ..., A_n)$ is an information operator in L and it is a total information operator because all $A_1, A_2, ..., A_n$ are total information operators.

Theorem is proved.

Proposition 7.3.1.14. If L is a linear space over the field of all real numbers R, then the set **Opt** L of all total information operators in L is a linear space over the field of all real numbers R.

Proof. Given information operators A and B in L, we define their sum $A + B$ as the information operator in L that acts on elements from L by the following rule:

$$(A + B)(x) = A(x) + B(x)$$

As information operators A and B are total, information operator $A + B$ is also total.

If a is a real number, then we define the information operator aA, which is the scalar product of a and A, by the following rule:

$$(aA)(x) = a(A(x))$$

It is possible to show that all axioms of a linear space over R are satisfied for the set **Opt** L with these two operations. Let us check the identities of a linear space (cf. Appendix D).

1. The associativity identity

$$A + (B + C) = (A + B) + C$$

is true for all total information operators A and B in L because for any x from L, we have

$$(A + (B + C))(x) = A(x) + (B(x) + C(x)) =$$
$$(A(x) + B(x)) + C(x) = (A + (B + C))(x)$$

2. The commutativity identity

$$A + B = B + A$$

is true for all total information operators A and B in L because for any x from L, we have

$$(A + B)(x) = A(x) + B(x) = B(x) + A(x) = (B + A)(x)$$

3. Addition has an identity element:

Taking the total information operator $\mathbf{0}$ such that $\mathbf{0}(x) = 0$ for any x from L, we have

$$A(x) + \mathbf{0}(x) = A(x)$$

for all A from **Opt** L.

4. Addition has inverse element because for any total information operator A, the total information operator $-A$ is its additive inverse, i.e.,

$$A(x) + (-A(x)) = \mathbf{0}(x)$$

5. Scalar multiplication is distributive over addition in **Opt** L because for all of real numbers a and total information operators A and B in L, we have

$$(a \cdot (A + B))(x) = a \cdot (A(x) + B(x)) =$$
$$a \cdot A(x) + a \cdot B(x) = (a \cdot A)(x) + (a \cdot B)(x)$$

6. Scalar multiplication is distributive over addition of real numbers because for all of real numbers a, b and any total information operator A, we have

$$((a + b) \cdot A)(x) = (a + b) \cdot A(x) = a \cdot A(x) + b \cdot A(x) = (a \cdot A)(x) + (b \cdot Ax)x)$$

7. Scalar multiplication is compatible with multiplication of real numbers because for all of real numbers a, b and any total information operator A, we have

$$(a \cdot (b \cdot A))(x) = a \cdot (b \cdot A(x)) = (a \cdot b) \cdot A(x) = ((a \cdot b) \cdot A)(x)$$

8. The number 1 also is an identity element for scalar multiplication: For all total information operators A, we have $1 \cdot A(x) = A(x)$

All axioms are verified. Thus, **Opt**$_U$ L is a linear space and Proposition 7.3.1.14 is proved.

Proposition 7.3.1.15. If L is a linear space over the field of all real numbers R, then the set **LOpt**$_U$ L of all total linear information operators in L is a linear space over the field of all real numbers R.

Proof. The sum $A + B$ of two linear operators A and B is a linear operator in L, as well as the scalar product aA of a and A (Dunford and Schwartz, 1957). Thus, **LOpt**$_U$ L is a subspace of the linear space **Opt**$_U$ L and hence, a linear space over the field of all real numbers R.

Corollary 7.3.1.4. All closed in the same representation domain information operators form a linear space.

Corollary 7.3.1.5. All closed in the same representation domain linear information operators form a linear algebra over the field F.

Let us assume that the representation space L for an infological system $IF(R)$ of a system R is a universal algebra of a type Ω.

Theorem 7.3.1.4. The set $\mathbf{Opt}_U L$ of all total in L and closed in U information operators is a universal algebra of the type Ω, which is a subalgebra of the algebra $\mathbf{Opt}\, L$.

Proof. Taking an operation ω from the set Ω with arity n and n closed in U information operators $A_1, A_2, \ldots, A_n$ in L, we know that $\omega_l(A_1, A_2, \ldots, A_n)$, as it is defined in the proof of Theorem 7.3.1.3, is an information operators in L. Now suppose that x is an element from the subalgebra U. As all $A_1, A_2, \ldots, A_n$ are closed in U information operators, all elements $A_1(x), A_2(x), \ldots, A_n(x)$ belong to U. As U is a subalgebra in L, the element $\omega(A_1(x), A_2(x), \ldots, A_n(x))$ also belongs to L. As ω is an arbitrary operation from the set Ω and x is an arbitrary element from the subalgebra U, the information operator $\omega_l(A_1, A_2, \ldots, A_n)$ is closed in U.

By construction, $\mathbf{Opt}_U L$ is a subalgebra of the algebra $\mathbf{Opt}\, L$.

Theorem is proved.

Definition 7.3.1.15. Operations ω_1 with arity n and ω_2 with arity m from the set Ω commute with one another if the following identity is true

$$\omega_1(\omega_2(x_{11}, x_{12}, \ldots, x_{1m}), \omega_2(x_{21}, x_{22}, \ldots, x_{2m}), \ldots, \omega_2(x_{n1}, x_{n2}, \ldots, x_{nm})) =$$
$$\omega_2(\omega_1(x_{11}, x_{21}, \ldots, x_{n1}), \omega_1(x_{12}, x_{22}, \ldots, x_{n2}), \ldots, \omega_1(x_{1m}, x_{2m}, \ldots, x_{nm}))$$

for any elements $x_{11}, x_{21}, \ldots, x_{n1}, \ldots, x_{1m}, x_{2m}, \ldots, x_{nm}$ from L.

Example 7.3.1.26. Let us take a linear space L over the field $\mathbf{R}$ of all real numbers. Then addition in L commutes with itself. Indeed,

$$(x + y) + (z + v) = (x + z) + (y + v) \tag{7.3.1.6}$$

It is easier to understand commutativity of operations when it is presented in the matrix form. Let us consider operations ω_1 with arity n and ω_2 with arity m from the set Ω and the following matrix M with elements from a universal algebra L.

$$\begin{pmatrix} x_{11}, x_{12}, \ldots, x_{1m} \\ x_{21}, x_{22}, \ldots, x_{2m} \\ \cdots\cdots\cdots\cdots \\ x_{n1}, x_{n2}, \ldots, x_{nm} \end{pmatrix}$$

Applying the operation ω_2 to the rows of M, we obtain n elements $\omega_2(x_{11}, x_{12}, \ldots, x_{1m})$, $\omega_2(x_{21}, x_{22}, \ldots, x_{2m})$, ..., $\omega_2(x_{n1}, x_{n2}, \ldots, x_{nm})$ from L. Then it is possible to apply the operation ω_1 to these elements and to obtain the element $\omega_1(\omega_2(x_{11}, x_{12}, \ldots, x_{1m}), \omega_2(x_{21}, x_{22}, \ldots, x_{2m}), \ldots, \omega_2(x_{n1}, x_{n2}, \ldots, x_{nm}))$ from L.

At the same time, applying the operation ω_1 to the columns of M, we obtain m elements $\omega_1(x_{11}, x_{21}, \ldots, x_{n1})$, $\omega_1(x_{12}, x_{22}, \ldots, x_{n2})$, ..., $\omega_1(x_{1m}, x_{2m}, \ldots, x_{nm})$ from L. Then it is possible to apply the operation ω_2 to these elements and to obtain the element $\omega_2(\omega_1(x_{11}, x_{21}, \ldots, x_{n1}), \omega_1(x_{12}, x_{22}, \ldots, x_{n2}), \ldots, \omega_1(x_{1m}, x_{2m}, \ldots, x_{nm}))$ from L.

Commutativity of operations ω_1 and ω_2 means that the results of both procedures are equal to one another.

Taking addition in linear algebras (cf. Appendix), we see that commutativity of addition with itself, which is described by the formula (7.3.1.6), corresponds to the following matrix

$$\begin{bmatrix} x & y \\ z & v \end{bmatrix}$$

Indeed, the matrix definition of commutativity of addition implies the following identity:

$$(x + z) + (y + v) = (x + y) + (z + v)$$

Taking $v = x = 0$, we have the conventional commutative law for addition:

$$z + y = y + z$$

Let us assume that the representation space L is a universal algebra of a type Ω, a set of operations Σ in L is a subset of Ω, and all operations from Ω commute with all operations from Σ.

Theorem 7.3.1.5. a) The set $\mathbf{Opt}^{\Sigma} L$ of all total coherent with respect to Σ information operators in L is a universal algebra of the type Ω and is a subalgebra of the algebra **Opt** L.

b) If all operations from Ω commute with one another, then the set **COpt** L of all total coherent information operators in L is a universal algebra of the type Σ.

Proof. a) Let us take an operation ω from the set Ω and assume that ω has arity n, i.e., $\omega: L^n \to L$ is a mapping from L^n into the representation space L (cf. Appendix D). We show that the operation ω induces the corresponding operation $\omega_t: (\text{Opt } L)^n \to \text{Opt } L$ in the set of total information operators Opt L.

Taking total information operators $A_1, A_2, \ldots, A_n$, we define the corresponding operation ω_t by the following formula:

$$\omega_t(A_1, A_2, \ldots, A_n)(x) = \omega(A_1(x), A_2(x), \ldots, A_n(x))$$

for any x from L

Let us assume that total information operators $A_1, A_2, \ldots, A_n$ are coherent with respect to Σ and σ is an m-ary operation from Σ. Coherence of information operators $A_1, A_2, \ldots, A_n$ with respect to Σ means that for each operator A_i, we have

$$A_i(\sigma(x_1, x_2, \ldots, x_m)) = \sigma(A_i(x_1), A_i(x_2), \ldots, A_i(x_m))$$

for any $x_1, x_2, \ldots, x_m$ from L and σ from Σ.

Then we have

$$\omega_t(A_1, A_2, \ldots, A_n)(\sigma(x_1, x_2, \ldots, x_m)) =$$
$$\omega(A_1(\sigma(x_1, x_2, \ldots, x_m)), A_2(\sigma(x_1, x_2, \ldots, x_m)), \ldots, A_n(\sigma(x_1, x_2, \ldots, x_m))) =$$
$$\omega(\sigma(A_1(x_1), A_1(x_2), \ldots, A_1(x_m)), \sigma(A_2(x_1), A_2(x_2), \ldots, A_2(x_m)),$$
$$\ldots, \sigma(A_n(x_1), A_n(x_2), \ldots, A_n(x_m)))$$

because all information operators $A_1, A_2, \ldots, A_n$ are coherent with respect to Σ. Then as all operations from Ω commute with all operations from Σ, we have

$$\omega(\sigma(A_1(x_1), A_1(x_2), \ldots, A_1(x_m)), \sigma(A_2(x_1), A_2(x_2), \ldots, A_2(x_m)),$$
$$\ldots, \sigma(A_n(x_1), A_n(x_2), \ldots, A_n(x_m))) =$$
$$\sigma(\omega(A_1(x_1), A_2(x_1), \ldots, A_n(x_1)), \omega(A_1(x_2), A_2(x_2), \ldots, A_n(x_2)),$$
$$\ldots, \omega(A_1(x_m), A_2(x_m), \ldots, A_n(x_m))) =$$
$$\sigma(\omega_t(A_1, A_2, \ldots, A_n)(x_1), \omega_t(A_1, A_2, \ldots, A_n)(x_2),$$
$$\ldots, \omega_t(A_1, A_2, \ldots, A_n)(x_m)))$$

This means that $\omega_I(A_1, A_2, \ldots, A_n)$ is a coherent with respect to σ information operator. As σ is an arbitrary operation from Σ, $\omega_I(A_1, A_2, \ldots, A_n)$ is a coherent with respect to Σ information operator.

As ω is an arbitrary operation from Ω and $A_1, A_2, \ldots, A_n$ are arbitrary total information operators from the set $\mathbf{Opt}^\Sigma\ L$, the statement (a) from the theorem is proved.

b) Taking $\Sigma = \Omega$, we see that the statement (b) is a direct corollary of the statement (a).

Besides, by construction, $\mathbf{Opt}^\Sigma\ L$ is a subalgebra of the algebra $\mathbf{Opt}\ L$.

Theorem is proved.

Let us assume that L is a universal algebra of a type Ω and all operations from Ω commute with one another.

Corollary 7.3.1.6. The set $\mathbf{COpt}\ L$ of all total coherent information operators in L is a universal algebra of the type Ω and is a subalgebra of the algebra $\mathbf{Opt}\ L$.

Let us assume that L is a universal algebra of a type Ω from a variety V and all operations in are mutually commutative.

Theorem 7.3.1.6. The set $\mathbf{Opt}\ L$ of all total information operators in L is a universal algebra from the variety V, i.e., algebras L and $\mathbf{Opt}\ L$ have the same operations and all identities in L have their counterparts in $\mathbf{Opt}\ L$.

Proof. Let us take an operation ω from the set Ω and assume that ω has arity n, i.e., $\omega: L^n \to L$ is a mapping from L^n into the representation space L (cf. Appendix D). We show that the operation ω induces the corresponding operation $\omega_I: (\mathrm{Opt}\ L)^n \to \mathrm{Opt}\ L$ in the set of total information operators Opt L.

Taking information operators $A_1, A_2, \ldots, A_n$, we define the corresponding operation ω_I by the following formula:

$$\omega_I(A_1, A_2, \ldots, A_n)\ (x) = \omega(A_1(x), A_2(x), \ldots, A_n(x))$$

for any x from L.

For instance, assume that the representation space L is a linear (vector) space over the field of all real numbers $\boldsymbol{R}$, and a is a real number. In this case, such operations with information operators are defined: the sum $A + B$, difference $A - B$, and multiplication aA by an element a from F. Then as it is proved in Proposition 7.3.1.14, all identities in L are also true in the set **Opt** L of all total information operators in L.

To prove Theorem 7.3.1.6 in a general case, we consider the general form of identities in universal algebras. As it is known (cf., for example, (Cohn, 1965)), any such an identity has the following form:

$$F(\omega_1, \omega_2, \ldots, \omega_k; x_1, x_2, \ldots, x_n) = G(\omega_1, \omega_2, \ldots, \omega_k; x_1, x_2, \ldots, x_n)$$

$$(7.3.1.7)$$

Here F and G are well-formed (correctly formed) expressions built from operations $\omega_1, \omega_2, \ldots, \omega_k \in \Omega$ and free variables $x_1, x_2, \ldots, x_n$. Note that it is not necessary that all operations $\omega_1, \omega_2, \ldots, \omega_k$ and all variables $x_1, x_2, \ldots, x_n$ belong to both expressions F and G.

Now let us assume that (7.3.1.7) is an identity in the algebra L and $A_1, A_2, \ldots, A_n$ are arbitrary total information operators from the set **Opt** L. Then for any x from L, we have

$$F(\omega_1, \omega_2, \ldots, \omega_k; A_1(x), A_2(x), \ldots, A_n(x)) =$$
$$G(\omega_1, \omega_2, \ldots, \omega_k; A_1(x), A_2(x), \ldots, A_n(x)) \qquad (7.3.1.8)$$

As $A_1, A_2, \ldots, A_n$ are arbitrary total information operators from the set **Opt** L, (7.3.1.8) is an identity in the algebra **Opt** L. Thus, all identities in the algebra L induce corresponding identities in the algebra **Opt** L.

Theorem is proved.

Corollary 7.3.1.7. The set **COpt**$_R$ L of all total and closed information operators in L is a universal algebra of the type Ω.

Corollary 7.3.1.8. All closed in the same representation domain information operators form a universal algebra of a type Ω.

Corollary 7.3.1.9. All closed in the same representation domain linear information operators form a universal algebra of a type Ω.

Let us assume that L is a universal algebra of a type Ω.

Theorem 7.3.1.7. The set **Op** L of all information operators in L is a universal algebra of the type Ω.

Proof. Let us take an operation ω from the set Ω and assume that ω has arity n, i.e., $\omega: L^n \to L$ is a mapping from L^n into the representation space L (cf. Appendix D). We show that the operation ω induces the corresponding operation $\omega_P: (\mathbf{Op}\ L)^n \to \mathbf{Op}\ L$ in the set of total information operators Opt L.

Let us select an element a from L. Taking n information operators $A_1, A_2, ..., A_n$ in L, we define the corresponding operation ω_P by the following formulas:

$$\omega_P(A_1, A_2, ..., A_n)\,(x) = \omega(A_1(x), A_2(x), ..., A_n(x))$$

when all $A_1, A_2, ..., A_n$ are defined for the element x from L.

Otherwise, i.e., when, at least, one information operator A_i ($i \in \{1, 2, 3, ..., n\}$) is undefined for the element x from L, we put

$$\omega_P(A_1, A_2, ..., A_n)\,(x) = a$$

By construction, $\omega_P(A_1, A_2, ..., A_n)$ is a total information operators in L.

At the same time, it is possible to assume that the information operator $\omega_P(A_1, A_2, ..., A_n)$ is undefined for x when, at least, one information operator A_i ($i \in \{1, 2, 3, ..., n\}$) is undefined for the element x from L. In this case, the mapping $\omega_P(A_1, A_2, ..., A_n)$ is not total if, at least, one operator A_i is not total. This shows that there are different ways to define a universal algebra in set **Op** L.

Theorem is proved.

Theorems 7.3.1.3 and 7.3.1.7 imply the following result.

Corollary 7.3.1.10. The set **Opt** L of all total information operators in L is a subalgebra of the algebra of the **Op** L.

Let us assume that L is a universal algebra of a type Ω and U is a subalgebra in L.

Theorem 7.3.1.8. The set **Op**$_U$ L of all closed in U information operators is a universal algebra of the type Ω, which is a subalgebra of the algebra **Op** L.

The proof of this theorem is based on the following lemma.

Lemma 7.3.1.1. If all operations ω in an identity $u = v$ that is true in L are substituted by their counterparts ω_I, then the obtained expression is an identity in the set of all closed in U information operators.

Theorems 7.3.1.6 and 7.3.1.8 imply the following results.

Corollary 7.3.1.11. The set **Op**$_U$ L of all closed in U total information operators in L is a universal algebra from the variety V.

One more type of operations on information operators is *inherited* from operations on systems represented in the space L and operations with representation spaces. For instance, we can consider the disjoint union of two systems or inclusion (integration) of one system into another system.

Definition 7.3.1.16. The *disjoint union* $A \amalg B$ of two systems A and B is defined only when these systems do not intersect, i.e., they do not have common elements or other parts, and is the system that consists of all elements from A and B and relations (ties) between these elements.

If L is the state space of A and H is the state space of B, then the Cartesian product $L \times H$ of spaces L and H is the state space of the system $A \amalg B$. Indeed, a state of $A \amalg B$ is a pair (a, b) where a is a state of A and b is a state of B. When spaces L and H are both linear spaces, the Cartesian product is called sum and denoted by $L \oplus H$.

Example 7.3.1.27. Taking two people X and Y and their brains as systems A and B, it is possible to define the disjoint union $A \amalg B$ because brains of different people do not intersect. Then it is possible to represent any state of the system $A \amalg B$ as a pair (a, b) where a is a state of A and b is a state of B.

Definition 7.3.1.17. The *sum* $I \oplus J$ of two information operators $I: L \to L$ and $J: H \to H$ is the mapping $I \oplus J: L \times H \to L \times H$ defined by mappings I and J.

As $L \times H$ is a direct product (or simply, product) in the category of all spaces with the same structure as spaces L and H have and their homomorhisms as its mappings, the mapping $I \oplus J$ is uniquely defined by mappings I and J (cf. (Herrlich and Strecker, 1973)).

Operations with representation spaces also give birth to operations with information operators. For instance, if representation spaces are linear (vector) spaces, then the operation of the sum (Cartesian product) $L \oplus H$ of two spaces L and H determines the corresponding operation that combines an information operator A in L with an information operator B in H, giving as a result an information operator $A \oplus B$ in $L \oplus H$.

Let us assume that L and H are universal algebras of the same type. Then another operation with information operators is induced by the free product $L * H$ (cf. (Cohn, 1965)) of algebras L and H.

Definition 7.3.1.18. The *free composition* $I * J$ of two information operators $I: L \to L$ and $J: H \to H$ is the mapping $I * J: L * H \to L * H$ defined by mappings I and J.

As $L * H$ is a free product (or coproduct) in the category of all spaces with the same structure as spaces L and H have and their homomorhisms as its mappings, the mapping $I * J$ is uniquely defined by mappings I and J (cf. (Herrlich and Strecker, 1973)).

Both the sum $I \oplus J$ and free composition $I * J$ of information operators I and J are their parallel compositions.

Proposition 18.16 from (Herrlich and Strecker, 1973) allows one to prove the following result.

Proposition 7.3.1.16. The sum and free product of covering (injective or bijective) information operators is a covering (injective or bijective, correspondingly) information operator.

Properties of direct products allow one to prove the following result.

Proposition 7.3.1.17. The sum of coherent information operators is a coherent information operator.

The union of two representation spaces defines a partial operation with information operators.

Let $A: L \to L$ and $B: H \to H$ be (partial) information operators such that the following condition is true.

(Condition C$_2$). For any a from $L \cap H$, we have $A(a) = B(a)$.

Definition 7.3.1.19. The union $A \cup B$ of information operators A and B is an information operator in the space $L \cup H$ defined by the following formula

$$A \cup B = \begin{cases} A(a) \text{ when } A(a) \text{ is defined} \\ \\ B(a) \text{ when } B(a) \text{ is defined} \end{cases}$$

Condition C$_2$ provides correctness of this definition.

Proposition 7.3.1.18. If an information operator A is closed in the space D and an information operator B is closed in the space C, then the information operator $A \cup B$ is closed in the space $C \cup D$.

It is also possible to define union for an arbitrary set of information operators.

Let $A_i: L_i \to L_i$ ($i \in I$) be (partial) information operators such that the following condition is true.

(Condition C). For any a from $L_i \cap L_j$, we have $A_i(a) = A_j(a)$ for all $i, j \in I$.

Definition 7.3.1.20. The union $\bigcup_{i \in I} A_i$ of information operators A_i is an information operator in the space $\bigcup_{i \in I} L_i$ defined by the following formula

$$(\bigcup_{i \in I} A_i(a))(a) = A_i(a) \text{ when } A_i(a) \text{ is defined}$$

Condition C provides correctness of this definition.

Theorem 7.3.1.9. Any general information operator I in **L** is the union of particular information operators.

Proof. Let us consider a general information operator A in **L**

$$A: \bigcup_{i \in I} L_i \to \bigcup_{i \in I} L_i$$

Then its restrictions $A_{ij}: L_i \to L_j$ on the pairs of spaces $L_i \to L_j$ for all $i, j \in I$ allow us to define mappings $A_{ij}^{\circ}: L_i \cup L_j \to L_i \cup L_j$. Namely, if a belongs to $L_i \cup L_j$ and $A_{ij}(a)$ is defined, then $A_{ij}^{\circ}(a) = A_{ij}(a)$. Otherwise, $A_{ij}^{\circ}(a)$ is not defined. All A_{ij}° are particular information operators.

Taking two information operators $A_{ij}: L_i \to L_j$ and $A_{kl}: L_k \to L_l$, we see that for any a from $L_i \cap L_k$, we have $A_{ij}(a) = A_{kl}(a)$ for all $i, j, k, l \in I$ as both operators are restrictions of one and the same information operator A. Consequently, for any a from $(L_i \cup L_j) \cap (L_i \cup L_j)$, we have $A_{ij}^{\circ}(a) = A_{kl}^{\circ}(a)$ for all $i, j, k, l \in I$, i.e., the system of information operators A_{ij}° satisfy Condition C. Thus, it is possible to take their union $\bigcup_{i,j \in I} A_{ij}$.

By Definition 7.3.1.17, $A = \bigcup_{i,j \in I} A_{ij}$. Indeed, $\bigcup_{i,j \in I} (L_i \cup L_j) = \bigcup_{i \in I} L_i$, and if a belongs to the space $\bigcup_{i \in I} L_i$, then a belongs to the space L_i for some $i \in I$. Consequently, we have

$$A(a) = A_{ij}(a) = A_{ij}^{\circ}(a) = (\bigcup_{i \in I} A_i(a))(a)$$

Theorem is proved.

Let R be a system, R_L be its representation in the system space L, and $A \in A(C, c)$

Definition 7.3.1.21. Application of A to R_L is called *informing R by C through c* or reception of the message (C, c, A).

Application of A to R_L models changes caused by the reception of the message (C, c, A) by the system R.

In many cases, the system space L is partially ordered. For instance, when L is a knowledge space, the inequality $a \leq b$ means that knowledge a is a part of knowledge b. It can be an actual part, i.e., $a \subseteq b$, or a potential part, i.e., knowledge a is deducible from knowledge b. When L is a linear space over the field of real numbers, it is also partially ordered. One of the possible partial orderings on L is given by the following rule: $a \leq b$ if $a = \alpha b$ and $|\alpha| \leq 1$.

Definition 7.3.1.22. An information operator A in L is called *increasing* (*decreasing*) [in a domain $U(R)$] if for any x [$x \in U(R)$], we have $x \leq A(x)$ ($A(x) \leq x$).

Proposition 7.3.1.19. The sequential composition (product) of increasing (decreasing) information operators is an increasing (decreasing) information operator.

Proof. Let us consider the sequential composition AB of two increasing information operators $A: L \to L$ and $B: L \to L$. Then for any x [$x \in U(R)$], we have $x \leq A(x)$. As B is also an increasing information operator, we have $A(x) \leq B(A(x))$. As any order relation is transitive (cf. Appendix A), $x \leq B(A(x))$.

For decreasing information operators, the proof is similar.

Proposition is proved.

Orderings in spaces L and H determine an ordering in the space $L \oplus H$. This allows one to prove the following result.

Proposition 7.3.1.20. The sum of increasing (decreasing) information operators is an increasing (decreasing) information operator.

Proof. Let us consider the sum $A \oplus B$ of two increasing information operators $A: L \to L$ and $B: H \to H$. It is the mapping $A \oplus B: L \times H \to L \times H$ defined by mappings A and B. Taking two pairs (x, y) and (u, v) such that $x, u \in L$ and $y, v \in H$, we see that by the definition of the order in the direct product $L \times H$, $(x, y) \leq (u, v)$ if and only if $x \leq u$ in L and $y \leq v$ in H.

Then for any $x \in L$, we have $x \leq A(x)$ and for any $y \in H$, we have $y \leq B(y)$. Consequently, $(x, y) \leq (A \oplus B)(x, y) = (A(x), B(y))$ for any pair $(x, y) \in L \times H$. It means that the sum $A \oplus B$ is an increasing information operator.

For decreasing information operators, the proof is similar.

Proposition is proved.

Definition 7.3.1.23. An information operator A in L_1 is called *monotone (antitone)* [in a domain U(R)] if for any x and y [$x, y \in$ U(R)], the inequality $x \leq y$ implies the inequality $A(x) \leq A(y)$ ($A(y) \leq A(x)$).

Proposition 7.3.1.21. The sequential composition (product) of monotone (antitone) information operators is a monotone (antitone) information operator.

Proof. Let us consider the sequential composition AB of two monotone information operators $A: L \to L$ and $B: L \to L$. Then the inequality $x \leq y$ implies the inequality $A(x) \leq A(y)$. In turn, the inequality $A(x) \leq A(y)$ implies the inequality $B(A(x)) \leq B(A(y))$. Consequently, the inequality $x \leq y$ implies the inequality $B(A(x)) \leq B(A(y))$. It means that the sequential composition AB is a monotone information operator.

For antitone information operators, the proof is similar.

Proposition is proved.

Corollary 7.3.1.3. If information operators A and B are closed and monotone (antitone) in the domain U(R), then their sequential composition AB is closed and monotone (antitone) in the domain U(R).

Orderings in spaces L and H determine an ordering in the space $L \oplus H$. This allows one to prove the following result.

Proposition 7.3.1.22. The sum of monotone (antitone) information operators is a monotone (antitone) information operator.

Proof. Let us consider the sum $I \oplus J$ of two monotone information operators $I: L \to L$ and $J: H \to H$. It is the mapping $I \oplus J: L \times H \to L \times H$ defined by mappings I and J. Taking two pairs (x, y) and (u, v) such that $x, u \in L$ and $y, v \in H$, we see that by the definition of the order in the direct product $L \times H$, $(x, y) \leq (u, v)$ if and only if $x \leq u$ in L and $y \leq v$ in H.

Then the inequality $(x, y) \leq (u, v)$ implies inequalities $x \leq u$ in L and $y \leq v$ in H. As I and J are monotone information operators, we have $I(x) \leq I(u)$ and $J(y) \leq J(v)$. Consequently, $(I(x), J(y)) \leq (I(u), J(v))$. It means that the sum $I \oplus J$ is a monotone information operator.

For antitone information operators, the proof is similar.

Proposition is proved.

Let $\otimes$ be an operation in L and $\otimes_I$ be the corresponding induced operation in the algebra A(C, c).

Proposition 7.3.1.23. If $\otimes$ is a monotone (antitone) operation, then the $\otimes_I$-composition of monotone (antitone) information operators is a monotone (antitone) information operator, i.e., the set of monotone (antitone) information operators is closed with respect to $\otimes_I$.

Proof is similar to the proof of Proposition 7.3.1.22.

Remark 7.3.1.4. In a general case, it is possible that the carrier and channel are also changed in the process of communication. This situation is modeled by a generalized information operator, which is a morphism (Burgin, 1990) of the substantial communication triad space (H, K, L) where H is the carrier system space and K is the channel system space. For instance, in the microworld, observations do not simply measure the system, but interfere with the system in a way that impacts observations. So, it is possible that in the process of information transaction (reception) the carrier and/or channel are changing and we need generalized information operators.

It is evident that a system can change itself when it receives information in a broad sense because energy is a kind of such information. Examples are interactions of subatomic particles, mutations of living organisms and transformations of chemical elements.

However, a system can change itself or/and the channel even during the time when the system receives information in the strict (proper) sense. Examples are given by self-transforming programs and algorithms (cf., for example, (Burgin, 1993a; Yip and Zhou, 2004)). Programs and algorithms are carriers of information that tells computers how and what to compute.

Transformations (changes) of channels for information transmission occur more frequently. For instance, we switch (change) channels when we watch TV or listen to the radio.

These situations bring us to the concept of information flow operator. To define it formally, we consider the following structures:

- a system space H for carriers of information;
- a system space L for receivers/receptors of information;
- interactions of carriers with receivers/receptors.

As before, it is assumed that carriers are represented in the system space H and receivers/receptors are represented in the system space L.

Interactions are mathematically represented by binary relations (in some cases, by functions) from the carrier system space H to receiver/receptor system space L. The set (system) of all tentative interactions is denoted by **V**.

Definition 7.3.1.24. An *information flow operator Fl* is a mapping of **V** into itself.

It is possible to represent interactions of carriers with receivers/receptors by triads (C, act, R). Then there is a projection of the space of all interactions (from some class) on the space of all receivers/receptors (from the corresponding class). Namely, the triad (C, act, R) is projected on the system R. Thus, any mapping of interactions induce a mapping of receivers/receptors. This allows one to build a projection of any information flow operator onto some information operator.

Developed mathematical models of information and its flow allow us to formalize ontological principles of the general theory of information. The first is the Weak Representation Postulate.

Postulate IP1. For any infological system $IF(R)$ of any system R, there is a system space L in which $IF(R)$ can be faithfully represented.

Postulate IP1 is a formalization of the Ontological Principle O1.

Let us consider a type **K** of infological system. For instance, cognitive, affective and effective are natural infological system types. Then we have the following typed representation postulate, which is stronger than Postulate IP1.

Postulate IP$_K$1. There is a system space L_u in which any infological system $IF(R)$ of the type **K** can be faithfully represented.

If we represent portions of information by information operators, it means that we have a function that corresponds information operators to portions of information.

Definition 7.3.1.25. A set Q of portions of information I is *faithfully represented* (in the space L_l) if different portions of information are corresponded to different information operators (in L).

Postulate IP2. Any portion of information can be faithfully represented by an information operator.

Postulate IP2 is a formalization of the Ontological Principle O2.

Note that Postulate IP2 demands that any portion of information has a representation as an information operator. However, it is not necessary that any information operator represents some portion of information.

Postulate IP3. Any information operator is a projection of some information flow operator.

Postulate IP3 is a formalization of the Ontological Principles O3 and O4.

The Ontological Principle O5 is a consequence of the faithfulness of representation of portions of information implied by Postulate IA2. The Ontological Principle O6 is a consequence of the fact that in a general case, there are different interactions of one carrier with one receptor/receiver.

Postulate IP1–IP3 provide a grounded interpretation for algebras of information operators and allow one to use these algebras for studying information processes and designing information systems.

When we consider information of some type and corresponding infological systems, additional postulates are necessary to reflect specific features of this type.

The mathematical stratum of the general theory of information has also axioms, which determine properties of mathematical structures used in this theory. For instance, if we develop a special information theory in which information is considered with respect to infological systems the state space of which is a linear space, then the axioms of the linear space become axioms of this special information theory. Namely, we have the following axioms (cf. Appendix D):

1. *Addition is associative*:

    For all x, y, z from L, we have $x + (y + z) = (x + y) + z$.

2. *Addition is commutative*:

    For all x, y from L, we have $x + y = y + x$.

3. *Addition has an identity element*:

    There exists an element **0** from L, called the *zero vector*, such that $x + \mathbf{0} = x$ for all x from L.

4. *Addition has an inverse element*:

    For any from L, there exists an element z from L, called the *additive inverse* of x, such that $x + z = \mathbf{0}$.

5. *Scalar multiplication is distributive over addition in L:*
   For all numbers a from R and vectors y, w from L, we have

   $$a(y + w) = a y + a w.$$

6. *Scalar multiplication is distributive over addition in R:*
   For all numbers a, b from R and any vector y from L, we have

   $$(a + b) y = a y + b y.$$

7. *Scalar multiplication is compatible with multiplication in R:*
   For all numbers a, b from R and any vector y from L, we have

   $$a(b y) = (ab) y.$$

8. *The number 1 is an identity element for scalar multiplication:*
   For all vectors x from L, we have $1x = x$.

Axioms of the general theory of information allow one to further develop information algebra, build information logic and use this theory for various practical purposes. For instance, let us consider a system of information portions that are represented by information operators. The developed algebraic setting allows us to select a set of basic portions of information and represent any information portion as an algebraic expression, e.g., a linear combination, of basic portions of information. Applying this approach to emotions, we can take portions of direct affective information that cause basic emotions and have unit intensity (cf. Section 2.6) as the basis and express any portion of direct affective information as an algebraic expression, e.g., a linear combination, of these unit portions.

Dealing with propositional cognitive information, we can represent information portions by logical information operators that add to and/or remove propositions from logical systems, e.g., logical calculi, which represent knowledge. By Theorem 7.3.1.7, these operators also form the same logical system. In the case when the representation space is a propositional calculus, logical information operators also form a propositional calculus. In such a way, we obtain a propositional logic of cognitive information.

It is possible to restrict utilized cognitive information operators. One possibility to do this is to consider only operations of deduction or inference as cognitive information operators. Another option is to allow adding new propositions to the initial system pending that these propositions are consistent with the system. One more option is to allow removing propositions from the initial system pending that the system does not become too weak. For instance, it is possible to demand that after all transformations the system always contains the formal arithmetic.

Representing knowledge by predicates, it is possible to build predicate calculi of cognitive information operators.

7.3.2. *Abstract information algebra*

> *Mathematics is a game played*
> *according to certain simple rules*
> *with meaningless marks on paper.*
> David Hilbert (1862–1943)

Information algebras were introduced by Kohlas (2003) as a mathematical model for situations when information that comes from different sources is combined from different elements and when parts of information that are relevant to a specific question are extracted. In this context, these two operations, combination and information extraction, are considered as basic modes of information processing. In addition, information algebras grasp similarities in various formal structures used in mathematics and computer science: relational databases, multiple systems of formal logic, and numerical methods in linear algebra.

Researchers study two types of information algebras: labeled information algebras and unlabeled information algebras (Kohlas and Stärk, 2007). The former are also called information algebras.

A labeled information algebra, or simply *information algebra*, is a two-sorted algebra $A = (\Phi, D)$ with six operations:

Combination/composition	$\otimes: \Phi \times \Phi \to \Phi$
Focussing/marginalization	$\Rightarrow: \Phi \times D \to \Phi$
Meet	$\wedge: D \times D \to D$
Join	$\vee: D \times D \to D$
Labeling	$d: \Phi \to D$
Unit selection	$e: D \to \Phi$

The set Φ is considered (interpreted) as a set of *pieces* (*portions*) of information. The set D is considered (interpreted) as a set of *frames* or *domains* for information. Each piece of information φ concerns a certain *domain* $d(\varphi)$ from D, which is attached to φ as a *label* or *mark*, i.e., a mapping $d: \Phi \to D$ is given. This unary operation formalizes a natural property of information.

If φ and ψ are portions of information, i.e., $\varphi, \psi \in \Phi$, and $x \in D$, then *composition*, or *combination*, of φ and ψ is denoted by $\varphi \otimes \psi$, or simply, by $\varphi\psi$, the unit element corresponding to $x \in D$ is denoted by $e(x) = e_x$, and *focusing*, or *marginalization*, of φ on x is denoted by $\varphi^{\Rightarrow x}$. Combination $\varphi \otimes \psi$ of information portions φ and ψ in the system Φ represents aggregation of information. Focusing shows that each portion of information refers to, at least, one domain (question) in D, or in other words, any portion of information has a *support*, i.e., $\varphi = \varphi^{\Rightarrow x}$ for some $x \in D$.

A partial order of information is introduced in the set Φ by defining $\varphi \leq \psi$ if $\varphi \otimes \psi = \psi$. This means that information φ is less informative than information ψ if φ adds no new information to ψ. In such a way, the set Φ with the operation of composition is turned into a semilattice relative to this order if we define $\varphi \vee \psi = \varphi \otimes \psi$. It is also possible to introduce a partial order in the set Φ relative to any question x from the domain D by defining $\varphi \leq_x \psi$ if $\varphi^{\Rightarrow x} \leq \psi^{\Rightarrow x}$. This means that information φ is less informative than information ψ relative to the question x if φ adds no new information to ψ with respect to x.

A (*labeled*) *information algebra* satisfies the following conditions (axioms):

LIA 1. The system D is a lattice with respect to meet (intersection) and join (union).

LIA 2. *Associativity* of composition: for any portions of information $\varphi, \psi, \phi \in \Phi$, $\varphi \otimes (\psi \otimes \phi) = (\varphi \otimes \psi) \otimes \phi$.

LIA 3. *Commutativity* of composition: for any portions of information $\varphi, \psi \in \Phi$, $\varphi \otimes \psi = \psi \otimes \varphi$.

LIA 4. *Distributivity* of labeling: for any portions of information $\varphi, \psi \in \Phi$, $d(\varphi \otimes \psi) = d(\varphi) \vee d(\psi)$

LIA 5. *Identity* information on each topic (question) x: $d(e_x) = x$ for all frames $x \in D$.

LIA 6. *Neutrality* of e_x on each topic (question) x: for any frame $x \in D$ and any portion of information $\varphi \in \Phi$, if $d(\varphi) = x$, then $\varphi \otimes e_x = \varphi$.

LIA 7. *Invertibility*: for any frame $x \in D$ and any portion of information $\varphi \in \Phi$, if $x \leq d(\varphi)$, then $d(\varphi^{\Rightarrow x}) = x$.

LIA 8. *Transitivity* of focusing: for any frames $x, y \in D$ and any portion of information $\varphi \in \Phi$, if $x \leq y \leq d(\varphi)$, then $(\varphi^{\Rightarrow y})^{\Rightarrow x} = \varphi^{\Rightarrow x}$

LIA 9. *Idempotency* of combination with respect to focusing: for any frame $x \in D$ and any portion of information $\varphi \in \Phi$, if $x \leq d(\varphi)$, then $\varphi \otimes \varphi^{\Rightarrow x} = \varphi$

LIA 10. *Stability*: for any frames $x, y \in D$, if $x \leq y$, then $d(e_y^{\Rightarrow x}) = e_x$.

LIA 11. *Combination* of focusing: for any frames $x, y \in D$ and for any portions of information $\varphi, \psi \in \Phi$, if $d(\varphi) = x$ and $d(\psi) = y$, then $(\varphi \otimes \psi)^{\Rightarrow x} = \varphi \otimes (\psi^{\Rightarrow x \wedge y})$.

Axioms of a labeled information algebra have the following interpretation:

- Axioms LIA 1 and LIA 2 imply that the system Φ is a commutative semigroup with respect to combination.
- The partial order in the lattice D of domains (related to questions) reflects the granularity of domains, while focusing represents extraction of information related to a given domain.
- Distributivity (Axiom LIA 11) means that to focus the combination of a portion of information φ on a domain x with another portion of information ψ on a domain y is the same as to combine the portion of information φ with the second portion of information ψ focused on $x \wedge y$.
- Transitivity (Axiom LIA 8) means that to focus a piece of information φ on a smaller frame x and then on a larger frame y is the same as to focus information φ only on x.
- Idempotency (Axiom LIA 9) means that a piece of information combined with a part of itself gives nothing new.
- Element e_x denotes the empty portion of information on the domain x.
- Axiom LIA 6 implies that adding the empty portion of information on a domain x, we do not change any portion of information related to the same domain.

Axioms of labeled information algebras imply the following properties.

Proposition 7.3.2.1 (Kohlas and Stärk, 2007).

a) If $d(\varphi) = x$, then $\varphi^{\Rightarrow x} = \varphi$.

b) For any portion of information $\varphi \in \Phi$, $\varphi \otimes \varphi = \varphi$.

c) For any frames $x, y \in D$, if $x \leq y$, then $e_x \otimes e_y = e_y$.

Proof. a) Let us assume that $d(\varphi) = x$. Then we have $\varphi^{\Rightarrow x} = (\varphi \otimes e_x)^{\Rightarrow x}$ (by Axiom LIA 6) $= \varphi \otimes (e_x)^{\Rightarrow x}$ (by Axiom LIA 11) $= \varphi \otimes e_x$ (by Axiom LIA 10) $= \varphi$ (by Axiom LIA 6).

b) For any portion of information $\varphi \in \Phi$, $\varphi \otimes \varphi = \varphi \otimes (\varphi)^{\Rightarrow x}$ (by Proposition 7.3.2.1.a) $= \varphi$ (by Axiom LIA 9).

c) For any frames $x, y \in D$, if $x \leq y$, then $e_x \otimes e_y = e_y^{\Rightarrow x} \otimes e_y$ (by Axiom LIA 10) $= e_y$ (by Axioms LIA 2 and 9).

An unlabeled information algebra is a two-sorted algebra $A = (\Phi, D)$ with five operations:

Combination/composition	$\otimes$:	$\Phi \times \Phi \rightarrow \Phi$
Focussing/marginalization	$\Rightarrow$:	$\Phi \times D \rightarrow \Phi$
Meet	$\wedge$:	$D \times D \rightarrow D$
Join	$\vee$:	$D \times D \rightarrow D$
Unit selection	$e \in \Phi$	

If $\varphi, \psi \in \Phi$ and $x \in D$, then combination of φ and ψ is denoted by $\varphi \otimes \psi$ and focusing of φ on x is denoted by $\varphi^{\Rightarrow x}$. If φ is a portion of information Φ and $\varphi = \varphi^{\Rightarrow x}$, then the frame $x \in D$ is called a *support* of φ.

An *unlabeled information algebra* satisfies the following conditions (axioms):

The system D is a lattice with respect to meet (intersection) and join (union).

UIA 1. *Associativity* of composition: for any portions of information $\varphi, \psi, \phi \in \Phi$, $\varphi \otimes (\psi \otimes \phi) = (\varphi \otimes \psi) \otimes \phi$.

UIA 2. *Commutativity* of composition: for any portions of information $\varphi, \psi \in \Phi$, $\varphi \otimes \psi = \psi \otimes \varphi$.

UIA 4. *Neutrality* of e: for any portion of information $\varphi \in \Phi$, $\varphi \otimes e = \varphi$.

UIA 5. *Transitivity* of focusing: for any frames x, $y \in D$ and any portion of information $\varphi \in \Phi$, $(\varphi^{\Rightarrow y})^{\Rightarrow x} = \varphi^{\Rightarrow x \wedge y}$.

UIA 6. *Idempotency* of combination with respect to focusing: for any frame $x \in D$ and any portion of information $\varphi \in \Phi$, $\varphi \otimes \varphi^{\Rightarrow x} = \varphi$

UIA 7. *Combination* of focusing: for any frame $x \in D$ and for any portions of information φ, $\psi \in \Phi$, $(\varphi^{\Rightarrow x} \otimes \psi)^{\Rightarrow x} = (\varphi^{\Rightarrow x}) \otimes (\psi^{\Rightarrow x})$.

UIA 8. Any portion of information $\varphi \in \Phi$, has a support, i.e., there is a frame $x \in D$ such that $\varphi = \varphi^{\Rightarrow x}$.

Axioms of a unlabeled information algebra imply the following properties:

- Axioms UIA 1 and UIA 2 imply that the system Φ is a commutative semigroup with respect to combination and has a neutral elements e (representing vacuous portion of information) for each element $x \in D$.

- Axiom UIA 5 implies that focusing is transitive.

- Axiom UIA 6 implies that the combination of information portions is idempotent with respect to focusing.

- Transitivity (Axioms LUA 5) means that to focus information φ on x and on y is the same as to focus information φ on $x \wedge y$.

- Idempotency (Axiom UIA 6) means that information combined with a part of itself gives nothing new.

- Element e denotes the empty portion of information.

- Axiom UIA 4 implies that adding the empty portion of information, we do not change any portion of information.

Proposition 7.3.2.2 (Kohlas and Stärk, 2007).

    a) The frame x is the support of $\varphi^{\Rightarrow x}$.

    b) If a frame x is a support of φ, then x is a support of $\varphi^{\Rightarrow y}$.

    c) If a frame x is a support of φ and $x \leq y$, then y is a support of φ.

    d) If frames x and y are supports of φ, then $x \wedge y$ is a support of φ.

    e) If a frame x is the support of φ and ψ, then x is the support of $\varphi \otimes \psi$.

Proposition 7.3.2.3 (Kohlas and Stärk, 2007). a) Each (labeled) information algebra generates an unlabeled information algebra. b) It is possible to convert any unlabeled information algebra into a labeled information algebra.

There are many classes of algebras and algebraic constructions studied by different authors that are special cases of information algebras. For instance, module algebras (Bergstra, Heering and Klint 1990; de Lavalette 1992) are information algebras. In constraint systems (Jaffar and Maher 1994), constraints form an information algebra. Preserving in a relational algebra (Codd, 1990) only such operations as join and projections, we obtain a labeled information algebra, in which join plays the role of combination.

Many systems in logic induce information algebras (Wilson and Mengin, 1999). For instance, reducts of cylindrical algebras (Henkin, Monk and Tarski, 1971) and polyadic algebras related to predicate logic (Halmos, 2000) are information algebras.

It addition, information algebras are related to other algebraic systems and are used for different applications. For instance, when we exclude the idempotency axiom from the definition of information algebra, we obtain valuation algebras. The latter were introduced by Shenoy and Shafer (1990) to generalize *local computation schemes* (Lauritzen and Spiegelhalter 1988) from Bayesian networks to more general formalisms, including belief function, possibility potentials, etc. (Kohlas and Shenoy, 2000). Systems of linear equations or linear inequalities induce information algebras (Kohlas 2003). Random variables with values in information algebras are used to represent *probabilistic argumentation systems* (Haenni, Kohlas and Lehmann, 2000).

In addition, there are relations between information algebras and logical inference. The standard way to determine logical inference uses deduction rules or entailment relations (cf., for example, (Shoenfield, 1967)). It is also possible to use consequence operators for inference in logic (Kohlas and Stärk, 2007). A logical language L of well-formed logical sentences is considered and portions of information are represented by subsets of L. This is a standard way of logical representation of semantic information (cf. Section 4.2).

Definition 7.3.2.1. A *consequence operator* is a mapping $C: 2^L \to 2^L$ of the power set 2^L of L into itself that satisfies the following conditions:

1. For any $X \in L$, we have $X \subseteq C(X)$.
2. For any $X \in L$, we have $C(C(X)) = C(X)$.
3. For any $X, Y \in L$, if $X \subseteq Y$, then $C(X) \subseteq C(Y)$.

In other words, a consequence operator is a topological closure operator on the set L (cf. (Kuratowski, 1966)).

It is demonstrated (Kohlas and Stärk, 2007) that any compact consequence operator satisfying the interpolation and deduction properties induces an information algebra where portions of information are represented by closed subsets of the language L and any finitary information algebra can be obtained from a consequence operator by the same procedure.

7.3.3. *Information geometry*

> *Inspiration is needed in geometry,*
> *just as much as in poetry.*
> Alexander Pushkin (1799–1837)

Taking a natural or artificial system, it is possible to assume that this system has a set of states and probabilities of these states give knowledge about this system. Taken together, these probabilities form a probability distribution, information about which gives knowledge. Thus, probability distributions represent information about different systems, situations, and the world as a whole. This approach is especially popular in economics (cf., for example, (Marschak, 1954; 1959; MacQueen and Marschak, 1975; Hirshliefer and Rilley, 1979; Lawrence, 1999)). A principal concern of information economics is the effect which information has upon a person's expectations and the resulting decisions she or he may contemplate or enact. Since these expectations are traditionally quantified by means of probability distributions assigned to future events (states of the world, including nature, actions of competitors, etc.), the relationship between information and probability distributions becomes crucial.

Information geometry studies probability distributions by methods of differential geometry, providing powerful tools for proving results about statistical models by considering them as well-behaved geometrical objects (Amari, 1985; Amari and Nagaoka, 2000). Probability distributions of random variables give information about these variables and related stochastic processes. As a result, such information or more exactly, its representation as a system of distributions forms a parametric manifold called info-manifold with the Fisher information measure $I(\varphi)$ as its Riemannian metric. Such manifolds naturally emerge in the theory of statistical estimation.

In 1920's when he studied estimation procedures, Fisher introduced the mathematical concept of *information* (more exactly, of information measure) in the form

$$I(\theta) = \mathbf{E} \left[[(\partial/\partial \theta) \ln f(X; \theta)]^2 \right]$$

He demonstrated that the variance of any estimator cannot be smaller than the inverse of his information. This information measure was used by mathematical statisticians Rao and Cramer to express the idea of an information manifold in terms of Riemannian geometry.

To build an information manifold, an n-parameter family $\mathbf{M} = \{\rho_\eta; \eta \in \mathbf{R}^n\}$ of probability distributions $\rho_\eta(x)$ is considered. In the context of the statistical theory of estimation, Fisher constructed the following information matrix

$$G^{ij} = \int \rho_\eta(x) \, (\partial \log \rho_\eta(x)/\partial \eta_i) \, (\partial \log \rho_\eta(x)/\partial \eta_j) \, dx$$

Later Rao (1945) demonstrated that this matrix provides a Riemannian metric for the manifold $\mathbf{M}$. This metric naturally arises in the statistical theory of estimation in the following way. A random n-dimensional variable X is considered the distribution ρ of which is believed or hoped to be one of those in the parametric family $\mathbf{M}$. Statisticians (or scientists, or economists, etc.) estimate the value of η by measuring X independently m times and getting the data $x_1, x_2, x_3, \ldots, x_m$. An estimator f is a function of $x_1, x_2, x_3, \ldots, x_m$ taking values in $\mathbf{R}^n$ that is used for this estimation. In this context, Fisher (1925) asserted and later Cramer proved that the variance V of an unbiased estimator f obeys the inequality $V \geq G^{-1}$ where $G = (G^{ij})$ is the Fisher information matrix. This inequality expresses that, given the family $\mathbf{M} = \{\rho_\eta; \eta \in \mathbf{R}^n\}$ of probability distributions, there is a limit to the reliability with which it is possible to estimate η.

The importance of studying statistical structures as geometrical structures lies in the fact that geometric structures are invariant under coordinate transformations. Information geometry allows one to find similar invariance for related families of distributions. For instance, a family of probability distributions, such as Gaussian distributions, may be transformed into another family of distributions, such as log-normal distributions, by a change of variables. However, the fact of it being an exponential family is not changed in this transformation because the latter is a geometric property. The distance between two distributions defined through Fisher metric is also preserved.

An important result in information geometry was obtained by Chentsov (1972) who proved that there is a unique intrinsic metric on the space of probability distributions on a sample space containing at least three points, and it is the Fisher information metric. This metric is the only one (up to a constant multiple) that is reduced by any stochastic mapping. In addition, Chentsov demonstrated that on the same space, there exists a unique one-parameter family of affine connections called α-affine connections.

There are classical and quantum information manifolds (Streater, 2000). The quantum version information manifolds was introduced by Chentsov (1972) and developed by Petz and his collaborators (Petz and Toth, 1993; Petz and Sudar, 1996). To do this, they considered algebras of $n \times n$ matrices used in quantum mechanics instead of random variables used in the classical case. Probability distributions were replaced by density matrices, which represented states of the quantum system. It was demonstrated that in contrast to the classical case, there are several metrics that make **M** a metric manifold (Chentsov, 1972).

Information geometry has been applied to different problems. For instance, in (Malagò, et al, 2008), information manifolds are used for providing a geometrical representation of estimation operators of distribution algorithms, a recent metaheuristic used in genetics-based machine learning to solve combinatorial and continuous optimization problems. Such a geometric representation provides a better understanding of the underlying behavior of this family of algorithms. It is demonstrated (Gibilisco and Isola, 2006) that many inequalities related to the uncertainty principle in quantum mechanics have a geometric interpretation in terms of quantum Fisher information. As an efficient way to study the theory of estimation in statistics, information geometry has been also introduced into the study of thermodynamics, neural nets, evolutionary processes, quantum physics, cosmology, cryptography, genetics, and finance.

Another approach to information geometry is suggested by Khrennikov (1999). Taking the ring Z_p of p-adic integers where p is a prime number (cf. (Van der Waerden, 1971)), he calls it or its Cartesian power Z_p^n, by the name *information space*, or *I-space*, and uses these spaces for a study of information processes. Information processes are realized by *information transformers*, or *I-transformers*. They have *inner states* that take values in an information space, and *internal clocks*, which also have inner states that take values in an information space. Inner states of internal clocks are called *information time*. Inner states are permanently changing. When inner states of an information transformer are vectors of p-adic numbers, their coordinates are called internal parameters of this transformer. Each information transformer is interacting with other information transformers and information fields and these interactions change internal parameters of interacting information transformers. A kind of Hamiltonian dynamics is developed on information spaces. Concepts of *information force*, *information velocity*, *information work*, and *information energy* are introduced and conservation laws for the total energy for I-transformers are obtained.

This approach stems from contemporary physics. It is well-known that results of any physical measurements are always expressed in terms of rational numbers since any measuring equipment works with finite precision, bounded, at least, by the quantum uncertainty principle. This brings many physicists to the conclusion that the field of real numbers is a theoretical abstraction valid only at large classical scales. Although the geometry of the Planck scales is unknown, there is a strong theoretical bias that the geometry of these scales should be non-Archimedian (cf., for example, (Brekke and Freund, 1993; Khrennikov, 1997)). The source of this bias is that finite non-Archimedian fields, i.e., p-adic fields, have a developed mathematical theory.

Khrennikov considers his model as an attempt to extend the p-adic model of physical reality by interpreting material objects as a particular class of transformers of information. At the same time, he applies his theory to social sciences, psychology, and cognitive sciences.

Another kind of geometry emerges in studies of information spaces related to representation of information by data and knowledge (Fabrikant, 2000; Fabrikant and Buttenfield, 1999; 2001). In this context, an *information space* is a type of information design in which representations of information objects are situated in a geometrical space. In such a space, location and direction have meaning, so that mapping and navigation become possible. Usually, elements of information spaces are data. For instance, Fabrikant (2000) writes, the use of spacial metaphor to depict complex database content has become popular within the information visualization community. The goal is to deal with rapidly growing volumes of data. Such spatialized database descriptions are called *information spaces*, or *information worlds*.

A more general approach treats any structured system of data as an information spaces. As Benyon and Höök (1977) make clear, the concept of an 'information space' is becoming increasingly important in our time of large computers, global and local networks, and exponentially growing storage of data. Information spaces are manifested in such areas as multiple interacting databases where data mining is a major issue, and in large hypermedia systems such as the Internet. It is possible to treat functioning of traditional information retrieval systems as navigation in information spaces.

Actually any representation space of an infological system (cf. Section 7.3.1) is an information space, geometry in this space is information geometry and any information space is a representation space for a properly chosen infological system.

People live, work and relax in information spaces. At one level of description all our multifarious interactions with the experienced world and other people are effected through the discovery, exchange, organization and manipulation of information. In computer systems, there is a range of information spaces, which have different characteristics. Hyperspace, typified by hypertext or hypermedia systems, provides a structure of an information space, in which the information nodes are linked and functions provided facilitating following these links to achieve intended goals. However, information spaces are not always related to computers and databases. They are central to the everyday experience of people in various areas. Finding a rout through an airport, a hotel or a city involves traveling in an information space. Paper documents represent another type of information spaces. Readers get quite different information from books, from newspapers and from magazines. Similarly they find diverse information in timetables, plans, vocabularies, guides, instruction manuals and maps.

There are three knowledge/data oriented approaches to building information spaces: *geographic*, *semantic*, and *cognitive* approaches. In the first approach, basic geographic concepts, such as identity, location, direction, magnitude and time, are used. Second approach is based on semantics constructs that preserve properties of mapped entities and represent functional relations between these entities in the information space called the Benedectine space (Benedict, 1991; Fabrikant, 2000). In the third, cognitive approach, the user-oriented view of information space design is adopted. Geographic spaces include affordances, experiential properties and socially constructed meanings. The cognitive approach focusses on how people explore available information, construct semantic relations, and are able to infer meaning encapsulated in spatialized views of knowledge discovery and data mining.

Chapter 8

Conclusion

Thus, we have seen that information is not merely an indispensable adjunct to personal, social and organizational functioning, a body of facts, data and knowledge applied to solutions of problems or to support actions. Rather it is a central and defining characteristic of all life forms, manifested in genetic transfer, in stimulus-response mechanisms, in the communication of signals and messages and, in the case of humans, in the intelligent acquisition of knowledge, understanding and achieving wisdom.

Different branches of the big tree called information theory, which grows in the forest of science, are described in this book. The trunk of this tree is the general theory of information. The roots of the tree are all works in information science that contributed to our knowledge of information, information processes and information systems. The branches of the tree are information theories created by numerous scientists. Some of these theories are new. Others appeared more than fifty years ago.

The general theory of information (GTI) is a new theory and when a new theory is created, an important question is what problems it solves.

The general theory of information solved the following problems

1. A unification mechanism, namely, an *infological system*, is constructed for achieving a common understanding of those multiple kinds and types of information that people use in society, observe in nature, process in technology, and study in various branches of science.
2. The main result of the general theory of information is the development of the unified definition information that encompasses

the enormous diversity of types, forms and classes of information. A new definition of information is based on the unification mechanism provided by the concept of an infological system. It is demonstrated that all considered and used by people kinds of information can be reduced to this concept. As the reader can see in Introduction, many researchers assumed and even tried to prove that the problem of information unification is unsolvable. However, such a unified concept of information exists and is based on ideas of many researchers who worked in this area. In addition, the new understanding of information does not exclude the everyday usage of the word *information*. It only connects this usage to a more exact meaning.

Three types of concepts of information are introduced, reflecting the main facets of the phenomenon:

- *information in a broad sense*, which includes different kinds of energy;
- *information as a relative concept*, which depends on the choice of an infological system;
- *information in a strict sense*, which acts on structures, such as knowledge, data, beliefs, and ideas.

Information in a broad sense is differentiated according to the global structure of the world expressed in the form of the Existential Triad. This gives us three classes:

- *information* in the Physical World is energy;
- *information* in the Mental World is mental energy;
- *information* in the Structural World is information *per se* or information in the strict sense, such as cognitive information.

3. A strict distinction is elaborated between information, which is an existing essence, and measures of information, which are functions that usually correspond numbers to portions of information. Measures of information are used to estimate information, to evaluate information sources, and to assess information carriers, representations and channels.

4. Different types of information, such as accessible information, available information, acceptable information, pseudoinformation, misinformation, and disinformation are separated and modeled.

5. New types of information, such as direct emotional/affective information and direct effective/regulative information, are discovered. Experimental data are provided to demonstrate their existence and distinction from such already known types of information as cognitive emotional information and cognitive effective information. Relations between information, emotions, motivations, intentions, and will are established.

6. A strict distinction is elaborated between information, data and knowledge, making it possible to explicate relations between these concepts. Confusion related to these concepts is so big that some researchers even suggested excluding the term information from the scientific usage (cf., for example, (Furner, 2004)). The general theory of information explains and eliminates many of existing misconceptions, contradictions and distortions in such fields as knowledge management, data mining and information studies.

7. A unification of existing information theories and directions is achieved in the context of the general theory of information.

8. A mathematical model of information is developed and studied. It explains, in particular, why the most adequate models in microphysics (e.g., in quantum mechanics, quantum electrodynamics, and quantum field theory) are based on the concept of operator.

9. The general theory of information explains in what sense and how computers process information and knowledge and not only data. Like people, computers are power plants for information as a structural energy. In the same way as conventional power plants change energy forms, computers and people often change symbolic information forms (information carriers) to a more efficient representation, for example, they perform data mining and derive knowledge from data.

10. Explanation of the essence of information and its functioning is given, as well as answers to the following questions:
    - how information is related to knowledge and data;
    - how information is modeled by mathematical structures;
    - how these models are used to better understand computers and Internet, cognition and education, communication and computation.

We have seen that the general theory of information encompasses such kinds of information as syntactic information, semantic information, and pragmatic information, allowing one to extend results of those theories that study these kinds of information. The general theory of information unifies these theories in the general theory of cognitive information (GTCI). Besides, all semantic information theories based on individual knowledge system transformations, such as theories of MacKay (1956, 1961, 1969), Shreider (1963; 1965; 1967), Brookes (1980), Mizzaro (1996; 1998; 2001) and Gackowski (2004) (cf. Section 4.3) study information in a local setting, i.e., as state transformations in a form of local information operators acting on knowledge systems, such as cognitive agents, databases and knowledge bases. In addition, the general theory of information extends the local setting of these theories to the global environment, studying global information operators. They generate connected systems of knowledge system transformations, unifying pointwise and local knowledge system transformations.

It is also demonstrated how the general theory of information explains relations between meaning and information, making possible to comprehend such phenomena as genuine information, false information, disinformation, and pseudoinformation. Types of information that are separated help us to understand and explain how information functions in society, as well as what are properties of information that is produced, extracted, transmitted, stored and processed by people and machines.

In addition, the general theory of information solves the controversy whether information exists only in society or there is information in nature and in technological artifacts, such as buildings, towers, pyramids, bridges, machines and mechanisms. In reality, according to ontological principles the answer to this question depends entirely on our choice of the class of infological systems and carriers of information. If we consider only infological systems such as the human brain, then we come to the situation where information is only what people accept on their mental level. At the same time, if we assume that possible carriers of information and infological systems are results of human activity (such as different books, documents, and databases) and existence (such as intelligence and memory of society), then it is possible to consider only information in society. However, if there are no restrictions on infological systems, we come to the conclusion that information exists everywhere.

When Capuro and Hjorland (2003) write, "anything can be information", or when Buckland (1991) underlines, "… we are unable to say confidently of anything that it could not be information", it sounds confusing. For instance, we can ask whether a chair or stone is information or if it can only carry information. In contrast to this, when the general theory of information asserts that anything can contain information, it provides exact meaning and mathematical models for this assertion.

It is necessary to remark that people in general and contemporary researchers, in particular, consider only cognitive information. Extending horizons of our knowledge, the general theory of information made possible the discovery of other types of information: direct emotional/affective information and direct effective/regulative information.

At the same time, the general theory of information does not completely eliminate common understanding of the word information. This theory allows one to preserve common usage in a modified and refined form. For instance, when people say and write that *information is knowledge of a specific event or situation* (The American Heritage Dictionary, 1996), the general theory of information suggests that it is more adequate to say and write that *information gives knowledge of a specific event or situation*. When people say and write that *information is a collection of facts or data* (The American Heritage Dictionary, 1996), the general theory of information suggests that it is more adequate to say and write that *a collection of facts or data contains information*.

Results obtained in the general theory of information and presented in this book open new directions for the future research in information studies:

1. It would be useful to build specific mathematical models for direct emotional/affective information and direct effective/regulative information.
2. Theory of Mizzaro spaces initiated by Mizzaro (1996; 1998; 2001) and further developed in Section 4.3 of this book opens new

perspectives for knowledge management. Thus, it is important to expand this theory and apply it to problems of knowledge management.

3. In Section 7.3.1, various operations with information operators are introduced and studied. It would be interesting to consider other operations with information operators, such as information fusion and information integration used in distributed information retrieval systems and databases. Note that information integration is a more general operation than information fusion.

4. In Section 7.3.1, a functional mathematical model for the general theory of information is developed. One more interesting problem for future research is building a categorical model for the general theory of information.

5. Bar-Hillel and Carnap (1952; 1958), Hintikka (1970; 1971; 1973; 1973a) and their followers developed semantic information theory based on the possible world representation using systems of propositions (cf. Section 4.2). However, there are other representations of possible worlds. For instance, Cohen and Levesque (1990) treat a possible world as a time-line representing a sequence of events, temporally extended into the past and the future. Rao and Georgeff (1991) consider a possible world as a time tree, which includes optional courses of events that an agent may choose in a particular world. This gives dynamic representations of possible worlds. Thus, it would be interesting to develop semantic information theory based on dynamic representations of possible worlds.

6. It might be attractive to extend semantic information theories based on logical calculi to similar theories based on logical varieties (Burgin, 1991b; 1997f; 2004c) and to apply this extension to problems of information integration and evaluation.

7. In pragmatic information theories, information quality and information value are studied only for cognitive information. It would be beneficial to apply these estimates to direct emotional/affective information and direct effective/regulative information.

8. One more problem for future research is connected to the theory of Mizzaro spaces (cf. Section 4.3). Cognitive information changes not only knowledge but also beliefs, ideals and other elements of a

cognitive infological system described in Section 2.2. Thus, it would be natural to include all these elements into Mizzaro spaces and develop a corresponding theory in a broader context. Moreover, it is possible to consider transformations of the individual's *Image* or *world-view*, the concept of which was introduced by Boulding (1956), instead of knowledge state transformations.

9. A synthesis of operator information theory presented in Sections 7.2 and 7.3.1 with information algebra theory presented in Section 7.3.2 looks very promising from the point of view of theoretical modeling of information systems and processes.

10. It might be interesting to apply mathematical tools developed in Section 7.3.1 to information geometry, considering information manifolds (cf. Section 7.3.3) and other information spaces as representation spaces of infological systems. For instance, a database is an infological system and all data in it form an information space. There are different kinds of estimates determined for these data. One of them is truthfulness considered in Section 2.4. Another estimate is a probability distribution considered in Section 3.2. One more estimate is truthfulness considered in Section 2.4. Data quality (cf. Section 6.2) provides a quantity of other data characteristics. Information entropy considered in Section 3.2 is also an important characteristic of information and data, reflecting their uncertainty. All these characteristics and estimates form a multidimensional functional space.

    Information operators studied in Section 7.3.1 act on this space if it is the state space of the chosen infological system. Thus, it would be useful for practice and theory of databases to study such database operators. Note that conventional operations with data in databases, e.g., projection, selection, intersection, set difference and join in relational databases (cf., for example, (Elmasri and Navathe, 2000)), are projections of special types of database operators.

11. The next step in the development of the mathematical stratum of the general theory of information is to study generalized information operators.

12. One more interesting direction is to study algebras of information operators from the general theory of information with operations used in databases and knowledge bases.

13. We see that there is a diversity of information quality characteristics and measures. In Chapter 6, three basic types of such characteristics and measures: (1) information representation characteristics and measures; (2) information source characteristics and measures; and (3) information channel characteristics and measures.
   Thus, it is important to form a unified system of information quality characteristics and measures and study their properties in the context of the general theory of information.

14. Computers have become the main technical devices for information processing. Thus, it is important to formalize and study operator spaces that represent operations with information performed by computers.

This problem is related to the following one.

15. Build an algorithmic theory of information operators in the general theory of information, i.e., a theory where information operators are computed by algorithms.

16. Cockshott and Michaelson (2005) suggested a computational approach to the value of information based on the computational complexity of extraction or production of a portion of information. This brings us to the natural problem of measuring information value using dual measures to time complexity, such as Kolmogorov complexity or information size, in appropriate classes of algorithms.

This problem is related to the following one.

17. Find relations between information quality characteristics, information quality measures, and complexity of algorithms.

18. In Section 4.3, the theory of Mizzaro spaces, which is a knowledge base oriented semantic theory of cognitive information, is represented and studied in a general form. At the same time, it is interesting to have specific versions of this theory where knowledge has some standard form. This brings us to the following problem.
   Develop a theory of logical Mizzaro spaces, i.e., spaces in which elements of knowledge are propositions and/or predicates.

This problem is related to the following one.

19. While propositions and predicates give descriptional knowledge, another basic type of knowledge is procedural knowledge (Burgin and Kuznetsov, 1991; 1992; 1993; 1994). Thus, it is important to study information that gives and transforms this type of knowledge. It gives us to the following problem.

    Develop a theory of algorithmic Mizzaro spaces, i.e., spaces in which elements of knowledge are algorithms, programs, scenarios, and/or procedures.

This problem is related to the following one where it is assumed that knowledge transformations are constructive.

20. Develop a theory of algorithmically powered Mizzaro spaces, i.e., spaces in which knowledge is obtained and transformed by algorithms.

21. Biological and, in particular, genetic information is of the first importance to people. The general theory of information gives efficient means for studying such information. Thus, we have an important problem.

    Specify the operator approach of the general theory of information for biological information in general and genetic information, in particular.

22. Build a model of mind as an infological system.

23. In the context of the operator approach of the general theory of information, find conditions for accessibility, availability, acceptability and validity of other properties of information.

24. Build a theory of information flow operators introduced in Section 7.3.1.

25. In the book, operator representation is developed and studied for information portions. At the same time, another important type of information quantization is a stream of information, which reflects information expansion in time. There are different ways to represent information streams in the operator model of the general theory of information.

    One way is to correspond a discrete (or may be, a continuous) dynamical operator system to each information stream.

Another way is to take an infological representation space L with time as one of its coordinates, e.g., space-time manifold in physics, and represent information streams as operators in L.

Thus, we have the following problem.

Develop a theory of information streams in the context of the general theory of information.

This problem is connected to the following one.

26. Study dynamical systems in operator spaces of the general theory of information considered in Section 7.3.1.

27. Find differences between intelligent information processing and mechanical information processing.

28. As we have seen in Chapter 2, intelligence is related to information. Thus, we have the following problem.

    Build a theoretical (mathematical) model of intelligence in the context of the general theory of information.

29. As we have seen in Chapter 2, emotions and other affective states are related to information. Thus, we have the following problem.

    Build a theoretical (mathematical) model of emotions in the context of the general theory of information.

30. As we have seen in Chapter 2, instinct is related to information. Thus, we have the following problem.

    Build a theoretical (mathematical) model of instinct in the context of the general theory of information.

31. As we have seen in Chapter 2, will of a person is related to information. Thus, we have the following problem.

    Build a theoretical (mathematical) model of will in the context of the general theory of information.

32. Develop logic of information based on the operator approach of the general theory of information.

33. Study topological properties of operator spaces in the general theory of information.

Appendix

Mathematical Foundations of Information Theory

What memory has in common with art
is the knack for selection, the taste for detail ...
Joseph Brodsky (1940–1996)

Some mathematical concepts, in spite of being basic and extensively used, have different interpretation in different books. In a similar way, different authors use dissimilar notation for the same things, as well as the same notation for distinct things. For this reason and to make our exposition self-consistent, we give here definitions and denotation for basic mathematical concepts used in this book.

Appendix A. Set Theoretical Foundations

Appendix B. Elements of the Theory of Algorithms

Appendix C. Elements of Logic

Appendix D. Elements of Algebra and Category Theory

Appendix E. Elements of Probability Theory

Appendix F. Numbers and Numerical Functions

Appendix G. Topological, Metric and Normed Spaces

561

Appendix A

Set Theoretical Foundations

Mathematics is the art of giving the same name to different things.

Henri Poincare (1854–1912)

$\emptyset$ is the *empty set*.

If X is a set, then $r \in X$ means that r belongs to X or r is a member of X.

If X and Y are sets, then $Y \subseteq X$ means that Y is a *subset* of X, i.e., Y is a set such that all elements of Y belong to X.

The *union* $Y \cup X$ of two sets Y and X is the set that consists of all elements from Y and from X.

The *intersection* $Y \cap X$ of two sets Y and X is the set that consists of all elements that belong both to Y and to X.

The *union* $\bigcup_{i \in I} X_i$ of sets X_i is the set that consists of all elements from all sets X_i, $i \in I$.

The *intersection* $\bigcap_{i \in I} X_i$ of sets X_i is the set that consists of all elements that belong to each set X_i, $i \in I$.

The *difference* $Y \setminus X$ of two sets Y and X is the set that consists of all elements that belong to Y but does not belong to X.

If X is a set, then 2^X is the *power set* of X, which consists of all subsets of X. The *power set* of X is also denoted by $\mathbf{P}(X)$.

If X and Y are sets, then $X \times Y = \{(x, y); x \in X, y \in Y\}$ is the direct or Cartesian product of X and Y, in other words, $X \times Y$ is the set of all pairs (x, y), in which x belongs to X and y belongs to Y.

Y^X is the set of all mappings from X into Y.

$$X^n = \underbrace{X \times X \times \ldots X \times X}_{n}.$$

Elements of the set X^n have the form $(x_1, x_2, \ldots, x_n)$ with all $x_i \in X$ and are called *n*-tuples, or simply, tuples.

A fundamental structure of mathematics is *function*. However, functions are special kinds of binary relations between two sets.

A *binary relation* T between sets X and Y is a subset of the direct product $X \times Y$. The set X is called the *domain* of T ($X = \text{Dom}(T)$) and Y is called the *codomain* of T ($Y = \text{CD}(T)$). The *range* of the relation T is $\text{Rg}(T) = \{y; \exists\, x \in X\, ((x, y) \in T)\}$. The *domain of definition* of the relation T is $\text{DDom}(T) = \{x; \exists\, y \in Y\, ((x, y) \in T)\}$. If $(x, y) \in T$, then one says that the elements x and y are in relation T, and one also writes $T(x, y)$.

Binary relations are also called multivalued functions (mappings or maps).

A family J of subsets of X, is called a *filter* if it satisfies the following conditions:

1. If $P \in J$ and $P \subseteq Q$, then $Q \in J$.
2. If $P, Q \in J$, then $P \cap Q \in J$.

It is possible to read more on set theory, for example, in (Bourbaki, 1960; Kuratowski and Mostowski, 1967).

A *preorder* (also called *quasiorder*) on a set X is a binary relation Q on X that satisfies the following axioms:

1. Q is reflexive, i.e. xQx for all x from X.
2. Q is transitive, i.e., xQy and yQz imply xQz for all $x, y, z \in X$.

A *partial order* is a preorder that satisfies the following additional axiom:

3. Q is antisymmetric, i.e., xQy and yQx imply $x = y$ for all $x, y \in X$.

A *strict partial order* is a preorder that is not reflexive, is transitive and satisfies the following additional axiom:

4. Q is asymmetric, i.e., only one relation xQy or yQx is true for all $x, y \in X$.

An *equivalence* on a set X is a binary relation Q on X that is reflexive, transitive and satisfies the following additional axiom:

5. Q is symmetric, i.e., xQy implies yQx for all x and y from X.

A *function* (also called a *mapping* or *map* or *total function* or *total mapping*) f from X to Y is a binary relation between sets X and Y in which there are no elements from X which are corresponded to more than one element from Y and to any element from X, some element from Y is corresponded. Often total functions are also called everywhere defined functions. Traditionally, the element $f(a)$ is called the image of the element a and denotes the value of f on the element a from X. At the same time, the function f is also denoted by $f: X \to Y$ or by $f(x)$. In the latter formula, x is a variable and not a concrete element from X.

A *partial function* (or *partial mapping*) f from X to Y is a binary relation between sets X and Y in which there are no elements from X which are corresponded to more than one element from Y. Thus, any function is also a partial function. Sometimes, when the domain of a partial function is not specified, we call it simply a function because any partial function is a total function on its domain.

A *multivalued function* (or *mapping*) f from X to Y is any binary relation between sets X and Y.

$f(x) \equiv a$ means that the function $f(x)$ is equal to a at all points where $f(x)$ is defined.

Two important concepts of mathematics are the domain and range of a function. However, there is some ambiguity for the first of them. Namely, there are two distinct meanings in current mathematical usage for this concept. In the majority of mathematical areas, including the calculus and analysis, the term "domain of f" is used for the set of all values x such that $f(x)$ is defined. However, some mathematicians (in particular, category theorists), consider the domain of a function $f: X \to Y$ to be X, irrespective of whether $f(x)$ is defined for all x in X. To eliminate this ambiguity, we suggest the following terminology consistent with the current practice in mathematics.

If f is a function from X into Y, then the set X is called the *domain* of f (it is denoted by Dom f) and Y is called the *codomain* of T (it is denoted by Codom f). The *range* Rg f of the function f is the set of all elements from Y assigned by f to, at least, one element from X, or formally, Rg $f = \{ y; \exists x \in X (f(x) = y) \}$. The *domain of definition* DDom f of the function f is the set of all elements from X that related by f to, at least, one element from Y is or formally, DDom $f =$

$\{x; \exists\, y \in Y\, (f(x) = y)\}$. Thus, for a partial function $f(x)$, its domain of definition DDom f is the set of all elements for which $f(x)$ is defined.

Taking two mappings (functions) f: $f: X \to Y$ and $g: Y \to Z$, it is possible to build a new mapping (function) $gf: X \to Z$ that is called *composition* or *superposition* of mappings (functions) f and g and defined by the rule $gf(x) = g(f(x))$ for all x from X.

For any set S, $\chi_S(x)$ is its *characteristic function*, also called *set indicator function*, if $\chi_S(x)$ is equal to 1 when $x \in S$ and is equal to 0 when $x \notin S$, and $C_S(x)$ is its partial characteristic function if $C_S(x)$ is equal to 1 when $x \in S$ and is undefined when $x \notin S$.

If $f: X \to Y$ is a function and $Z \subseteq X$, then the restriction $f|_Z$ of f on Z is the function defined only for elements from Z and $f|_Z(z) = f(z)$ for each element z from Z.

If U is a correspondence of a set X to a set Y (a binary relation between X and Y), i.e. $U \subseteq X \times Y$, then $U(x) = \{y \in Y; (x, y) \in U\}$ and $U^{-1}(y) = \{x \in X; (x, y) \in U\}$.

An *n*-ary relation R in a set X is a subset of the nth power of X, i.e., $R \subseteq X^n$. If $(a_1, a_2, \ldots, a_n) \in R$, then one says that the elements $a_1, a_2, \ldots, a_n$ from X are in relation R.

Let X be a set. An *integral operation* W on the set X is a mapping that given a subset of X, corresponds to it an element from X, and for any $x \in X$, $W(\{x\}) = x$.

Examples of integral operation are: sums, products, taking minimum, taking maximum, taking infimum, taking supremum, integration, taking the first element from a given subset, taking the sum of the first and second elements from a given subset, and so on.

Examples of finite integral operations defined for numbers are: sums, products, taking minimum, taking maximum, taking average, weighted average, taking the first element from a given subset, and so on.

As a rule, integral operations are partial, that is, they assign values, e.g., numbers, only to some subsets of X.

Proposition A.1. Any binary operation in X generates a finite ordinal integral operation on X.

It is possible to read more about integral operations and their applications in (Burgin and Karasik, 1976; Burgin 2004d).

Set theory is correctly considered the base of the major part of contemporary mathematics. For a long time, the word *set* was used in mathematics as an informal notion. Only at the end of the 19th century and at the beginning of the 20th century this notion was formalized in the process of the set theory development. In turn, set theory provided rigorous foundations for the whole mathematics. However, applications of mathematical structures to real life phenomena demonstrated limitations of sets. As a result various generalizations of sets have been suggested. The most popular of these generalizations are fuzzy sets and multisets.

A *multiset* is similar to a set, but can contain indiscernible elements or different copies of the same elements. It is possible to read more about multisets in (Aigner, 1979; Knuth, 1997; 1998).

A *fuzzy set A* in a set U is the triad $(U, \mu_A, [0,1])$, where $[0,1]$ is an interval of real numbers, $\mu_A: U \to [0,1]$ is a membership function of A, and $\mu_A(x)$ is the degree of membership in A of $x \in U$. It is possible to read more about fuzzy sets, for example, in (Klir and Folger, 1988; Zimmermann, 2001).

Named sets as the most encompassing and fundamental mathematical construction encompass all generalizations of ordinary sets and provide unified foundations for the whole mathematics (Burgin, 2004c).

A *named set* (also called a *fundamental triad*) has the following graphic representation (Burgin, 1990; 1991; 1997):

$$\textbf{connection}$$
$$\textbf{Entity 1} \xrightarrow{\hspace{3cm}} \textbf{Entity 2} \qquad (A.1)$$

or

$$\textbf{correspondence}$$
$$\textbf{Essence 1} \xrightarrow{\hspace{3cm}} \textbf{Essence 2} \qquad (A.2)$$

In the fundamental triad (1) or (2), Entity 1 (Essence 1) is called the *support*, the Entity 2 (Essence 2) is called the *reflector* (also called the set or component of names) and the connection (correspondence) between Entity 1 (Essence 1) and Entity 2 (Essence 2) is called the *reflection* (also called the *naming correspondence*) of the fundamental triad (1) (respectively, (2).

In the symbolic form, a *named set* (*fundamental triad*) **X** is a triad (X, f, I) where X is the *support* of **X** and is denoted by S(**X**), I is the *component of names* (also called *set of names* or *reflector*) of **X** and is denoted by N(**X**), and f is the *naming correspondence* (also called *reflection*) of the named set **X** and is denoted by n(**X**). The most popular type of named sets is a named set **X** = (X, f, I) in which X and I are sets and f consists of connections between their elements. When these connections are set theoretical, i.e., each connection is represented by a pair (x, a) where x is an element from X and a is its name from I, we have a *set theoretical named set*, which is binary relation. Even before the concept of a fundamental triad was introduced, Bourbaki in their fundamental monograph (1960) had also represented binary relations in a form of a triad (named set).

Using the term *triad*, it is necessary to distinguish it from the notion of a *triplet*. A triad is a system that consists of three parts (elements or components), while a triplet is any three objects. Thus, any triad is a triplet, but not any triplet is a triad. In a triad, there are ties and/or relations between all three parts (objects from the triad), while for a triplet, this is not necessary.

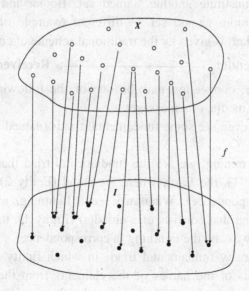

Figure A1. A set theoretical named set **X** = (X, f, I)

There are many named sets that are not set theoretical. For instance, an algorithmic named set $\mathbf{A} = (X, A, Y)$ consists of an algorithm A, the set X of inputs and the set Y of outputs. Let us take as X and Y the set of all words in some alphabet Q and all Turing machines that work with words in the alphabet Q as algorithms. Then theory of algorithms tells us that there are much more algorithmic named sets than relations (set theoretical named sets) because several different algorithms (e.g., Turing machines) can define the same function or relation. Thus, algorithmic named sets are different from set theoretical named sets.

Mereological named sets are essentially different from set theoretical named sets (Leśniewski, 1916; 1992; Leonard and Goodman, 1940; Burgin, 2004c). Categorical named sets (fundamental triads) also are different from set theoretical named sets. For instance, an arrow in a category is a fundamental triad but does not include sets as components (Herrlich and Strecker, 1973). Named sets with physical components, such as (a woman and her name), (a paper and its title), (a book and its title) and many others, are far from being set theoretical.

People meet fundamental triads (named sets) constantly in their everyday life. People and their names constitute a named set. Cars and their owners constitute another named set. Books and their authors constitute one more named set. A different example of a named set (fundamental triad) is given by the traditional scheme of communication:

Sender ⎯⎯⎯⎯⎯⎯⎯⎯⎯⟶ Receiver

In this case, connection may be one of the following: a channel, communication media, or a message.

People can even see some fundamental triads (named sets). Here are some examples.

When it is raining, we see the fundamental triad that consists of a cloud(s) (Entity 1), the Earth where we stand (Entity 2) and flows of water (a correspondence). When we see a lightning, we see another fundamental triad that consists of a cloud(s) (Entity 1), the Earth where we stand (Entity 2) and the lightning (a correspondence).

There are many fundamental triads in which Entity 1 is some set, Entity 2 consists of the names of the elements from the Entity 1 and elements are connected with their names by the naming relation. This explains the name "named set" that was for an essential time applied to this structure (Burgin, 1990; 1991). A standard model of a named set is a set of people who constitute the carrier, their names that form the set of names, and the naming relation consists of the correspondences between people and their names.

Many mathematical systems are particular cases of named sets or fundamental triads. The most important of such systems are fuzzy sets (Zadeh, 1965; 1973; Zimmermann, 2001), multisets (Hickman, 1980; Blizard, 1991; Knuth, 1997), graphs and hypergraphs (Berge, 1973), topological and fiber bundles (Husemoller, 1994; Herrlich and Strecker, 1973). Moreover, any ordinary set is, as a matter of fact, some named set, and namely, a singlenamed set, i.e., such a named set in which all elements have the same name (Burgin, 2004c). It is interesting that utilization of singlenamed sets instead of ordinary sets allows one to solve some problems in science, e.g., Gregg's paradox in biology (Ruse, 1973; Burgin, 1983).

It is possible to find many named sets in physics. For instance, according to particle physics, any particle has a corresponding antiparticle, e.g., electron corresponds to positron, while proton corresponds to antiproton. Thus, we have a named set with particles as its support and antiparticles as its set of names. A particle and its antiparticle have identical mass and spin, but have opposite value for all other non-zero quantum number labels. These labels are electric charge, color charge, flavor, electron number, muon number, tau number, and barion number. Particles and their quantum number labels form another named set with particles as its support and quantum number labels as its set of names.

When we study information and information processes, fundamental triads become extremely important. Each direction in information theory has fundamental concepts that are and models of which are either fundamental triads or systems built of fundamental triads. Indeed, the relation between information and the receiver/recipient introduced in the Ontological Principle O1 is a fundamental triad. The Ontological Principles O4 and O4a introduce the interaction and second communication triads (Chapter 2). The first communication triad is basic in statistical information theory (Chapter 3). A special case of named sets are Chu spaces used in information studies, introduced by Barr (1979) and theoretically developed by Chu.

A *Chu space C* over a set B is a triad (A, r, X) with $r: A \times X \to B$. The set A is called the *carrier* and X the *cocarrier* of the Chu space C. Matrices are often used to represent Chu spaces. Note that in the form (A, r, X), Chu space is a triad but not a fundamental triad. At the same time, in the more complete form $(A \times X, r, B)$, Chu space is a particular case of fundamental triads.

One more example of named sets is given by operands from the multidimensional structured model of computer systems and computations (Burgin and Karasik, 1975; 1976; 1976a; Burgin, 1980; 1982a; 1984).

An operand Q is a triad (Z, r, C) where $Z = Z_1 \times Z_2 \times ... \times Z_n$, each Z_i is a subset of the set Z of all integer numbers, $r: Z \times Z \to C$ and C is a set. The set Z is called the *support* of the operand Q.

Taking an operand with $n = 2$ and arbitrary sets Z_1 and Z_2, we get the concept of a valued relation.

An operator A in the multidimensional structured model of computer systems and computations is a mapping

$$A: Q^l \to Q^h$$

where Q^k is the set of all k-tuples $(Q_1, Q_2, ... Q_k)$ and Q_i all are operands $(i = 1, 2, 3, ..., k; k \in \{l, h\})$.

Many constructions, such as valued relations, fuzzy relations (Salii, 1965), classifications in the sense of (Bairwise and Seligman, 1997), and Chu spaces, are particular cases of such operands.

Named sets are explicitly used in many areas: in models of computers and computation (Burgin and Karasik, 1975; 1976), artificial intelligence (Burgin and Gladun, 1989; Burgin and Gorsky, 1991; Burgin and Kuznetsov, 1992), mathematical linguistics (Burgin and Burgina, 1982; 1982a), software engineering (Browne, et al, 1995) and Internet technology (Terry, 1986; Sollins, 2003; Balakrishnan, et al, 2004; Cunnigham, 2004). Set-theoretical named sets are very popular in databases and knowledge engineering due to the fact that both binary relations and hierarchical structures are specific kinds of named sets. Even images of data and knowledge structures and operations in such structures are presented as named sets. For instance, Figures 3.3, 3.4, 4.1, and 4.2 from (Martin, 1977), Figures 3.3.1, 3.3.3, and 3.3.4 from (Tsichritzis and Lochovsky, 1982), and Figures 3.9–3.12 and 6.1–6.6 from (Elmasri and Navathe, 2000) are similar to Figure A1 given above because they give graphical images of named sets. The reason for such a popularity of named sets in these areas is the fact that it is possible to represent any data structure by named sets and their chains (Burgin, 1997; Burgin and Zellweger, 2005; Burgin, 2008).

There are different formal mathematical definitions of named sets/fundamental triads: in categories (Burgin, 1990), in set theory (Burgin, 2004c), and by axioms (Burgin, 2004b). Axiomatic representation of named sets shows that named set theory, as a formalized mathematical theory, is independent from set theory and category theory. When category theory is build independently from set theory, then categorical representations of named sets are also independent from set theory. It is also necessary to emphasize that physical fundamental triads (named sets), i.e., fundamental triads that are structures of physical objects, are independent from set theory.

It is possible to read more about named sets in (Burgin, 1990; 1991a; 1992b; 1997; 2004c).

Appendix B

Elements of the Theory of Algorithms

> *The gem cannot be polished without friction,*
> *nor man perfected without trials.*
> Chinese proverb

The theory of algorithms is the most abstract part of the theory of algorithms, computation and automata. Abstract algorithms and automata work with symbolic data, usually, in the form of words.

An alphabet is a set of symbols, e.g., $A = \{1, 0\}$ or $B = \{a, b, c\}$.

A string is a sequence of alphabet symbols, e.g., 101000.

A word is a string that belongs to some language.

A formal language is a set of words in a fixed alphabet.

The most general structure in the theory of abstract automata is *state transition machine*.

A *state transition machine* (STM), also called a *state transition system*, **A** consists of three structures and can be represented as **A** = (L, S, δ):

- The *linguistic structure* L = (A_I, Q, A_O) where A_I is a alphabet of *input symbols*, Q is a *set of states*, and A_O is a set of *output symbols* of the STM **A**;
- The *state structure* S = (Q, q_0, F) where q_0 is an element from Q called the *initial* or *start state* and F is a subset of Q called the set of *final* states of the STM **A**;
- The *action structure* δ traditionally called the *transition function*, or more exactly, *transition relation* of the STM **A**; the transition relation δ determines how input and the current state determine the next state and output, i.e.,

$$\delta: A_I \times Q \rightarrow Q \times A_O$$

δ is, in general, a multivalued function. It can be represented by two relations (functions):

The state transition relation (function):

$$\delta_{tr}: A_I \times Q \rightarrow Q$$

and the output relation (function):

$$\delta_{out}: A_I \times Q \rightarrow A_O$$

Examples of state transition machines are finite automata, pushdown automata, Turing machines, inductive Turing machines, limit Turing machines, Petri nets, neural networks, and cellular automata (cf. (Burgin, 2005)).

The structure of an inductive Turing machine, as an abstract automaton, consists of three components called *hardware*, *software*, and *infware*. Infware is a description and specification of information that is processed by an inductive Turing machine. Computer infware consists of data processed by the computer. Inductive Turing machines are abstract automata working with the same symbolic information in the form of words as conventional Turing machines. Consequently, formal languages with which inductive Turing machines works constitute their infware.

Computer hardware consists of all devices (the processor, system of memory, display, keyboard, etc.) that constitute the computer. In a similar way, an inductive Turing machine M has three abstract devices: a *control device A*, which is a finite automaton and controls performance of M; a *processor* or *operating device H*, which corresponds to one or several *heads* of a conventional Turing machine; and the *memory E*, which corresponds to the *tape* or tapes of a conventional Turing machine. The memory E of the simplest inductive Turing machine consists of three linear tapes, and the operating device consists of three heads, each of which is the same as the head of a Turing machine and works with the corresponding tape.

The *control device A* is a finite automaton that regulates: the state of the whole machine M, the processing of information by H, and the storage of information in the memory E.

The *memory E* is divided into different but, as a rule, uniform cells. It is structured by a system of relations that organize memory as a, well-structured system and provide connections or ties between cells. In particular, *input* registers, the *working* memory, and *output* registers of M are separated. Connections between cells form an additional structure K of E. Each cell can contain a symbol from an alphabet of the languages of the machine M or it can be empty.

In a general case, cells may be of different types. Different types of cells may be used for storing different kinds of data. For example, binary cells, which have type B, store bits of information represented by symbols 1 and 0. Byte cells (type BT) store information represented by strings of eight binary digits. Symbol cells (type SB) store symbols of the alphabet(s) of the machine M. Cells in conventional Turing machines have SB type. Natural number cells, which have type NN, are used in random access machines (Aho, et al, 1976). Cells in the memory of quantum computers (type QB) store q-bits or quantum bits (Deutsch, 1985). Cells of the tape(s) of real-number Turing machines (Burgin, 2005), have type RN and store real numbers. When different kinds of devices are combined into one, this new device has several types of memory cells. In addition, different types of cells facilitate modeling the brain neuron structure by inductive Turing machines.

It is possible to realize an arbitrary structured memory of an inductive Turing machine M, using only one linear one-sided tape L. To do this, the cells of L are enumerated in the natural order from the first one to infinity. Then L is decomposed into three parts according to the input and output registers and the working memory of M. After this, nonlinear connections between cells are installed. When an inductive Turing machine with this memory works, the head/processor is not moving only to the right or to the left cell from a given cell, but uses the installed nonlinear connections.

Such realization of the structured memory allows us to consider an inductive Turing machine with a structured memory as an inductive Turing machine with conventional tapes in which additional connections are established. This approach has many advantages. One of them is that inductive Turing machines with a structured memory can be treated as multitape automata that have additional structure on their tapes. Then it is conceivable to study different ways to construct this structure. In addition, this representation of memory allows us to consider any configuration in the structured memory E as a word written on this unstructured tape.

If we look at other devices of the inductive Turing machine M, we can see that the processor H performs information processing in M. However, in comparison to computers, this operational device performs very simple operations. When H consists of one unit, it can change a symbol in the cell that is observed by H, and go from this cell to another using a connection from K. This is exactly what the head of a Turing machine does.

It is possible that the processor H consists of several processing units similar to heads of a multihead Turing machine. This allows one to model in a natural way various real and abstract computing systems by inductive Turing machines. Examples of such systems are: multiprocessor computers; Turing machines with several tapes; networks, grids and clusters of computers; cellular automata; neural networks; and systolic arrays.

We know that programs constitute computer software and tell the system what to do (and what not to do). The *software* R of the inductive Turing machine M is also a program in the form of simple rules:

$$q_h a_i \rightarrow a_j q_k \qquad \text{(B.1)}$$

$$q_h a_i \rightarrow c q_k \qquad \text{(B.2)}$$

$$q_h a_i \rightarrow a_j q_k c \qquad \text{(B.3)}$$

Here q_h and q_k are states of A, a_i and a_j are symbols of the alphabet of M, and c is a type of connection in the memory E.

Each rule directs one step of computation of the inductive Turing machine M. The rule (B.1) means that if the state of the control device A of M is q_h and the processor H observes in the cell the symbol a_i, then the state of A becomes q_k and the processor H writes the symbol a_j in the cell where it is situated. The rule (B.2) means that the processor H then moves to the next cell by a connection of the type c. The rule (B.3) is a combination of rules (B.1) and (B.2).

Like Turing machines, inductive Turing machines can be deterministic and nondeterministic. For a *deterministic* inductive Turing machine, there is at most one connection of any type from any cell. In a *nondeterministic* inductive Turing machine, several connections of the same type may go from some cells, connecting them with (different) other cells. If there is no connection of the prescribed by an instruction type that goes from the cell that is observed by H, then H stays in the same cell. There may be connections of a cell with itself. Then H also stays in the same cell. It is possible that H observes an empty cell. To represent this situation, we use the symbol ε. Thus, it is possible that some elements a_i and/or a_j in the rules from R are equal to ε in the rules of all types. Such rules describe situations when H observes an empty cell and/or when H simply erases the symbol from some cell, writing nothing in it.

The rules of the type (B.3) allow an inductive Turing machine to rewrite a symbol in a cell and to make a move in one step. Other rules (B.1) and (B.2) separate these operations. Rules of the inductive Turing machine M define the transition function of M and describe changes of A, H, and E. Consequently, these rules also determine the transition functions of A, H, and E.

A general step of the machine M has the following form. At the beginning of any step, the processor H observes some cell with a symbol a_i (for an empty cell the symbol is Λ) and the control device A is in some state q_h.

Then the control device A (and/or the processor H) chooses from the system R of rules a rule r with the left part equal to $q_h a_i$ and performs the operation prescribed by this rule. If there is no rule in R with such a left part, the machine M stops functioning. If there are several rules with the same left part, M works as a nondeterministic Turing machine, performing all possible operations. When A comes to one of the final states from F, the machine M also stops functioning. In all other cases, it continues operation without stopping.

For an abstract automaton, as well as for a computer, three things are important: how it receives data, process data and obtains its results. In contrast to Turing machines, inductive Turing machines obtain results even in the case when their operation is not terminated. This results in essential increase of performance abilities of systems of algorithms.

The computational result of the inductive Turing machine M is the word that is written in the output register of M: when M halts while its control device A is in some final state from F, or when M never stops but at some step of computation the content of the output register becomes fixed and does not change although the machine M continues to function. In all other cases, M gives no result.

The memory E is called *recursive* if all relations that define its structure are recursive.

Here recursive means that there are some Turing machines that decide/build all naming mappings and relations in the structured memory.

Inductive Turing machines with recursive memory are called *inductive Turing machines of the first order*.

The memory E is called *n-inductive* if all relations that define its structure are constructed by an inductive Turing machine of order n.

Inductive Turing machines with n-inductive memory are called *inductive Turing machines of the order $n + 1$*.

Limit Turing machines have the same structure (hardware) as inductive Turing machines. The difference is in a more general way of obtaining the result of computation. To obtain their result, limit Turing machines need some topology in the set of all words that are processed by these machines.

Let a limit Turing machine L works with words in an alphabet A and in the set A^* of all such words, a topology T is defined. While the machine L works, it produces in the output tape (memory) words $w_1, w_2, \ldots, w_n, \ldots$. Then the result of computation of the limit Turing machine L is the limit of this sequence of words in the topology T.

When the set A^* has the discrete topology, limit Turing machines coincide with inductive Turing machines.

Turing machines are special cases of inductive Turing machines. The difference is in a more general way of obtaining the result of computation. To obtain their result, Turing machine gives the result if and only if its control device A comes to a final state from F. Besides, the memory of a Turing machine consists of one (or several) potentially infinite one-dimensional (or multidimensional) tapes.

It is possible to read more on algorithms in (Sipser, 1997; Hopcroft, J.E. et al, 2001; Burgin, 2005).

Appendix C

Elements of Logic

Logic must look after itself.
Ludwig Wittgenstein (1889–1951)

Basic structures of logic are logical calculi and logical varieties. They are constructed using different logical languages.

A *language* in a constructive representation/definition is a triad (a named set) of the form $L = (X, R, L)$ where X is the alphabet, R is the set of constructive algorithms/rules for building well-formed (correct) expressions (e.g., words or texts) from L, and L is the set of words of the language L.

Logical languages are a special kind of artificial languages developed intentionally within a culture. The typical feature of logical languages is that their structure (inner relations) and grammar (formation rules) are intended to express the logical information within linguistic expressions in a clear and effective way. Languages used in logic have, as a rule, constructive definitions in a form of production rules. Elements of logical languages are logical expressions or formulas. To emphasize that these formulas are constructed in a proper way, they are often called *well-formed formulas*.

Elements of the language of the classical propositional or sentential logic/calculus give a formal representation of propositions. Propositional variables are denoted by letters, which are considered as atomic formulas and form a part of the alphabet of the language.

Another part is formed by logical operations:
- *negation* is denoted by ⌐ or by ~,

- *conjunction* also called logical "and" is denoted by ∧ or by & or by ·,
- *disjunction* also called logical "or" is denoted by ∨,
- *implication* is denoted by → or by ⇒ or by ⊃,
- *equivalence* is denoted by ↔ or by ≡ or by ⇔.

If $P(x)$ and $Q(x)$ are some propositions or predicates, then:

- Their *conjunction* is equal to $P(x)$ & $Q(x)$ or $P(x) \wedge Q(x)$.
- Their *disjunction* is equal to $P(x) \vee Q(x)$.
- The *negation of $P(x)$* is equal to $\rceil P(x)$.

It is possible to use fewer operations (and thus, a smaller alphabet) by expressing some of these operations by mean of others, e.g., $P \rightarrow Q$ is equivalent to $\rceil P \vee Q$. For example, Church (1956) uses only one logical operation ⊃. In addition, the left and right parentheses, (and) or the left and right brackets [and] are included in the alphabet.

Elements of the language L_P of the classical propositional (or sentential) logic/calculus are called well-formed formulas (*wff*s) and built by the following rules:

1. Letters of the alphabet that denote propositions are *wff*s from L_P.
2. If φ is a *wff*, then $\rceil$ φ is a *wff* from L_P.
3. If φ and ψ are *wff*s, then (φ ∧ ψ), (φ ∨ ψ), (φ → ψ), and (φ ↔ ψ) are *wff*s from L_P.

These rules form the set of algorithms R that build the language L_P.

Elements of the language L_{CPC} of the classical predicate logic/calculus of the first order give a formal representation of binary properties. The predicate calculus language has a developed alphabet and elaborated symbolic notation. Lower-case letters a, b, c, ..., x, y, z, ... are used to denote individuals. Upper-case letters M, N, P, Q, R, ... are used to denote predicates.

The alphabet L_{CPC} of the language L_{CPC} consists of six parts:

- A set **F** of function symbols (common examples include + and ·);
- A set **P** of predicate symbols (e.g., P_1 or P_a);
- A set **C** of logical connectives (usually, it is: $\rceil$, ∧, ∨, →, and ↔);
- A set **S** of punctuation symbols (usually, it is: (,), : and ,);
- A set **Q** of quantifiers (usually, they are ∀, which means "for all" or "for any", and ∃, which means "there exists" or "there is");
- A set **V** of variables.

Every function symbol, predicate symbol, and connective is associated with an arity. Namely, an n-ary function has the form $f: X^n \rightarrow X$ and an n-ary predicate has the form $P(x_1, x_2, ..., x_n)$. As a rule, 0-ary predicates and/or 0-ary functions are called constants. Another way to deal with constants is to include their names in the alphabet of the language.

The language L_{CPC} of the classical predicate calculus, as a general structure, encompasses the language L_P of the classical propositional calculus because propositions may be formed by juxtaposition of a predicate with an individual.

Elements of the language L_{CPC} of the classical predicate logic/calculus are also called well-formed formulas (*wffs*) and built by the following rules:

1. Letters of the alphabet A of the language L_{CPC} are *wffs* from L_{CPC}.
2. Expressions $P(x_1, x_2, ..., x_n)$ where P is an n-ary predicate symbol is a *wff* from L_{CPC}.
3. If φ is a *wff*, then $\rceil \varphi$ is a *wff* from L_{CPC}.
4. If φ and ψ are *wffs*, then $(\varphi \wedge \psi)$, $(\varphi \vee \psi)$, $(\varphi \rightarrow \psi)$, and $(\varphi \leftrightarrow \psi)$ are *wffs* from L_{CPC}.
5. If $H(x_1, x_2, ..., x_n)$ is a *wff* containing a free variable x, then $\exists x H(x_1, x_2, ..., x_n)$ and $\forall x H(x_1, x_2, ..., x_n)$ are *wffs* from L_{CPC}.

Here a variable is free if it is not related to a quantifier. Consequently, the rule (4) makes any instance of x bound (that is, not free) in the formulas $\exists x H(x_1, x_2, ..., x_n)$ and $\forall x H(x_1, x_2, ..., x_n)$. In logic, symbol $\exists$ means "there is" or "there are" and symbol $\forall$ means "for any" or "for all."

If $A = \{a_i; I \in I\}$ is an infinite set, then the expression "a predicate $P(x)$ is *true for almost all* elements from A", or "*almost all* elements from A have a property P" or in logical notation $\forall\forall x P(x)$, means that $P(x)$ can be untrue only for a finite number of elements from A. For instance, if $A = \omega$, then almost all elements of A are bigger than 10, or another example is that conventional convergence of a sequence l to x means that any neighborhood of x contains almost all elements from l.

Let L be a logical language or a language of well-formed formulas and R be an algorithmic language, procedural language or a language of rules of inference in L.

A (*syntactic* or *deductive*) *logical calculus*, usually called *calculus*, in the pair of languages (L, R) is a triad (a named set) of the form $C = (A, H, T)$ where $H \subseteq R$, A, $T \subseteq L$, A is the set of axioms, H consists of inference rules and the set of theorems T is obtained by applying algorithms/procedures/rules from H to elements from A.

Different systems of the classical propositional calculus have been devised to achieve consistency, completeness, and independence of axioms. All these systems are logically equivalent. For instance, Kleene (2002) suggests the following list of axioms (axiom schemas) for the classical propositional calculus:

$$\varphi \to (\chi \to \varphi) \tag{C.1}$$

$$(\varphi \to (\chi \to \psi)) \to ((\varphi \to \chi) \to (\varphi \to \psi)) \tag{C.2}$$

$$\varphi \to (\chi \to (\varphi \wedge \chi)) \tag{C.3}$$

$$\varphi \to \varphi \vee \chi \tag{C.4}$$

$$\chi \to \varphi \vee \chi \tag{C.5}$$

$$\varphi \wedge \chi \to \varphi \tag{C.6}$$

$$\varphi \wedge \chi \to \chi \tag{C.7}$$

$$(\varphi \to \psi) \to ((\chi \to \psi) \to (\varphi \vee \chi \to \psi)) \tag{C.8}$$

$$(\varphi \to \chi) \to ((\varphi \to \neg\chi) \to \neg \varphi) \tag{C.9}$$

$$\neg\neg \varphi \to \varphi \tag{C.10}$$

Usually the system of inference rules has only one rule called *modus ponens*:

$$\varphi, \varphi \to \psi \vdash \psi$$

or in the natural/programming language notation, we have

If φ and $\varphi \to \psi$, then ψ.

Other rules are derived from *modus ponens* and then used in formal proofs to make proofs shorter and more understandable. These rules serve to directly introduce or eliminate connectives, e.g., "**If φ and χ, then $\varphi \wedge \chi$**" (or $\varphi, \chi \vdash \varphi \wedge \chi$) or "**If φ, then $\varphi \vee \chi$**" (or $\varphi \vdash \varphi \vee \chi$).

A standard transformation rule is substitution. This rule is necessary because axiom schemas demand substitution to become axioms and be applied.

Shoenfield (1967) suggests the following list of axiom schemata of the classical first-order predicate calculus:

Propositional Axiom: $\varphi \vee \rceil \varphi$

Identity Axiom: $x = x$

Substitution Axiom: $\varphi_x[a] \rightarrow \exists x \, \varphi$

Equality Axioms: a) If F is a symbol of an n-ary function from **F**, then

$$x_1 = y_1 \wedge x_2 = y_2 \wedge \, ... \, \wedge x_n = y_n \rightarrow F(x_1, x_2, ..., x_n) = F(y_1, y_2, ..., y_n);$$

b) If P is a symbol of an n-ary predicate from **P**, then

$$x_1 = y_1 \wedge x_2 = y_2 \wedge \, ... \, \wedge x_n = y_n \rightarrow P(x_1, x_2, ..., x_n) = P(y_1, y_2, ..., y_n)$$

Shoenfield (1967) also suggests the following inference rules:

Extension rule: φ implies $\varphi \vee \psi$

Cancellation rule: $\varphi \wedge \psi$ implies φ

Associative rule: $(\varphi \vee \psi) \vee \chi = \varphi \vee (\psi \vee \chi)$

Cut rule: $(\varphi \vee \psi)$ and $(\rceil \varphi \vee \chi)$ imply $(\psi \vee \chi)$

$\exists$-introduction rule: $\varphi \rightarrow \psi$ implies $\exists x \, \varphi \rightarrow \psi$ if x is not a free variable in ψ.

One of the most important mathematical ways of reasoning is the, so-called, *proof from contradiction*. The essence of this proof is that trying to prove that some system (object) A has a property P, we make an assumption that A does not have this property P. Then we show that this contradicts the initial conditions. This allows us to conclude that our assumption was not true and due to the Principle or Law of Excluded Middle, the system (object) A has the property P.

The *Principle* or *Law of Excluded Middle*, or in Latin, *tertium non datur* (a third is not given) is formulated in traditional logic as "A has a property B or A does not have a property B".

Syntactic logical calculi provide functional formalization to the notion of a formal theory (Smullyan, 1962). In turn, a formal theory formalizes some source theory from a scientific discipline (e.g., from mathematics, physics or economics). In order to specify a formal theory, predicates, functions and relations, which are regarded as basic for a given field of study, are chosen. These predicates delimit the scope of the formal theory and are the primitives of the theory and together with logical and punctuation symbols form the alphabet of the theory language.

The logical symbol ⊨ is usually interpreted as *truth-functional entailment* (cf., for example, (Bergman, et al, 1980) or (Kleene, 1967)). That is, if *a* and *b* are some propositions or predicates, then $a \models b$ means "if *a* is true in some interpretation, then *b* is true in the same interpretation." ⊨*b* means that *b* is always true.

It is possible to read more about classical logical systems and structures, for example, in (Kleene, 2002; Shoenfield, 1967; Bergman, et al, 1980; Manin, 1991; Mendelson, 1997).

However, classical logical systems have definite limitations that bound their applications in many cases. For instance, the principal condition for any logical calculus is its consistency. At the same time, knowledge about large object domains (in science or in practice) is essentially inconsistent (Burgin, 1991b; Mestdash, et al, 1991; Nguen, 2008). From this perspective, Partridge and Wilks (1990) write, "because of privacy and discretionary concerns, different knowledge bases will contain different perspectives and conflicting beliefs. Thus, all the knowledge bases of a distributed AI system taken together will be perpetually inconsistent". Consequently, when conventional logic is used for formalization of such knowledge, it is possible to represent only small fragments of the object domain. Otherwise, contradictions appear.

To eliminate these limitations in a logically correct way, the concepts of a logical prevariety and variety were introduced (Burgin, 1997f; 2004c). They generalize in a natural way the concept of a logical calculus and are more advanced systems of logic. They include calculi as the simplest case. There are different types and kinds of logical varieties and prevarieties:

1. Deductive or syntactic varieties and prevarieties.
2. Functional or semantic varieties and prevarieties.
3. Model or pragmatic varieties and prevarieties.

Syntactic varieties and prevarieties are built from logical calculi as building blocks. Semantic logical varieties and prevarieties are formed by separating those parts that represent definite semantic units. In contrast to syntactic and semantic varieties, model varieties are essentially formal structures.

Let U be some class (universe) of objects, **M** be an abstract class of partially ordered sets, i.e., such a class that with any partially ordered set contains all partially ordered sets isomorphic to it, and $L \in$ **M**.

An abstract property P of objects from the universe U is a named set $P = (U, p, L)$ where $p : U \to L$ is a partial function (L-predicate) called the evaluation function $Ev(P)$ of P. L is called the scale $Sc(P)$ of the abstract property P. Abstract properties are used as mathematical models of real properties.

For instance, when you want to know something about a computer, you look into a list of its specifications, which contains many properties of a computer. Here is one of such specifications:

System memory installed: 128MB

Hard disk capacity: 30GB

Monitor diagonal size: 18 in.

Graphics memory amount: 64MB

Graphics chipset: nVidia GeForce II GTS

Supported operating systems: Windows 98 SE

Graphics card: Guillemot 3D Prophet II GTS

Monitor type: CRT

Processor model: Athlon

Processor clock speed: 1000 MHz,

Some properties of computers are now incorporated into their names. For instance, when you see the name "*Dell Dimension L866r Pentium III 866MHz*", you know that this computer has the processor clock frequency 866MHz. In turn, the processor clock frequency determines how many instructions it can execute per second.

The most popular property in logic is *truth* defined for logical expressions. In classical logics only one such property as *truth* with the scale $L = \{T, F\}$ is considered Thus, predicates and proposition take one of two value T (true) and F (false) dering on them are used as the scales for this properly. For modal logics, which have only one truth property that is determined for logical expressions, modalities are expressed by means of modal operators. Another possibility to express modality is to determine different modal truth properties: *"truth"*, *"necessary truth"*, and *"possible truth"*,

An abstract property $P = (U, p, L)$ is equivalent to a system $\mathbf{Z}$ of abstract properties $\{P_i = (U, p_i, L_i); I \in I\}$ if the validity of the inequality $P_i(a) \neq P_i(b)$ for some $I \in I$ implies $P(a) \neq P(b)$ for any $a, b \in U$, and vice versa.

Composition of properties makes possible to prove the following result (cf. (Burgin, 1997)).

Theorem C1. For any system $\mathbf{Z} = \{P_i = (U, p_i, L_i); I \in I\}$ of abstract properties, there is a property $P = (U, p, L)$ equivalent to $\mathbf{Z}$.

The mathematical theory of abstract properties includes logic as its subtheory (Burgin, 1989).

Appendix D

Elements of Algebra and Category Theory

Algebra reverses the relative importance
of the factors in ordinary language.
Alfred North Whitehead

An *algebraic system* is a structure $A = (X, \Omega, R)$ that consists of: a non-empty set X called the *carrier* or the *underlying set* of A and elements of which are called the elements of A; a family Ω of algebraic operations, which are mappings $\omega_i: X^{n_i} \to X$ ($i \in I$); and a family R of relations $r_j \subseteq X^{m_j}$ ($j \in J$) defined on X. The non-negative integers n_i and m_j are called the *arities* of the respective operations ω_i and relations r_j. When ω is an operation from Ω with arity n, then the image $\omega(a_1, a_2, ..., a_n)$ of the element $(a_1, a_2, ..., a_n)$ from X^n under the mapping $\omega: X^n \to X$ is called the *value* of the operation ω for elements $a_1, a_2, ..., a_n$.

In a general case, some operations from Ω may be partial mappings. For instance, division in such a universal algebra (algebraic system) as a field is not defined for 0, i.e., it is impossible to divide by 0.

The operations from Ω and the relations from R are called *basic* or *primitive*. Using basic operations, it is possible to build many different derivative operations in the algebra A. The pair of families ($\{n_i; i \in I\}$; $\{m_j; j \in J\}$) is called the *type* of the algebraic system A. Two algebraic systems A and A' have the *same type* if $I = I'$, $J = J'$, and $n_i = n_i'$ and $m_j = m_j'$ for all $i \in I, j \in J$.

An algebraic system A is called finite if the set X is *finite* and it is called of finite type if the set $I \cup J$ is finite. An algebraic system $A = (X, \Omega, R)$ is called a *universal algebra* or simply, an *algebra* if the set R of its basic relations is empty, and it is called a *model* (in logic) or a *relational system* if the set Ω of basic operations is empty. Classical algebraic systems are groups, rings, linear (vector) spaces, linear algebras, lattices, ordered sets, ordered groups, etc. Groups, rings, modules, fields, linear spaces, linear algebras, lattices, and semigroups are classical universal algebras.

For instance, a *linear space* or a *vector space L over the field F*, e.g., the field R of real numbers, elements of which are often called *vectors*, has two operations:

- *addition*: $L \times L \rightarrow L$ denoted by $x + y$ where x and y belong to L;
- *scalar multiplication*: $F \times L \rightarrow L$ denoted by ax where $a \in F$ and $x \in L$.

These operations satisfy the following axioms:

1. Addition is associative:
   For all x, y, z from L, we have $x + (y + z) = (x + y) + z$.

2. Addition is commutative:
   For all x, y from L, we have $x + y = y + x$.

3. Addition has an identity element:
   There exists an element 0 from L, called the *zero vector*, such that $x + 0 = x$ for all x from L.

4. Addition has an inverse element:
   For any x from L, there exists an element z from L, called the *additive inverse* of x, such that $x + z = 0$.

5. Scalar multiplication is distributive over addition in L:
   For all elements a from F and vectors y, $\mathbf{w}$ from L, we have
   $$a\,(\mathbf{y} + \mathbf{w}) = a\,\mathbf{y} + a\,\mathbf{w}.$$

6. Scalar multiplication is distributive over addition in F:
   For all element elements a, b from F and any vector y from L, we have
   $$(a + b)\,y = a\,y + b\,y.$$

7. Scalar multiplication is compatible with multiplication in F:
   For all elements a, b from F and any vector y from L, we have
   $$a\,(b\,y) = (ab)\,y.$$

8. The identity element 1 from the field F also is an identity element for scalar multiplication:
   For all vectors x from L, we have $1x = x$.

Vectors x_1, x_2, ..., x_n from L are called *linearly dependent* in L if there is an equality $\Sigma_{i=1}^{n} a_i x_i = 0$ where a_i are elements from R (from F, in a general case) and not all of them are equal to 0. When there are no such an equality, vectors $x_1, x_2, ..., x_n$ are called *linearly independent*.

A system B of linearly independent vectors from L is called a *basis* of L if any element x from L is equal to a sum $\Sigma_{i=1}^{n} a_i x_i$ where n is some natural number, x_i are elements from B and a_i are elements from R (from F, in a general case).

The number of elements in a basis is called the *dimension* of the space L. It is proved that all bases of the same space have the same number of elements.

The space R is a one-dimensional vector (linear) space over itself. The space R^n is an n-dimensional vector (linear) space over R.

A subsystem of a universal algebra A closed with respect to basic operations is called a *subalgebra* of A. A subsystem of a model is called a submodel. The concept of a subalgebra essentially depends on the set of operations of the algebra under consideration. In contrast to this, any non-empty subset of a model is a submodel.

A set V of universal algebras of a type Ω is called a variety if there is a system of identities such that V consists of all universal algebras of a type Ω that satisfy these identities. A variety of universal algebras may be characterized as a non-empty class of algebras closed under taking quotient algebras, subalgebras and direct products.

It is possible to find other structures from algebra and their properties, for example, in (Kurosh, 1963; 1974; Malcev, 1970; Van der Varden, 1971; Giles, 2000).

There are two approaches to the mathematical structure called a *category*. One approach treats categories in the framework of the general set-theoretical mathematics. Another approach establishes categories independently of sets and uses them as a foundation of mathematics different from set theory. It is possible to build the whole mathematics in the framework of categories. For instance, such a basic concept as a binary relation is frequently studied in categories (cf., for example, (Burgin, 1970)). Toposes allow one to reconstruct set theory as a subtheory of category theory (cf., for example, (Goldblatt, 1979)).

According to the first approach, we have the following definition of a category.

A *category* **C** consists of two collections Ob **C**, the objects of **C**, and Mor **C**, the morphisms of **C** that satisfy the following three axioms:

A1. For every pair A, B of objects, there is a set $\text{Mor}_C(A, B)$, also denoted by $H_C(A, B)$ or $\text{Hom}_C(A, B)$, elements of which are called morphisms from A to B in **C**. When f is a morphism from A to B, it is denoted by $f: A \rightarrow B$.

A2. For every three objects A, B and C from Ob **C**, there is a binary partial operation, which is a partial function from pairs of morphisms that belong to the direct product $\text{Mor}_C(A, B) \times \text{Mor}_C(B, C)$ to morphisms in $\text{Mor}_C(A, C)$. In other words, when $f: A \rightarrow B$ and $g: B \rightarrow C$, there is a morphism $g \circ f: A \rightarrow C$ called the composition of morphisms g and f in **C**. This composition is associative, that is, if $f: A \rightarrow B$, $g: B \rightarrow C$ and $h: C \rightarrow D$, then $h \circ (g \circ f) = (h \circ g) \circ f$.

A3. For every object A, there is a morphism 1_A in $\text{Mor}_C(A, A)$, called the identity on A, for which if $f: A \rightarrow B$, then $1_B \circ f = f$ and $f \circ 1_A = f$.

Examples of categories:

1. The category of sets **SET**: objects are arbitrary sets and morphisms are mappings of these sets.
2. The category of groups **GRP**: objects are arbitrary groups and morphisms are homomorphisms of these groups.
3. The category of topological spaces **TOP**: objects are arbitrary topological spaces and morphisms are continuous mappings of these topological spaces.

Mapping of categories that preserve their structure are called functors. There are functors of two types: covariant functors and contravariant functors.

A *covariant functor* **F**: **C** → **K**, also called a *functor*, from a category **C** to a category **K** is a mapping that is stratified into two related mappings F_{ObC}: Ob **C** → Ob **K** and F_{MorC}: Mor **C** → Mor **K**, i.e., F_{ObC} associates an object $F(A)$ from the category **K** to each object A from the category **C** and F_{MorC} associates a morphism $F(f): F(A) \rightarrow F(B)$ from the category **K** to each morphism $f: A \rightarrow B$ from the category **C**. In addition, F satisfies the following two conditions:

1. $F(1_A) = 1_{F(A)}$ for every object A from the category **C**;
2. $F(f \circ g) = F(f) \circ F(g)$ for all morphisms f and g from the category **C** when their composition $f \circ g$ exists.

That is, functors preserve identity morphisms and composition of morphisms.

A *contravariant functor* **F: C → K** from a category **C** to a category **K** consists of two mappings F_{ObC}: Ob **C** → Ob **K** and F_{MorC}: Mor **C** → Mor **K**, i.e., F_{ObC} associates an object $F(A)$ from the category **K** to each object A from the category **C** and F_{MorC} associates a morphism $F(f)$: $F(B) \to F(A)$ from the category **K** to each morphism f: $A \to B$ from the category **C**, that satisfy the following two conditions:

3. $F(1_A) = 1_{F(A)}$ for every object A from the category **C**;
4. $F(f \circ g) = F(g) \circ F(f)$ for all morphisms f and g from the category **C** when their composition $f \circ g$ exists.

It is possible to define a contravariant functor as a covariant functor on the dual category $\mathbf{C}^{\text{op}}$.

A functor from a category to itself is called an *endofunctor*.

There is an approach to the definition of a category in which a category **C** consists only of the collection Mor **C** of the morphisms (also called *arrows*) of **C** with corresponding axioms. Objects of **C** are associated with identity morphisms 1_A. It is possible to do this because 1_A is unique in each set Mor$_C(A, A)$, and uniquely identifies the object A. In any case, morphism is the central concept in a category. But what is a morphism?

If f: $A \to B$ is a morphism, then it is a one-to-one named set $(\{A\}, f, \{B\})$. Thus, the main object of a category is a named set, and categories are built from these named sets. Besides, a construction or separation of a category begins with separation of all elements into two sets and calling all elements from one of these sets by the name "objects" and elements from the other sets by the name "morphisms". In such a way, two named sets appear. In addition, composition of morphisms, as any algebraic operation, is also represented by a named set (Mor$_C(A, B)$ × Mor$_C(B, C)$, $\circ$, Mor$_C(A, C)$). This shows that the informal notion of a named set is prior both to categories and sets. As a result, we come to the conclusion that any category is built of different named sets. Moreover, functors between categories, which are structured mappings of categories (Herrlich and Strecker, 1973), are morphisms of those named sets.

It is possible to read more about categories, functors and their properties, for example, in (Goldblatt, 1979; Herrlich and Strecker, 1973).

Appendix E

Elements of Probability Theory

The essence of mathematics lies in its freedom.
Georg Cantor

Probability theory is nowadays an important tool in physics and information theory, engineering and industry. As a mathematical foundation for statistics, probability theory is essential to many human activities that involve quantitative analysis of large data sets. Methods of probability theory also apply to description of complex systems given only partial knowledge of their state, as in statistical mechanics. A great discovery of twentieth century physics was the probabilistic nature of physical phenomena at microscopic scales, described in quantum mechanics and quantum field theory. Common applications of probability theory occur in our daily lives and are usually associated with risk assessments in such areas as business ventures, health decisions, insurance, sports, product reliability, gambling, etc. Governments typically apply probabilistic methods in environmental regulation where it is called "pathway analysis", often measuring well-being using methods that are stochastic in nature, and choosing projects to undertake based on statistical analyses of their probable effect on the population as a whole. Another significant application of probability theory in everyday life is reliability. Many consumer products, such as automobiles and consumer electronics, utilize reliability theory in the design of the product in order to reduce the probability of failure. The probability of failure is also closely associated with the product's warranty.

Probability theory is usually defined as the branch of mathematics concerned with analysis of random phenomena, for which probability is the central concept. However, there is no unique understanding of the term *probability* and there is no single mathematical system that formalizes the term *probability*, but rather a host of such systems.

An interpretation of the concept of probability is based on a choice of some class of events (or statements) and an assignment of some meaning to probability claims about those events (or statements). An *event* is usually considered as a collection of tentative results or outcomes of an experiment, procedure or process. A *simple event* is an even that cannot be broken down into simpler events. A simple event usually is one result or outcome of an experiment.

Usually three main interpretations of the probability of a random event are considered: the frequency interpretation, the belief interpretation, and the support interpretation. The classical example of a random event is a toss of a fair coin. A coin is considered fair if heads and tails are equally likely in a toss. However, other interpretations, such as the logical interpretation or the propensity interpretation, also exist. A sample space consists of all possible simple events.

In the traditional (relative) frequency interpretation of probability, it is assumed the *same random experiment* is indefinitely repeated and we keep track of the relative frequency with regard to a particular event A in the following ratio called the relative frequency:

$$\frac{\text{Number of times that the event } A \text{ occurs during a sequence } Q \text{ of trials}}{\text{Total number of trials in } Q}$$

When the number of trials in Q is N, the idea is that as $N \to \infty$ the preceding ratio approaches the *probability $p(A)$* of event A.

In practice, three types of probabilities are used:

- *Objective probability* is defined as a numerical property of relative frequency sequences associated with an event in natural, social or technical phenomena.
- *Subjective probability* is defined (exists) as a belief or measure of confidence in an outcome of a certain event.

- *Combined probability* is defined (exists) as a supported by both observational and experimental evidence belief (measure of confidence) in an outcome of some event.

Examples of probability types:

1. When there are 10 blue balls, 5 green balls and 5 red balls in a box, and these balls are well mixed, then the objective probability to draw a blue ball from this box is ½. To come to this value, we assume that for any two balls in the box, there are equal chances that any of them will be drawn. Thus the probability to draw a blue ball is the same as the probability to draw not a blue ball as there are 10 blue balls and 10 balls that are not blue. As the probability of drawing a ball is equal to one, we have that.

2. In a similar way, when we believe that the ratio of blue balls to all balls in a box is ½, then the subjective probability of drawing a blue ball from this box is ½.

3. When we conjecture that the ratio of blue balls to all balls in a box is ½, we tested this hypothesis and got some experimental evidence in support of it, then the combined probability of drawing a blue ball from this box is ½.

Problems with probability interpretations and necessity to have sound mathematical foundations brought forth an axiomatic approach in probability theory. Based on ideas of Fréchet and following the axiomatic mainstream in mathematics, Kolmogorov developed his famous axiomatic exposition of probability theory (1933). According to this axiomatization, probability is defined in the following way.

A set Ω is taken with a set field (or set algebra) $\mathbf{F}$ on Ω, i. e., a set of subsets of Ω that has Ω as a member, and that is closed under complementation (with respect to Ω) and union. Elements from $\mathbf{F}$, i.e., subsets of Ω that belong to $\mathbf{F}$, are called *random events*.

A function from $\mathbf{F}$ to the set $\mathbf{R}$ of real numbers is called a *probability function*, if it satisfies the following axioms:

P 1. (*Non-negativity*) $P(A) \geq 0$, for all $A \in \mathbf{F}$.

P 2. (*Normalization*) $P(\Omega) = 1$.

P 3. (Finite additivity)

$$P(A \cup B) = P(A) + P(B)$$

for all sets $A, B \in \mathbf{F}$ such that

$$A \cap B = \emptyset.$$

The triad $(\Omega, \mathbf{F}, P)$ is called a *probability space*.

If $A \in \mathbf{F}$, then $P(A)$ is called the *probability* of the event A.

Eventually one more axiom called *axiom of continuity* was also added to this list to comply with the traditional approach to probability.

P 4. (*Continuity*) If

$$A_1 \supseteq A_2 \supseteq A_3 \supseteq \dots \supseteq A_i \supseteq \dots$$

is a decreasing sequence of events from $\mathbf{F}$ such that

$$\cap_{i=1}^{\infty} A_i = \emptyset,$$

then

$$\lim_{i \to \infty} P(A_i) = 0.$$

It is possible to read more about probability, for example, in (Kolmogorov, 1933; Feller, 1950; Carnap, 1950)).

The *expectation* (also called, the *expected value* or *expectation value* or *mathematical expectation* or *first moment*) of a function $f(x)$ of a random variable x with the probability distribution $p(x)$ is equal to the sum:

$$\mathrm{E} f(x) = \sum_{i=1}^{n} f(x_i)\, p(x_i) = f(x_1)\, p(x_1) + f(x_2)\, p(x_2) + \dots + f(x_n)\, p(x_n)$$

in the finite case when $x \in \{ x_1, x_2, x_3, \dots, x_n \}$,

is equal to the sum:

$$\mathrm{E} f(x) = \sum_{i=1}^{\infty} f(x_i)\, p(x_i)$$

in the discrete case when $x \in \{x_1, x_2, x_3, \dots, x_n, \dots\}$,

and is equal to the integral:

$$\mathrm{E} f(x) = \int_{-\infty}^{\infty} f(x) p(x)$$

in the continuous case.

When $f(x) = x$, then its expectation is called the *expectation* (or *expected value* or *expectation value* or *mathematical expectation* or *first moment*) of the random variable x.

Appendix F

Numbers and Numerical Functions

> Everything is a number.
>
> *Pythagoras*

N is the set of all natural numbers 1, 2, ..., n, ...

ω is the sequence of all natural numbers.

N_0 is the *set of all whole numbers* 0, 1, 2, ..., n, ...

Z is the set of all integer numbers or of integers.

Q is the set of all rational numbers.

R is the *set of all real numbers* or of *reals*. The geometric form of the set **R** is called the *real line*.

C is the set of all complex numbers.

∞ is the *positive infinity*, −∞ is the *negative infinity*. Usually, these elements are added to the set **R**. The new set is denoted by $R_\infty = R \cup \{\infty, -\infty\}$.

If a is a real number, then $|a|$ or $\|a\|$ denotes its *absolute value* or *modulus*. Thus, $|a| = a$ when a is non-negative and $|a| = -a$ when a is negative. Two more important constructions are the *integral part* or *integral value* [a] (also denoted by $\lfloor a \rfloor$) of a, which is equal to the largest integer number that is less than a, and]a[(also denoted by $\lceil a \rceil$), which is equal to the least integer number that is larger than a. For instance, [1.2] = 1, [7.99] = 7 and [−1.2] = −2. The difference a − [a], also denoted by a mod 1, or $\{a\}$, is called the *fractional part* of a. All these construction define real-valued functions $|x|$, [x], $\{x\}$, and]x[where]x[is sometimes called the *ceiling function* and [x] is sometimes called the *floor function*.

The symbol $\approx$ denotes the relation "approximately equal". For instance, we have $5 \approx 5.001$.

Axioms for operations with real and complex numbers:

1. Commutativity of addition: $a + b = b + a$;
2. Associativity of addition: $(a + b) + c = a + (b + c)$;
3. Commutativity of multiplication: $a \cdot b = b \cdot a$;
4. Associativity of multiplication: $(a \cdot b) \cdot c = a \cdot (b \cdot c)$;
5. Distributivity of multiplication with respect to addition:

$$a \cdot (b + c) = a \cdot b + a \cdot c$$

6. Zero is a neutral element with respect to addition:

$$a + 0 = 0 + a = a$$

7. One is a neutral element with respect to multiplication:

$$a \cdot 1 = 1 \cdot a = a.$$

A function with **R** as its range is called a *real function* or a *real-valued function*.

A function with **C** as its range is called a *complex function* or a *complex-valued function*.

Operations with numbers induce similar operations with functions:
- Addition of functions $(f + g)(x) = f(x) + g(x)$.
- Subtraction of functions $(f - g)(x) = f(x) - g(x)$.
- Multiplication of functions $(f \cdot g)(x) = f(x) \cdot g(x)$.
- Scalar multiplication of functions $(k \cdot f)(x) = k \cdot g(x)$ by a number k.

Logarithm $\log_a x$ is an important numerical function used in information theory. It has the following properties:

$\log_a b = c$ if and only if $a^c = b$ (this is the definition of logarithms).

$\log_a a^c = c$

$\log_a b + \log_a c = \log_a bc$.

$\log_a b - \log_a c = \log_a b/c$.

$\log_a b^c = c \cdot \log_a b$.

$\log_e b = \ln b$

$\log_{10} b = \log b$

Appendix G

Topological, Metric and Normed Spaces

> Space is the breath of art.
> *Frank Lloyd Wright* (1867–1959)

A *topology* in a set X is a system $O(X)$ of subsets of X that are called open subsets and satisfy the following axioms:

T1. $X \in O(X)$ and $\varnothing \in O(X)$.

T2. For all A, B, if A, $B \in O(X)$, then $A \cap B \in O(X)$.

T3. For all A_i, $i \in I$, if all $A_i \in O(X)$, then $\bigcup_{i \in I} A_i \in O(X)$.

A set X with a topology in it is called a *topological space*.

Topology in a set can be also defined by a system of neighborhoods of points from this set. In this case, a set is *open* in this topology if it contains a standard neighborhood of each of its points. For instance, if a is a real number and $t \in R^{++}$, then an open interval $O_t a = \{x \in R;\ a - t < x < a + t\}$ is a standard neighborhood of a.

If X is a subset of a topological space, then $Cl(X)$ denotes the closure of the set X.

In many interesting cases, topology is defined by a metric.

A *metric* in a set X is a mapping $\mathbf{d}: X \times X \to R^+$ that defines distances between points of X and satisfies the following axioms:

M1. $\mathbf{d}(x, y) = 0$ if and only if $x = y$,

i.e., the distance between an element and itself is equal to zero, while the distance between any different elements is equal to a positive number.

M2. $\mathbf{d}(x, y) = \mathbf{d}(y, x)$ for all $x, y \in X$,

i.e., the distance between x and y is equal to the distance between x and y.

M3. The triangle inequality:

$\mathbf{d}(x, y) \le \mathbf{d}(x, z) + \mathbf{d}(z, y)$ for all $x, y, z \in X$.

That is, the distance from x through z to y is never less than the distance directly from x to y, or the shortest distance between any two points is a straight line

A set X with a metric $\mathbf{d}$ is called a *metric space*. The number $\mathbf{d}(x, y)$ is called the *distance* between x and y in the metric space X.

For instance, in the set R of all real numbers, the distance $\mathbf{d}(x, y)$ between numbers x and y is the absolute value $|x - y|$, i.e., $\mathbf{d}(x, y) = |x - y|$. This metric defines the following topology in R. If a is a point from R, then a standard neighborhood of a has the form $O_r(a) = \{y; |a - y| < r$ with $r \in R^{++}$. A set is open in this topology if it contains a standard neighborhood of each of its points.

Mappings of metric (topological) spaces that preserve the metric (topological) structure are called continuous.

It is possible to find other structures from topology and their properties, for example, in (Bers, 1957; Kelly, 1957; Kuratowski, 1966; 1968).

$\sum_{i=1}^{n} c_i$ denotes the sum $c_1 + c_2 + c_3 + \ldots + c_n$.

If U is a subset of a metric space, then the diameter $D(U)$ of the set U is equal to $\sup\{\mathbf{d}(x, y); x, y \in U\}$.

If $l = \{a_i \in M; i = 1, 2, 3, \ldots\}$ is a sequence, and $f: M \to L$ is a mapping, then $f(l) = \{f(a_i); i = 1, 2, 3, \ldots\}$.

$a = \lim l$ means that a number a is a limit of a sequence l.

It is possible to introduce a natural metric in the space R^n: if $x, y \in R^n$, $x = \sum_{i=1}^{n} a_i x_i$, and $y = \sum_{i=1}^{n} b_i x_i$ where x_i are elements from a basis B of the space R^n, then $\mathbf{d}(x, y) = \sqrt{(a_1 - b_1)^2 + (a_2 - b_2)^2 + \ldots + (a_n - b_n)^2}$. It is called the *Euclidean metric* in the space R^n. The space R^n with the Euclidean metric is called an *Euclidean space* and often denoted by E^n.

Another natural metric in R^n is the Manhattan distance, where the distance between any two points, or vectors, is the sum of the distances between corresponding coordinates, i.e., $\mathbf{d}(x, y) = |a_1 - b_1| + |a_2 - b_2| + \ldots + |a_n - b_n|$.

It is possible to consider any non-empty set X as a metric space with the distance $\mathbf{d}(x, y) = 1$ for all x not equal to y and $\mathbf{d}(x, y) = 0$ otherwise. It is a discrete metric.

A *norm* in a linear space L over the field $\mathbf{R}$ is a mapping $\| \ \| : L \to \mathbf{R}^+$ that satisfies the following axioms:

N1. $\|x\| = 0$ if and only if $x = 0$,

i.e., the zero vector has zero length, while any other vector has a positive length.

N2. For any positive number a from $\mathbf{R}$, we have $\|ax\| = a\|x\|$,

i.e., multiplying a vector by a positive number has the same effect on the length.

N3. The triangle inequality:

$$\|x + y\| \le \|x\| + \|y\|$$

i.e., the norm of a sum of vectors is never larger than the sum of their norms.

This implies:

$$\|x\| - \|y\| \le \|x + y\|$$

A linear space L with a norm is called a *normed linear space* or simply, a *normed space*.

The space $\mathbf{R}^n$ is a normed linear space with the following norm: if $x \in \mathbf{R}^n$ and $x = \sum_{i=1}^{n} a_i x_i$ where x_i are elements from a basis B of the space $\mathbf{R}^n$, then $\|x\| = = \sqrt{a_1^2 + a_2^2 + \ldots + a_n^2}$.

Proposition D.1. Any normed linear space is a metric space.

Indeed, we can define $\mathbf{d}(x, y) = \|x - y\|$ and check that all axioms of metric are valid for this distance.

Another natural metric in a normed linear space is the British Rail metric (also called the Post Office metric or the SNCF metric) on a normed vector space, given by $\mathbf{d}(x, y) = \|x\| + \|y\|$ for distinct vectors x and y, and $\mathbf{d}(x, x) = 0$.

A *Hilbert space* is an abstract linear (vector) space over the field $\mathbf{R}$ of real numbers or the field $\mathbf{C}$ of complex numbers with an inner product and complete as a metric space.

An *inner product* in a vector space V is a function $\langle \cdot, \cdot \rangle$: $V \times V \rightarrow R$ (or in the complex case $\langle \cdot, \cdot \rangle$: $V \times V \rightarrow C$) that satisfies the following properties for all vectors x, y, and x from V and number a:

Conjugate symmetry:

$$\langle x, y \rangle = \langle y, x \rangle$$

Linearity in the first argument:

$$\langle ax, y \rangle = a\, x, y \rangle$$

$$\langle x + z, y \rangle = \langle x, y \rangle + \langle z, y \rangle$$

Positive-definiteness:

$$\langle x, y \rangle > 0$$

for all $x \neq 0$ from V.

A *fiber bundle* B (also called fibre bundle) is a triad (E, p, B) where the topological space E is called the *total space* or simply, *space* of the fiber bundle B, the topological space B the *base space* or simply, *base* of the fiber bundle B, and p is a topological projection of E onto B such that every point in the base space has a neighborhood U such that $p^{-1}(b) = F$ for all points b from B and $p^{-1}(U)$ is homeomorphic to the direct product $U \times F$. The topological space F is called the *fiber* of the fiber bundle B. Informally, a fiber bundle is a topological space which looks locally like a product space $U \times F$.

Fiber bundles are special cases of topological bundles. A *topological bundle* D is a triad (named set) (E, p, B) where E and B are topological spaces and p is a topological (i.e., continuous) projection of E onto B.

It is possible to find introductory information on topological manifolds, fiber and topological bundles, and other topological structures in (Gauld, 1974; Husemoller, 1994; Lee, 2000).

Bibliography

Abercrombie, N., Hill, S. and Turner, B.S. (2000) Social Structure, in *The Penguin Dictionary of Sociology*, Penguin, London, pp. 326–327

Ackoff, R.L. (1989) From Data to Wisdom, *Journal of Applied Systems Analysis*, v. 16, pp. 3–9

Aczél, J.D. and Daróczy, Z. (1975) *On Measures of Information and Their Generalizations*, Academic Press, New York

Aczel, J., Forte B. and Ng, C.T. (1974) Why the Shannon and Hartley entropies are 'natural', *Adv. Appl. Prob.*, v. 6, pp. 131–146

Adami, C. and N.J. Cerf, (1999) What information theory can tell us about quantum reality, *Lect. Notes in Comp. Sci.*, v. 1509, pp. 1637–1650

Adams, S. (2003) Information Quality, Liability, and Corrections, *Online*, v. 27, No. 5, pp. 16–22

Agre, P. (1995) Institutional Circuitry: Thinking about the Forms and Uses of Information, *Information Technologies and Libraries*, December, pp. 225–230

Agre, P.E. (2000) The Market Logic of Information, *Knowledge, Technology, and Policy*, v. 13, No. 3, pp. 67–77

Alber, G., Beth, T., Horodecki, M., Horodecki, P., Horodecki, R., Rötteler, M., Weinfurter, H., Werner, R. and Zeilinger, A. (2001) *Quantum Information: An Introduction to Basic Theoretical Concepts and Experiments*, Springer Verlag, New York/Berlin/Heidelberg

Alexander, J. and Tate, M.A. (1999) *Web wisdom: How to evaluate and create information on the web*, Erlbaum, Mahwah, NJ

Al-Hakim, L. (Ed.) (2006) *Challenges of managing information quality in service organizations*, Idea Group Pub., Hershey, PA

Allo, P. (2005) Being Informative, in *Proceedings of the Second International Workshop on Philosophy and Informatics*, Kaiserslautern, Germany, pp. 579–586

Allo, P. (2007) Informational content and information structures: a pluralist approach, in *Proceedings of the Workshop on Logic and Philosophy of Knowledge, Communication and Action*, The University of the Basque Country Press, pp. 101–121

Allo, P. (2007) Formalising Semantic Information: Lessons from Logical Pluralism, in *Computation, Information, Cognition — The Nexus and The Liminal*, Cambridge Scholars Publishing, Cambridge, pp. 41–53

Alter, S. (2002) *Information Systems Foundation of e-business*, Prentice Hall, Englewood Cliffs

Alter, S. (2006) Goals and Tactics on the Dark Side of Knowledge Management, Proc. of the *39th Hawaii Internat. Conf. on System Sciences*, IEEE Press

Amari, S. (1985) *Differential-geometrical methods in statistics*, Lecture notes in statistics, Springer-Verlag, Berlin

Amari, S. and Nagaoka, H. (2000) *Methods of information geometry*, Transactions of mathematical monographs, v. 191, American Mathematical Society

American Heritage Dictionary of the English Language (1996) Third Edition, Laurel, New York

Anderson, J.R. and Bower, G.H. (1973) *Human Associative Memory*, V.H. Winston and Sons, Washington

Anderson, R. and Joshi, G.C. (2006) *Interpreting Mathematics in Physics: Charting the Applications of* SU(2) *in* 20th *Century Physics*, Preprint History of Physics physics.hist-ph/0605012 (electronic edition: http://arXiv.org)

Antonov, A.V. (1988) *Information: Comprehension and Understanding*, Naukova Dumka, Kiev (in Russian)

Arbib, M. (1992) Schema Theory, in *The Encyclopedia of AI*, Wiley-Interscience, New York, pp. 1427–1443

Arrow, K.J. (1979) The economics of information, in *The computer age: A twenty year view* (M.L. Dertouzos & J. Moses, Eds.), MIT Press, Cambridge, MA, pp. 306–317

Arrow, K. J. (1980/81) Jacob Marschak's contributions to the economics of decision and information, *Math. Social Sci.*, v. 1, No. 4, pp. 335–338

Arrow, K.J. (1984) *The economics of information*, Collected papers of Kenneth J. Arrow, v. 4, The Belknap Press of Harvard University Press, Cambridge, MA

Arrow, K.J. (1984a) *Information and economic behavior*, A presentation to the Federation of Swedish Industries, Stockholm, 1973 (reprinted in Arrow, 1984, pp. 136–152)

Arrow, K.J., Marschak, J. and Harris, T. (1951) Economics of information and organization: Optimal inventory policy, *Econometrica*, v.19, No. 3, pp. 250–272

Artandi, S. (1973) Information concepts and their utility. *J. Am. Soc. Inform. Sci.*, v. 24, pp. 242–245

Arruda, A.I. (1980) A survey of paraconsistent logic, in *Mathematical Logic in Latin America*, North-Holland, pp. 3–41

Ash, R. (1965) *Information Theory*, Interscience Publ., New York

Atanasov, K. *Intuitionistic Fuzzy Sets: Theory and Applications*, Physica-verlag, Heidelberg/New York, 1999

Atkinson, R.L., Atkinson, R.C., Smith E.E. and Bem, D.J. (1990) *Introduction to Psychology*, Harcourt Brace Jovanovich, Inc., San Diego/New York/Chicago

Atmanspacher, H. and Weidenmann, G. (Eds.) (1991) *Information Dynamics*, Plenum, New York

Attneave, F. (1974) Informationstheorie in der Psychologie: Grundbegriffe, in *Techniken Ergebnisse*, Verlag Hans Huber, Bern/Stuttgart/Wien

Aumann, R.J. (1974) Subjectivity and correlation in randomized strategies, *Journ. of Math. Economy*, v. 1, pp. 67–96

Autar, R. (1965), On Relative Information Function, *Indian J. Pure and Applied Math*, v. 6, pp. 1449–1454

Baba M.L. (1999) Dangerous liaisons: Trust, distrust, and information technology in American work organizations, *Human Organization*, v. 58, pp. 331–346

Babcock, B.A. (1990) The Value of Weather Information in Market Equilibrium. *American Journal of Agricultural Economics*, No. 2, pp. 63–72

Bækgaard, L. (2006) Interaction in information systems — beyond human-computer interaction, in ALOIS'06 — *Action in Language, Organisation and Information Systems*, Borås, Sweden, 12 p.

Balakrishnan, H., Lakshminarayanan, K., Ratnasamy, S., Shenker, S., Stoica, I. and Walfish, M. (2004) A Layered Naming Architecture for the Internet, *SIGCOMM '04*, Portland, Oregon, pp. 343–352

Balcazar, J.L., Diaz, J. and Gabarro, J. (1988) *Structural Complexity*, Springer-Verlag, Berlin/Heidelberg/New York

Baldoni, M., Giordano, L. and Martelli, A. (1998) A modal extension of logic programming: Modularity, beliefs and hypothetical reasoning, *Journal of Logic and Computation*, v. 8, pp. 597–635

Baldwin, J.F. (1986) Support Logic Programming, *Int. J. of Intelligent Systems*, v. 1, pp. 73–104

Ballou, D. and Pazer, H. (1985) Modeling Data and Process Quality in Multi-Input, Multi-Output Information Systems, *Management Science*, v. 31, No. 2, pp. 150–162

Ballou, D. and Tayi, G. (1989) Methodology for Allocating Resources for Data Quality Enhancement, *Communications of the ACM*, v. 32, No. 3, pp. 320–331

Banathy, B.A. (1995) The 21st century Janus: The three faces of information, *Systems Research*, v. 12, No. 4, pp. 319–320

Barbieri, M. (2003) *The Organic Codes: An Introduction to Semantic Biology*, Cambridge University Press, Cambridge

Bar-Hillel, Y. (1952) Semantic information and its measures. In *Transactions of the Tenth Conference on Cybernetics*, J. Macy Foundation, New York, pp. 33–48

Bar-Hillel, Y. (1955) An examination of information theory, *Philos. Sci.*, v. 22, pp. 86–105

Bar-Hillel, Y. (1964) *Language and Information, Selected Essays on their Theory and Application*, Addison-Wesley Publishing Company, Reading, Massachusetts

Bar-Hillel, Y. and Carnap, R. (1952) *An Outline of a Theory of Semantic Information*, Technical Report No. 247, Research Laboratory of Electronics, MIT, Cambridge,

MA (reproduced in Bar-Hillel (1964) (Ed.), *Language and Information*, Addison-Wesley, Reading, MA, pp. 221–274)

Bar-Hillel, Y. and Carnap, R. (1958) Semantic Information, *British J. of Philosophical Sciences*, v. 4, No. 3, pp. 147–157

Barlow, J.P. (1999) *Selling Vine without Bottles*: The Economy of Mind, (electronic edition: http://www.eff.org/pub/Publications/John_Perry_Barlow/HTML/idea_economy_article.html)

Barnouw, E. and Kirkland, C. (1989) Entertainment, in "*The International Encyclopedia of Communications*", Oxford University Press, New York

Barr, M. (1979) *-autonomous categories, LNM 752, Springer, Berlin/Heidelberg/New York

Barwise, J. and Perry, J. (1983) *Situations and Attitudes*, MIT Press, Cambridge, Massachusetts, and London, England

Barwise, J. (1993) Constraints, channels, and the flow of information, in *Situation Theory and its application*, Stanford CSLI, Stanford, CA, pp. 3–27

Barwise, J. and Seligman, J. (1997) *Information Flow*: The Logic of Distributed Systems, Cambridge Tracts in Theoretical Computer Science 44, Cambridge University Press

Barzdin, Ja. M. (1968) Complexity of programs which recognize whether natural numbers not exceeding *n* belong to a recursively enumerable set, *Dokl. Akad. Nauk USSR*, v. 182, pp. 1249–1252

Bates, M.J. (1999) The invisible substrate of information science, *Journal of the American Society for Information Science*, v. 50, No. 12, pp. 1043–1050

Bateson, G. (1972) *Steps to an Ecology of Mind*, Ballantine Books, New York

Bateson, G. (2000) *Steps to an Ecology of Mind*, University of Chicago Press, Chicago

Bateson, G. and Ruesch, J. (1951) *Communication*: The Social Matrix of Psychiatry, W.W.Norton and Co., New York

Bechara, A. Damasio H. and Damasio, A.R. (2000) Emotion, Decision Making and the Orbitofrontal Cortex, *Cerebral Cortex*, v. 10, No. 3, pp. 295–307

Beck, J. (1995) *Cognitive Therapy*: Basics and Beyond, Guilford, New York

Bednarz, J. (1988) Information and meaning: Philosophical remarks on some cybernetic concepts, *Grundlagenstud. Kybernetik-Geistenwissenschaft*, v. 29, No. 1, pp. 17–25

Beedie, C., Terry, P. and Lane, A. (2005) Distinctions between emotion and mood, *Cognition and Emotion*, v. 19, No. 6, pp. 847–878

Bekenstein, J.D. (1981) Universal upper bound on the entropy-to-energy ratio for bounded systems, *Phys. Rev. D, Part. Fields*, v. 23, pp. 287–298

Bekenstein, J.D. (2003) Information in the holographic universe, *Scientific American*, 289, No. 2, pp. 58–65

Belkin, N.J. (1978) Information Concepts for Information Science, *Journal of Documentation*, v. 34, pp. 55–85

Belkin, N. and Robertson, S. (1976) Information science and the phenomenon of information, *J. Am. Soc. Inform. Sci.*, v. 27, pp. 197–204

Bell, A.J. (2003) The co-information lattice, Proceedings of the *Fifth International Workshop on Independent Component Analysis and Blind Signal Separation* (ICA 2003), pp. 921–926

Bell, D. (1973) *The coming of the post-industrial society: A venture in social forecasting*, Basic Books, New York

Bell, D. (1980) The Social Framework of Information Society, in *"The Microelectronic Revolution"* (Forester, ed.), Oxford

Bell, E.T. *Men of Mathematics*, Simon and Shuster, New York/London/Toronto, 1965

Bellinger, G., Castro, D. and Mills, A. (1997) *Data, Information, Knowledge, and Wisdom*, (*http://www.outsights.com/systems/dikw/dikw.htm*)

Bem, D.J. (1970) *Beliefs, Attitudes, and Human Affairs*, Brooks/Cole P.C., Belmont, California

Benci, V., Bonanno, C., Galatolo, S., Menconi, G. and Virgilio, M. (2002) *Dynamical systems and computable information*, Preprint in Physics cond-mat/0210654 (electronic edition: http://arXiv.org)

Benioff, P. (1980) The computer as a physical system: a microscopic quantum mechanical model of computers as represented by Turing machines, *J. Stat. Physics*, v. 22, pp. 563–591

Beniof, P.A. (1982) The Thermodynamics of Computation, *Int. Journal of Theoretical Physics*, v. 21, pp. 905–940

Bennett, C.H. and DiVincenzo, D.P. (2000) Quantum information and computation. *Nature*, v. 404, pp. 247–255

Benyon, D.R. (1998) Cognitive Ergonomics as Navigation in Information Space, *Ergonomics*, v. 41, No. 2, pp. 153–156

Benyon, D.R. (2001) The new HCI? Navigation of Information Space, *Knowledge-based Systems*, v. 14, No. 8, pp. 425–430

Benyon, D. and Höök, K. (1997), Navigation in information space: supporting the individual, in *Proceedings of INTERACT'97*, Chapman and Hall, Sydney, pp. 39–46

Benci, V., Bonanno, C., Galatolo, S., Menconi, G. and Ponchio, F. (2001) Information, complexity and entropy: a new approach to theory and measurement methods, (http://arXiv.org/abs/math/0107067)

Berge, C. (1973) *Graphs and Hypergraphs*, North Holland P.C., Amsterdam/New York

Bergman, M., Moor, J. and Nelson, J. (1980) *The Logic Book*, Random House, New York

Bergstra, J.A., J. Heering and P. Klint (1990), "Module algebra", *J. of the assoc. for Computing Machinery* , v. 73, No. 2, pp. 335–372

Berkeley, G. (1901) *The Works of George Berkeley*, Clarendon Press, Oxford, UK

Berland, G.K., Elliot, M.N., Morales, L.S., Algazy, J.I., Kravitz, R.I., Broder, M.S., Kanouse, D.E., Munos, J.A., Puyol, J.-A., Lara, M., Watkins, K.E., Yang, H. and

McGlynn, E.A. (2001) Health information on the Internet: Accessibility, quality, and readability in English and Spanish, *JAMA*, v. 285, No. 20, pp. 2612–2621

Berlyant, A.M. (1986) *The Image of the Space: Map and Information*, Mysl, Moscow (in Russian)

Bernecker, S. and Dretske, F. (Eds.) (2000) *Knowledge: Readings in Contemporary Epistemology*, Oxford University Press, Oxford/New York

Bernknopf, R.L., Brookshire, D. S., McKee, M. and Soller, D. R. (1997) Estimating the Social Value of Geologic Map Information: A Regulatory Application, *Journal of Environmental Economics and Management*, v. 32, pp. 204–218.

Berthiaume, A., van Dam, W. and Laplante, S. (2001) Quantum Kolmogorov Complexity, *Journal of Computer and Systems Sciences*, v. 63, No. 2, pp. 201–221

Beth, T. *Quantum Information*, Springer Verlag, New York/Berlin/Heidelberg

Bialynicki-Birula, I. (2006) Formulation of the uncertainty relations in terms of the Rényi entropies, *Phys. Rev.*, A 74, No. 5, 052101.1– 052101.6

Bjorken, J.D. and Drell, S.D. (1964) *Relativistic Quantum Mechanics*, McGraw-Hill Book Company

Bjorken, J.D. and Drell, S.D. (1965) *Relativistic Quantum Fields*, McGraw-Hill Book Company

Blizard, W.D. (1991) The Development of Multiset Theory, *Modern Logic*, v. 1, No. 4, pp. 319–352

Bloch, A.S. (1975) Graph-Schemes and their Application, Vysheishaya Shkola, Minsk, Belaruss (in Russian)

Bloch, M. (1949) *Apologie pour l'Histoire ou Métier d'Historien*, Armand Colin, Paris

Blum, M. (1967) On the Size of Machines, *Information and Control*, v. 11, pp. 257–265

Blum M. (1967a) A Machine-independent Theory of Complexity of Recursive Functions, *Journal of the ACM*, v. 14, No 2, pp. 322–336

Blum, M. and S. Kannan, S. (1995) Designing programs that check their work, *Journal of the ACM*, v. 42, No. 1, pp. 269–291

Boisot, M.H. (1995) *Information Space: A Framework for Learning in Organizations, Institutions and Cultures*, Routledge, London

Boisot, M.H. (1998) *Knowledge Assets: Securing Competitive Advantage in the Information Economy*, Oxford University Press, Oxford

Boisot, M. (2002) Information, Space, and the Information-Space: A Conceptual Framework, (www.uoc.edu/in3/gnike/eng/docs/dp_02_boisot.doc)

Boisot, M. (2002) The Structuring and Sharing of Knowledge, in *Strategic Management of Intellectual Capital and Organization Knowledge* (Chun Wei Choo and N. Bontis, Eds) Oxford University Press, Oxford, pp.65–77

Boisot, M. and Canals, A. (2004) Data, Information, and Knowledge: Have we got it right, *Journal of Evolutionary Economics*, v. 14, pp. 43–67

Bojowald, M. and Skirzewski, A. (2008) Effective theory for the cosmological generation of structure *Advanced Science Letters*, v. 1, pp. 92–98

Bongard, M.M. (1963) On the concept "useful information", *Problems of Cybernetics*, v. 9 (in Russian)

Bongard, M.M. (1967) The Problem of Recognition, Nauka, Moscow (in Russian)

Bonnevie, E. (2001) Dretske's semantic information theory and meta-theories in library and information science, *Journal of Documentation*, v. 57, pp. 519–534

Borel, E. (1927) *Leçons sur la théorie des fonctions*, Gauthier-Villars, Paris

Borel, E. (1952) *Les nombre inaccessible*, Gauthier-Villiars, Paris

Borgmann, A. (1984) *Technology and the Character of contemporary life*, University of Chicago Press, Chicago/London

Borgmann, A. (1999) *Holding on to reality*: *The nature of information at the turn of the millenium*, University of Chicago Press, Chicago/London

Borgman, C.L. (2000) *From Gutenberg to the Global Information Infrastructure*: *Access to Information in the Networked World*, Cambridge University Press, Cambridge, MA

Borko, H. (1968) Information science: What is it? *Journal of the American Society for Information Science*, v. 19, pp. 3–5

Born, M. (1953) Physical Reality, *The Philosophical Quarterly*, v. 3, No. 11, pp. 139–149

Bosma, H. (1983) Information Quality rather then Information Quantity, in *Information Policy and Scientific Research*, Elsevier, Amsterdam, pp. 99–106

Bosmans, A. and Baumgartner, H. (2005) Goal-Relevant Emotional Information: When Extraneous Affect Leads to Persuasion and When It Does Not, *Journal of Consumer Research*, volume, v. 32, pp. 424–434

Bougnoux, D. (1993) *Sciences de l'information et de la communication*, Larousse, Paris

Bougnoux, D. (1995) *La communication contre l'information*, Hachette, Paris

Boulding, K.E. (1956) *The Image*, University of Michigan Press, Ann Arbor

Bourbaki, N. (1948) L'architecture des mathématiques, Legrands courants de la pensée mathématiques, *Cahiers Sud*, pp. 35–47

Bourbaki, N. (1957) *Structures*, Hermann, Paris

Bourbaki, N. (1960) *Theorie des Ensembles*, Hermann, Paris

Bower, G.H. and Hilgard, E. R. (1975) *Theories of Learning*, Prentice-Hall, Englewood Cliffs, NJ

Bowker, G. (1994) Information Mythology, in *Information Acumen*: *The Understanding and Use of Knowledge in Modern Business*, Routledge, 1994

Boyce, B.R., Meadow, C.T. and Kraft, D.H. (1994) *Measurement in Information Science*, Academic Press, San Diego

Bradford, D.F. and Kelejian, H.H. (1977) The Value of Information for Crop Forecasting in a Market System, *Bell Journal of Economics*, v. 9, pp. 123–144

Brading, K. and Landry, E. Scientific Sructuralism: Presentation and Representation, *Pholosophy of Science*, v. 73, pp. 571–581

Braman, S. (1989) Defining information: An approach for policymakers, *Telecommunications Policty*, v. 13, No. 1, pp. 233–242

Bratman, M.E. (1987) *Intentions, Plans, and Practical Reason*. Harvard University Press, Cambridge, MA,

Brekke, L. and Freund, P. (1993) *p*-adic numbers in physics, *Phys. Rep*, v. 231, pp. 1–66

Bremer, M.E. (2003) Do logical truths carry information, *Minds and Machines*, v. 13, pp. 567–575

Bremer, M.E. and Cohnitz, D. (2004) *Information and Information Flow: An Introduction*, Ontos Verlag, Frankfurt.

Brendal V. (1939) Linguistique structurale, *Acta linguistica*, v. 1, No. 1, pp. 2–10

Brier, S. (2004) Cybersemiotics and the Problem of the Information-Processing Paradigm as a Candidate for a Unified Science of Information Behind Library and Information Science, *Library Trends*, Vol. 52, No. 3, pp.629–657

Brillouin, L. (1956) *Science and Information Theory*, Academic Press Inc. Publishers, New York

Brooks, K. (Ed.) (1967) *The Communicative Arts and Sciences of Speech*, Merrill, Columbus, Ohio

Brooks, K. (1977) The developing cognitive viewpoint in information science, In *Proc. International Workshop on the Cognitive Viewpoint*, University of Ghent, Ghent, pp. 195–203

Brookes, B.C. (1980) The foundations of information science, pt. 1, Philosophical aspects, *Journal of Information Science*, v. 2, pp. 125–133

Brooks, D. (2007) The Outsourced Brain, *The New York Times*, October 26

Browne, S., Dongarra, J., Green, S., Moore, K., Pepin, T., Rowan, T. and Wade, R. (1995) Location-Independent Naming for Virtual Distributed Software Repositories, *ACM SIGSOFT Software Engineering Notes, Proceedings of the 1995 Symposium on Software reusability*, v. 20, pp. 179–185

Bruce, B. (2000) Credibility of the Web: Why we need dialectical reading, *Journal of Philosophy of Education*, v. 34, No. 1, pp. 97–109

Brudno, A.A. (1983) Entropy and the complexity of the trajectories of a dynamical system, *Trans. Moscow Math. Soc.*, v. 2, pp. 127–151

Brukner, C. and Zeilinger, A. (1999) Operationally Invariant Information in Quantum Mechanics, *Physical Review Letters*, v. 83, No. 17, pp. 3354

Brukner, C. and Zeilinger, A. (2001) Conceptual Inadequacy of the Shannon Information in Quantum Measurements, *Physical Review* A, v. 63, pp. 022113

Brush, S. G. (1967) Boltzmann's "Eta Theorem": Where's the Evidence? *American Journal of Physics*, v. 35, pp. 892

Bruza, P.D. (1993) *Stratified Information Disclosure: A Synthesis between Hypermedia and Information Retrieval*, Doctoral Dissertation, University of Nijmegen, Nijmegen, the Netherlands

Buckland, M. (1991) *Information and Information Systems*, Praeger, New York

Bunge, M. and Ardila, R. (1987) *Philosophy of psychology*, Springer-Verlag, New York

Burgin, M. (1970) Categories with involution and relations in γ-categories, *Transactions of the Moscow Mathematical Society*, v. 22, pp. 161–228 (translated from Russian)

Burgin, M. (1977) *Non-classical Models of Natural Numbers*, Russian Mathematical Surveys, v. 32, No. 6, pp. 209–210 (in Russian)

Burgin, M. (1980) Functional equivalence of operators and parallel computations, *Programming and Computer Software*, v. 6, No. 6, pp. 283–294

Burgin, M. (1982) Generalized Kolmogorov complexity and duality in theory of computations, *Notices of the Russian Academy of Sciences*, v. 25, No. 3, pp.19–23

Burgin, M. (1982a) Products of Operators in a Multidimensional Structured Model of Systems, *Mathematical Social Sciences*, No. 2, pp. 335–343

Burgin, M. (1983) Inductive Turing Machines, *Notices of the Academy of Sciences of the USSR*, v. 270, pp. 1289–1293 (translated from Russian, v. 27, No. 3)

Burgin, M. (1983a) On the Greg's paradox in taxonomy, *Abstracts presented to the American Mathematical Society*, v. 4, No. 3, p. 303

Burgin, M. (1984) Composition of operators in a multidimensional structured model of parallel computations and systems, *Cybernetics and System Analysis*, v. 19, No. 3, pp. 340–350

Burgin, M. (1989) Numbers as properties, *Abstracts presented to the American Mathematical Society*, v. 10, No. 1, p. 134

Burgin, M. (1990) Generalized Kolmogorov Complexity and other Dual Complexity Measures, *Cybernetics*, No. 4, pp. 21–29 (translated from Russian)

Burgin, M. (1990a) Abstract Theory of Properties and Sociological Scaling, in *Expert Evaluation in Sociological Studies*, Kiev, pp. 243–264

Burgin, M. (1990b) Theory of Named Sets as a Foundational Basis for Mathematics, in *Structures in Mathematical Theories*, San Sebastian, pp. 417–420

Burgin, M. (1991) What the Surrounding World is Build of, *Philosophical and Sociological Thought*, No. 8, pp. 54–67 (in Russian and Ukrainian)

Burgin, M. (1991a) Named Set Compositions in Categories, in *Problems of Group Theory and Homological Algebra*, Yaroslavl, pp. 39–52 (in Russian)

Burgin, M. (1991b) Logical Methods in Artificial Intelligent Systems, *Vestnik of the Computer Science Society*, No. 2, pp. 66–78 (in Russian)

Burgin, M. (1992) Universal limit Turing machines, *Notices of the Russian Academy of Sciences*, v. 325, pp. 654–658 (translated from Russian)

Burgin, M. (1992a) Complexity measures in the axiomatic theory of algorithms, in *Methods of design of applied intellectual program systems*, Kiev, pp. 60–67 (in Russian)

Burgin, M. (1992b) Algebraic Sructures of Multicardinal Numbers, in *Problems of group theory and homological algebra*, Yaroslavl University Press, Yaroslavl, pp. 3–20 (in Russian)

Burgin, M. (1992c) Reflexive Calculi and Logic of Expert Systems, in *Creative Processes Modeling by Means of Knowledge Bases,* Sofia, pp. 139–160 (in Russian)

Burgin, M. (1993) Information triads, *Philosophical and Sociological Thought*, No. 7/8, pp. 243–246 (in Russian and Ukrainian)

Burgin, M. (1993a) Reflexive Turing Machines and Calculi, *Vychislitelnyye Systemy* (*Logical Methods in Computer Science*), No. 148, pp. 94–116, 175–176 (in Russian)

Burgin, M. (1994) Evaluation of Scientific Activity in the Dynamic Theory of Information, *Science and Science of Science,* No. 1, pp. 124–131

Burgin, M. (1995) Algorithmic Approach in the Dynamic Theory of Information, *Notices of the Russian Academy of Sciences*, v. 342, No. 1, pp. 7–10

Burgin, M. (1995a) Information Measurement in Information Retrieval Systems, in *Library and Bibliographic Classifications and Information Retrieval Systems*, Kiev, pp. 25–29 (in Ukrainian)

Burgin M. (1995b) Named Sets as a Basic Tool in Epistemology, *Epistemologia*, v. XVIII, pp. 87–110

Burgin, M. (1995c) Logical Tools for Inconsistent Knowledge Systems, *Information*: *Theories & Applications*, v. 3, No. 10, pp. 13–19

Burgin, M. (1996) Information as a Natural and Technological Phenomenon, *Informatization and New Technologies*, No. 1, pp. 2–5 (in Russian)

Burgin, M. (1996a) What is the Cost of Information, *Information and Market*, No. 5–6, pp. 35–36 (in Ukrainian)

Burgin, M. (1996b) Information Needs and Resources, in *"Information Resources*: *Creation, Integration, and Utilization"*, Kiev, pp. 24–28 (in Russian)

Burgin, M. (1996c) Methodological Analysis of Informatization, *Informatization and New Technologies*, No. 3, pp. 48–53 (in Russian)

Burgin, M. (1996d) *The Structural Level of Nature*, Kiev, Znannya (in Russian)

Burgin, M. (1996e) Flow-charts in programming: arguments pro et contra, *Control Systems and Machines*, No. 4–5, pp. 19–29 (in Russian)

Burgin, M. (1997) *Fundamental Structures of Knowledge and Information*, Academy for Information Sciences, Kiev (in Russian)

Burgin, M. (1997a) The Extentional Model of Personality, in *"Ananyev Readings,"* S.-Petersburg, pp. 114–115 (in Russian)

Burgin, M. (1997b) Information Algebras, *Control Systems and Machines*, No. 6, pp. 5–16 (in Russian)

Burgin, M. (1997c) Time as a Factor of Science Development, *Science and Science of Science*, No. ½, 45–59

Burgin, M. (1997d) *Non-Diophantine Arithmetics or is it Possible that 2+2 is not Equal to 4?* Ukrainian Academy of Information Sciences, Kiev (in Russian, English summary)

Burgin, M. (1997e) Duality between Information and Energy, in *Creation, integration, utilization of information resources for innovative development*, Kiev, pp. 101–106 (in Russian)

Burgin, M. (1997f) Logical Varieties and Covarieties, in *Methodological and Theoretical Problems of Mathematics and Information and Computer Sciences*, Kiev, 1997, pp. 18–34 (in Russian)

Burgin, M. (1998) *On the Nature and Essence of Mathematics*, Academy of Information Sciences of Ukraine, Kiev (in Russian)

Burgin, M. (1998a) *Intellectual Components of Creativity*, International Academy "Man in Aerospace systems", Kiyv (in Ukrainian)

Burgin, M. (1998/1999) Information and Transformation, *Transformation*, No. 1, pp. 48–53 (in Polish)

Burgin, M. (1999) Super-recursive Algorithms as a Tool for High Performance Computing, *Proceedings of the High Performance Computing Symposium*, San Diego, pp. 224–228

Burgin, M. (2001) *Entertainment and Information*, UCMA, Los Angeles

Burgin, M. (2001a) Information in the Context of Education, *The Journal of Interdisciplinary Studies*, v. 14, pp. 155–166

Burgin, M. (2001b) *Diophantine and Non-Diophantine Aritmetics: Operations with Numbers in Science and Everyday Life*, Preprint Mathematics GM/0108149, (electronic edition: http://arXiv.org)

Burgin, M. (2002) Information, Organization, and System Functioning, in *Proceedings of the 6th World Multiconference on Systemics, Cybernetics and Informatics*, v. 2, Orlando, Florida, pp. 155–160

Burgin, M. (2002a) Knowledge and Data in Computer Systems, in Proceedings of the ISCA 17th International Conference *"Computers and their Applications"*, International Society for Computers and their Applications, San Francisco, California, pp. 307–310

Burgin, M. (2002b) Theory of Hypernumbers and Extrafunctions: Functional Spaces and Differentiation, *Discrete Dynamics in Nature and Society*, v. 7, No. 3, pp. 201–212

Burgin, M. (2003) Information Theory: A Multifaceted Model of Information, *Entropy*, v. 5, No. 2, pp. 146–160

Burgin, M. (2003a) Information: Problems, Paradoxes, and Solutions, *TripleC*, v. 1, No.1, pp. 53–70

Burgin, M. (2004) Data, Information, and Knowledge, *Information*, v. 7, No. 1, pp. 47–57

Burgin, M. (2004a) Algorithmic Complexity of Recursive and Inductive Algorithms, *Theoretical Computer Science*, v. 317, No. 1/3, pp. 31–60

Burgin, M. (2004b) *Named Set Theory Axiomatization*: T_{NZ} Theory, Elsevier, Preprint 0402024, (electronic edition: http://www.mathpreprints.com/math/Preprint/)

Burgin, M. (2004c) *Unified Foundations of Mathematics*, Preprint in Mathematics LO/0403186 (electronic edition: http://arXiv.org)

Burgin, M. (2004d) Logical Tools for Program Integration and Interoperability, in *Proceedings of the IASTED International Conference on Software Engineering and Applications*, MIT, Cambridge, pp. 743–748

Burgin, M. (2005) *Super-recursive Algorithms*, Springer, New York/Berlin/Heidelberg

Burgin, M. (2005a) Measuring Power of Algorithms, Programs, and Automata, in *Artificial Intelligence and Computer Science*, Nova Science Publishers, New York, pp. 1–61

Burgin, M. (2005b) Is Information Some Kind of Data? *Proceedings of the Third Conference on the Foundations of Information Science* (FIS 2005), Paris, France, July, pp. 1–31 (electronic edition: http://www.mdpi.net/fis2005/proceedings.html)

Burgin, M. (2005c) Grammars with Prohibition and Human-Computer Interaction, in *Proceedings of the Business and Industry Simulation Symposium*, Society for Modeling and Simulation International, San Diego, California, pp. 143–147

Burgin, M. (2005d) Recurrent Points of Fuzzy Dynamical Systems, *Journal of Dynamical Systems and Geometric Theories*, v. 3, No. 1, pp.1–14

Burgin, M. (2006) Operational and Program Schemas, in *Proceedings of the 15th International Conference on Software Engineering and Data Engineering* (SEDE-2006), ISCA, Los Angeles, California, pp. 74–78

Burgin, M. (2006a) Algorithmic Control in Concurrent Computations, in *Proceedings of the 2006 International Conference on Foundations of Computer Science*, CSREA Press, Las Vegas, pp. 17–23

Burgin, M. (2006b) Mathematical Schema Theory for Modeling in Business and Industry, *Proceedings of the 2006 Spring Simulation MultiConference* (SpringSim '06), Huntsville, Alabama, pp. 229–234

Burgin, M. (2006c) *Nonuniform Operations on Named Sets*, 5th Annual International Conference on Statistics, Mathematics and Related Fields, 2006 Conference Proceedings, Honolulu, Hawaii, January, pp. 245–271

Burgin, M. (2007) Universality, Reducibility, and Completeness, *Lecture Notes in Computer Science*, v. 4664, pp. 24–38

Burgin, M. (2007a) Algorithmic Complexity as a criterion of unsolvability, *Theoretical Computer Science*, v. 383, No. 2/3, pp. 244–259

Burgin, M. (2007b) *Elements of Non-Diophantine Arithmetics*, 6th Annual International Conference on Statistics, Mathematics and Related Fields, 2007 Conference Proceedings, Honolulu, Hawaii, pp. 190–203

Burgin, M. (2008) Structural Organization of Temporal Databases, in Proceedings of the 17th *International Conference on Software Engineering and Data Engineering* (SEDE-2008), ISCA, Los Angeles, California, pp. 68–73

Burgin, M. and Burgina, E. (1982) Information retrieval and multi-valued partitions in languages, *Cybernetics and System Analysis*, No. 1, pp. 30–42

Burgin, M. and Burgina, E.(1982a) Partitions in Languages and Parallel Computations, *Programming, Programming and Computer Software*, v. 8, No. 3, pp. 112–120

Burgin, M. and Chelovsky, Yu. A. (1984) Program Quality Estimation Based on Structural Characteristics, International Conference "*Software*", Kalinin, pp. 79–81 (in Russian)

Burgin, M. and Chelovsky, Yu. A. (1984a) Quality of the Automated Control System Software, in *"Perspectives of the Automated Control System Development"*, Kiev, pp. 58–66 (in Russian)

Burgin, M. and Debnath, N.C. (2003) Complexity of Algorithms and Software Metrics, in Proceedings of the ISCA 18th International Conference *"Computers and their Applications"*, International Society for Computers and their Applications, Honolulu, Hawaii, pp. 259–262

Burgin, M. and Debnath, N. (2003a) Hardship of Program Utilization and User-Friendly Software, in "Proceedings of the International Conference 'Computer Applications in Industry and Engineering' ", Las Vegas, Nevada, pp. 314–317

Burgin, M. and Debnath, N.C. (2006) Software Correctness, in Proceedings of the ISCA 21st International Conference *"Computers and their Applications"*, ISCA, Seattle, Washington, pp. 259–264

Burgin, M. and Debnath, N. (2007) Correctness In The Software Life Cycle, in *Proceedings of the 16th International Conference on Software Engineering and Data Engineering (SEDE*-2007*)*, ISCA, Las Vegas, Nevada, pp. 26–31

Burgin, M. and Eggert, P. (2004) Types of Software Systems and Structural Features of Programming and Simulation Languages, in *Proceedings of the Business and Industry Simulation Symposium*, Society for Modeling and Simulation International, Arlington, Virginia, pp. 177–181

Burgin, M. and Gladun, V. (1989) Mathematical Foundations of the Semantic Networks Theory, *Lecture Notes in Computer Science*, v. 364, pp. 117–135

Burgin, M. and Gorsky, D. (1991) Towards the construction of general theory of concept, in *"The Opened Curtain"*, Oulder/San Francisco /Oxford, pp. 167–195

Burgin, M. and Karasik, A. (1975) A Study of an Abstract Model of Computers, *Programming and Computer Software*, v. 1, No.1, pp. 72–82

Burgin, M. and Karasik, A. (1976) Operators of Multidimensional Structured Model of Parallel Computations, *Automation and Remote Control*, v. 37, No. 8, pp. 1295–1300

Burgin, M. and Karasik, A. (1976a) On a Construction of Matrix Operators, *Problems of Radio-Electronics*, No. 8, pp. 9–25 (in Russian)

Burgin, M. and Kavunenko, L. (1994) *Measurement and Evaluation in Science*, STEPS Center, Kiev (in Russian)

Burgin, M. and Kuznetsov, V. (1991) *Axiological Aspects of Scientific Theories*, Kiev, Naukova Dumka (in Russian)

Burgin, M. and Kuznetsov, V. (1992) Nomological Structures Modeling in Intelligent Systems, in *Intelligent Systems and Methodology*, Novosibirsk, pp. 116–133 (in Russian)

Burgin, M. and Kuznetsov, V. (1993) Properties in science and their modeling, *Quality & Quantity*, v. 27, pp. 371–382

Burgin, M. and Kuznetsov, V. (1994) *Introduction to Modern Exact Methodology of Science*, International Science Foundation, Moscow (in Russian)

Burgin, M., Liu, D. and Karplus, W. (2001) The Problem of Time Scales in Computer Visualization, in "*Computational Science*", *Lecture Notes in Computer Science*, v. 2074, part II, pp.728–737

Burgin, M.S. and Milov, Yu. (1998) *Grammatical Aspects of Language in the Context of the Existential Triad Concept*, in *On the Nature and Essence of Mathematics*, Ukrainian Academy of Information Sciences, Kiev, pp. 136–142

Burgin, M. and Milov, Yu. (1999) Existential Triad: A Structural Analysis of the Whole, *Totalogy*, v. 2/3, pp. 387–406 (in Russian)

Burgin, M.S. and Neishtadt, L.A. (1993) *Communication and discourse in teachers professional activity*, Daugavpils Pedagogical Institute, Daugavpils (in Russian)

Burgin, M. and Schumann, J. (2006) Three Levels of the Symbolosphere, *Semiotica*, v. 160, No. 1/4, pp. 185–202

Burgin, M. and Simon, I. (2001) *Information, Energy, and Evolution*, Preprint in Biology 2359, Cogprints (electronic edition: http://cogprints.ecs.soton.ac.uk)

Burgin, M. and Zellweger, P. (2005) *A Unified Approach to Data Representation*, in Proceedings of the 2005 International Conference on Foundations of Computer Science, CSREA Press, Las Vegas, pp. 3–9

Burton, D.M. (1997) *The History of Mathematics*, The McGrow Hill Co., New York,

Bush, S.F. and Todd, H. (2005) On the Effectiveness of Kolmogorov Complexity Estimation to Discriminate Semantic Types, LANL, Computer Science cs/0512089, (electronic edition: http://arXiv.org)

Bush, S.F. and Kulkarni, A.B. (2001) *Active Networks and Active Virtual Network Management Prediction: A Proactive Framework*, Kluwer Academic/Plenum Publishers

Cabanac, M., Cabanac, R.A. and Hammel, H.T. (1999) The Fifth Influence. In *Proceedings The 43rd meeting of the Internation Society for the Systems Sciences*, Asilomar, Ca., pp. 1–10

Cahill, L., Babinsky, R., Markowitsch, H.J. and McGaugh, J.L. (1995) The amygdala and emotional memory, *Nature*, v. 377, pp. 295–296

Cahill, R.T. (2002) Process Physics, *Nature*, v. 377, pp. 295–296

Cahill, R.T. (2002a) Process Physics: Inertia, Gravity and the Quantum, *General Relativity and Gravitation*, v. 34, pp. 1637–1656

Cahill, R.T. (2002b) *Process Physics: From Quantum Foam to General Relativity*, (electronic edition: http://xxx.lanl.gov/abs/gr-qc/0203015)

Calude, C.S. (1988) *Theories of Computational Complexity*, Annals of Discrete Mathematics 35, North-Holland, Amsterdam/New York/Oxford/Tokyo

Calude, C.S. (1991) Algorithmic complexity: A topological point of view, *Singularité*, v. 2, No. 10 pp. 28–29

Calude, C.S. (1996) The finite, the unbounded and the infinite, *J. UCS* 2, pp. 242–244

Calude, C.S. (1996a) Algorithmic information theory: Open problems, *J. UCS*, v. 2, pp. 439–441

Calude, C.S. (2002) *Information and Randomness: An Algorithmic Perspective*, (Texts in Theoretical Computer Science. An EATCS Series), Springer-Verlag, Berlin

Calude, C.S. (2004) Algorithmic randomness, quantum physics, and incompleteness, in M. Margenstern (ed.). *Proceedings of the Conference "Machines, Computations and Universality"* (MCU'2004), Lectures Notes in Comput. Sci., v. 3354, Springer, Berlin, pp. 1–17.

Campbell, D.G., Brundin, M., MacLean, G. and Baird, C. (2007) Everything old is new again: Finding a place for knowledge structures in a satisficing world, in *Proceedings North American Symposium on Knowledge Organization* 2007, v. 1, 21–30

Cantor, G. (1895) *Beiträge zur Begründung der transfiniten Mengenlehre*, Math. Ann., v. 46, pp. 481–512

Capurro, R. (1978) *Information. Ein Beitrag zur etymologischen und ideengeschichtlichen Begründung des Informationsbegriffs*, München

Capurro, R. (1991) Foundations of Information Science: Review and Perspectives, Proceedings of the *International Conference on Conceptions of Library and Information Science*, University of Tampere, Tampere, Finland, pp. 26–28

Capurro, R., Fleissner, P. and Hofkirchner, W. (1999) Is a Unified Theory of Information Feasible? In *The Quest for a unified theory of information*, Proceedings of the 2nd International Conference on the Foundations of Information Science, pp. 9–30

Capuro, R. and Hjorland, B. (2003) The Concept of Information, *Annual Review of Information Science and Technology*, v. 37, No. 8, pp. 343–411

Cardoso, A.I., Crespo, R.G. and Kokol, P. (2000) Two different views about software complexity, in *Escom 2000*, Munich, Germany, pp. 433–438

Carnap, R. (1950) *Logical Foundations for Probability*, University Chicago Press, Chicago

Carnap, R. (1952) *The Continuum of Inductive Methods*, University Chicago Press, Chicago

Carpenter, E. (1995) The New Languages, in *The New Languages: A Rhetorical Approach to the Mass Media and Popular Culture* (Ohlgren T.H. and Berk L.M., eds.), Prentice-Hall, Inc., Englewood Cliffs, pp. 5–13

Carr, N. (2008) Is Google Making us Stupid? *Atlantic Monthly*, July/August

Cartwright, N. (1983) *How the Laws of Physics Lie*, Clarendon Press, Oxford

Carlucci, L., Case, J. and Jain, S. (2007) Learning Correction Grammars, *Lecture Notes in Computer Science*, v. 4539, pp. 203–217

Castell, L., Drieschner, M. and von Weizsäcker, C.F. (Eds.) (1975–1986) *Quantum Theory and the Structures of Time and Space* (6 volumes), Hanser, Munich

Caws, P. (1988) *Structuralism: The Art of the Intelligible*, Humanities Press, Atlantic Highlands, NJ

Chaitin, G.J. (1966) On the Length of Programs for Computing Finite Binary Sequences, *J. Association for Computing Machinery*, v. 13, No. 4, pp. 547–569

Chaitin, G.J. (1975) A Theory of Program Size Formally Identical to Information Theory, *J. Association for Computing Machinery*, v. 22, pp. 329–340

Chaitin, G.J. (1976) Information theoretic characterizations of recursive infinite strings. *Theoretical Computer Science*, v. 2, pp. 45–48

Chaitin, G.J. (1977) Algorithmic information theory, *IBM Journal of Research and Development*, v. 21, No. 4, pp. 350–359

Chalmers, D. (1995) Facing Up to the Problem of Consciousness, *Journal of Consciousness Studies*, v. 2, No. 3, pp. 200–219

Chechkin, A.V. (1991) *Mathematical Informatics*, Nauka, Moscow (in Russian)

Chentsov, N.N. (1972) *Statistical Decision and Optimal Inference*, Nauka, Moscow (translated from Russian: American Math. Society, v. 53, 1982)

Chernavsky, D.S. (1990) *Synergetics and Information*, Znaniye, Moscow (in Russian)

Chisholm, R. (1989) *Theory of Knowledge*, Prentice Hall, Englewood Cliffs

Choo, C.W. (1998) *The Knowing Organization*, Oxford University Press, New York, NY

Choo, C.W., Detlor, B. and Turnbull, D. (2000) *Web Work: Information Seeking and Knowledge Work on the World Wide Web*, Kluwer Academic Publishers, Dordrecht

Church, A. (1956) *Introduction to Mathematical Logic*, Princeton University Press, Princeton

Clancey, W.J. (1997) *Situated Cognition: On Human Knowledge and Computer Representations*, Cambridge University Press, Cambridge, UK

Clarke, A. (1962) Hazards of Prophecy: The Failure of Imagination, in *Profiles of the Future*, Harper & Row

Clayton, A. (1982) Technology is not Enough, in "*Communications and the Future*", World Future Society, Bethesda, MD

Cleary, T. (1993) *No Barrier: Unloking the Zen Koan*, Harper Collins, New York/London

Cleveland, H. (1982) Information as a Resource, *The Futurist*, v. 16, No. 6, pp. 34–39

Cleveland, H. (1985) *The Knowledge Executive: Leadership in an Information Society*; Truman Talley.Books, New York

Clore, G.L. (1994) Why emotions are never unconscious, in *The nature of emotion*, Oxford University Press, Oxford, UK, pp. 285–290

Cockshott, P. and Michaelson, G. (2005) *Information, Work and Meaning*, (http://www.dcs.gla.ac.uk/~wpc/reports/infoworkmeaning.pdf)

Codd, E. (1990) Relational Model for Data Management, Addison-Wesley, Reading, MA

Colburn, T.R. (2000) *Philosophy and Computer Science*, M.E. Sharpe, Armonk, New York

Columbia Encyclopedia (2000) Sixth Edition, New York, Columbia University Press

Cohen, P.J. (1966) *Set Theory and the Contonuum Hypothesis*, Benjamin, New York

Cohen, P.R. and Levesque, H.J. (1990) Intention is Choice with Commitment, *Artificial Intelligence*, v. 42, No. 3, pp. 213 – 261

Cohn, P.M. (1965) *Universal Algebra*, Harper&Row, Publ., New York/Evanston/London

Collins, R. (1993) Emotional energy as the common denominator of rational action, *Rationality and Society*, v. 5 (http://rss.sagepub.com/cgi/content/abstract/5/2/203)

Connell, T.H. and Triple, J.E. (1999) Testing the accuracy of information on the World Wide Web using AltaVista search engine, *Reference and User Services Quaterly*, v. 38, No. 4, pp. 360–368

Cornford, F.M. (2003) *Plato's Theory of Knowledge: The Theaetetus and The Sophist*, Dover, New York

Cooley, M. (1987) *Architecture or Bee?* The Hogarth Press, London

Cooper, W.S. (1978) *Foundations of Logico-Linguistics*, D. Reidel P.C., Dordrecht, Holland/Boston

Cooper-Chen, A. (1994) *Games in the Global Vilage: A 50-Nation Study of Entertainment Television*, Bouling Green State University Popular Press, Bouling Green, OH

Cornelius, I. (2002) Theorizing information for information science, *Annual Review of Information Science and Technology*, v. 36, No. 1, pp. 392 – 425

Corry, L. (1996) *Modern Algebra and the Rise Mathematical Structures*, Birkhäuser, Basel/Boston/Berlin

Cory, G.A. (1999) *The reciprocal modular brain in economics and politics: shaping the rational and moral basis of organization, exchange, and choice*, Kluwer Academic/Plenum Publishers, New York

Cottrell, A., Cockshott, P. and Michaelson, G. (2007) *Information, Money and Value* (electronic edition: http://www.dcs.gla.ac.uk/~wpc/reports/info_book.pdf)

Cover, T.M. and Thomas, J.A. (1991) *Elements of Information Theory*, Wiley, New York

Crawford, S. and Bray-Crawford, K. (1995) Self-Determination in the Information Age, *The Internet Society 1995 International Networking Conference*, Honolulu, (electronic edition: http://www.hawaii-nation.org/sdinfoage.html#ref)

Crosby, S.A. and Wallach, D.S. (2003) *Denial of Service via Algorithmic Complexity Attacks*, Technical Report TR-03-416, Department of Computer Science, Rice University

Crutchfield, J.P. (1990) Information and its Metric, in *Nonlinear structures in Phyiscal Systems — Pattern Formation, Chaos and Waves*, Springer-Verlag, New York

Csanyi, V. (1989) *Evolutionary Systems and Society*, Duke University Press, Durham/London

Cunnigham, W. (2004) Objects, Patterns, Wiki and XP: All are systems of Names, OOPSLA, Vancuver, Canada, 2004 (http://www.oopsla.org/2004/)

Curras, E. (2002) *Towards a Theory of Information Science*, B.R. Pub., Delhi

Curtis, G. (1989) *Business information systems: Analysis, design and practice*, Addison-Wesley, Workingham

Cutting, J.C. and Ferreira, V.S. (1999) Semantic and phonological information flow in the production lexicon, *Journal of Exper. Psychol., Learning, Memory and Cognition*, v. 25, No. 2, pp. 318–44

Dahl, O.-J., Dijkstra, E.W. and Hoare, C.A.R. (1972) *Structured Programming*, Academic Press, London

Daley, R.P. (1973) Minimal-program complexity of sequences with restricted resources, *Information and Control*, v. 23, pp. 301–312

Dalkir, K. (2005) *Knowledge Management in Theory and Practice*, Elsevier, Amsterdam/London/New York

Damasio, A.R. (1994) *Descartes' error: emotion, reason, and the human brain*, Grosset/Putnam, New York

Damasio, A.R. (1999) *The feeling of what happens: Body and emotion in the making of consciousness*, Harcourt Brace & Company, New York, NY

Darwin, C. (1872) *The Expression of the Emotions in Man and Animals*, London, John Murray

Davenport, T. H. (1997) *Information Ecology*, Oxford University Press, New York

Davenport, T. H. and Prusak, L. (1998) *Working Knowledge: How organizations manage what they know*, Harvard Business School Press, Boston

Davenport, R.H., Eccles, R.G. and Prusak, L. (1996) Information politics, *Sloan Management Review*, pp. 53–65

Day, R.E. (2001) *The Modern Invention of Information*, Southern Illinois University Press

de Lavalette, Gerard R. Renardel (1992), Logical semantics of modularisation, in *CSL: 5th Workshop on Computer Science Logic*, Lecture Notes in Computer Science, v. 626, Springer, pp. 306–315

De Vey Mestdagh, C.N.J. (1998) Legal Expert Systems. Experts or Expedients? The Representation of Legal Knowledge in an Expert System for Environmental Permit Law, *The Law in the Information Society*, Conference Proceedings on CD-Rom, Firenze, 8 pp.

De Vey Mestdash, C.N.J., de Verwaard, V. and Hoepman, J. H. (1991) The logic of reasonable inference, in *Legal knowledge based systems: Model-based legal reasoning*, Koninlijke Vermande, Lelstad, pp. 60–76

DeArmond, S.J., Fusco, M.M. and Dewey, M. (1989) *Structure of the Human Brain: A Photographic Atlas*, Oxford University Press, New York

Debnath, N. and Burgin, M. (2003) *Software Metrics from the Algorithmic Perspective*, in Proceedings of the ISCA 18th International Conference *"Computers and their Applications"*, Honolulu, Hawaii, pp. 279–282

Debons, A. and Horne, E.E. (1997) NATO Advanced Study Institutes of Information Studies and Foundations, *J. Amer. Soc. Information Science*, v. 48, No. 9, pp. 794–803

Dedeke, A. (2000) A Conceptual Framework for Developing Quality Measures for Information Systems, *Proceedings of the 2000 Conference on Information Quality (IQ-2000)*, Cambridge, MA, pp.126–128

DeLone, W.H. and McLean, E.R. (1992) Information Systems Success: The Quest for the Dependent Variable, *Information Systems Research*, v. 3, No. 1, pp. 60–95

DeLone, W.H. and McLean, E.R. (2003) The DeLone and McLean model of information systems success: A ten-year update, *Journal of Management Information Systems*, v. 19, No. 4, pp. 9–30

Demri, S. and Orlowska, E. (1999) Informational representability of models for information logic, in *Logic at Work*, Physica-Verlag, Heidelberg, pp. 383–409

Dempster, A.P. (1967) Upper and Lower Probabilities Induced by Multivalued Mappings, *Ann. Math. Statist.*, v. 38, pp. 325–339

Demski, J.S. (1980) *Information Analysis*, Addison-Wesley, Massachussets

Denning, P.J. (2006) Infoglut, *Communications of the ACM*, v. 49, No. 7, pp. 15–19

Derr, R. (1985) The concept of information in ordinary discourse, *Inform. Process. Manage.* v. 21, No. 6, pp. 489–499

Derrida, J. (1982) *Positions*, University of Chicago Press, Chicago

Deutsch, D. (1985) Quantum theory, the Church-Turing principle, and the universal quantum Turing machine, *Proc. Roy. Soc.*, Ser. A, v. 400, pp. 97–117.

Deutsch, D. (1989) Quantum computational networks, *Proc. Roy. Soc.*, Ser. A, v. 425, pp. 73–90.

Devlin, K. (1991) *Logic and Information*, Cambridge University Press, Cambridge

Devlin, K. (2001) *Infosense*: *Turning Information into Knowledge*, W.H. Freeman and Co., New York

DeWeese, M.R. and Meister, M. (1999) How to measure the information gained from one symbol, *Network*: *Comput. Neural. Syst.*, v. 10, pp. 325–340

Dewey, T.G. (1996) The Algorithmic Complexity of a Protein, *Phys. Rev. E*, v. 54, pp. R39–R41

Dewey, T.G. (1997) Algorithmic Complexity and Thermodynamics of Sequence: Structure Relationships in Proteins, *Phys. Rev. E*, v. 56, pp. 4545–4552

Dijkstra, P., Prakken, H. and De Vey Mestdagh, C.N.J. (2007) *An implementation of norm-based agent negotiation.* Proceedings of the 11th International Conference on Artificial Intelligence and Law, Stanford, pp. 167–175

Dirac, P.A.M. (1931) Quantised Singularities in the Electromagnetic Field, *Proceedings of the Royal Society of London*, A113, pp. 60–70

Dodig-Crnkovic, G. (2005) System Modeling and Information Semantics, in *Proceedings of the Fifth Promote IT Conference*, Borlänge, Sweden

Dodig-Crnkovic, G. (2006) Semantics of Information and Interactive Computation, in *Investigations into Information Semantics and Ethics of Computing*, PhD Thesis, Mälardalen University Press

Doi, T. (1973) *The anatomy of dependence.* New York, NY: Kodansha International.

Dolan, R.J. (2002) Emotion, Cognition, and Behavior, *Science*, v. 298 (5596), pp. 1191–1194

Drathen, (1990) Grundlagen der Prozeßleittechnik, *Zuckerindustrie*, v. 115, No. 8, pp. 631–637

Dretske, F. I. (1981) *Knowledge and the Flow of Information*, Basil Blackwell, Oxford

Dretske, F. (1983) Précis of Knowledge and the Flow of Information, *Behavioral Brain Sciences*, v. 6, pp. 55–63

Dretske, F. (1988) *Explaining behavior*, MIT Press, Cambridge, MA

Dretske, F. (2000) *Perception, Knowledge and Belief: Selected Essays*, Cambridge University Press, Cambridge

Drikker, A. (2002) Deficiency of the Emotional Information in a Digital Format, *Cidoc Porto Alegre Conference*, International Council of Museums — ICOM, Brazil

Duff A.S. (2003) Four "e"pochs: the story of informatization, *Library Review*, v. 52, No. 2, pp. 58–64

Duncan, H.D. (1968) *Symbols in Society*, Oxford University Press, New York

Duncan, H.D. (1969) *Symbols and Social Theory*, Oxford University Press, New York

Dunford, N. and Schwartz, J.T. (1957) *Linear Operators*, Interscience Publishers, Inc., New York

Dussauchoy, R.L. (1982) Generalized Information theory and the Decomposability of Systems, *International Journal of General Systems*, v. 9, pp. 13–36

Duwell, A. (2008) Quantum information does exist, *Studies in History and Philosophy of Modern Physics*, v. 39, pp. 195–216

Dyer, R. (1992) *Only Entertainment*, Routledge, London/New York

Dzhunushaliev, V.D. (1998) Kolmogorov's algorithmic complexity and its probability interpretation in quantum gravity, *Classical and Quantum Gravity*, v. 15, pp. 603–612

Dzhunushaliev, V. and Singleton, D. (2001) *Algorithmic Complexity in Cosmology and Quantum Gravity*, Preprint in Physics gr-qc/0108038 (electronic edition: http://arXiv.org)

Easterbrook, J.A. (1959) The effect of emotion on cue utilization and the organization of behavior, *Psychol Rev.*, v. 66, pp. 183–201

Eaton, J.J. and Bawden, D. (1991) What kind of resource is information? *International Journal of Information Management*, v. 11, No. 2, pp. 156–165

Eddington, A.S. (1922) *The Theory of Relativity and its Influence on Scientific Thought*, Clarendon Press, Oxford

Edmonds, A. (1999) *Syntactic Measures of Complexity*, CPM Report No. 99–55, University of Manchester, Manchester, UK

Ekman, P., Friesen, W.V., O'Sullivan, M., Chan, A., Diacoyanni-Tarlatzis, I., Heider, K., Krause, R., LeCompte, W.A., Pitcairn, T. and Ricci-Bitti, P.E. (1987) Universals and Cultural Differences in the Judgments of Facial Expressions of Emotion, *Journal of Personality and Social Psychology*, v. 53, No. 4, pp. 712–717

Eliot, T.S. (1934) *The Rock*, Faber & Faber

Elmasri, R. and Navathe, S.B. (2000) *Fundamentals of Database Systems*, Addison-Wesley Publishing Company, Reading, Massachusetts

English, L.P. (1999) *Improving Data Warehouse and Business Information Quality*, Wiley & Sons, New York

Eppler, M.J. (2001) The concept of information quality: an interdisciplinary evaluation of recent information quality frameworks, *Studies in Communication Sciences*, v. 1 pp.167–182

Eppler, M. and Muenzenmayer, P. (2002) Measuring information quality in the web context: A survey of state-of-the-art instruments and an application methodology, Proceedings of the 7th *International Conference on Information Quality* (IQ-02), MIT, Cambridge, pp. 187–196

Eppler, M. and Wittig, D. (2000) Conceptualizing Information Quality: A Review of Information Quality Frameworks from the Last Ten years, *Proceedings of the 2000 Conference on Information Quality*, Boston, USA, pp. 83–96

Etzioni, O., Banko, M., Soderland, S. and Weld, D.S. (2008) Open Information Extraction from the Web, *Communications of the ACM*, v. 51, No. 12, pp. 68–74

Eves, H. (1983) *An Introduction to the History of Mathematics*, Saunders College Publishing, Philadelphia

Exploration in Education, http:// www.stsci.edu/exined/exin/; http:// www.merli-is.com

Fabrikant, S. (2000) The Geography of Semantic Information Spaces, Paper presented at *GIScience 2000*, Savannah, GA (http://www.cits.ucsb.edu/publication/the-geography-semantic-information-spaces)

Fabrikant, S.I. and Buttenfield, B.P. (2001) Formalizing Spaces for Information Access, *Annals of the Association of American Geographers*, v. 91, No.2, pp. 263–280

Farhi, E., Goldstone, J. Gutmann, S. and Sipser, M. (2000) *Quantum computation by adiabatic evolution*, E-print (http://arxiv.org/pdf/quant-ph/0001106)

Fairthorn, R.A. (1973) Information: One label, several bottles, in *Perspectives in Information Science*, pp. 65–73

Fallis, D. (2004) On Verifying the Accuracy of Information, *Library Trends*, v. 52, No. 3, pp. 463–487

Fano, R.M. (1952) *Lecture notes on statistical theory of information*, Massachusetts Institute of Technology, Cambridge, MA

Fano, R.M. (1961) *Transmission of information*: *A statistical theory of communications*, M.I.T. Press, Cambridge, MA

Farradane, J. (1979) The Nature of Information, *Journal of Information Science*, v. 1, pp. 13–17

Feinstein, A. (1954) A New basic theorem of information theory, *IEEE Transactions on Information Theory*, v. 4, No. 4, pp. 2–22

Feinstein, A. (1958) *Foundations of Information Theory*. McGraw-Hill, New York

Feller, W. (1950) *An Introduction to Probability Theory*, John Wiley and Sons, Inc., New York

Fetzer, J.H. (2004) Information, Misinformation, and Disinformation, *Minds and Machines*, v. 14, No. 2, pp. 223–229

Feynman, R.P. (1982) Simulating Physics with Computers, *Int. Journal of Theoretical Physics*, v. 21, pp. 467–488

Feynman, R.P. (1986) Quantum Mechanical Computers, *Foundations of Physics*, v. 16 (6), pp. 507–531

Feynman, R.P. (1986) Quantum Mechanical Computers, *Foundations of Physics*, v. 16 (6), pp. 507–531

Feynman, R.P. and Hibbs, A.R. (1965) *Quantum Mechanics and Path Integrals*, McGraw-Hill Companies

Fineman, S. (2004) Getting the Measure of Emotion — and the Cautionary Tale of Emotional Intelligence, *Human Relations*, v. 57, No. 6, pp. 719–740

Fisher, R.A. (1922) On the mathematical foundations of theoretical statistics. *Philos. Trans. Roy. Soc., London, Sec. A*, v. 222, pp. 309–368

Fisher, R.A. (1925) Theory of statistical estimation. *Proc. Cambridge Phil. Society*, v. 22, pp. 700–725

Fisher, R.A. (1950) *Contributions to Mathematical Statistics*, New York, Wiley

Fisher, E.L., Chengalur-Smith, S. and Wang, R.Y. (2006) *Introduction to Information Quality*, MITIQ Publication, Boston

Fisher, H.-D. and Melnik S. (1979) *Entertainment: A Cross-Cultural Examination*, Hastings House, New York

Fleissner, P. and Hofkirchner, W. (1995) Informatio revisited, Wider den dinglichen Informationsbegriff, *Informatik Forum*, v. 3, pp. 126–131

Flükiger, D.F. (1995) *Contributions towards a Unified Concept of Information*, Doctoral Thesis, University of Berne

Flükiger, D.F. (1999) Towards a unified concept of information: Presentation of a new approach, in *The quest for a unified theory of information. Proceedings of the Second International Conference on the Foundations of Information Science*, Gordon and Breach, Amsterdam, pp. 101–111

Fontana, D. (2001) *Discover Zen*, Chronicle Books, San Francisco

Fox, C. (1983) *Information and Misinformation: An Investigation of the Notions of Information, Misinformation, Informing, and Misinforming*, Greenwood Press, Westport, CT

Fraenkel, A.A. and Bar-Hillel, Y. (1958) *Foundations of Set Theory*, North Holland P.C., Amsterdam

Freedman, M.H. (2001) Quantum Computation and the Localization of Modular Functors, *Foundations Comput. Math.*, v. 1, No. 2, pp. 183–204

Freedman, M.H., Kitaev, A., Larsen, M.J. and Wang, Z. (2002) Topological Quantum Computation, *Bull. Amer. Math. Soc.*, v. 40, No. 1, pp. 31–38

Freedman, M.H., Larsen, M.J. and Wang, Z. (2002) A Modular Functor which is Universal for Quantum Computation, *Comm. Math. Phys.*, v. 227, No. 3, pp. 605–622

Freeman, A. and DeWolf, R. (1992) *The 10 Dumbest Mistakes Smart People Make and How to Avoid Them*, Harper Collins Publ., New York

Frege, G. (1891) *Funktion und Begriff*, Herman Pohle, Jena

Frege, G. (1892) Uber Begriff und Gegenstand, *Vierteljahrschrift fur wissenschaftliche Philosophie*, b. 16, pp. 192–205

Frege, G. (1892a) Uber Sinn und Bedeutung, in *Zeitschrift fur Philosophie und philosophische Kritik*, b. 100, pp. 25–50

Freud, S. (1949) *The Ego and the Id*, The Hogarth Press Ltd., London

Fricke, M. (2008) The Knowledge Pyramid: A Critique of the DIKW Hierarchy, Preprint (electronic edition: http://dlist.sir.arizona.edu/2327/)

Frieden, R.B. (1998) *Physics from Fisher Information*, Cambridge University Press, Cambridge

Friedman, J.P. (2000) *Dictionary of Business Terms*, Barron's Educational Series

Fuchs, C.A. (1995) *Distinguishability and Accessible Information in Quantum Theory*, PhD thesis, The University of New Mexico, Albuquerque, NM

Furner, J. (2004) Information studies without information, *Library Trends*, v. 52, No. 3, pp. 427–446

Furth, J. (1994) The Information Age in Charts, *Fortune International*, April

Fyffe, G. (1998) *There is no Absolute Measure of Information*, (electronic edition: http://www.geocities.com/CollegePark/9315/infocont.htm)

Gabbay, D.M., et al. (1994) Nonmonotonic Reasoning and Uncertain Reasoning, in *Handbook of Logic in Artificial Intelligence*, v. 3, Oxford University Press, Oxford

Gabbey, A. (1995) The Pandora's Box Model of the History of Philosophy, *Etudes maritainiennes* (*Maritain Studies*), No. 11, pp. 61–74

Gackowski, Z.J. (2004) What to Teach Business Students in MIS Courses about Data and Information, *Issues in Informing Science & Information Technology*; v. 1, pp. 845–867

Gács, P. (1974) On a Symmetry of Algorithmic Information, *Soviet Math. Dokl.*, v. 218, pp. 1265–1267

Gács, P. (1993) *Lecture Notes on Descriptional Complexity and Randomness*, Boston University, Boston

Gács, P. (2001) Quantum Algorithmic Entropy, (electronic edition: http://arXiv.org/quant-ph/0011046)

Gallager, R. G. (1968) *Information Theory and Reliable Communication*, Wiley, New York

Gammerman, A. and Vovk, V. (1999) Kolmogorov complexity: sources, theory and applications, *The Computer Journal*, v. 42, pp. 252–255

Gardner, H. (1975) *Frames of Mind: The Theory of Multiple Intelligencies*, Basic Books, New York

Gardner, H. (1975a) The Shattered Mind. New York: Knopf Vintage

Gardner, H. (2006) *Multiple Intelligences: New Horizons*, Basic Books, New York

Garfinkel, S. (2000) *Database Nation: The Death of Privacy in the 21st Century*, O'Reilly

Gatlin, L.L. (1972) *Information Theory and the Living System*, Columbia Univ. Press, New York

Gauld, D.B. (1974) Topological Properties of Manifolds, *The American Mathematical Monthly*, v. 81, No. 6, pp. 633–636

Gelfand, I.M., Kolmogorov, A.N. and Yaglom, A.M. (1956) On a General Definition of Information Quantity, *Notices of the Academy of Sciences of the USSR*, v. 111, No. 4, pp. 745–748

Gell-Mann, M. (1994) *The Quark and the Jaguar*, W.H. Freeman and Co., New York

Gell-Mann, M. (1995) Remarks on Simplicity and Complexity, *Complexity*, v. 1, pp. 16–19

Gell-Mann, M. and Lloyd, S. (1996) Information Measures: Effective Complexity and Total Information, *Complexity*, v. 2, No. 5, pp. 16–19

George, P. and McIllhagga, M. (2000) *The Communication of Meaningful Emotional Information for Children Interacting with Virtual Actors*. Affect and Interaction, Springer

Gernert, D. (1996) Pragmatic information as a unifying concept, *Information — New Questions to a Multidisciplinary Concept*, Akademie–Verlag, Berlin, pp. 147–162

Gershenfeld, N. (1992) Information in Dynamics, in *Workshop on Physics and Computation*, pp. 276–280

Gettier, E. (1963) Is Justified True Belief Knowledge?, Analysis, v. 23, pp. 121–123

Giachetti, R.I. (1999) A standard manufacturing information model to support design for manufacturing in virtual enterprises, *Journal of Intelligent Manufacturing*, v. 10, pp. 49–60

Gibilisco, P. and Isola, T. (2006) Uncertainty principle and quantum Fisher information, *Annals of the Institute of the Statistical Mathematics*, v. 59, No. 1, pp. 147–159

Gibbs, P.E. (1996) Principle of Event Symmetry, *International Journal of Theoretical Physics*, v. 35. 1037–1062

Gibbs, P.E. (1999) *Event-Symmetric Space-Time*, Weburbia Press

Giddens, A. (1984) *The Constitution of Society: Outline of the Theory of Structuration*, Polity Press, Cambridge

Gilligan, J. (1994) Patterns on glass: The language games of information, *Symposium on the Philosophical and Logical Aspects of Information Systems*, University of the West of England, Bristol, pp. 20–22

Ginsberg, Matthew L. (Ed.) (1987) Readings in Nonmonotonic Reasoning. Morgan Kaufmann Pub. Inc.

Giuliano, V.E. (1983) The United States of America in the Information Age, in *Information Policy and Scientific Research*, Elsevier, Amsterdam, pp. 59–76

Gödel, K. (1938) The consistency of the axiom of choice and of the generalized continuum hypothesis, Proc. National Acad. of Sciences of USA, v. 24, pp. 556–557

Godin, B. (2008) The Information Economy: the history of a concept through its measurement, 1949–2005, *History and Technology*, v. 24, No. 3, pp. 255–287

Goeleven, E., De Raedt, R., Baert, S. and Koster, E. Deficient inhibition of emotional information in depression, *Journal of Affective Disorders*, v. 93, No. 1–3, pp. 149–157

Goguen, J.A. (1997) Towards a Social, Ethical Theory of Information, In *Social Science Research, Technical Systems and Cooperative Work*, Erlbaum, pp. 27–56

Gofman, W. (1970) Information Science: Discipline or Disappearence? *Aslib Proc.*, v. 22, pp. 589–585

Gold, E.M. (1967) Language Identification in the Limit, *Information and Control*, v. 10, pp. 447–474

Goldblatt, R. (1979) *Topoi: The Categorical analysis of Logic*, North-Holland P.C., Amsterdam

Goldman, A.I. (1967) A Causal Theory of Knowledge, *The Journal of Philosophy*, v. 64, pp. 357–372

Goguen, J.A. (1997) Towards a Social, Ethical Theory of Information, in *Social Science Research, Technical Systems and Cooperative Work*, Erlbaum, pp. 27–56

Goleman, D. (1995) *Emotional Intelligence*, Bantam Books, New York/Toronto/London

Goleman, D. (1998) *Working with emotional intelligence*, Bantam Books, New York/Toronto/London

Good, I.J. (1952) Rational Decisions, *J. Royal Stat. Soc.*, Ser. B, v. 14, No. 1, pp. 107–115

Gould, J.P. (1974) Risk, stochastic preference and the value of information, *Journ. Econom. Theory*, v. 48, pp. 64–84

Grant, R.M. (1996) Toward a Knowledge-Based Theory of the Firm, *Strategic Management Journal*, v. 17, pp. 109–122

Griffith, B.C. (Ed.) (1980) *Key papers in information science*, Knowledge Industry Publications, New York

Grossmann, R. (1990) *The Fourth Way: A Theory of Knowledge*, Indiana University Press, Bloomington/Inianapolis

Grothendieck, A. (1957) Sur quelques points d'algèbre homologique, *Tôhoku Math Journal*, Second Series, v. 9, No. 2, 3, pp. 119–121

Gundry, J. (2001) Knowledge Management. (electronic edition: http://www.knowab.co.uk/kma.html)

Haenni, R., J. Kohlas and N. Lehmann (2000) Probabilistic argumentation systems, written at Dordrecht, in J. Kohlas and S. Moral, *Handbook of Defeasible Reasoning and Uncertainty Management Systems*, v. 5: Algorithms for Uncertainty and Defeasible Reasoning, Kluwer, pp. 221–287

Haken, H. (1988) *Information and Self-Organisation: a macroscopic approach to complex systems*, Springer-Verlag, Berlin

Halliday, M.A.K. (1967) Notes on Transitivity and Theme in English, *Journal of Linguistics*, v. 3, pp. 199–244

Halmos, P.R. (2000) An autobiography of polyadic algebras, *Logic Journal of the IGPL*, v. 8 No. 4, pp. 383–392

Halpern, J.Y. and Moses, Y. (1985) Towards a theory of knowledge and ignorance, in *Logics and Models of Concurrent Systems*, Springer-Verlag, New York, pp. 459–476

Halstead, M.H. (1977) *Elements of Software Science*, Elsevier, New York

Hammer, D. (1998) *Complexity inequalities*, Wissenschaft & Technik Verlag, Berlin

Hammer, D., Romashchenko, A.E., Shen, A. and Vereshchagin, N.K. (2000) Inequalities for Shannon entropies and Kolmogorov complexities, *Journal of Computer and Systems Sciences*, v. 60, pp. 442–464

Han, T.S. (2003) *Information-Spectrum Methods in Information Theory*, Springer-Verlag, Berlin

Hansen, S. (2002) Excessive Internet usage or 'Internet Addiction'? The implications of diagnostic categories for student users, *Journal of Computer Assisted Learning*, v. 18, No. 2, pp.232–236

Harding, M.E. (1973) *Psychic Energy: Its Source and Its Transformation*, Princeton University Press, Princeton

Harkevich, A. (1960) On the value of information, *Problems of Cybernetics*, v. 4, pp. 2–10 (in Russian)

Harmuth, H.F. (1992) *Information Theory Applied to Space-Time Physics*, World Scientific, Singapore

Harrah, D. (1967) *Communication: A Logical Model*, Cambridge University Press, Cambridge

Hartley, R.V. (1928) Transmission of information. *Bell Sys. Tech. Journal*, v. 7, pp. 335–363

Hartmanis, J. and Hopcroft, J.E. (1971) An Overview of the Theory of Computational Complexity, *J. Association for Computing Machinery*, v. 18, pp. 444–475

Hausdorff, F. (1927) *Mengenlehre*, Berlin and Leipzig, Walter de Gruyter

Hayakava, S.I. (1949) *Language in Thought and Action*, Hardcourt, Brace and Co., New York

Heckermann, D., Horvitz, E. and Middleton, B. (1993) An Approximate Nonmyoptic Computation for Value of Information, *IEEE Transactions on Pattern Analysis and Machine Intelligence*, v. 15, No. 3, pp. 292–298

Heidegger, M. and Fink, E. (1970) *Heraklit*, Klostermann, Frankfurt am Main, Germany

Heisenberg, W. (1958) The Representation of Nature in Contemporary Physics, *Daedalus*, v. 87, pp. 95–108

Henkin, L., Monk, J.D. and A. Tarski (1971) *Cylindric Algebras*, North-Holland, Amsterdam

Herbert, N. (1987) *Quantum Reality: Beyond the New Physics*, Anchor Books, New York

Hernon, P. (1995) Disinformation and misinformation through the Internet: Findings of an exploratory study, *Government Information Quaterly*, v. 12, No. 2, pp. 133–139

Herold, K. (Ed.) (2004) The Philosophy of Information, *Library Trends*, v. 52, No. 3, pp. 373–670

Herrlich, H. and Strecker, G.E. (1973) *Category Theory*, Allyn and Bacon Inc., Boston

Herrmann, N. (1988) *The Creative Brain*, Brain Books

Heyderhoff, P. and Hildebrand, T. (1973) *Informationsstrukturen, eine Einführung in die Informatik*, B. I.-Wissenschaftsverlag, Mannheim/Wien/Zürich

Hickman, J.L. (1980) A Note on the Concept of Multiset, *Bull. of the Australian Mathematical Society*, v. 22, pp. 211–217

Higashi, M. and Klir, G.J. (1983) Measures of uncertainty and information based onpossibility distributions, *Internat. Journ. General Systems*, v. 9, No. 1, pp. 43–58

Hilbert, D. (1899) *Die Grundlagen der Geometrie*, Berlin, Teubner

Hill, G. (2004) An information-theoretic model of customer information quality, *Proceedings of the Decision Support Systems Conference*, Prato, Italy, pp 359–371

Hilton, R.W. (1981) The Determinants of Information Value: Synthesizing Some General Results, *Management Science*, v. 27, No. 1, pp. 57–64

Hintikka, J. (1964) Distributive normal form and deductive interpolation, *Zeitschrift fur matematische Logic and Grundlagen der Matematik*, v. 10, pp. 185–191

Hintikka, J. (1968) The Varieties of Information and Scientific Explanation, in *Logic, Methodology and Philosophy of Science* (Rootselaar, B. van and Staal, J. F., Eds.), Amsterdam, v. III, pp. 311–331

Hintikka, J. (1970) Surface Information and Depth Information, in *Information and Inference*, Synthese Library, Humanities Press, New York, pp. 263–97

Hintikka, J. (1971) On Defining Information. *Ajatus*, v. 33, pp. 271–273.

Hintikka, J. (1973) Surface semantis: Definition and its motivation, in *Truth, Syntax, and Modality*, pp.127–147

Hintikka, J. (1973a) *Logic, Language-Games and Information*, Clarendon, Oxford.

Hintikka, J. (1984) Some varieties of information, *Inf. Process. Management*, v. 20, No. 1–2, pp. 175–181

Hintikka, J. (2005) *Knowledge and Belief*, Kings College Publications

Hintikka, J. and Suppes, P. (Eds) (1970) *Information and Inference*, Kluwer Academic Publishers Group, Dordrecht/Boston, Mass.

Hirshliefer, J. and Rilley, J.G. (1979) The analytics of uncertainty and information: An expository survey, *Journ. of Economic Literature*, v. 17, pp. 1375–1421

Hjalmars, S. (1977) Evidence for Boltzmann's *H* as a capital eta, *American Journal of Physics*, v. 45, pp. 214–215

Hjelmslev L. (1958) Dans quelle mesureles significations des mots peuvent'elles etre consideres corn me formant une structure? In *Proceedings of the 8th International Congress of Linguistics*, Oslo, pp. 636–654

Hjørland, B. (1998) Theory and metatheory of information science: A new interpretation. *Journal of Documentation*, v. 54, No. 4, pp. 606–621

Hjørland, B. (2002) Epistemology and the Socio-Cognitive Perspective in Information Science. *Journal of the American Society for Information Science and Technology*, v. 53, No. 4, pp. 257–270

Hobart, M.E. and Schiffman, Z.S. (2000) *Information ages: Literacy, numeracy, and the computer revolution*, The John Hopkins University Press, Baltimore, MD

Hodgson, G.M. and Knudsen, T. (2007) Information, Complexity, and Generative Replication, *Biol. Philosophy*, v. 43, No. 1, pp. 47–65

Hofkirchner, W. (1995) Information Science: An Idea Whose Time Has Come, *Informatik Forum*, v. 3, pp. 99–106

Hofkirchner, W. (Ed.) (1999) *The Quest for a Unified Theory of Information.* Proceedings of the Second International Conference on the Foundations of Information Science, Gordon and Breach Publ.

Hopcroft, J.E. Motwani, R. and Ullman, J.D. (2001) *Introduction to Automata Theory, Languages, and Computation*, Addison Wesley, Boston/San Francisco/New York

Hjørland, B. (1998) Theory and metatheory of information science: A new interpretation. *Journal of Documentation*, v. 54, No. 4, pp. 606–621

Horn, P. (2001) *Announcement of the Autonomous Computing Program*, www.research.ibm.com/autonomic/

Hornby, A.S. (1980) *Oxford Advanced Learner's Dictionary of Current English*, Oxford University Press, Oxford

Hromkovic, J. (1997) *Communication Complexity and Parallel Computing*, Springer, New York

Hu, W. and Feng, J. (2006) Data and Information Quality: An Information-theoretic Perspective, in Proceedings of the 2nd *International Conference on Information Management and Business* (IMB), Sydney, Australia, pp. 482–491

Hu, X. and Ye, Z. (2006) Generalized quantum entropy, *Journal of Mathematical Physics*, v. 47, No. 2, pp.023502–1 – 023502–7

Huang K., Lee Y. W. and Wang R. Y. (1999) *Quality Information and Knowledge*, Prentice Hall PTR, New Jersey

Husemoller, D. (1994) *Fibre Bundles*, Springer Verlag, Berlin/New York

Ingversen, P. (1992) *Information Retrieval Interaction*, Taylor Graham Publishing, London/Los Angeles

Irwin, A. (1995) Facing up to emotion, *Time Higher Education Supplement*, 21 July

Israel, D. and Perry, J. (1990) What is information? in *Information, Language and Cognition*, University of British Columbia Press, Vancouver, pp. 1–19

Jaffar, J. and M.J. Maher (1994), Constraint logic programming: A survey, *J. of Logic Programming*, v. 19/20, pp. 503–581

Jakobson, R. (1971) *Selected Writings, Word and Language*, v. II, Mouton de Gruyter, Berlin, Germany

Janich, P. (1992) *Grenzen der Naturwissenschaft*, Beck, Munich, Germany

Jeffreys, H. (1946) An invariant form for the prior probability in estimation problems, *Proc. Royal Society*, Ser. A, v. 186, pp. 453–461

Johnson, S.R. and Holt, M.T. (1986) The Value of Climate Information, in *Policy Aspects of Climate Forecasting.* Washington, DC, pp. 53–78

Joshi, H. and Bayrak, C. (2004) *Semantic Information Evolution, in* Proceedings of *Artificial Neural Networks in Engineering Conference 2002*, St. Louis, Missouri, pp. 519–525

Juedes, D.W. and Lutz, J.H. (1992) Kolmogorov Complexity, Complexity Cores and the Distribution of Hardness, in *Kolmogorov Complexity and Computational Complexity*, Springer-Verlag, Berlin/Heidelberg/New York

Jumarie, G.M. (1986) *Subjectivity, Information, Systems*, Gordon and Breach Science Publishers, New York/London/Paris

Jumarie, G. (1990) *Relative Information: Theories and Applications*, Springer Verlag, New York

Jung, C.G. (1969) *The Structure and Dynamics of the Psyche*, Princeton University Press, Princeton

Jung, C.G. (1969a) On Psychic Energy, in *On the Nature of the Psyche*, Princeton University Press, Princeton

Kagan, M.S. (1984) Art and Communication, in *Art and Communication*, Leningrad University Press, Leningrad, pp. 15–36 (in Russian)

Kahn, K., Strong, D. and Wang, R. (2002) Information Quality Benchmarks: Product and Service performance, *Communications of the ACM*, v. 45, No. 4, pp. 184–193

Kahre, J. (2002) *The Mathematical Theory of Information*, (Kluwer International Series in Engineering and Computer Science, v. 684) Kluwer Academic Publishers, Dordrecht

Kalfoglou, Y. and Schorlemmer, M. (2003) IF-Map: an ontology-mapping method based on information-flow theory, *Journal on Data Semantics*, v. 1, pp. 98–127

Kampis, G. (1988) Information, Computation and Complexity, in *Nature, Cognition and Systems* (Carvallo, M.E., Ed.), Kluer Academic Publishers, Dordrecht, pp. 313–320

Kampis, G. (1991) *Self-Modifying Systems in Biology and Cognitive Science*, Pergamon Press, Oxford

Kantor, F.W. (1977) *Information Mechanics*, John Wiley & Sons, NewYork

Kantor, F.W. (1982) An Informal Partial Overview of Information Mechanics, *Intenat. Journ. of Theoretical Physics*, v. 21, No. 6/7, pp. 525–535

Kary, M. (1990) Information theory and the*Treatise*: Towards a new understanding, in *Studies on Mario Bunge's Treatise*, Rodopi, Amsterdam, pp. 263–280

Katerattanakul, P. and Siau, K. (1999) Measuring information quality of web sites: Development of an instrument, *Proceedings of the 20th international conference on Information Systems*, Charlotte, North Carolina, USA, pp. 279–285

Katseff, H.P. and Sipser, M. (1981) Several results on program-size complexity, Theoretical Computer Science, v. 15, pp. 391–408

Kaye, D. (1995) The nature of information, Library Review, v. 44, No. 8, pp. 37–48

Keane, S.P. and Parrish, A.E. (1992) The role of affective information in the determination of intent, *Developmental psychology*, v. 28, No. 1, pp. 159–162

Kellogg, P. *Introduction: What is Information?* (electronic edition: http://www.patrickkellogg.com/school/papers/)

Kelly, J.R. (1956) New interpretation of information rate, *Bell Syst. Techn. Journ.*, v. 35, pp. 917–926

Keltner, J.W. (1970) *Interpersonal Speech-Communication*: *Elements and Structures*, Wadsworth P.C., Belmont, CA

Kent, R.E. (2000) The Information Flow Foundation for Conceptual Knowledge Organization, in *Proceedings of the Sixth International ISKO Conference on Advances in Knowledge Organization*, v. 7, pp. 111–117

Kerridge, D.F. (1961) Inaccuracy and inference, *Journ. Royal Statist. Society*, Ser. B, v. 23, pp. 184–194

Kessler, G.E. (1998) *Voices of Wisdom*, Wadsworth Publishing Company, Boston/New York/Paris

Ketelaar, E. (1997) Can We Trust Information? *The International Information & Library Review*, v. 29, No. 3–4, pp. 333–338

Ketelaar, T. and Clore, G.L. (1997) Emotions and Reason: The Proximate Effects and Ultimate Functions of Emotions, in *"Personality, Emotion, and Cognitive Science"*, Advances in Psychology Series, Amsterdam, Elsevier Science Publ., pp. 355–396

Ketelaar, T. and Goodie, A.S. (1998) The Satisficing Role of Emotions in Decision-making, *Psykhe*, v. 7, pp. 63–77

Khrennikov, A. (1997) *Non-Archimedean Analysis*: *Quantum Paradoxes, Dynamical Systems and Biological Models*, Kluwer, Boston/Dordrecht/London

Khrennikov, A. (1999) Classical and quantum mechanics on information spaces with applications to cognitive, psychological, social, and anomalous phenomena, Foundations of Physics, v. 29, No. 7, pp. 1065–1097

Kiernan, V. (April 28 1995) Gravitational constant is up in the air, *The New Scientist*, p. 18

Kieu, T.D. (2002) Quantum hypercomputation, *Minds and Machines*, v. 12, pp. 541–561.

Kieu, T.D. (2003) Computing the noncomputable, *Contemporary Physics*, v. 44, pp. 51–77.

Kirk, J. (1999) Information in organizations: Directions for information management, *Information Research*, v. 4, No. 3 (http://informationr.net/4-3/paper57.html)

Kleene, S.C. (2002) *Mathematical Logic*, Courier Dover Publications, New York

Klein, B.D. (2002) When do users detect information quality problems on the World Wide Web? *American Conference on Information Systems*, Dallas, TX, pp. 1101–1103

Kline M. (1967) *Mathematics for Nonmathematicians*, Dover Publications, New York

Kline, M. (1980) *Mathematics*: *The Loss of Certainty*, Oxford University Press, New York

Klir, G.J. (1991) Generalized Information Theory, *Fuzzy Sets and Systems*, v. 40, pp. 127–142

Klir, G.J. and Folger, T.A. (1988) *Fuzzy Sets, Uncertainty and Information*, Prentice Hall, New Jersey

Klir, G.J. and Mariano, M. (1987) On the uniqueness of possibilistic measure of uncertainty and information, *Fuzzy Sets and Systems*, v. 24, No. 2, pp. 197–219

Klir, G.J. and Wang, Z. (1993) *Fuzzy Measure Theory*, Kluwer Academic Publishers, Boston/Dordrecht/London

Knapska, E., Nikolaev, E., Boguszewski, P., Walasek, G., Blaszczyk, J., Kaczmarek, L. and Werka, T. (2006) Between-subject transfer of emotional information evokes specific pattern of amygdala activation, *Proc. National Academy of Sciences*, v. 103, No. 10, pp. 3858–3862

Knight, S.-A. and Burn, J. (2005) Developing a framework for assessing information quality on the World Wide Web, *Informing Science Journal*, v. 8, pp. 159–172

Knuth, D. (1997) *The Art of Computer Programming*, v.2: *Seminumerical Algorithms*, Addison-Wesley, Reading, MA

Kogut, B. and Zander, U. (1992) Knowledge of the Firm: Combinative Capabilities, and the Replication of Technology, *Organization Science*, v. 3, No. 3, pp. 383–397

Kohlas, J. (2003) *Information Algebras: Generic Structures for Inference*, Springer-Verlag, New York

Kohlas, J. and P.P. Shenoy (2000) Computation in valuation algebras, in J. Kohlas and S. Moral, *Handbook of Defeasible Reasoning and Uncertainty Management Systems,* v. 5: *Algorithms for Uncertainty and Defeasible Reasoning*, Kluwer, Boston/Dordrecht/London, pp. 5–39

Kohlas, J. and Stärk, R.F. (2007) Information Algebras and Consequence Operators, *Logica Universalis*, v. 1, No. 1, pp. 139–165

Kolb, B. and Whishaw, I.Q. (1990) *Fundamentals of Human Neuropsychology*, W.H. Freeman and Co., New York

Kolmogorov, A.N. (1933) *Grundbegriffe der Wahrscheinlichkeitrechnung*, Ergebnisse Der Mathematik (English translation: (1950) *Foundations of the Theory of Probability*, Chelsea P.C.)

Kolmogorov, A.N. (1953) On the Concept of Algorithm, *Russian Mathematical Surveys*, v. 8, No. 4, pp. 175–176

Kolmogorov, A.N. (1957) Theory of Information Transmission, in *Session of the National Academy of Sciences of the USSR on problems of automation*, AN USSR, Moscow, pp. 66–99

Kolmogorov, A.N. (1965) Three approaches to the definition of the quantity of information, *Problems of Information Transmission*, No. 1, pp. 3–11

Kolmogorov, A.N. (1968) Logical basis for information theory and probability theory, *IEEE Trans. Inform. Theory,* v. IT-14, pp. 662–664

Kononenko, I. and Bratko, I. (1991) Information-Based Evaluation Criterion for Classifier's Performance, *Machine Learning*, v. 6, pp. 67–80

Kornwachs, K. (1996) Pragmatic Information and System Surface, in *Information. New questions to a multidisciplinary concept*, Akademie Verlag, Berlin, Germany, pp. 163–185

Kornwachs, Klaus (1998) Pragmatic Information and the Emergence of Meaning, in Van de Vijver, G./Salthe, S./Delpos, M. (Eds.) *Evolutionary Systems*, Kluwer, Boston/Dordrecht/London, pp. 181–196

Kraus, S., Lehmann, D. and M. Magidor, (1990) Nonmonotonic reasoning, preferential models and cumulative logics, *Artificial Intelligence*, v. 44, pp. I67–207

Kraut, R., M. Patterson, V. Lundmark, Kiesler, S., Mukopadhyay, T. and Scherlis, W. (1998) Internet Paradox: A Social Technology That Reduces Social Involvement and Psychological Well-Being? *American Psychologist*, v. 53, pp. 1017–1031.

Kreinovich, V. (1999) Coincidences Are Not Accidental: A Theorem, *Cybernetics and Systems: An International Journal*, v. 30, No. 5, pp. 429–440

Kreinovich, V. and Kunin, I. (2004) A. *Application of Kolmogorov Complexity to Advanced Problems in Mechanics*, University of Texas at El Paso, Computer Science Department Reports, UTEP-CS-04-14

Kshirsagar, S. and Magnenat-Thalmann, N. (2002) A multilayer personality model, Proceedings of the 2nd international symposium on Smart graphics, ACM Press pp. 107–115

Kullback, S. (1959) *Information Theory and Statistics*, John Wiley and Sons, Inc. New York

Kullback, S. and Leibler, R.A. (1951) On Information and Sufficiency, *Ann. Math. Statistics*, v. 22, pp. 79–86

Kuratowski, K. (1966) *Topology*, v. 1, Academic Press, Waszawa

Kuratowski, K. (1968) *Topology*, v. 2, Academic Press, Waszawa

Kuratowski, K. and Mostowski, A. (1967) *Set Theory*, North Holland P.C., Amsterdam

Kurosh, A.G. (1963) *Lectures on general algebra*, Chelsea P. C., New York

Kushilevitz, E. and Nisan, N. (1997) *Communication Complexity*, Cambridge University Press, Cambridge

Lacan J. (1966) *Ecrits*, Paris

Lacan, J. (1977) *The Four Fundamental Concepts of Psychoanalysis*, Hogarth, London

Laffont, J. J. (1989) The economics of uncertainty and information (J. P. Bonin and H. Bonin, Transl.), MIT Press, Cambridge, MA

Lakatos, I. (1976) *Proofs and Refutations*, Cambridge University Press, Cambridge

Land, F., Land, N., Sevasti M. and Amjad, U. (2007) Knowledge management: the darker side of KM. *Ethicomp journal*, v. 3, No. 1 (http://www.ccsr.cse.dmu.ac.uk/journal/do_previous.php?prev=View+Papers&id=5)

Landauer, C. (1998) Data, Information, Knowledge, Understanding: Computing Up the Meaning Hierarchy, in *Proceedings of the 1998 IEEE International Conference on Systems, Man, and Cybernetics* (SMC'98), San Diego, California, pp. 2255–2260

Landauer, R. (2002) Information is Inevitably Physical, in *Feynmann and Computation: Exloring the limits of computers*, Westview Press, Oxford, pp. 76–92

Landry, E. (1999) Category Theory as a Framework for Mathematical Structuralism, *The 1998 Annual Proceedings of the Canadian Society for the History and Philosophy of Mathematics*, p. 133–142

Langefors, B. (1963) Some approaches to the theory of information systems, *BIT*, v. 3, pp. 229–254

Langefors B. (1966) *Theoretical Analysis of Information Systems*, Studentlitteratur, Lund, Sweden

Langefors, B. (1974) Information systems, in *Proc. IFIP Congress* 74, North-Holland, Amsterdam, pp. 937–945

Langefors, B. (1977) Information systems theory, *Information systems*, v. 2, pp. 207–219

Langefors, B. (1980) Infological models and information user views, *Information systems*, v. 5, pp. 17–32

Lancaster, F.W. (1968) *Information Retrieval Systems*, John Wiley & Sons, New York/London/Sydney

Laudisa, F. and Rovelli, C. Relational Quantum Mechanics, *Stanford Encyclopedia of Philosophy* (electronic edition: http://plato.stanford.edu/entries/qm-relational/)

Laudon, K.C. (1996) *Information Technology and Society*, Wadsworth P.C., Belmont, California

Lauritzen, S.L. and Spiegelhalter, D.J. (1988) Local computations with probabilities on graphical structures and their application to expert systems, *J. Royal Statis. Soc.*, B 50, pp. 157–224

Lawrence, D. (1999) *Economic Value of Information*, Springer-Verlag, New York

Lawson, H.W. (2002) Rebirth of Computer Industry, *Communications of the ACM*, v. 45, pp. 25–29

Lazarus, R.S. (1991) *Emotion and Adaptation*, Oxford University Press, New York

Lazarus, R.S. and Lazarus, B.N. (1994) *Passion & Reason: Making Sense of Our Emotions*, Oxford University Press, New York

LeDoux, J.E. (1986) The neurobiology of emotion, in *Mind and Brain: dialogues in cognitive neuroscience*, Cambridge Review, New York, pp. 301 – 327

LeDoux, J.E. (1989) Cognitive-emotional interactions in the brain, *Cognition and Emotion*, v. 3, pp. 267–289

LeDoux J.E. (1996) *The Emotional Brain*, Simon and Schuster, New York

LeDoux, J.E. (2000) Emotion circuits in the brain, *Annual Review of Neuroscience*, v. 23, pp. 155–184

Lee, P.M. (1964) On the Axioms of Information Theory, *Ann. Math. Statist.*, v. 35, No. 1, pp. 415–418

Lee, J.M. (2000) *Introduction to Topological Manifolds*, Graduate Texts in Mathematics 202, Springer, New York

Lee, Y.W., Pipino, L., Funk, J.D. and Wang, R. Y. (2006) *Journey to Data Quality*, MIT Press

Lenski, W. (2004) Remarks on a Publication-Based Concept of Information, in *New Developments in Electronic Publishing AMS/SMM Special Session*, ECM4 Satellite Conference, Stockholm, pp. 119–135

Leonard, H.S. and N. Goodman, (1940) *The Calculus of Individuals and Its Uses*, Journal of Symbolic Logic, v. 5, pp. 45–55

Lerner, V. (1993) Macroeconomics Analysis of Information Dynamic Model, *Cybernetics and Systems*, v. 24, No. 6, pp. 591–633

Lerner, V.S. (1996) The Information model of mutual learning in Artificial Intelligence, in Proceedings the 6th IEEE International Conference on Electronics, Communications and Computers, Pueblo, pp. 350–355

Lerner, V.S. (1997) Information Macrodynamic Approach for Modeling in Biology and Medicine, *Journal of Biological Systems*, v. 5, No. 2, pp. 215–264

Lerner, V.S. (1998) Informational Systemic Approach to Education Process, in Proceedings of the 26th International Conference of the International Council for Innovation in Higher Education, Costa Mesa, pp. 54–55

Lerner V.S. (1999) *Information Systems Analysis and Modeling: An Informational Macrodynamics Approach*, Kluwer Academic Publishers, Boston/Dordrecht/London

Lerner V.S. (2003) *Variation Principle in Informational Macrodynamics*, Kluwer Academic Publishers, Boston/Dordrecht/London

Lerner, V.S. (2006) Macrodynamic cooperative complexity of biosystems, *Journal of Biological Systems*, v. 14, No. 1, pp.131–68

Lerner, V.S. (2006a) About the biophysical conditions initiating the cooperative complexity, letter to the editor, *Journal of Biological Systems*, v. 14, No. 2, pp. 315–322

Lerner, V.S. (2006b) An elementary information macrodynamic model of market economic system, *J. Information Sciences*, v. 176, No. 23, pp. 3556–3590

Lerner, V.S. (2006c) Macrodynamic Cooperative Complexity of Biosystems, *Journal of Biological Systems*, v. 14, No. 1, pp. 1–35

Lerner, V.S. (2007) Information Systems Theory and Informational Macrodynamics: Review of the Main Results, *IEEE Transactions on Systems, Man, and Cybernetics*, v. 37, No. 6, pp. 1–17

Lerner, V.S. and Talyanker, M.I. (2000) Informational Geometry for Biology and Environmental Applications, in Proceedings of the *International Conference on Environmental Modeling and Simulation*, San Diego, pp. 79–84

Leśniewski, S. (1916) *Podstawy ogólnej teoryi mnogosci. I*, Prace Polskiego Kola Naukowego w Moskwie, Sekcya matematyczno-przyrodnicza (Eng. trans. by D.I. Barnett: 'Foundations of the General Theory of Manifolds I', in S. Leśniewski, *Collected Works*, ed. S.J. Surma, J. Srzednicki, D.I. Barnett, and F.V. Rickey, Kluwer, Dordrecht, v. 1, 1992, pp. 129–173)

Levin, L.A. (1973) On the notion of a random sequence, *Soviet Math. Dokl.*, v. 14, pp. 1413–1416 (translated from Russian)

Levin, L.A. (1974) Laws of information (nongrowth) and aspects of the foundation of probability theory, *Problems of Information Transmission*, v. 10, pp. 206–210

Lévi-Strauss, C. (1963) *Structural Anthropology*, Basic Books, New York

Lévi-Strauss, C. (1987) *Anthropology and Myth: Lectures, 1951–1982*, Blackwell, Oxford

Lewis, G.N. (1916) The atom and the molecule, *Journal Amer. Chem. Soc.*, v. 38, pp. 762–785

Lewis, J.P. (2001) Limits to Software Estimation, *Software Engineering Notes*, v. 26, pp. 54–59

Lewis, M.A. and Newton, J.T. (2006) An evaluation of the quality of commercially produced patient information leaflets, *British Dental Journal*, v. 201, pp. 114–117

Lewis, P. (1991) The Decision Making Basis for Information Systems: The contribution of Vicker's concept of appreciation to a soft system perspective, *J. Inf Systems*, v. 1, No. 1, pp. 33–43

Li, M. and Vitanyi, P. (1997) *An Introduction to Kolmogorov Complexity and its Applications,* Springer-Verlag, New York

Li Weihua and Li Shixian, (2005) Using information-flow theory to support information interoperability in the grid, in *The 3rd International Conference on Information Technology and Applications* (ICITA 2005), v. 1, pp. 272–275

Liebenau, J. and Backhouse, J. (1990) *Understanding Information: An Introduction,* Macmillan, London

Lindley, D.V. (1956) On the measure of information provided by an experiment, *The Annals of Mathematical Statistics*, v. 27, No. 4, pp. 986–1005

Lindsay, R.B. (1971) *Basic Concepts of Physics*, Van Nostrand Reinhold Co., NewYork/Toronto/London

Loewenstein, W.R. (1999) *The Touchstone of Life*: *Molecular Information, Cell Communication, and the Foundation of Life*, Oxford University Press, Oxford/New York

Loewenstein, G.F., Weber, E.U., Hsee, C.K. and Welch, E.S. (2001) Risk as feelings, *Psychological Bulletin*, v. 127, No. 2, pp. 267–286

Lombardi, O. (2004) What is Information? *Foundations of Science*, v. 9, No. 2, pp. 105–134(30)

Losee, R.M. (1997) A Discipline Independent Definition of Information, *J. of the American Society for Information Science*, v. 48, No. 3, pp. 254–269

Loveland, A.W. (1969) A variant of the Kolmogorov concept of complexity, *Information and Control*, v. 15, pp. 510–526

Lucadou, W. V. (1987) The Model of Pragmatic Information, *Proceedings of the 30th Parapsychological Association Convention*, pp. 236–254

Luck, S. J., Vogel, E. K. and Shapiro, K. L. (1996) Word meanings can be accessed but not reported during the attentional blink, *Nature*, 383, pp. 616–618

Lyon, D. (1988) *The Information Society*: *Issues and Illusions*, Polity, Cambridge

Lyons, J. (1972) *Introduction to Theoretical Linguistics*, Cambridge University Press, Cambridge

Lyre, H. (1997) Time and Information, in *Time, Temporality, Now: Experiencing Time and Concepts of Time in an Interdisciplinary Perspective*, Springer, Berlin.

Lyre, H. (1998) *Quantentheorie der Information*, Springer, Vienna

Lyre, H. (2002) *Quanten der Informationtheorie*: *Eine philosophisch-naturwissenschaftliche Einfürung*, Fink, Munich

Lyre, H. (2003) C.F. von Weizsäcker's reconstruction of physics: Yesterday, Today, Tomorrow, in *Time, Quantum and Information*, Springer, Berlin

Macauley, M.K. (1997) Some Dimensions of the Value of Weather Information: General Principles and a Taxonomy of Empirical Approaches, in *The Social and Economic Impacts of Weather*, National Center for Atmospheric Research, Boulder, CO

Macgregor, G. (2005) The nature of information in the twenty-first century: Conundrums for the informatics community?, *Library Review*, v. 54, No. 1, pp. 10–23

Machlup, F. and Mansfield, U. (Eds.) (1983) *The Study of Information: Interdisciplinary Messages*, Wiley, New York

Machlup, F. (1983) Semantic quirks in studies of information, in *The study of information: Interdisciplinary Messages*, John Wiley & Sons, New York/London/Sydney, pp. 641–671

MacKay, D.M. (1955) Complementary measures of scientific information-content. *Methodos*, v. 7, pp. 63–89

MacKay, D. (1956) The place of 'meaning' in the theory of information. In Cherry, C. (ed.), *Information Theory: Third London Symposium*, Butterworths, London, pp. 215–225

MacKay, D. (1961) The informational analysis of questions and commands, in *Information Theory: Fourth London Symposium*, Butterworths, London, pp. 469–476

MacKay, D.M. (1969) *Information, Mechanism and Meaning*, The MIT Press, Cambridge, Massachusetts

MacLean, P.D. (1973) *A Triune Concept of the Brain and Behavior*, University of Toronto Press, Toronto

MacLean, P.D. (1982) On the Origin and Progressive Evolution of the Triune Brain, in *Primate Brain Evolution*, Plenum Press, New York

MacQueen, J. and Marschak, J. (1975) Partial knowledge, entropy, and estimation, *Proc. Nat. Acad. Sci. U.S.A.*, v. 72, No. 10, pp. 3819–3824

Magnée, M. J.C.M., de Gelder, B., van Engeland, H. and Kemner, C. Facial electromyographic responses to emotional information from faces and voices in individuals with pervasive developmental disorder, *Journal of Child Psychology and Psychiatry*, v. 48, No. 11, pp. 1122–1130

Malagò, L., Matteucci, M. and Bernardo Dal Seno, B. (2008) An information geometry perspective on estimation of distribution algorithms: boundary analysis, in Proceedings of the *2008 GECCO conference companion on Genetic and evolutionary computation*, Atlanta, GA, pp. 2081–2088

Malcev, A.I. (1970) *Algebraic systems*, Nauka, Moscow (in Russian)

Malik, G. (2004) An Extension of the Theory of Information Flow to Semiconcept and Protoconcept Graphs, in *ICCS 2004*, pp. 213–226

Malinowski, K. (2008) *On the Information Society and the danger of digital divide*, Polish Academy of Sciences (http://www.pte.pl/pliki/2/12/Infor-Society-present-12-06-2008.pdf)

Manin, Yu I. (1991) *Course in Mathematical Logic*, Springer-Verlag, New York

Mansilla, R. (2001) Algorithmic Complexity in Real Financial Markets (electronic edition: http://arXiv.org/cond-mat/0104472)

Manten, A.A. and Timman, T. (Eds), (1983) *Information Policy and Scientific Research*, Elsevier, Amsterdam

Marchand, D.A. and Forest W.H., Jr. (1986) *Infotrends: Profiting from your Information Resources*; John Wiley & Sons, New York

Markov, A.A. (1964) On Normal Algorithms that Compute Boolean Functions, *Notices of the Academy of Sciences of the USSR*, v. 157, No. 2, pp. 262–264 (translated from Russian)

Markov, K., Ivanova, K. and Mitov, I. (2003) Basic structure of the general information theory, *Information Theories and Applications*, v. 14, pp. 5–19

Marr, D. (1982) *Vision*, Freeman, San Francisco

Marschak, J. (1954) Towards an economic theory of organization and information, in *Decision Processes*, New York, pp. 187–220

Marschak, J. (1959) *Remarks on the economics of information*, Cowles Foundation Discussion Papers, No 70, Yale University

Marschak, J. (1964) Problems in information economics, in *Economic Information, Decision, and Prediction: Selected Essays*, v. 2, pp. 63–76

Marschak, J. (1971) Economics of information systems, *Journ. Amer. Statist. Assoc.*, v. 66, pp. 192–219

Marschak, J. (1972) Optimal systems for information and decision, in *Techniques of optimization* (Fourth IFIP Colloq. Optimization Techniques, Los Angeles, Calif., 1971), Academic Press, New York, pp. 355–370

Marschak, J. (1974) Value and cost of information systems, in *Production theory* (Proc. Internat. Sem., Univ. Karlsruhe, Karlsruhe, 1973), Lecture Notes in Economics and Math. Systems, v. 99, Springer, Berlin, pp. 335–358

Marschak, J. (1976) Guided soul-searching for multi-criterion decisions. *Multiple criteria decision making (Papers, 22nd Internat. Meeting Inst. Management Sci. (TIMS), Kyoto, 1975)*, Lecture Notes in Econom. and Math. Systems, v. 123, Springer, Berlin, pp. 1–16

Marschak, J. (1980) Economic information, decision, and prediction, Sellected essays, v. I, Economics of decision, *Theory and Decision Library*, D. Reidel Publishing *Co.*, Dordrecht/Boston, Mass.

Marschak, J. (1980a) Economic information, decision, and prediction, Selected essays. Vol. II, Economics of information and organization, *Theory and Decision Library*, D. Reidel Publishing Co., Dordrecht/Boston, Mass.

Marschak, J. and Miyasawa, K. (1968) Economic comparability of information systems, International Economic Review, v. 9, No. 2, pp. 137–174.

Martin, J. (1977) *Computer Database Organization*, Prentice-Hall, Englewood Cliffs, NJ

Martin, W.J. (1995) *The Global Information Society*, Aslib/Gover,Aldershot

Martin, W. (1999) *Structures: Theory and Analysis*, Palgrave Macmillan

Mason, R.O. (1978) Measuring Information Output: A communication Systems Approach, *Information and Management*, v. 1, pp. 219–234.

Mason, R.O. (1986) Four Ethical Issues of the Information Age, *Management Information Systems Quarterly*, v. 10, No. 1, (http://www.misq.org/archivist/vol/no10/issue1/vol10no1mason.html)

Mattessich, R. (1993) On the nature of information and knowledge and the interpretation in the economic sciences, *Library Trends*, v. 41, No. 4, pp. 567–93

Maturana, H. and Varela, F. (1980) *Autopoesis and Cognition: The Realization of the Living*, Reidel, Dordrecht, The Netherlands

Mazur, M. (1970) *Jakosciowa Teoria Informacji*, PAN, Warszawa (in Polish)

McCabe, T.J. (1976) A Complexity Measure, *IEEE Transaction on Software Engineering*, SE-2, pp. 308–320

McCarthy, J. (1956) Measures of the value of information, *Proc. Nat Acad. Sci.*, v. 42, pp. 654–655

McCrae, R.R. and Allik, J. (Eds.). (2002) *The Five-Factor Model across Cultures*, Kluwer Academic/Plenum Publishers, New York

McGee, J. and Prusak, L. (1993) *Managing Information Strategically*, The Ernst & Young Information Management Series, John Wiley & Sons, Inc., New York

McGraw-Hill Encyclopedia of Science and Technology (2005) 5th Edition, McGraw-Hill Companies

McLuhan, M. (1964) *Understanding Media: The Extensions of Man*, MIT Press, Cambridge, MA

McMillan, B. (1953) The basic theorems of information theory, *Ann. Math. Stat.*, v. 24, pp. 196–219

McPherson, P.K. (1994) Accounting for the value of information, *Aslib Proceedings*, v. 46, No. 9, pp. 203–215

Meadow, C.T. (2006) *Messages, Meanings, and Symbols: The Communication of Information*, Scarecrow Press, Lanham, MD

Meadow, C.T. and Yuan, W. (1997) Measuring the impact of information: Defining the concepts, *Information Processing and Management*, v. 33, No. 6, pp. 697–714

Meadow, C.T. (1996) A proposed method of measuring the utility of individual information retrieval tools, *Canadian Journal of Information and Library Science*, v. 21, No. 1, pp. 22–34.

Melik-Gaikazyan, I.V. (1997) *Information processes and reality*, Nauka, Moscow (in Russian, English summary)

Menant, C. (2002) Information and Meaning, *Electronic Conference on Foundations of Information Science* (FIS 2002) (electronic edition: http://www.mdpi.net)

Mendelson, E. (1997) *Introduction to Mathematical Logic*, Chapman & Hall, London/Glasgow/ New York

Mendelsohn, H. (1966) *Mass Entertainment*, College and UP, New Haven

Mensky, M.B. (1990) The self-measurement of the universe and the concept of time in quantum cosmology, *Class. Quantum Grav.*, v. 7, pp. 2317–2329

Merton, R.K. (1968) *Social Theory and Social Structure*, The Free Press, New York

Mesarovic, M.D. and Takahara, Y. (1975) *General Systems Theory: Mathematical Foundations*, Academic Press, New York/London/San Francisco

Mey, H. (1986) Der Eintritt in das Informationszeitalter, in *Denken über die Zukunft*, Ringier AG, Zürich

Meyrowitz, J. (1985) *No Sense of Place: The Impact of Electronic Media on Social Behavior*, Oxford University Press, Oxford/New York

Miller, G.A. (1953) What is information measurement? *American Psychologist*, v. 8, pp. 3–11

Miller, G.R. (1966) Speech Communication: A Behavioral Approach, Bobbs-Merril, Indianapolis, IN

Miller, G.L. (1988) The concept of information: A historical perspective on modern theory and technology. In Ruben, Brent D. (Ed). *Information and behavior*, v. 2, Transaction Publishers, New Brunswick, NJ, pp. 27–53

Mingers, J.C. (1995) Information and meaning: foundations for an intersubjective account, *Information Systems Journal*, v. 5, pp. 285–306

Mingers, J.C. (1996) An Evaluation of Theories of Information with Regard to Semantic and Pragmatic Aspects of Information Systems, *Systems Practice*, v. 9, No. 3, pp. 187–209

Minsky, M. (1962) Problems of formulation for artificial intelligence, *Proc. Symp. on Mathematical Problems in Biology*, American Mathematical Society, Providence, pp. 35–46

Minsky, M. (1974) *A framework for knowledge representation*, AI Memo No. 306, MIT, Cambridge

Minsky, M. (1986) *The Society of Mind*, Simon and Schuster, New York

Minsky, M. (1998) The mind, artificial intelligence and emotions, *Brain & Mind Magazine*, No. 7 (http://www.cerebromente.org.br/n07/opiniao/minsky/minsky_i.htm)

Mizzaro, S. (1996) On the Foundations of Information Retrieval, in *Atti del Congresso Nazionale AICA'96* (Proceedings of AICA'96), Roma, IT, pp. 363–386.

Mizzaro, S. (1998) How many relevances in information retrieval? *Interacting with computers*, v. 10, pp. 303–320

Mizzaro, S. (2001) Towards a theory of epistemic information. Information Modelling and Knowledge Bases, *IOS Press*, v. 12, Amsterdam, pp. 1–20

Moles, A. (1994) Information Theory, in: *Encyclopedic Dictionary of Semiotics*, Mouton de Gruyter, Berlin/New York, pp. 349–351

Montanet, L., *et al*, (1994) Review of Particle Properties, *Phys. Rev.*, D 50, pp. 1173–1826

Monte, C.F. (1980) *Beneath the Mask: An Introduction to Theories of Personality*, Holt, Rinehart & Winston, New York

Mora, C. and Briegel, H.J. (2004) *Algorithmic Complexity of Quantum States*, LANL, Quantum Physics, quant-ph/0412172, 2004 (electronic edition: http://arXiv.org)

Morgan, G. (1998) *Images of Organization*, Sage, Beverly Hills, CA

Morris, C.W. (1938) Foundation of the Theory of Signs, in *International Encyclopedia of Unified Science*, v. 1, No. 2

Morris, C.W. (1946) *Signs, Language and Behavior*, Prentice-Hall, Inc., New York

Morris, C.W. (1964) *Signification and Significance*: *A Study of the Relations of Signs and Values*, MIT Press, Cambridge, Massachusetts

Morris, P.E. and Conway, M.A. (Eds) (1993) *The psychology of memory*, vol. 1–3, Edward Elgar, Aldershot

Motl, L.L. *Is Space-Time Discrete*? (electronic edition: http://www.karlin.mff.cuni.cz/~motl/Gibbs/discrete.htm)

Mowshowitz, A. and Kumar, N. (2009) And Then There Were Three, *Computer*, v. 42, No. 2, pp. 108, 106–107

Muchnik, An. and Vereshchagin, N. (2006) Shannon Entropy vs. Kolmogorov Complexity, *Lecture Notes in Computer Science*, v. 3967, pp. 281–291

Muller, S.J. (2007) *Asymmetry*: *The Foundation of Information*, Springer, Berlin/Heidelberg/New York

Müller, B., Reinhardt, J. and Strickland, M.T. (1995) *Neural Networks*: *An Introduction*, Springer-Verlag, Berlin/Heidelberg/New York

Munindar P.S. and Nicholas M.A. (1993) A logic of intentions and beliefs, *Journal of Philosophical Logic*, v. 22, pp. 513–544

Murray, M. and Rice, J. (1993) *Differential geometry and statistics*, Monographs on Statistics and Applied Probability, v. 48, Chapman and Hall, London/Glasgow/New York

Nariman, H.N. (1993) *Soap Operas for Social Change*, Praeger, Westport/London

Nauta, D. Jr. (1970) *The Meaning of Information*, Mouton, The Hague/Paris

Neisser, U. (1976) *Cognition and Reality*: *Principles and Implications of Cognitive Psychology*, San Fransisco

Nguyen, N.T. (2008) *Advanced Methods for Inconsistent Knowledge Management*, Springer, New York

Nguen, N.T. (2008a) Inconsistency of knowledge and collective intelligence, *Cybernetics and Systems*, v. 39, No. 6, pp. 542–562

Niedenthal, P.M., Krauth-Gruber, S. and Ric, F. (2006) *Psychology of Emotion Interpersonal, Experimental, and Cognitive Approaches*, Psychology Press, New York, NY

Nielson, H.R. and Nielson, F. (1995) *Semantics with Applications, A Formal Introduction*, John Wiley & Sons, Chicester, England

Nielsen, M.A. and Chuang, I.L. (2000) *Quantum Computation and Quantum Information*, Cambridge University Press, Cambridge

Nilsson, N. (1998) *Artificial Intelligence*: *A New Synthesis*, Morgan Kaufmann, San Francisco

Nonaka, I. (1996) *Knowledge Has to Do with Truth, Goodness, and Beauty, an interview with Professor Nonaka*, (http://www.dialogonleadership.org/Nonaka-1996.html)

Nonaka, I. and Takeuchi, H. (1995) *The knowledge-creating company: how Japanese companies create the dynamics of innovation*, Oxford University Press, New York, NY

North, C. (2008) *The key to patient compliance – Effective information at the right time* (electronic edition: http://social.eyeforpharma.com/content/patient-compliance/key-patient-compliance-%E2%80%93-effective-information-right-time)

Nyquist, H. (1924) Certain factors affecting telegraph speed, *Bell System Technical Journal*, v. 3, pp. 324–346

Nyquist, H. (1924) Certain topics in telegraph transmission theory, *AIEE Trans.*, v. 47, pp. 617–625

Oakley, K. and Jenkins, J.M. (1996) *Understanding Emotions*, Blackwell Publishers

O'Brien, J.A. (1995) *The Nature of Computers*, The Dryden Press, Philadelphia/San Diego

O'Brien, J.A. (2004) *Management of Information Systems*, McGraw-Hill, Irvine

Ochsner, K. N., Bunge, S. A., Gross, J. J. and Gabrieli, J. D. E. (2002) Rethinking Feelings: An fMRI Study of the Cognitive Regulation of Emotion, *J. Cogn. Neurosci.*, v. 14, No. 8, pp. 1215–1229

Ogden, C.K. and Richards, I.A. (1953) *The Meaning of Meaning*, Routledge and Kegan, London

Olson, J.E. (2003) *Data Quality: The Accuracy Dimension*, The Morgan Kaufmann Series in Data Management Systems

Ore, O. (1935) On the foundation of abstract algebra, I, *Ann. of Math.*, v. 36, pp. 406–437

Ore, O. (1936) On the foundation of abstract algebra, II, *Ann. of Math.*, v. 37, pp. 265–292

Ortony, A., Clore, G.L. and Collins, A. (1988) *The Cognitive Structure of Emotions*, Cambridge University Press, Cambridge, UK

Ortony, A. and Turner, T.J. (1990) What's basic about basic emotions? *Psychological Review*, v. 97, pp. 315–331

Ossowski, M. and Ossowski, S. The Science of Science, *Organon*, Warszawa, No. 1

Osuga, S. (1989) *Knowledge Processing*, Moscow, Mir, (Russian translation from the Japanese)

Osuga, S. and Saeki I. (Eds.) (1990) *Knowledge Acquisition*, Moscow, Mir (Russian translation from the Japanese)

Otten, K. (1975) Information and communication: A conceptual model as framework for development of theories of information. In Debbons, A. and Cameron, W. (eds.), *Perspectives in Information Science*, Noordhof, Leyden, pp. 127–148

Oussalah, M. (2000) On the Qualitative/Necessity Possibility Measure, I, *Information Sciences*, v. 126, pp. 205–275

Page, T. (1957) The value of information in decision-making, in Proc. of the First Internat. Conf. on Operations Research, Oxford, pp. 306–314

Parahonsky, B.A. (1988) *The Language of Culture and Genesis of Knowledge*, Kiev, Naukova Dumka (in Russian)

Parker, E.B. (1974) Information and Society, in *Library and information service needs of the nation*, National Commission on Libraries and Information Science, Washington, DC, pp. 9–50

Parker, M.B., Moleshe, V., De la Harpe, R. and Wills, G.B. (2006) An evaluation of information quality. frameworks for the World Wide Web, in Proceedings of the 8th Annual Conference on World Wide Web Applications, Bloemfontein, Cape Town (electronic edition: http://eprints.ecs.soton.ac.uk/12908/1/WWW2006_MParker.pdf)

Parker, N. J. (2001) Student learning as information behaviour: exploring Assessment Task Processes, *Information Research*, v. 6, No. 2 (electronic edition: http://InformationR.net/ir/6-2/ws5.html)

Partridge, D. and Wilks, Y. (1990) *The Foundations of Artificial Intelligence*, Cambridge University Press, Cambridge

Payne, W.L. (1983/1986) *A study of emotion: developing emotional intelligence; self integration; relating to fear, pain and desire*, Dissertation Abstracts International, 47

Pattee, H.H. (2006) The Physics of Autonomous Biological Information, *Biological Theory: Integration, Development, Evolution, and Cognition*, v. 1, No. 3, pp. 224–226

Peirce C.S. (1931–1935) *Collected papers*, v. 1–6, Cambridge University Press, Cambridge, England

Penrose R. (1994) *Shadows of the Mind*, Oxford University Press, Oxford/New York

Peres, A. (1993) *Quantum Theory: Concepts and Methods*, Kluwer Academic Publishers, Dordrecht

Peters, J.D. (1988) Information: Notes toward a critical history, *Journal of Communication Inquiry*, v. 12, No. 2, pp. 9–23

Petz, D. (2001) Entropy, von Neumann and von Neumann entropy, in *John von Neumann and the Foundations of Quantum Physics*, Kluwer Academic Publishers, Dordrecht

Petz, D. and Toth, G. (1993) The Bogolubov inner product in quantum statistics, *Letters in Mathematical Physics*, v. 27, pp. 205–216

Petz, D. and Sudar, C. (1996) Entropy, von Neumann and von Neumann entropy, *Mathematical Physics*, v. 37, pp. 2662–2673

Pfeifer, R. (1988) Artificial Intelligence Models of Emotion, in *Cognitive Perspectives on Emotion and Motivation*, Proceedings of the NATO Advanced Research Workshop, Kluwer, Dordrecht, pp. 287–320

Phanzagl, J. (1971) *Theory of Measurement*, Physica-Verlag, Würzburg/Wien

Philippot, P. and Feldman, R.S. (Eds.) (2004) *The regulation of emotion*, Lawrence Erlbaum, New York

Phlips, L. (1988) *The economics of imperfect information*: Cambridge University Press, Cambridge, England

Piaget, J. (1971) *Structuralism*, Routledge and Kegan Paul, London

Piattini, M.G., Calero, C. and Genero, M.F. (2002) *Information and Database Quality*, Kluwer Academic Publishers,

Peirce C.S., *Collected papers*, v. 1–6, Cambridge, 1931–1935

Pierce, E., Kahn, B.K. and Melkas, H. (2006) A Comparison of Quality Issues for Data, Information and Knowledge, in *International Conference of the Information Resource Management Association*, pp. 60–63

Picard, R.W. (1997) *Affective Computing*, MIT Press, Cambridge

Picard, R.W. (1998) Toward agents that recognize emotion, Actes Proceedings IMAGINA, Monaco, pp. 153–165

Pierce, J.R. (1961) *An Introduction to Information Theory: Symbols, Signals, and Noise*, Dover Publications, New York

Pinson G. (1995) Cognitive Information Theory, *Proc.* 14th *Intern. Congress on Cybernetics,* Ass. Int. de Cybernétique, Namur, Belgique

Pinson, G. (1995a) Une théorie cognitive de l'information, *Rev. Int. Systémique*, v. 9, No. 1, pp. 27–66

Pinson G. (1999) Beyond the "Information Wall" with discrete Information Processing, *Analyse de systèmes*, v. XXV, No. 4, pp. 42/1–42/42

Pipino, L., Lee, Y.W. and Wang, R.Y. (2002) Data quality assessment. Commun. ACM, v. 45, No. 4, pp. 211–218

Redman, T.C. (2001) *Data quality: the field guide*, Boston, MA: Digital Press.

Plotkin, B.I. (1966) *Automorphism Groups of Algebraic Systems*, Nauka, Moscow (in Russian)

Plotkin, B. (1994) *Universal Algebra, Algebraic Logic and Databases*, Kluwer Academic Publishers, Dordrecht

Pollock, J.L. and Cruz, J. (1999) *Contemporary Theories of Knowledge*, Rowman and Littlefield, Lanham/New York

Popper, K.R. (1965) *The Logic of Scientific Discovery*, New York

Popper, K.R. (1974) Replies to my critics, in *The Philosophy of Karl Popper* (P.A. Schilpp, Ed.), Open Court, La Salle, IL, pp. 949–1180

Popper, K.R. (1979) *Objective knowledge: An evolutionary approach*, Oxford University Press, New York

Porat, M.U. and Rubin, M.R. (1977) *The Information Economy*, Office of Telecommunications, US Department of Commerce, Washington DC

Porpora, D.V. (1989) Four Concepts of Social Structure, *Journal for the Theory of Social Behavior*, v. 19, No. 2, pp. 195–211

Poster, M. (1990) *The Mode of Information: Post-structuralism and Social Contexts*, University of Chicago Press, Chicago

Prigogine, I. (1947) *Etude Thermodynamique des Phénomènes irréversibles*, Desoer, Liège

Prigogine, I. (1961) *Introduction to Thermodynamics of Irreversible Processes*, Interscience Publishers, Inc., New York

Probst, G., Raub, S. and Romhard, K. (1999) *Managing Knowledge*, Wiley, London

Quigley, E.J. and Debons, A. (1999) Interrogative Theory of Information and Knowledge, in *Proceedings of SIGCPR '99*, ACM Press, New Orleans, pp. 4–10

Qvortrup, L. (1993) The controversy over the concept of information: An overview and a selected and annotated bibliography, *Cybernetics & Human Knowing 1*(4), pp. 3–24

Raatikainen, P. (1998) *Complexity and information — A critical evaluation of algorithmic information theory*, Reports from the Department of Philosophy, University of Helsinki, No 2 (http://www.mv.helsinki.fi/home/praatika/information.pdf)

Raber, D. (2003) *The Problem of Information: An Introduction to Information Science*, Scarecrow, Lanham, MD

Raber, D. and Budd, J.M. (2003) Information as sign: semiotics and information science. *Journal of Documentation*, v. 59, No. 5, pp. 507–522

Raymond, M. (2005) The Concept of Structure in IR: Toward a Theoretical Synthesis? *Paper presented at the annual meeting of the International Studies Association, Hilton Hawaiian Village, Honolulu, Hawaii, Online* 2008-03-11 (electronic edition: http://www.allacademic.com/meta/p72132_index.html)

Raisbeck, G. (1963) Information Theory: An Introduction for Scientists and Engineers, MIT Press, Cambridge, Massachusetts

Rankin, T.L. (1988) When is Reasoning Nonmonotonic? in *Aspects of Artificial Intelligence*, Kluwer Academic Publishers, Boston/Dordrecht/London, pp. 289–308

Rao, C.R. (1945) Information and accuracy attainable in the estimation of statistical parameters, *Bull. Calcutta Math. Soc.*, v. 37, pp. 81–91

Rapoport, A. (1953) What Is Information? *ETC: A Review of General Semantics*, v. 10, No. 4, pp. 247–260

Rapoport, A. (1956) The Promise and Pitfalls of Information Theory, *Behavioral Science*, v. 1, pp. 303–309

Rauterberg, M. (1995) About a Framework for Information and Information Processing of Learning systems, in Information System Concepts: Towards a Consolidation of Views, Proc. *IFIP Intern. Working Conference on Information Systems Concepts*, Chapman &Hall, London/Glasgow/ New York, pp. 54–69

Reading, A. (2006) The Biological Nature of Meaningful Information, *Biological Theory*, v. 1, No. 3, pp. 243–249

Reeves, A.M., Beamish, N.L., Anderson, R.B. and Buel, J.W. (1906) *The Norse Discovery of America*, Norroena Society, London/Stockholm/Copenhagen/Berlin/New York

Reinagel, P. (2000) Information theory in the brain, *Current Biology*, v. 10, No. 15, pp. 542–544

Rényi, A. (1960) Some fundamental questions of information theory, *Magyar Tud. Akad. Mat. Fiz. Oszt. Közl*, v. 10, pp. 251–282

Rényi, A. (1961) On measures of entropy and information, *Proc. 4th Berk. Symp. Math. Stat. and Prob.*, v. 1, pp. 547–561

Restall, G. (1996) Information Flow and Relevant Logics, in *Logic, Language and Computation*, v. 1, CSLI, Stanford, pp. 463–478

Ridley, M. (2000) *Genome: The Autobiography of a Species*, Perennial

Rifkin, J. (1995) *The End of Work: The Decline of the Global Laborforce and the Dawn of the Post-Market Era*, New York, Tarcher/Putnam's Sons

Rikur, P. (1995) *Hermeneutics, Ethics, Politics*, ACADEMIA,Moscow (in Russian)

Roberts, F.S. (1979) *Measurement Theory: with applications to decisionmaking, utility, and the social sciences*, Encyclopedia of Mathematics and its Applications, v. 7, Addison-Wesley, Reading, MA

Roberts, C. (1996) Information Structure in Discourse: Towards an Integrated Formal Theory of Pragmatics, in *OSU Working Papers in Linguistics*, pp. 91–136

Robinson, A. (1963) *Introduction to Model Theory and Metamathematics of Algebra*, North-Holland, Amsterdam/New York

Rochester, J.B. (1996) *Using Computers and Information*, Education and Training, Indianapolis

Roederer, J.G. (2002) On the Concept of Information and its Role in Nature, *Electronic Conference on Foundations of Information Science* (FIS 2002) (electronic edition: http://www.mdpi.net)

Roederer, J.G. (2005) *Information and its Role in Nature*, Springer, New York

Rogers, H. (1987) *Theory of Recursive Functions and Effective Computability*, MIT Press, Cambridge Massachusetts

Roget's II (1995) *The New Thesaurus*, Third Edition, Bantam Books, New York/London

Rolls E.T. (1999) *The brain and emotion*, Oxford University Press, Oxford/ New York

Rosenfeld, A. (1979) Digital topology, *American Mathematical Monthly*, v. 86, pp. 621–630

Rosenfeld, A. (1986) Continuous functions on digital pictures, *Pattern Recognition Letters*, v. 4, pp. 177–184

Rovelli, C. (1996) Relational quantum mechanics, *International Journal of Theoretical Physics*, v. 35, pp. 1637–1678

Rowley, J. (1998) What is Information? *Information Services and Use*, v. 18, pp. 243–254

Rowley, J. (2007) The wisdom hierarchy: representations of the DIKW hierarchy, *Journal of Information Science*, v. 33, No. 2, pp. 163–180

Ruben, B.D. (1992) The Communication-Information Relationship in System-Theoretic Perspective, *Journ. Amer Society for Information Science*, v. 43, No. 1, pp. 15–27

Rucker, R. (1987) *Mind Tools: The Five Levels of Mathematical Reality*, Houghton Mifflin Co., Boston

Ruse, M. (1973) *The Philosophy of Biology*, Hutchinson University Library, London

Russell, B. (1926) Theory of Knowledge, in *Encyclopedia Britannica*

Russell, P. (1992) *The Brain Book*, Penguin Books, London

Russell, S. and Norvig, P. (1995) *Artificial Intelligence*: *A Modern Approach*, Prentice-Hall, Englewood Cliffs, NJ

Ruževičius, J. and Gedminaitė, A. (2007) Peculiarities of the Business Information Quality Assessment, *Vadiba/Management*, v. 14, No. 1, pp. 54–60

Ryckman, R. (1993) *Theories of Personality*, Brooks/Cole P.C., Pacific Groves, California

Salii, V.N. (1965) Binary *L*-relations, *Izv. Vysh. Uchebn. Zaved.*, *Matematika*, v. 44, No.1 pp. 133–145 (in Russian)

Salovey, P. and Mayer, J.D. (1990) Emotional intelligence, *Imagination, Cognition, and Personality*, v. 9, No. 3, pp. 185–211

Salustri, F.A. and Venter, R.D. (1992) An Axiomatic Theory of Engineering Design Information, *Engineering with Computers*, v. 8, pp. 197–211

Samuelson, L. and Swinkels, J. (2006) Information, Evolution and Utility, *Theoretical Economics*, v. 1, pp. 119–142

Sanders, G. (1985) The Perception and Decoding of Expressive Emotional Information by Hearing and Hearing-Impaired Children, *Early Child Development and Care*, v. 21, No. 1–3, pp. 11–26

Sanfey, A.G. and Cohen, J.D. (2004) Is knowing always feeling? Proc. of the National Academy of Sciences, v. 101, No. 48, pp. 16709 – 16710

de Saussure, F. (1916) Nature of the Linguistics Sign, in: Bally, C. and Sechehaye, A. (Ed.), *Cours de linguistique générale*, McGraw Hill Education

de Saussure, F. (1916a) *Cours de linguistique générale*, ed. C. Bally and A. Sechehaye, with the collaboration of A. Riedlinger, Payot, Lausanne and Paris (English transl. W. Baskin, *Course in General Linguistics*, Fontana/Collins, Glasgow, 1977)

Scarrott, G.G. (1989) The Nature of Information, *Computer Journal*, v. 32, No. 3, pp. 262–266

Schank R.C. and Abelson, R.P. (1975) Scripts, plans and knowledge, in *Proceedings of the IJCAI Conference*, Tbilisi, v. 1, pp. 151–157

Scharffe, F. and Ding, Y. (2006) Three Levels of Knowledge Structuration for the Web, Citeulike, (http://www.citeulike.org/user/Sulpicus/article/1092431)

Schepanski, A. and Uecker, W. (1983) Toward a positive theory of information evaluation, *The Accounting Review*, v. 58, No. 2, pp. 259–283

Scheutz, M. (2002) Agents with or without Emotions? in Proc. of the *FLAIRS Conference*, Pensacola, Florida, pp. 89–93

Scheutz, M. (2002a) The Evolution of Affective States and Social Control, in Proceedings of International Workshop on Self-Organisation and Evolution of Social Behaviour, Swiss Federal Institute of Technology, Monte Verita, Switzerland, pp. 358–367

Scheutz, M. and Sloman, A. (2001) Affect and Agent Control: Experiments with Simple Affective States, in Proceedings of *IAT-01*, World Scientific Publisher, pp. 200–209

Schmidhuber, J. (2002) Hierarchies of Generalized Kolmogorov Complexities and Nonenumerable Universal Measures Computable in the Limit, *International Journal of Foundations of Computer Science*, v. 3, No. 4, pp. 587–612

Schnorr, C.-P. (1973) Process complexity and effective random tests, Fourth Annual ACM Symposium on the Theory of Computing, *J. Comput. System Sci.*, v. 7, pp. 376–388

Schrader, A.M. (1983) *Toward a Theory of Library and Information Science*, Doctoral dissertation, Indiana University

Schrer, K. (2005) What are emotions and how they can be measured? *Social Science Information*, v. 44, No. 4, pp. 695–729

Schrödinger, E. (1967) *What is Life? The Physical Aspects of the Living Cell*, Cambridge University Press, Cambridge

Severino, S.K., Bucci, W. and Creelman, M.L. (1989) Cyclical Changes in Emotional Information Processing in Sleep and Dreams, *Journal of American Academy of Psychoanalysis*, v. 17, pp. 555–577

Seiffert, H. (1968) *Information über die Information, Verständigung im Alltag, Nachrichtentechnik, wissenschaftliches Verstehen, Informationssoziologie, das Wissen des Gelehrten*, Verlag C.H. Beck, München

Senge, P.M. (1990) *The Fifth Discipline: The Art and Practice of the Learning Organization*, Doubleday, New York

Senn, J.A. (1990) *Information Systems in Management*, Wadsworth Publishing, Belmont, CA

Sequoiah-Grayson, S. (2008) The scandal of deduction: Hintikka on the Information Yield of Deductive Inferences, *Journal of Philosophical Logic*, v. 37, No. 1, pp. 67–94

Shafer, G. (1976) *A Mathematical Theory of Evidence*, Princeton University Press, Princeton

Shanks, G. and Darke, P. (1998) Understanding Data Quality in Data Warehousing: a Semiotic Approach, Proc. of the *MIT Conference on Information Quality*, Boston, pp. 247–264

Shannon, C.E. (1948) The Mathematical Theory of Communication, *Bell System Technical Journal*, v. 27, No. 1, pp. 379–423; No. 3, pp. 623–656

Shannon, C.E. (1993) *Collected Papers*, (N.J.A. Sloane and A.D. Wyner, Eds) IEEE Press, New York

Shannon, C.E. and Weaver, W. (1949) *The Mathematical Theory of Communication*. Univ. of Illinois Press

Shanks, G. and Corbitt, B. (1999) Understanding Data Quality: Social and Cultural Aspects, in *Proceedings of the 10th Australasian Conference on Information Systems*

Shanks, G. and Darke, P. (1998) Understanding Data Quality in Data Warehousing: a Semiotic Approach, in Proc. *MIT Conference on Information Quality*, Boston, pp. 247–264

Shanteau, J. and N.H. Anderson, (1972) Integration Theory Applied to Judgments of the Value of Information, *Journal of Experimental Psychology*, v. 2, pp. 266–275

Sharma, N. (2005) *The Origin of the "Data Information Knowledge Wisdom" Hierarchy* (electronic edition: http://www-personal.si.umich.edu/~nsharma/dikw_origin.htm)

Sharma, B.D. and Taneja, I.J. (1974) On Axiomatic Characterization of Information-Theoretic Measures, *Journ. Stat. Physics*, v. 10, No. 4, pp. 337–346

Shaw F.Z., Chen R.F., Tsao H.W. and Yen C.T. (1999) Algorithmic complexity as an index of cortical function in awake and pentobarbital-anesthetized rats, *J. Neurosci Methods*, v. 93, No. 2, pp. 101–110

Shenker, O.R. and Hemmo, M. (2007) Von Neumann's entropy does not correspond to thermodynamic entropy, *Philosophy of Science*, v. 73, No. 2, pp. 153–174

Shenoy, P.P. and Shafer, G. (1990) Axioms for probability and belief-function proagation, in *Uncertainty in Artificial Intelligence* 4: *Machine intelligence and pattern recognition*, v. 9, pp. 169–198

Shih, Y-S. (1999) Families of Splitting Criteria for Classification Trees, *Statistics and Computing*, v. 9, pp. 309–315

Shiller, R.J. (2000) Irrational Exuberance, Prinston University Press, Princeton, NJ

Shoenfield, J.R. (1967) *Mathematical Logic*, Addison-Wesley, Reading, Massachussets

Sholle, D. (1999) What is Information? The Flow of Bits and the Control of Chaos, *MIT Commucation Forum* (http://web.mit.edu/comm-forum/papers/sholle.html)

Shor, P.W. (1994) Algorithms for quantum computation: Discrete logarithms and factoring, in Proc. of the *35th Annual Symposium on Foundations of Computer Science*, Los Alamitos, CA, pp. 124–134

Shreider, Yu.A. (1963) On Quantitave Characteristics of Semantic Information, *Scientific and Tecnological Information*, No. 10, pp. 33–38 (in Russian)

Shreider, Yu.A. (1965) On the Semantic Characteristics of Information, *Information Storage and Retrieval*, v. 2, pp. 221–233

Shreider, Yu.A. (1967) On Semantic Aspects of Information Theory, *Information and Cybernetics*, Moscow, Radio, pp. 15–47 (in Russian)

Sibson, R. (1969) Information Radius, Z. *Wahrs. und verw Geb.*, v. 14, pp. 149–160

Siegfried, T. (2000) *The Bit and the Pendulum: From Quantum Computing to M-theory — the New Physics of Information*, John Wiley, New York

Simon, H.A. (1967) Motivational and emotional controls of cognition, *Psychology Review*, v. 74, pp. 29–39

Simon, V.M. (1998) Emotion participation in decision-making, *Psychology in Spain*, v. 2, No. 1, pp. 100–107

Simonov, P. (1986) *The emotional brain*. New York-London: Plenum Press.

Simonov, P.V. (1998) Psychophysiology of emotions, in *Principles of psychophisiology*, Moscow: "INFRA-M" Press

Singh, J. (1966) *Great Ideas in Information Theory, Language and Cybernetics*, Dover, New York

Sipser, M. (1985) A topological view of some problems in complexity theory, Theory of algorithms (Pécs, 1984), *Colloq. Math. Soc. János Bolyai*, v. 44, pp. 387–391

Sipser, M. (1997) *Introduction to the Theory of Computation*, PWS Publishing Company

Skagestad, P. (1993) Thinking with machines: Intelligence augmentation, evolutionary epistemology, and semiotics, *Journal of Social and Evolutionary Systems*, v. 16, pp. 157–180

Skinner, Q. (1969) Meaning and understanding in the history of ideas, *History and Theory*, v. 8, pp. 3–53

Sloman, A. (2001) Beyond shallow models of emotions, *Cognitive Processing*, v. 2, No. 1, pp. 177–198

Sloman, A. (2002) Architecture-based conception of mind, in *In the Scope of Logic, Methodology, and Philosophy of Science*, Kluwer, Dordrecht, v. 2, pp. 403–427

Smith, J.M. and Szathmary, E. (1998) *The Major Transitions in Evolution*, Oxford University Press, Oxford

Smith, J.M. and Szathmary, E. (1999) *The Origins of Life: From the Birth of Life to the Origin of Language*, Oxford University Press

Smolin, L. (1999) *The Life of the Cosmos*, Oxford University Press, Oxford/ New York

Smullyan, R.M. (1962) *Theory of Formal Systems*, Princeton University Press, Princeton

Sneed, J.D. (1979) *The Logical Structure of Mathematical Physics*, D. Reidel, Dordrecht

Sober, E. (1991) *Core Questions in Philosophy*, MacMillan P.C., New York

Sollins, K.R. (2003) *Effective Design of Naming Systems for Networks*, (electronic edition: http://krs.lcs.mit.edu/regions/docs/naming.pdf)

Solomonoff, R.J. (1960) *A Preliminary Report on a General Theory of Inductive Inference,* Technical Report ZTB-138, Zator Company, Cambridge, MA

Solomonoff, R.J. (1964) A Formal Theory of Inductive Inference, *Information and Control*, v. 7, No. 1, pp. 1–22; No. 2, pp. 224–254

Solomonoff, R.J. (1995) The discovery of algorithmic probability: A guide for the programming of true creativity, *Lecture Notes in Computer Science*; v. 904 Proceedings of the Second European Conference on Computational Learning Theory, pp. 1–22

Sowa, J. F. Semantic networks, in *Encyclopedia of Artificial Intelligence*, Wiley, New York, 1992

Spek, R. v.d. and Spijkervet, A. (1997) *Knowledge Management: Dealing Intelligently with Knowledge*, CIBIT, Utrecht

Spence, S.A. (2000) Between will and action, Journal of Neurology, Neurology and Psychiatry, v. 69, No. 5, p. 702

Sperber, D. and Wilson, D. (1986) *Relevance: Communication and Cognition*, Blackwell, Oxford

Stamper, R. (1973) *Information in Business and Administrative Systems*, Wiley, London

Stamper, R. (1985) Towards a Theory of Information: Mystical fluid or subject for scientific enquiry? *Computer Journal*, v. 28, No. 3, pp. 195–199

Stamper, R. (1987) Semantics, in *Critical Issues in Information Systems Research*, Wiley, London

Steedman, M. (2000) Information Structure and the Syntax-Phonology Interface, *Linguistic Inquiry*, v. 31, pp. 649–689

Stenmark, D. (2002) The Relationship between Information and Knowledge and the Role of intranets in Knowledge Management, in *Proceedings of the 35th Annual Hawaii International Conference on System Sciences* (*HICSS-35*), v. 4, IEEE Press, Hawaii (http://csdl2.computer.org/comp/proceedings/hicss/2002/1435/04/14350104b.pdf)

Stegmüller, W. (1969) *Main Currents in Contemporary German, British and American Philosophy*, D. Reidel P. C., Dordrecht, Holland

Stenmark, D. (2001) The Relationship between Information and Knowledge, in *Proceedings of IRIS 24*, Ulvik, Norway

Stenmark, D. (2002) Information vs. Knowledge: The Role of intranets in Knowledge Management, in *Proceedings of HICSS-35*, IEEE Press, Hawaii

Stepanov, Yu.S. (1975) *Foundations of General Linguistics*, Prosveshchenie, Moscow (in Russian)

Stepanov, Yu.S. (1985) *In the Three-dimensional Language Space*: *Semiotic Problems of Linguistics, Philosophy, and Art*, Nauka, Moscow (in Russian)

Stephanidis, C. (2000) Challenges towards Universal Access in the Information Age, *ERCIM News*, No. 40 (electronic edition: http://www.ercim.org/publication/Ercim_News/enw40/stephanidis.html)

Steup, M. (1997) The Analysis of Knowledge, in: E.N. Zalta, *Stanford Encyclopedia of Philosophy,* Center for the Study of Language and Information (CSLI), Stanford University, Stanford, CA, (electronic edition: http://plato.stanford.edu/entries/church-turing/)

Stigler, G.J. (1961) The economics of information, *Journal of Political Economy*, v. 69, No. 3, pp. 213–225

Stockwell, F. (2000) *A History of Information Storage and Retrieval*, McFarland & Company, Jefferson, NC

Stomer, T. (1990) *Information and the Internal Structure of the Universe*, Springer, London

Stone, A.R. (1997) *The War of Desire and Technology at the Close of the Mechanical Age*, The MIT Press, Cambridge, Massachussetts

Stonier, T. (1990) *Information and the Internal Structure of the Universe*: *An Exploration into Information Physics*, Springer, New York/London

Stonier, T. (1991) Towards a new theory of information, *Journal of Information Science*, v. 17, pp. 257–263

Stonier, T. (1992) *Beyond Information*: *The Natural History of Intelligence*, Springer Verlag, London

Stonier, T. (1996) Information as a basic property of the universe, *Bio Systems*, v. 38, pp. 135–140

Stonier, T. (1997) *Information and meaning*: *An evolutionary perspective*, Springer, London

Stonier, T. (1999) The Emerging Global Brain, in Hofkirchner, W. (Ed.), *The quest for a unified theory of information. Proceedings of the Second International Conference on the Foundations of Information Science*, Gordon and Breach, Amsterdam, pp. 561–578

Strauss, A.G.H. (1986) *Mead on Social Psychology*, University of Chicago Press, Chicago

Streater, R.F. (2000) *Classical and Quantum Info-manifolds*, Preprint in Mathematical Physics, math-ph/0002050 (electronic edition: http://www.arXiv.org)

Strong, D., Lee, Y. and Wang, R. (1997) Data Quality in Context, *Communications of the ACM*, v. 40, No. 5, pp. 103–110

Strong, D.M., Lee, Y.W. and Wang, R.Y. (1997a) 10 potholes in the road to information quality, *Computer*, v. 30, No. 8, pp. 38–46

Suber, P. (2000) *Truth of Statements, Validity of Reasoning* (electronic edition: http://www.earlham.edu/~peters/courses/log/tru-val.htm)

Sugeno, M. (1974) *Theory of fuzzy integrals and its application*, Doctoral Thesis, Tokyo Institute of Technology

Sugeno, M. (1977) *Fuzzy measures and fuzzy integrals — a survey*, in Fuzzy Automata and Decision Processes, North-Holland, New York, pp. 89–102

Sundgren, B. (1974) Conceptual foundation of the infological approach to databases, in *Database management*, North-Holland, Amsterdam, pp. 61–96

Suppe, F. (Ed.) (1979) *The Structure of Scientific Theories*, University of Illinois Press, Urbana

Suppes, P. (1967) What is a scientific theory? in *Philosophy of Science Today*, basic Books, New York, pp. 55–67

Suppes, P. and Han B. (2000) Brain-wave representation of words by superposition of a few sine waves, *Proceedings of the National Academy of Sciences*, v. 97, pp. 8738–8743

Suppes, P., Han, B., Epelboim, J. and Lu Z.-L. (1999) Invariance between subjects of brain wave representations of language, *Proceedings of the National Academy of Sciences*, v. 96, pp. 12953–12958

Suppes, P., Han, B., Epelboim, J. and Lu Z.-L. (1999a) Invariance of brain-wave representations of simple visual images and their names. *Proceedings of the National Academy of Sciences*, v. 96, pp. 14658–14663

Sveiby, K.-E. (1994/1998) What is information? What is information? (http://www.sveiby.com/portals/0/articles/Information.html)

Svozil, K. (1993) *Randomness and Undecidability in Physics*, World Scientific

Svozil. K. (1996) Quantum Algorithmic Information Theory, *Journal of Universal Computer Science*, v. 2, pp. 311–346

Szpankowski, W. and Konorski, J. (2008) What is Information? in *Festschrift in Honor of Jorma Rissanen*, pp. 154–172

Tassey, G. (2002) *The Economic Impacts of Inadequate Infrastructure for Software Testing*, NIST Report 7007.011

Taurinus, F.A. *Theorie der Parallellinien*, Cologne, 1825

Tegmark, M. (1996) Does the Universe in Fact Contain almost no Information? *Found. Phys. Lett.*, v. 9, pp. 25–42

Tegmark, M. (2007) *The Mathematical Universe*, Preprint in Physics GR-QC/07040646 (electronic edition: http://arXiv.org)

Teisberg, T.J. and Weiher, R.F. (2000) Valuation of Geomagnetic Storm Forecasts: An Estimate of the Net Economic Benefits of a Satellite Warning System. *Journal of Policy Analysis and Management*, v. 19, No. 2, pp. 329–334

Terry, D. (1986) Structure-free name management for evolving distributed environments, 6th *International Conference on Distributed Computing Systems*, pp. 502–508

Thayse, A., Gribomont, P., Louis, G., Snyers, D., Wodon, P., Gochet, P., Grégoire, E., Sanchez, E. and Delsarte, P. (1988) *Approche Logique de lÍntelligence Artificielle*, Bordas, Paris

Thibodeau, P. Buggy software costs users, vendors nearly $60B annually, *Computerworld*, Washington, 2002

Thompson, F. (1968) The organization is the information, *Am. Document*, v. 19, pp. 305–308

Thorndike, R.L. (1936) Factor analysis of social and abstract intelligence, *Journal of Educational Psychology*, v. 27, pp. 231–233

Thorndike, R.L. and Stein, S. (1937) An evaluation of the attempts to measure social intelligence, *Psychological Bulletin*, v. 34, pp. 275–284

Timpson, C.G. (2003) The Applicability of Shannon Information in Quantum Mechanics and Zeilinger's Foundational Principle, *Philosophy of Science*, v. 70, pp. 1233–1244

Titze, H. (1971) Ist Information ein Prinzip?, in *Monographien zur philosophischen Forschung*, Verlag Anton Hain, Meisenheim am Glan

Tohm, R. (1975) *Structural Stability and Morphogenesis*, Benjamin, New York

Traub, J.F., Wasilkowski, G.W. and Wozniakowski, H. (1983) *Information, Uncertainty, Complexity*, Addison-Wesley, Reading, MA

Traub, J.F., Wasilkowski, G.W. and Wozniakowski, H. (1988) *Information-Based Complexity*, Academic Press, London

Tribus, M. (1961) *Thermostatics and Thermodynamics*, D. van Nostrand Company, Inc., Princeton, NJ

Tribus, M. and McIrvine, E.C. (1971) Energy and information, *Scientific American*, v. 225, No. 3, pp. 179–188

Tsai, B.-S. (2003) Information landscaping: information mapping, charting, querying and reporting techniques for total quality knowledge management, *International Journal of Information Processing and Management*, v. 39, No. 4, pp. 639–664

Tsichritzis, D.C. and Lochovsky, F.H. (1982) *Data Models*, Prentice-Hall, Englewood Cliffs, NJ

Tuomi, I. (1999) Data is More Than Knowledge: Implications of the Reversed Knowledge Hierarchy for Knowledge Management and Organizational Memory, *Journal of Management Information Systems*, Vol. 16, No. 3, pp. 107–121

Ueno, H., Koyama, T., Okamoto, T., Matsubi, B. and Isidzuka, M. (1987) *Knowledge Representation and Utilization*, Moscow, Mir (Russian translation from the Japanese)

Ursul, A.D. (1971) *Information*, Nauka, Moscow (in Russian)

Van der Waerden, B.L. (1971) *Algebra*, Springer-Verlag, New York/Berlin

van Frassen, B. (2000) The semantic approach to scientific theories, in *The Nature of Scientific Theory*, pp. 175–194

van Rijsbergen, C.J. (1986) A Nonclassical Logic for Information Retrieval, *Computer Journal*, v. 29, pp. 481–485

van Rijsbergen, C.J. (1989) Towards an information logic, in *Proceedings of the 12th annual international ACM SIGIR conference on Research and development in information retrieval*, Cambridge, Massachusetts, pp. 77–86

van Rijsbergen, C.J. and Laimas, M. (1996) Information Calculus for Information Retrieval, *Journal of the American Society for Information Science*, v. 47, No. 5, pp. 385–398

Vegas, S. and Basili, V. (2005) A characterization schema for software testing technique, *Empirical Software Engineering*, v. 10, pp. 437–466

Veldhuizen, T.L. (2005) Software Libraries and Their Reuse: Entropy, Kolmogorov Complexity, and Zipf's Law, in Proceedings of the *First International Workshop on Library-Centric Software Design* (*LCSD* '05), pp. 11–23

Veltman, F. (1996) Defaults in Update Semantics, *Journal of Philosophical Logic*, v. 25, pp. 221–261.

Venda, V.F. (1990) *Systems with hybrid intelligence*, Mashinostroyeniye, Moscow (in Russian)

Vetschera, R. (2000) Entropy and the Value of Information. *Central European Journal of Operations Research*, v. 8, pp. 195–208.

Vitanyi, P.M.B. (1999) Three approaches to the quantitative definition of information in an individual pure quantum state, 15th *IEEE Conference on Computational Complexity*, pp. 263–270

Vitanyi, P.M.B. (2001) Quantum Kolmogorov Complexity Based on Classical Descriptions, *IEEE Transactions on Information Theory*, v. 47, No. 6, pp. 2464–2479

Voishvillo (1984) Logical entailment and semantics of generalized state descriptions, in *Modal and Intentional Logics and Application to the Methodology of Science*, pp. 183–191 (in Russian)

Voishvillo (1988) *Philosophical and Methodological Aspects of Relevant Logic*, MGU, Moscow (in Russian)

von Bayer, H.C. (2004) *Information: The New Language of Science*, Harvard University Press, Harvard von Foerster, H. (1980) Epistemology of Communication, In K. Woodward (Ed.), *The myths of information*, Baumgartner, Sun Prairie, WI, pp. 18–27

von Foerster, H. (1984) *Observing systems*, Intersystems Publications, Seaside, CA

von Neumann, J. (1927) Thermodynamik quantummechanischer Gesamheiten, *Göttingen Nachr.*, b. 1, pp. 273–291

von Neumann, J. (1932) *Mathematische Grundlagender Quantenmechanik*, Springer, Berlin

von Neumann, J. and Margenstern, O. (1944) *Theory of Games and Economic Behavior*, Princeton University Press, Princeton

von Weizsäcker, C.F. (1958) Die Quantentheorie der einfachen Alternative (Komplementarität und Logik, II), *Zeitschrift für Naturforschung*, v. 13, pp. 245–253

von Weizsäcker, C.F., Scheibe, E. and Süssmann, G. (1958) Komplementarität und Logik, III (Mehrfache Quantelung), *Zeitschrift für Naturforschung*, v. 13, pp. 705–721

von Weizsäcker, C.F. (1974) *Die Einheit der Natur*, Deutscher Taschenbuch Verlag, Munich, Germany

von Weizsäcker, C.F. (1985) *Aufbau der Physik*, Hanser, Munich, Germany (Eglish translation: *The Structure of Physics*, Springer, Berlin/Heidelberg/New York, 2006)

von Weizsäcker, C.F. (1992) *Zeit und Wissen*, Hanser, Munich, Germany

Vygotski, L.S. (1956) *Selected Psychological Works*, Moscow (in Russian)

Vyugin, V.V. (1981) Algorithmic Entropy (Complexity) of a Finite Object and its Application to the Definition of Randomness and Quantity of Information, *Semiotika and Informatika*, v. 16, pp. 14–43 (in Russian)

Walker, R. (Ed.) (1993) *AGI Standards Committee GIS Dictionary*, Association for Geographical Information

Wallace, C.S. and Boulton, D.M. (1968) An information measure for classification, *Computer Journal*, v. 11, No. 2, 1968, pp. 185–194

Wallace, C.S. and Dowe (1999) Minimum Message Length and Kolmogorov complexity, *Computer Journal*, v. 42, No. 4, pp. 270–283

Wand, Y. and Wang, R. (1996) Anchoring data quality dimensions in ontological foundations, *Communications of the ACM*, v, 39, No. 11, pp. 86–95

Wang, Y. and Junkang Feng, J. (2006) Semantic Databases: An Information Flow (IF) and Formal Concept Analysis (FCA) Reinforced Information Bearing Capability (IBC) Model, *Computing and Information Systems Journal*, v. 10, No. 1, pp. 15–26

Wang, R.Y., Ziad, M. and Lee Y.W. (2001) Data Quality, Kluwer Academic Publishers

Warner, T. (1996) *Communication Skills for Information Systems*, London, Pittman Publishing

Weber, M. (1947) *The Theory of Social and Economic Organization*, New York/London

Webster, F. (2002) *Theories of the Information Society*, Routledge, London

Webster's Revised Unabridged Dictionary (1998) MICRA, Inc. of Plainfield, NJ

Wechsler, D. (1940) Non-intellective factors in general intelligence, *Psychological Bulletin*, v. 37, pp. 444–445.

Weinberger, E.D. (2002) A theory of pragmatic information and its application to the quasi-species model of biological evolution, *Biosystems*, v. 66, No. 3, pp. 105–119

Westin, A.F. (1972) *Databanks in a Free Society: Computers, Record-Keeping and Privacy*, Quadrangle Books, New York

Wheeler, J.A. (1977) Include the Observer in the Wave Function? in *Quantum Mechanics, a Half Century Later* (Lopes, J.L. and M. Paty M., Eds.), Riedel, Dordrecht, pp. 1–18

Wheeler, J.A. (1979) From the Big Bang to the Big Crunch (an interview by M. R. Gearhart), *Cosmic Search*, v. 1, No. 4

Wheeler, J.A. (1990) Information, Physics, Quantum: The Search for Links, in *Complexity, Entropy, and the Physics of Information* (Zurek, W., ed.), Redwood City, CA: Addison-Wesley, pp. 3–28

Weiss, G. (1996) What is information? What is the truth of consciousness? *DEcon'95 symposium*, Baden-Baden , Germany, v. 106, pp. 65–69

Wersig, G. (1997) Information Theory, in Encyclopaedic Dictionary of of Library and information Science, Routledge, London, pp. 220–227

White, N.P. (1976) *Plato on Knowledge and Reality*. Hackett, Indianapolis.

Wiener, N. (1954) *The Human Use of Human Beings: Cybernetics and Society*, Free Association, London

Wiener, N. (1961) *Cybernetics, or Control and Communication in the Animal and the Machine*, 2nd revised and enlarged edition, New York and London: MIT Press and Wiley, New York, London

Wiig, K.M. (1993) *Knowledge Management Foundations: Thinking About Thinking — How People and Organizations Create, Represent, and Use Knowledge*, Schema Press, Arlington, TX

Wigner, E. (1960) The Unreasonable Effectiveness of Mathematics, *Communications in Pure and Applied Mathematics*, v. 13, pp. 1–14

Wilson, T. (1993), Trends and Issues in Information Science — a General Survey, in *Media, Knowledge and Power* (Boyd-Barrett, O. and Braham, P., Eds.), Groom Helm, London/Sydney, pp. 407–422

Wilson, N. and Mengin, J. (1999) Logical deduction using the local computation framework, in *Symbolic and Quantitative Approaches to Reasoning and Uncertainty*, ECSQARU'99, Lecture Notes in Computer Science,, v. 1638, Springer, pp. 386–396

Wing, J.M. (2008) Five Deep Questions in Computing, *Communications of the ACM*, v. 51, No. 1, pp. 58–60

Wirth, N. (1975) *Algorithms + Data Structures = Programs*, Prentice Hall,

Wirth, N. (1986) *Algorithms and Data Structures*, Prentice Hall,

Wittgenstein, L. (1922) *Tractatus Logico-Philosophicus*, (English translation by C.K. Ogden and F.P. Ramsey) Routledge and Kegan Paul, London.

Wolverton, R.W. (1974) The Cost of Developing Large-Scale Software, *IEEE Transactions on Computer*, C-23, pp. 615–636

Wooldridge, M. (2000) *Reasoning About Rational Agents*, The MIT Press

Worboys, M. F. (1994) A Unified Model of Spatial and Temporal Information, *Computer Journal*, v. 37, No. 1, pp. 26–34

Wormell, I. (Ed.) (1990) *Information quality: Definitions and dimensions*, Taylor Graham, Los Angeles

Wu, C.-H. (2004) Building knowledge structures for online instructional/learning systems via knowledge elements interrelations, *Expert Systems with Applications*, v. 26, No. 3, pp. 311–319

Yaglom, I.M. (1980) *Mathematical Structures and Mathematical Modeling*, Sov. Radio, Moscow (in Russian)

Yip, S. and Zhou, Q. (2004) Enhancing software protection with poly-metamorphic code, *The Computers and Law Journal*, Issue 56 (www.nswscl.org.au/journal/)

Yourdon, E. (1979) *Managing the Structured Techniques*, Prentice-Hall, Englewood Cliffs, NJ

Yovits, M.C. (1975) A theoretical framework for the development of information science, in *Theoretical problems in informatics* (FID 530), pp. 90–114

Yasin, E.G. (1970) To the problem of measurement of information quantity, content, and value, in *Economic Semiotic*, Nauka, Moscow, pp. 46–66 (in Russian)

Yurtsever, U. (2000) *Quantum Mechanics and Algorithmic Randomness*, Quantum Physics, quant-ph/9806059

Zadeh, L. (1965) Fuzzy Sets, *Information and Control*, v. 8, No. 3, pp. 338–353

Zadeh, L.A. (1973) *The Concept of a Linguistic Variable and its Application to Approximate Reasoning*, Memorandum ERL-M 411, Berkeley

Zadeh, L.A. (1978) *Fuzzy Sets as a Basis for a Theory of Possibility*, Fuzzy Sets and Systems, v. 1, pp. 3–28

Zajonc, R.B. (1980) Feelings and Thinking: Preferences Need No Inferences, *American Psychologist*, v. 35, No. 2, pp. 151–175

Zeeman, E.C. and Buneman, O.P. Tolerance spaces and the brain, in *Towards a Theoretical Biology*, v. 1, pp. 140–151

Zeist, R.H.J. and Hendriks, P.R.H. (1996) Specifying software quality with the extended ISO model, *Software Quality Management IV — Improving Quality*, BCS, pp. 145 –160

Zeleny, M. (1987) Management support systems: towards integrated knowledge management, Human Systems Management, v. 7, No.1, pp. 59–70

Zhu, X. and Gauch, S. (2000) Incorporating quality metrics in centralized/distributed information retrieval on World Wide Web, *Proceedings of the 23 International ACM SIGIR Conference on research and development in information retrieval*, Athens, Greece, pp. 288–295

Zimmermann, K.-J. (2001) *Fuzzy set theory and its applications*, Kluwer Academic Publishers, Boston/Dordrecht/London

Zurek, W.H. (1989) Algorithmic randomness and physical entropy, *Phys. Rev. A*, v. 40, No. 8, pp. 4731–4751

Zurek, W.H. (1991) Algorithmic information content, Church-Turing thesis, physical entropy, and Maxwell's demon. *Information dynamics* (Irsee, 1990), NATO Adv. Sci. Inst., Ser. B Phys., v. 256, *Plenum, New York* , pp. 245–259

Zuse, H. (1998) *History of Software Measurement*, Berlin

Zvonkin, A.K. and Levin, L.A. (1970) The Complexity of Finite Objects and the Development of the Concepts of Information and Randomness by Means of the Theory of Algorithms, *Russian Mathematics Surveys*, v. 256, No. 6, pp. 83–124

Subject Index